## Modern Small Antennas

If you're involved in designing and developing small antennas, this complete, cutting-edge guide covers everything you need to know. From fundamentals and basic theory to design optimization, evaluation, measurements, and simulation techniques, all the essential information is included. You'll also get many practical examples from a range of wireless systems, whilst a glossary is provided to bring you up to speed on the latest terminology. A wide variety of small antennas is covered, and design and practice steps are described for each type: electrically small, functionally small, physically constrained small, and physically small. Whether you are a professional in industry, a researcher, or a graduate student, this is your essential guide to small antennas.

**Kyohei Fujimoto** is a Professor Emeritus at the University of Tsukuba, Japan, and is also a Senior Scientist at the Foundation for Advancement of International Science. He has published extensively on the subject of antennas and has written several books including the classic *Small Antennas* (1987). He is a Life Fellow of the IEEE, a Fellow of the IEICE, and a member of Sigma Xi.

**Hisashi Morishita** is a Professor in the Department of Electrical and Electronic Engineering at the National Defense Academy, Japan, where he has worked since 1992. His research has long been concerned with small antennas and mobile communications, on which he has so far published many papers in major technical journals. He is a member of the IEICE and a senior member of the IEEE.

"*Modern Small Antennas* is a clearly written, comprehensive, yet practical treatment for the design and development of the physically small and compact antennas required by many of today's wireless systems. It is an indispensable day-to-day desk reference for the small antenna designer."

*Gary A. Thiele, University of Dayton*

"This book on small antennas is masterfully written. The wealth of design information on the state-of-the-art on small antenna configurations is unparallel to any existing book on this subject. Both the antenna practicing engineers and researchers should find the book very valuable."

*Yahya Rahmat-Samii, University of California, Los Angeles*

# Modern Small Antennas

KYOHEI FUJIMOTO
University of Tsukuba, Japan

HISASHI MORISHITA
National Defense Academy, Japan

Shaftesbury Road, Cambridge CB2 8EA, United Kingdom

One Liberty Plaza, 20th Floor, New York, NY 10006, USA

477 Williamstown Road, Port Melbourne, VIC 3207, Australia

314–321, 3rd Floor, Plot 3, Splendor Forum, Jasola District Centre, New Delhi – 110025, India

103 Penang Road, #05–06/07, Visioncrest Commercial, Singapore 238467

Cambridge University Press is part of Cambridge University Press & Assessment, a department of the University of Cambridge.

We share the University's mission to contribute to society through the pursuit of education, learning and research at the highest international levels of excellence.

www.cambridge.org
Information on this title: www.cambridge.org/9780521877862



First published 2013

*A catalogue record for this publication is available from the British Library*

*Library of Congress Cataloging-in-Publication data*
Fujimoto, K. (Kyohei), 1929–
Modern small antennas / Kyohei Fujimoto, University of Tsukuba, Japan; Hisashi Morishita, National Defense Academy, Japan.
pages cm
Includes bibliographical references and index.
ISBN 978-0-521-87786-2
1. Antennas (Electronics) 2. Miniature electronic equipment. I. Morishita, Hisashi. II. Title.
TK7871.6.F85 2013
621.382′4 – dc23 2012030785

ISBN 978-0-521-87786-2 Hardback

# Contents

# Preface

Small antenna, big impact!

More than 20 years have elapsed since the first book on small antennas was published in 1987. Small antennas have been used for a long time, from the earliest communications to present-day applications in various small wireless systems; typically mobile phones and other hand-held equipment that emerged into common use several decades ago. Publication of the first small-antenna book dates back to 1987, starting with its planning in 1981, when we, the late Professor J. R. James and I (Fujimoto), saw the necessity of a book on small antennas. At that time design techniques vital for development and successful fabrication of small antennas had not been systematically treated and were seldom even recorded in the literature. Hoping to improve that situation, we agreed to produce a book entitled "Small Antennas." Six years later, in 1987, Research Studies Press, UK, published the book as the first text devoted to small antennas. We intended the book to provide data and information that should assist with the analysis, design, and development of practical small antennas. We were pleased that the book was used worldwide and was recognized as useful by many people who had taken advantage of data and practical examples provided in the book.

Now, more than 20 years later, I recognized that the original book no longer matches current trends in the discipline, and it calls for extensive updating to reflect current trends in advanced antenna technology and acute requirements of these times.

Coincidentally, at the ISAP 2006, Singapore, in 2006, quite a few attendees there asked me about renewal of the book "Small Antennas." There was a common recognition of the expanding boundaries of the wireless world that had grown with newly evolved systems, for which advanced antennas were required. Then, the present authors (Fujimoto and Morishita) started planning a renewed book and gathering materials necessary for it. I was happy at that time for having received inquiries of publishing a book on small antennas by some publishing companies, among which was Cambridge University Press (CUP), in the United Kingdom. Then, I contacted Dr. Julie Lancashire, Engineering Editor of the CUP, and got agreement for publishing a renewed version of "Small Antennas." Dr. Lancashire was formerly a commissioning editor of Artech House, UK, who supported us (the late Professor J. R. James and me) in editing "Handbook of Antenna Systems for Mobile Communications."

Today, small antenna technology has advanced along with progress in various wireless systems, for which small antennas are required. The most typical wireless systems requiring small antennas were mobile phones, having progressed to date through at least four system generations. Mobile phones have evolved well beyond traditional voice-communication instruments to widely versatile systems, often referred to as "smart phones" and hand-held "tablets" that function as information terminals, being capable of handling high-rate data and video transmissions (both still and moving pictures) in addition to standard telephone capabilities. Besides mobile phones, numerous other types of wireless systems have been deployed. Representative systems are NFC (Near Field Communication), including RFID (Radio Frequency Identification), broadband wireless such as WLAN (Wireless Local Area Network) and WiMAX (Worldwide Interoperability for Microwave Access), wireless power transmission, radio control, body-centric communications, medical systems, and so forth. Most of these systems need small and compact antennas, as the equipment is generally very small and normally requires internal antennas to avoid damage in normal handling and to lend convenience for operators as well as better appearance.

This book is designed to provide readers with the latest data and information that would be useful for engineers and researchers to design and develop small antennas, placing particular emphasis on practical usefulness. With that emphasis on practicality, this book purposely does not aim to concern itself too deeply with underlying theory. Extensive bibliographic references accompany each chapter to guide readers needing to pursue details to a greater degree.

It is worth mentioning that this book is unique in treating four types of small antennas, differing from conventional small-antenna books so far published, in which only ESA (Electrically Small Antennas) are dealt with. The categories of small antennas here are based on functions, physically constrained size, and physically small dimensions, in addition to electrically small size in comparison with the wavelength. Hence, the book endeavors to cover these four types of small antennas which are practically employed in various wireless systems that require not only electrically small size, but also enhanced performances or improved characteristics, even with reduced size. This book provides design concepts of small antennas, and many examples based on them are given. Novel design methods such as applications of integration techniques, inclusion of environmental materials in the antenna design, applications of electromagnetic (EM) composite structure and the latest topical EM materials like metamaterials, are described along with many examples. Antennas today are no longer a single device, but constructed within a composite structure to perform sophisticated functions even with physically small dimensions. We the authors sincerely hope that our book is useful for readers who need to design, develop, and create novel and sophisticated small antennas.

In this book, some important small antennas applicable to human body communications and those employed in medical uses such as endoscopes and cancer treatment are omitted, although they are unquestionably interesting subjects and important antennas.

Small antennas have vital importance in small wireless equipment, as in some cases they will determine the limits of system performances; in turn they can promote further deployment and advancement of novel sophisticated wireless systems and modern information systems that are beneficial and ameliorative for human life and society. Thus, even though their size is small, quite big is the impact of small antennas on the human condition as well as on their technically related fields such as antennas, communications, and information concerns, to which they contribute greatly and significantly.

We say again "Small Antennas, Big Impact!"

The contents of this book, consisting of eleven chapters, and the persons in charge of each chapter are described briefly as follows.

## Chapter 1 Introduction (K. Fujimoto)

This chapter presents introductory remarks on small antennas, starting with a brief history of small antennas, followed by comments on the current status of practical small antennas and some related subjects such as fundamental limitations and so forth. Then, explanation of the concepts underlying small antennas is given, which is unique in categorizing in a wider sense into four types that include functionally small, physically constrained small, and simply physically small, in addition to electrically small that differs from conventional books. Then, variations of small antenna types are covered briefly, ranging from simple examples like dipoles to a variety of geometries constructed with basic shapes and structures, modified structures, composite designs, and designs integrated with materials, including metamaterials.

## Chapter 2 Small antennas (K. Fujimoto)

This chapter begins with a definition of small antennas categorized into four types from the viewpoint of function, physical dimensions, and partly constrained physical dimensions, in addition to dimensions small in comparison with the wavelength. The four types are: (1) FSA (Functionally Small Antenna), (2) PCSA (Physically Constrained Small Antenna), (3) PSA (Physically Small Antenna), and (4) ESA (Electrically Small Antenna). The significance of small antennas can be recognized as the essential and indispensable feature in a great many types of small wireless equipment and the fact that they may often determine the performance level of wireless systems; in turn, newly emerged wireless systems have been aided by small antennas in cases where they couldn't even operate properly without them.

## Chapter 3 Properties of small antennas (K. Fujimoto and Y. Kim)

This chapter starts with a discussion on specific characteristics of small antennas; typically impedance, antenna $Q$, bandwidth, and radiation efficiency. Impedance matching of small antennas, particularly when the size is only a small fraction of a wavelength, exhibits difficulty, because of the antenna's small radiation resistance and large reactance compared with the connected circuit impedance. Some useful methods for matching are introduced. Other issues specifically noted regarding small antennas are proximity effect that cannot be avoided in almost all cases due to the installation inside small equipment.

The proximity effects usually deteriorate antenna performances; however, in a way it can be turned into enhancement of the antenna performance, when materials near an antenna can be utilized as a part of the radiator.

## Chapter 4 Fundamental limitation of small antennas (K. Fujimoto)

Typical work on the fundamental limitation of small antennas is reviewed, beginning with that done by Wheeler, up to recent work by Thal. H. A. Wheeler was a pioneer who treated small antennas first in 1947, discussed performances of small antennas, and introduced the concept of the fundamental limitation of small antennas. Chu followed Wheeler's work in 1948 and derived the minimum possible $Q$ of an antenna for either TE or TM wave mode radiation that is known as Chu's limitation. After Chu, many researchers such as Hansen, Harrington, Collin, Rothchild, Fante, McLean, Folts, Thiel, Geyi, Best, and Yaghjian discussed and calculated antenna $Q$ by using each individual method. Best and Yaghjian uniquely gave relationships between $Q$, impedance, and bandwidth. Thal had shown the most verifiable $Q$ by calculating reactive energy inside Chu's sphere, a topic that had been ignored by Chu. Hansen later showed a new method for calculating $Q$, which gave a more rigorous value.

## Chapter 5 Subjects related with small antennas (K. Fujimoto)

This short chapter describes major subjects that concern small antennas; firstly investigation of fundamental characteristics of the small antenna, and secondly exploitation of methods or ideas of how to realize a practical small antenna. Discussed next are practical design problems that should be considered for small antennas; for instance, design issues for mobile terminals. Lastly, general topics are covered, such as problems in designing small antennas suitable for specific wireless systems, and the necessity of compromising between the theoretical designs and practical performances encountered when the antenna size is very small.

## Chapter 6 Principles and techniques for making antennas small (H. Morishita and K. Fujimoto)

This chapter first covers principles of making antennas small, and follows with techniques to realize those principles for four types of small antennas. Each type is introduced with examples. One of the significant methods is to use materials, especially recently developed metamaterials (MM), by which novel small antennas may be created. Use of materials composed with SNG (Single Negative) and DNG (Double Negative) materials, including NRI (Negative Refractive Index) TL (Transmission Line) MMs, is described. A review to solve matching problems in very small antennas with application of SNG materials in the near field of a radiator is given.

Later in the chapter, optimization techniques in designing are introduced. Small antenna designers often encounter difficulty when an antenna is either too small to treat or is located in a complicated environment. Optimization techniques may be useful to ease such difficulty in designing. Four typical optimization techniques, GA (Genetic Algorithm), PSO (Particle Swarm Optimization), TO (Topology Optimization), and VMO (Volumetric Material Optimization) are described along with some application examples.

## Chapter 7 Design and practice of small antennas I (K. Fujimoto)

This lengthy chapter along with the next (Chapter 8) intends to provide practical design methods and illustrative examples. The chapter consists of four sections corresponding to four types of small antenna: ESA, PCSA, FSA, and PSA. This chapter is the first part, where ESA is dealt with. Design methods based on the principles shown in Chapter 6 are discussed and then numerous design examples taken from some related journals are provided.

## Chapter 8 Design and practice of small antennas II (K. Fujimoto)

Following the previous chapter's treatment of ESA, this again-lengthy chapter describes design methods and practical examples for the remaining three types of small antennas, PCSA, FSA, and PSA. In FSA, methods to enhance antenna performances, typically wideband, multiband, and UWB (Ultra Wideband) operations, are described. Integration of function into antenna structure is added as an important method to produce FSA, which includes reconfigurable antennas. Typical methods to produce PCSA are applications of EM materials/structures such as HIS (High Impedance Surface), EBG (Electromagnetic Band Gap), and DGS (Defected Ground Surface) as well as PEC (Perfect Electric Conductor) ground plane, by which antennas of low profile, wide bandwidth, higher gain, and arrays with closely spaced elements are realized. Today we see various small widely deployed RFID devices, which employ very small antennas that are considered as the most representative PSA.

## Chapter 9 Evaluation of small antenna performance (H. Morishita)

At first, specific problems that must be considered for evaluation of small antenna performances are discussed. In small-antenna measurements, a prime important matter is balanced and unbalanced geometries in the antenna structure and feed line to achieve measurement errors as small as possible. Optical fiber systems can replace coaxial cable systems in the small-antenna measurements to avoid serious errors duc to unfavorable current flow on the feed cable. Recommended practices in measuring important antenna parameters such as impedance, radiation patterns, and efficiency are described.

## Chapter 10 Electromagnetic simulation (H. Morishita and Y. Kim)

Electromagnetic (EM) simulation plays an important role in designing an antenna and finding its characteristics, especially when the antenna is too small to deal with and/or it is employed in complicated environments so that conventional design or evaluation of antenna characteristics is almost impossible. This chapter explains concepts of EM simulation first and then describes typical EM simulators. The typical simulation methods considered here are the IE3D, FIDELITY, HFSS, and MW studio, which are based on the Method of Moments (MoM), FDTD (Finite-Difference Time-Domain), FEM (Finite Element Method), and FIT (Finite Integration Technique), respectively. Examples of simulation applied to practical antenna models for either design or evaluation of performances are described.

## Chapter 11 Glossary (K. Fujimoto and N. T. Hung)

This chapter gives a catalog of typical small antennas to provide readers with data and information for assisting design and development. Most of the antennas treated are covered in more detail elsewhere in the book, alongside a few antennas covered in other literature. The glossary list gives a brief view of each antenna, its antenna type, its main features, and applications for every antenna, along with references and the chapter/section number where the antenna is described.

# Acknowledgements

The authors would like to express their highest appreciation to a number of people who assisted them to complete this book, providing useful information, data, and materials related with small antennas, contributing to prepare manuscripts and endeavoring to produce the book through sincere effort.

Particular acknowledgement should go to Dr. J. R. Copeland, who supported us with his editing skill being devoted to our manuscripts, and improved our nonnative English, especially from a technical point of view as an experienced antenna researcher. (He was one of the members who did pioneering work on the active integrated antennas at the Ohio State University Antenna Laboratories in the early 1960s, where the first author was also one of the members.)

The book wouldn't have been completed without the continuous support and patience of Cambridge University Press editorial and production staff. Sarah Marsh and many other editorial staff assisted us to produce and arrange our manuscripts. We paticularly acknowledge the support of Dr Julie Lancashire, Publisher, Engineering, who at first showed interest in publishing this book, then promoted the book project, and encouraged us in the process of production. She is especially thanked for her patience at the time when the progress of the project stalled, by understanding of our situation, especially my health condition and unavoidable complications caused by the unexpected huge disaster in north-eastern Japan on March 11, 2011. The authors are also deeply indebted to Jon Billam for his kind review of the manuscripts in detail, by which the manuscript was greatly improved and refined.

Kyohei Fujimoto would personally like to thank many people for their assistance and contribution to preparing manuscripts. Among them, immense appreciation should go to Professor Hiroyoshi Yamada, Niigata University, Japan, who generously helped in producing figures and tables in Chapters 6, 7, and 8 with hearty cooperation of not fewer than 10 of his students who worked excellently under his instruction. We are also much obliged to Dr. Yongho Kim, Nguyen Tuan Hung, and Yuki Kobayashi, current and former graduate students of Professor H. Morishita, who assisted us by exhibiting their skillful work, in drawing figures in Chapters 1 to 4, improving poor quality figures in some few chapters, and producing Chapter 11. Last but not least, I must acknowledge my wife Machiko Fujimoto, who allowed me to continue to work on the book by sacrificing much domestic life, with patience for a couple of years until completion of the book.

Hisashi Morishita would like to thank all his graduate students of his laboratory in National Defense Academy, Japan for their wholehearted cooperation in producing figures and tables in Chapters 9 and 10. Especially, Dr. Y. Kim, who was his graduate student, who contributed to produce Chapter 10.

# 1 Introduction

The antenna first used in radio communication was a small antenna, which was a fan-type monopole (Figure 1.1), developed by G. Marconi, used for trans-Atlantic Ocean communication in 1901 [1]. The antenna appeared to be very large, as it was hung by two masts 48 meters high and 60 meters apart so that it could never be considered small [1]; however, since the dimensions were a small fraction of wavelength (about 1/6 of the operating wavelength, 366 meters), it was "electrically" small.

An Electrically Small Antenna – ESA – (the definition is described in 2.1) is an antenna of dimensions much smaller than the wavelength. In the classical sense, there are two types of ESA; one is an electric element, which couples to the electric field and is referred to as a capacitive antenna, and another is a magnetic element (electric loop), which couples to the magnetic field and is referred to as an inductive antenna. These are ESA categories; however, many practical antennas are some combination of these two types. It should be noted that small electric and magnetic elements in the forms of dipoles and loops have been used since 1887 when Hertz successfully generated and detected electromagnetic waves and verified Maxwell's prediction [2]. His invention of such small antennas certainly proved his success in demonstrating existence of electromagnetic waves. Although more than 120 years have passed since Hertz's experiment, his small dipoles and loops shown in Figure 1.2 (a) and (b) were so basic that essentially the same antennas are still being used.

Typical antennas in practical use for the early days of radio communications were ESAs. Examples of them are shown in Figure 1.3: (a) top-loaded monopole, (b) fan-type monopole, (c) multi-wire, (d) rectangular cage type, and (e) cage type, and so forth. Most of them were installed on ships and operated in frequency bands of LF and HF. Since then, small antennas have been used in various communication systems, particularly mobile systems, where small antennas were required. Antenna types used are not only linear but also planar and others such as composite and integrated types. Antenna technology has made steady progress along with the progress in communication systems and electronic devices, especially during World Wars I and II that provided the need for advanced antenna design and the surge of technology. Operating frequencies have gradually been raised to higher regions; from MF and HF bands to VHF, UHF, and SHF in recent years. Use of higher frequencies and smaller-sized electronic devices gave impetus to develop smaller antennas. One of the major trends in recent antenna technology is miniaturization of antenna systems, yet with improved functioning and further sophistication. The latest applications of small antennas are mainly to mobile communications and newly deployed

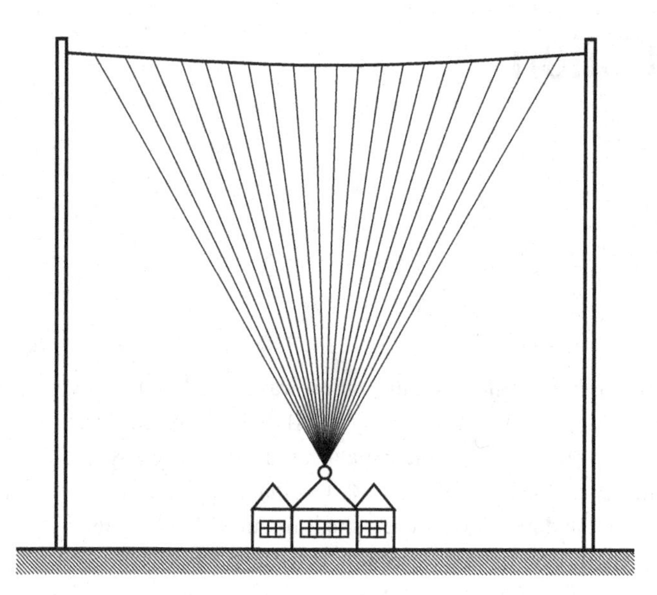

**Figure 1.1** The first radio communication antenna used for trans-oceanic communications developed by Marconi in 1901 [1].

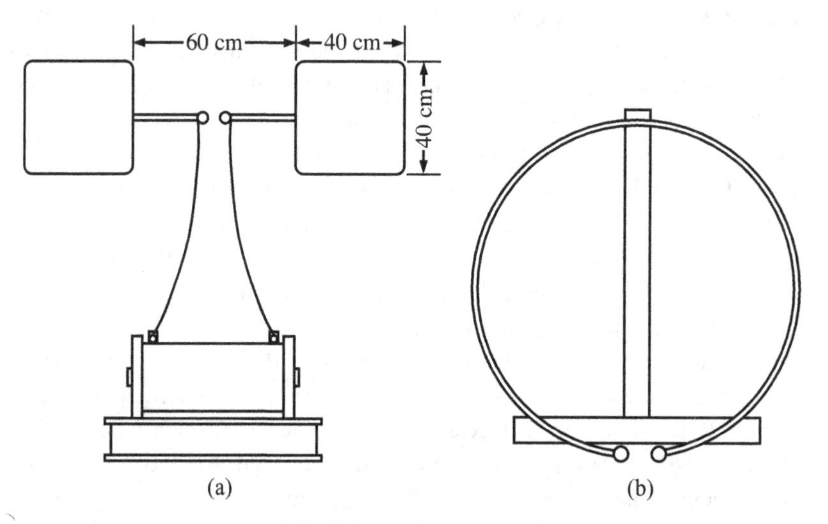


**Figure 1.2** Hertz antenna used for his experiment of the first radio wave transmission experiment [2, 3, 4].

various wireless systems. The typical mobile systems are mobile phone systems, which have evolved from the first generation systems in the early 1950s to the present fourth generation through the third-generation systems and further progressed wireless systems such as smart phones and handy tablets, where small antennas are indispensable. Recently emerged wireless systems also demand small antennas. These recent wireless systems are applied not only to communications, but also to control, sensing, identification, medical use, body communications, and data and video transmission. Typical systems

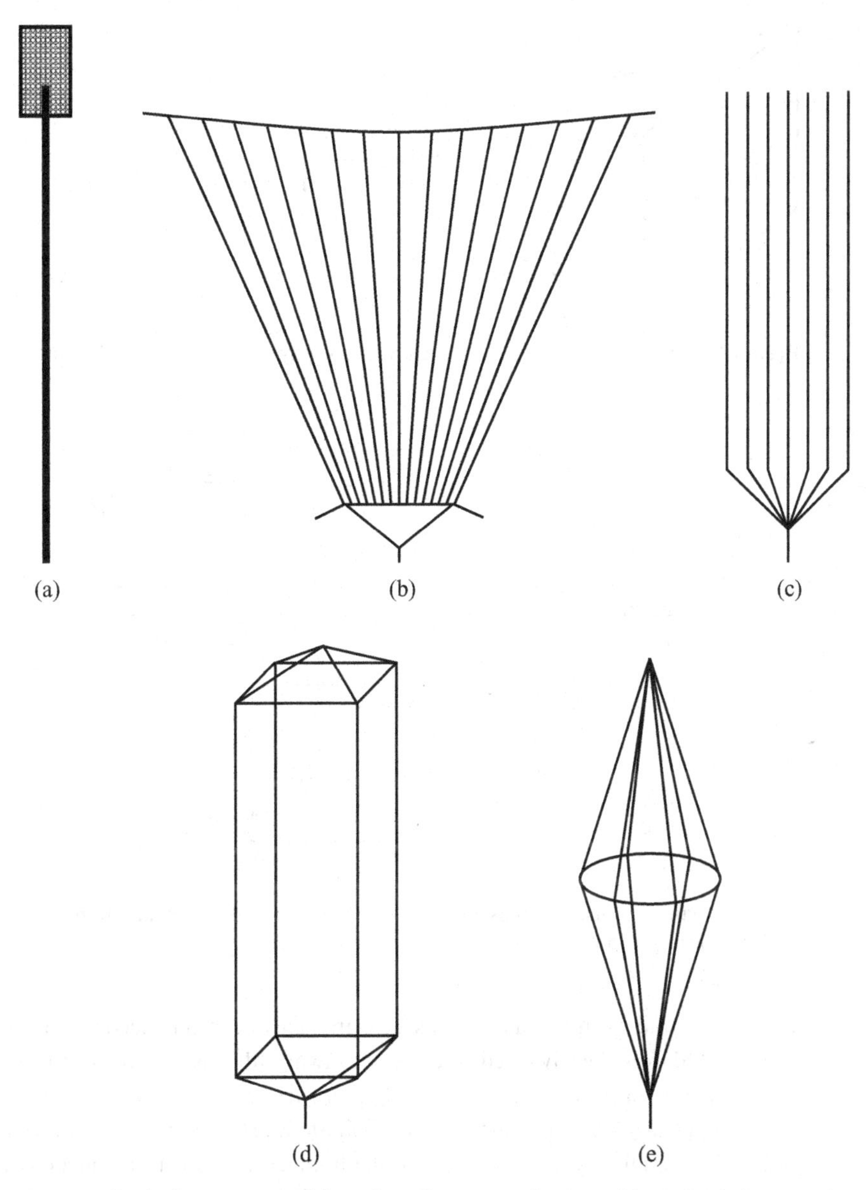


**Figure 1.3** Typical antennas used in early radio communications; (a) top-loaded monopole, (b) fan-type monopole, (c) multi-wire, (d) rectangular cage type, and (e) cage type [2, 3, 4].

for which small antennas are required, are NFC (Near Field Communication) systems, including RFID (Radio Frequency Identification), UWB (Ultra Wideband) systems, and wireless broadband systems such as WLAN (Wireless Local Area Network) systems and WiMAX (Worldwide Interoperability for Microwave Access) systems.

Requirements from various mobile systems have intensely stimulated research and study of small antennas and as a consequence have brought promotion of novel antenna systems development. They have also contributed to improved antenna performances

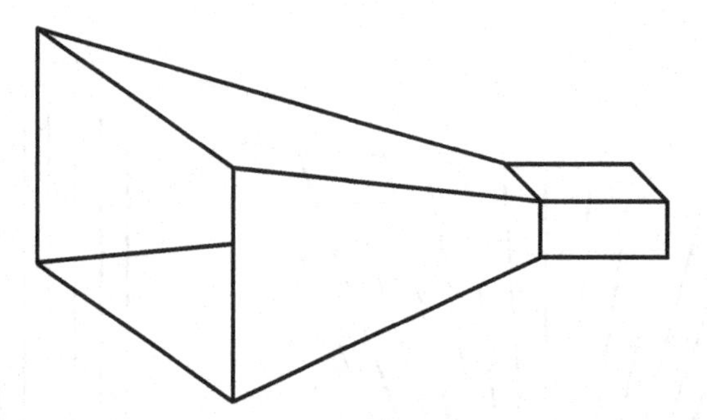

**Figure 1.4** A millimeter-wave horn antenna, which is physically small, but electrically large.

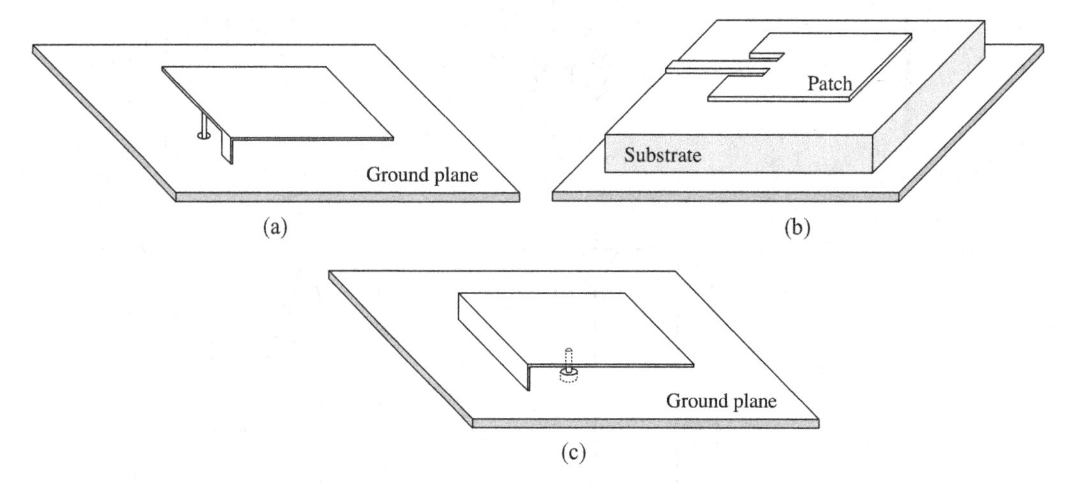


**Figure 1.5** Typical planar antennas; (a) PIFA (Planar Inverted-F Antenna), (b) patch antenna, and (c) MSA (Microstrip Antenna).

year by year and the antenna functions have gradually been enhanced to meet the requirements raised from the advanced systems. Meanwhile, discussions on the fundamentals of small antennas have continued ceaselessly ever since Wheeler's work, because of continued significance. Problems that concern limitations are still regarded as very interesting and controversial topics. Knowledge of the limitations informs antenna engineers how to approach the limitations with an antenna of given size.

There is some ambiguity in the expression "small antenna," because one can say, "This is a small antenna," when it looks physically small. For example, a millimeter-wave (MMW) horn antenna (Figure 1.4) of only the size of a human palm looks simply small and it might be called a "small" antenna. However, since a MMW antenna often has the aperture of a few or more wavelengths, the antenna dimension shouldn't be said to be "small," but in fact "electrically large" in terms of the operating wavelength. On the contrary, for an antenna having dimensions of a small fraction of the operating wavelength like the Marconi antenna mentioned earlier, the antenna is referred to as an "electrically small" antenna. Here, small antennas will be treated in a wider sense, in

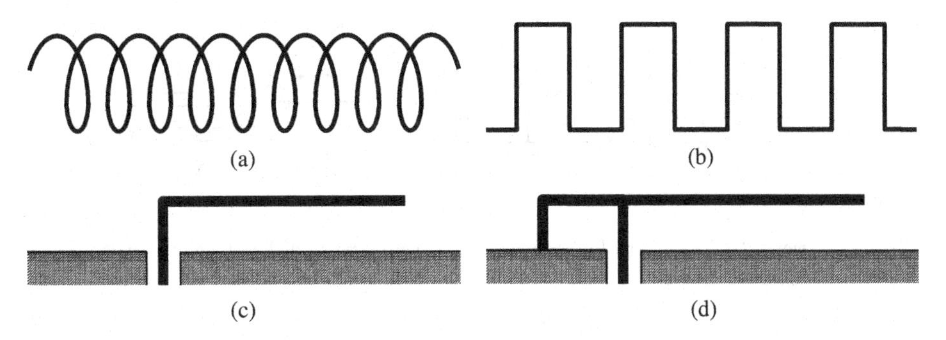


**Figure 1.6** Typical linear small antennas; (a) NMHA (Normal Mode Helical Antenna), (b) Meander Line Antenna, (c) ILA (Inverted-L Antenna), and (d) IFA (Inverted-F Antenna).

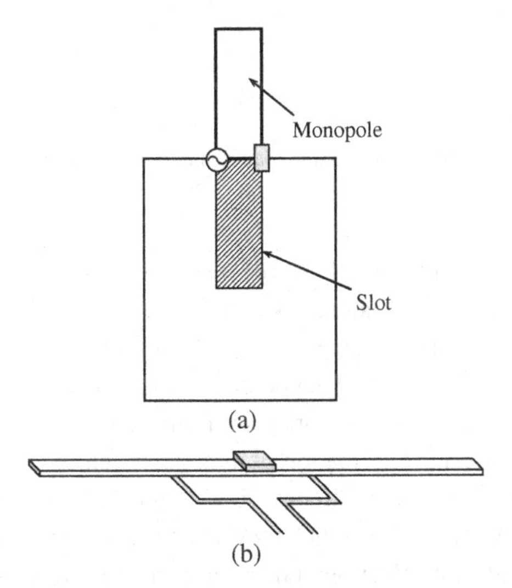


**Figure 1.7** Examples of composite antenna; (a) self complementary antenna and (b) active integrated antenna.

which antennas are defined four ways; first, in terms of simply physical size; second, in comparison with the operating frequency; third, in relation with function; and fourth, in the size-constrained structure. Small antennas defined in terms of the function may be unfamiliar in general; however, antennas so defined should be appreciated as a significant concept that comprehensively embraces the term "small."

Various types of small antennas have so far been evolved from simple dipoles or loops. Planar and low-profile antennas are typical ones. Practical examples are illustrated in Figure 1.5, which shows (a) Planar Inverted-F Antenna (PIFA), (b) patch antenna and (c) Microstrip Antenna (MSA). Meanwhile, examples of linear antennas are shown in Figure 1.6, which depicts (a) Normal Mode Helical Antenna (NMHA), (b) Meander

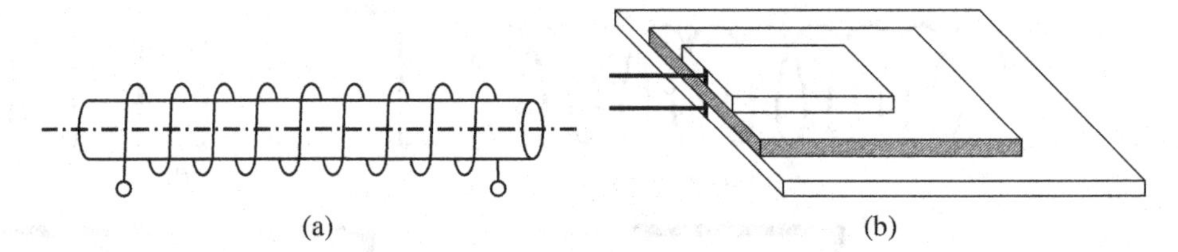


**Figure 1.8** Examples of loaded antenna: (a) a ferrite loaded coil, and (b) dielectric loaded patch antenna.

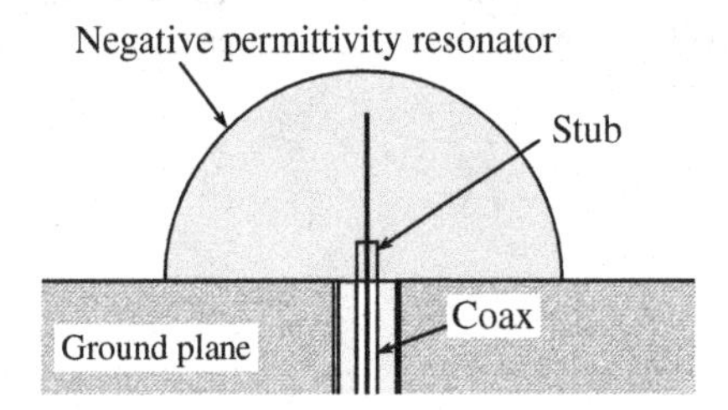


**Figure 1.9** Example of metamaterial (negative permittivity) loaded small antenna [5–7].

Line Antenna (MLA), (c) Inverted-L Antenna (ILA), (d) Inverted-F Antenna (IFA) and composite antenna. Examples of composite antennas are shown in Figure 1.7, which depicts (a) an example of complementary antennas and (b) an example of active integrated antennas. Integration technique, by which some device or circuitry is combined with an antenna structure so that the antenna works with enhanced performance or function, can be utilized for small-sizing. Application of materials such as ferrite and/or dielectrics is also a useful means to reduce antenna size. Examples are an antenna loaded with a ferrite and an MSA loaded with dielectric material as shown in Figure 1.8(a) and (b), respectively. Presently, introduction of metamaterials such as mu-negative or epsilon-negative materials (MNM or ENM) and double negative materials (DNG) into the antenna structure has been discussed extensively worldwide as one of the promising candidates for antenna miniaturization [5–13]. An example of application of negative-epsilon material to small antenna structure is depicted in Figure 1.9 [6]. Development of negative-mu materials will also contribute to new small antennas [14]. These antennas will be discussed in later sections.

## References

[1] J. Ramsay, Highlight of Antenna History, *IEEE Antennas and Propagation Society Newsletter*, December 1981, 8.

[2] S. Tokumaru, *Invitation to Radiowave Technology*, Kohdansha, Japan, 1978, pp. 48–50 (in Japanese).

[3] O. Lodge, *Signaling Across Space Without Wires*, Van Nostrand, 1900.

[4] G. Pierce, *Principles of Wireless Telegraphy*, McGraw-Hill, 1910.

[5] H. R. Stuart and A. Pilwerbetsky, Electrically Small Antenna Elements Using Negative Permittivity Resonator, *IEEE Transactions on Antennas and Propagation*, vol. 54, 2006, no. 6, pp. 1644–1653.

[6] F. Bilotti, A. Ali, and L. Vegni, Design of Miniaturized Metamaterial Patch Antennas with $\mu$-Negative Loading, *IEEE Transactions on Antennas and Propagation*, vol. 56, 2008, no. 6, pp. 1640–1647.

[7] P. Y. Chen and A. Alu, Sub-Wavelength Elliptical Patch Antenna Loaded with $\mu$-Negative Permittivity Resonator, *IEEE Transactions on Antennas and Propagation*, vol. 58, 2010, no. 1, pp. 13–25.

[8] N. Engheta and R. W. Ziolkowsky, eds., *Metamaterials-Physics and Engineering Explorations*, IEEE Press, 2006, John Wiley and Sons.

[9] A. Lai, C. Caloz, and T. Itoh, Composite Right/left-Handed Transmission Line Metamaterials, *IEEE Microwave Magazine*, September 2004, pp. 34–50.

[10] C. Caloz, T. Itoh, and A. Rennings, CRLH Metamaterial Leaky-Wave and Resonant Antennas, *IEEE Antennas and Propagation Magazine*, vol. 50, 2008, no. 5, pp. 25–39.

[11] G. V. Eleftheriades and K. G. Balmain (eds.), *Negative Refraction Metamaterials: Fundamental Principles and Applications*, Wiley-IEEE Press, 2005.

[12] J. L. Volakis, C.-C. Chen, and K. Fujimoto, *Small Antennas, Miniaturization Techniques and Applications*, McGraw-Hill, 2010, chapter 6.

[13] F. Capolino (ed.), *Applications of Metamaterials*, CRC Press, 2009, Part IV.

[14] T. Tsutaoka, *et al.*, Negative Permeability Spectra of Magnetic Materials, *IEEE iWAT* 2008, Chiba, Japan, P202, pp. 279–281.

# 2 Small antennas

## 2.1 Definition of small antennas

Here in this book, small antennas are treated with a concept that embraces not only electrically small antennas, but also other types of small antennas. Categories used to classify small antennas include functions as well as dimensions, because small antennas being used practically are not only what we call "Electrically Small Antennas," but also simply physically small antennas – antennas of partly electrically small dimensions and antennas equivalently small in terms of functions. Conventionally, ESA has been the main subject when small antennas are discussed; however, other types of small antenna have comparable significance with the ESA, depending on the situation of the practical applications. The categories used here are Electrically Small Antenna (ESA), Physically Constrained Small Antenna (PCSA), Functionally Small Antenna (FSA), and Physically Small Antenna (PSA) [1].

An ESA is an antenna conventionally defined as an electrically small-sized antenna; i.e., one having dimensions much smaller than the wavelength. However, this definition is unclear, since the dimensions are not described precisely. Wheeler defined the ESA as an antenna having the maximum size that can be circumscribed by a radian sphere, with a radius of one radian in length ($= \lambda/2\pi$) [2]. However, an antenna having the maximum dimension of a radian length may not necessarily be categorized as an ESA, because taking a dipole antenna as an example – which has the length of a radian length, $2 \times \lambda/2\pi$ ($= 0.32\lambda$) – it can no longer be called electrically small, as the size becomes no longer a small fraction of the wavelength. Hence classifying the radian-length dipole antenna as an ESA is not reasonable.

There was another definition of ESA – that was an antenna having dimension not greater than one eighth of a wavelength [3]. One more definition was given by King, who did not use the term ESA, but "very short antenna," that referred to an antenna having the length in terms of $ka \leq 0.5$ ($a$: a half length of a thin linear dipole and $k = 2\pi/\lambda$, ($\lambda$: wavelength)) [4]. Recently, Best followed King's definition that ESA is an antenna having $ka \leq 0.5$, the same dimensions as King's. This is more reasonable to define as ESA, so hereafter we will reserve the term ESA for small antennas having dimension $ka$ smaller than 0.5, where $a$ is the radius of sphere circumscribing the antenna.

A PCSA is an antenna, not having dimensions of ESA, but a part of which has dimensions corresponding to the ESA. A low-profile antenna, for example, with the height over the ground plane or the thickness of say $\lambda/50$ or so, like an Inverted-L antenna, planar antennas such as a patch antenna, a microstrip antenna (MSA), and so forth, can be classified into "Physically Constrained Small Antenna" (PCSA).

Another classification of small antennas relates to antenna functions. When an antenna can attain either additional functions or enhanced performances with the dimensions being kept unchanged, it can be said to have an equivalent reduction in the antenna size, because with normal antennas such enhanced function or improved performance could result only from enlargement of dimensions. An example is an antenna designed to have wideband, multiband performance, radiation pattern shaping, and array thinning without any change in the dimensions. This class of antenna is referred to as "Functionally Small Antenna" (FSA).

Among the FSA, there may be antennas that can be categorized also as ESA and/or PCSA at the same time, because their antenna dimensions are small enough to be comparable with a fraction of a wavelength, which is the criterion for ESA and/or PCSA.

A "Physically Small Antenna" (PSA) is an antenna, which differs from any of the above, but has physically small dimensions as measured. When the volume of an antenna is bounded by 30 cm or less, we may call the antenna a PSA. There is not any strict physical meaning in the definition, but only a sensual measure based on the common understanding of "smallness" in physical size. An example is a millimeter-wave horn antenna with an aperture size of, say 5 cm approximately. Many of the MSAs may be classified into either PCSA or PSA.

These definitions cannot be analytically formulated in practice, because an antenna may be differently classified depending on the practical situation. For instance, when a short monopole categorized in the ESA is put on a rectangular conducting plate of 1/4 wavelength, the antenna can no longer be classified into ESA, because the antenna dimensions are effectively enlarged with inclusion of the plate. The plate of finite size acts as a part of the radiator so that the monopole plus the plate constitutes an antenna system. In this case, the radiation performance of the monopole is enhanced by contribution of the radiation current on the plate induced by the monopole. A very small ceramic-chip antenna practically used in small mobile terminals is another example.

In any case the size of the antenna system must be defined with the maximum size, which includes the platform together with the antenna element.

Although an antenna may be classified into one of the four categories mentioned above, the categories are not necessarily distinct so that one class of antenna can be categorized also in another class. For instance, when a very thin patch antenna has the edge (radiating aperture) of 1/5 wavelength in its periphery, the antenna is referred to as both an ESA and a PCSA.

## 2.2 Significance of small antennas

The significance of the small antenna is recognized from its important role in various wireless systems, where small antennas are indispensable. In the early days of communications, when electronic devices such as thermionic vacuum tubes were not yet available, antenna technology was essential to extend the communication range. Development of tuning and matching concepts gave impetus to promote further increase in communication capability. In those days, most of the antennas used were ESA, because operating frequencies were mostly MF and HF bands. In the 1930s, antennas such as slot antennas and ferrite antennas were put to practical use. It was around this era when needs for small antennas furthered researchers' interest in studying the fundamentals of small antennas, and investigation of limitations of small antennas was inspired. In parallel with research, practical development of small antennas was advanced. Examples are normal mode helical antennas, MSAs, and patch antennas. Their practical applications extended to various mobile systems, particularly to land vehicles. Remarkable progress of antenna technology can be attributed to World Wars I and II. After the wars, and toward the end of the twentieth century, various composite antennas and integrated antennas appeared. Meanwhile, mobile communications drove development of small antennas, as mobile terminals needed small equipment, for which small-sized, compact, and lightweight antennas were required.

So far, creation and invention of small antennas have accelerated rapidly in response to worldwide demands raised urgently by the growth of mobile phone systems. Recent trends are increasing demands arising particularly for newly deployed wireless mobile systems – NFC (Near Field Communication) systems, including RFID (Radio Frequency Identification), UWB (Ultra-Wide Band), and wireless broadband systems such as WLAN (Wireless Local Area Networks) and WiMAX (Worldwide Interoperability for Microwave Access). Small antennas have already played an obscure but key role in such wireless mobile systems, as the systems do not work without these devices. Antennas in these systems have acted as a key component, since they almost determine the system performance, especially in small wireless equipment, because the systems are designed to exhibit almost the highest achievable electronic performance, and only the antenna can further improve the system performance. Since downsizing of electronic devices and their design technology as well has expedited progress in small wireless equipment, small antennas have become inevitable for these systems. As such, it can also be said that such newly deployed wireless systems owe much in their development to small antennas, and in turn small-antenna technology has advanced by being employed in these wireless systems. Recent urgent demand is application of small antennas to the MIMO (Multi-Input Multi-Output) systems. WiMAX already employs the MIMO system in order to transmit high data rate to the 4G (4th Generation) mobile systems, and other wireless broadband systems where high data rates will be needed will adopt the MIMO system. Small antennas are suitable for small MIMO equipment, where space to install antennas is limited to small areas, and close proximity of antenna elements is unavoidable.

Downsizing of an antenna without degrading the antenna performance imposes a severe design issue for antenna engineers – how to keep the antenna performance unchanged while the antenna size is reduced. The latest demands for small antennas are not only physically small, but also functional, typically wideband, multiband operation, or including signal processing capability.

Types of antennas required for modern wireless systems are not only ESA, but also PCSA, PSA, and FSA, depending on the system specifications. The major design parameters are the operating frequency, system specifications, dimension of the unit to which the antenna is installed, place and space to install the antenna inside the unit, and so forth. Selection of antenna type and the design of the antenna differ depending upon these parameters. It is most important to optimize the antenna design that meets the system requirements. There have been many small antennas used practically in current systems without well-optimized design. Such non-optimally designed antennas must be used only with understanding that if the system may not fully function to its maximum capability, it shouldn't bring any serious impediment. If the antennas were optimally designed, the antenna and thus the system performance would be improved. In a small mobile unit – for example a small RFID unit – an antenna is often used in poor impedance-matching condition, because the size of the antenna is too small to allow good matching. However, if good impedance matching were achieved, the system sensitivity would be increased so that the operating range could be extended. Another important advantage claimed by better matching is saving the battery life; the battery might be used for a longer time or could be replaced by a smaller one.

Increase in the bandwidth may also be expected by improving the matching condition. Proper design of matching circuitry is essential.

Although small antennas have been used widely in communications and other wireless systems, there are still many issues remaining that concern theory, design, and practical applications. Realization of an ESA is still a big, challenging problem for antenna engineers, and at the same time antenna engineers are urged to develop skills to develop practically useful small antennas.

Study of fundamentals of ESAs assists optimization of antenna design and development of new small antennas, augments the knowledge of essential properties and characteristics of small antennas, and provides a substantial clue to creating novel ESA implementations. In addition, knowledge of the radiation mechanism of small antennas, still being pursued, should add a significant key for giving birth to a unique small antenna when it is successfully mastered.

Small antennas offer the vital key to further development and progress of wireless systems, and in turn, these systems bestow an unparalleled opportunity for further advancement of small antennas.

## References


[1] K. Fujimoto *et al.*, *Small Antennas*, Research Studies Press, UK, 1987, p. 4.

[2] H. A. Wheeler, Fundamental Limitations of Small Antennas, *Proceedings of IRE*, vol. 35, Dec. 1947, pp. 1479–1484.

[3] S. A. Schelkunoff and H. T. Friis, *Antennas and Theory*, John Wiley and Sons, 1952, chapter 10.

[4] R. W. P. King, *The Theory of Linear Antennas*, Harvard University Press, 1956, p. 184.

# 3 Properties of small antennas

## 3.1 Performance of small antennas

Small antennas exhibit some specific characteristics that differ from antennas of a size comparable to the wavelength. The input impedance of a small electric (magnetic) dipole antenna is highly capacitive (inductive), while its resistive component is very small. Difficulty in the perfect matching to the load of 50 ohms is mainly due to this impedance characteristic.

The radiation pattern of a small antenna, as its size becomes smaller, tends to approach that of the classical elemental vertical dipole which is omnidirectional in the horizontal plane, with a figure-of-eight pattern in the vertical plane, and with a directivity approaching 1.5.

Small antennas, particularly ESAs, need specific design techniques.

### 3.1.1 Input impedance

A small dipole antenna of radius $d$ and the half length $a$ (Figure 3.1), when $ka$ is less than 0.5 ($k$: $2\pi/\lambda$), exhibits the input impedance $Z_a = R_a + \mathrm{j}\,X_a$ as follows [1];

$$R_a = 20(ka)^2 \tag{3.1}$$

$$X_a = 60(\Omega - 3.39)/ka \tag{3.2}$$

where $\Omega = \ln(2a/d)$.

Numerical values are illustrated in Figure 3.2. This is an excellent approximation of the correct value for $\beta a \leq 0.2$ and quite a good approximation for $\beta a \leq 0.5$. The figure shows loss-free cases, where $k = \beta$ with $\alpha = 0$ in $k = \alpha + \mathrm{j}\beta$.

Meanwhile, the radiation resistance $R_a$ of a small circular loop antenna of radius $a$ and wire radius $d$ (loop area: $A = a^2\pi$) for $\beta a \leq 1$ (Figure 3.3) [2][3] is

$$R_a = 20\,\beta^4 A^2. \tag{3.3}$$

The reactance $X$ can be found by calculating $\omega L$, where $L$, the inductance of a small circular loop, is given by [3]

$$L = a\mu\,[\ln(8a/d) - 2]. \tag{3.4}$$

Figure 3.4(a) gives $R_a$ with respect to $\beta a$, and Figure 3.4(b) depicts $L$ against $a/d$.

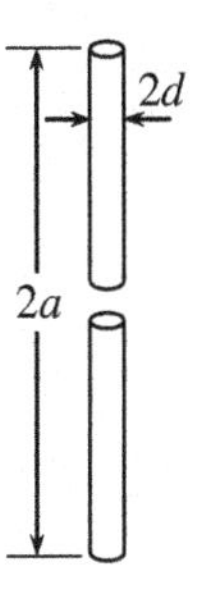


**Figure 3.1** A small dipole [1].

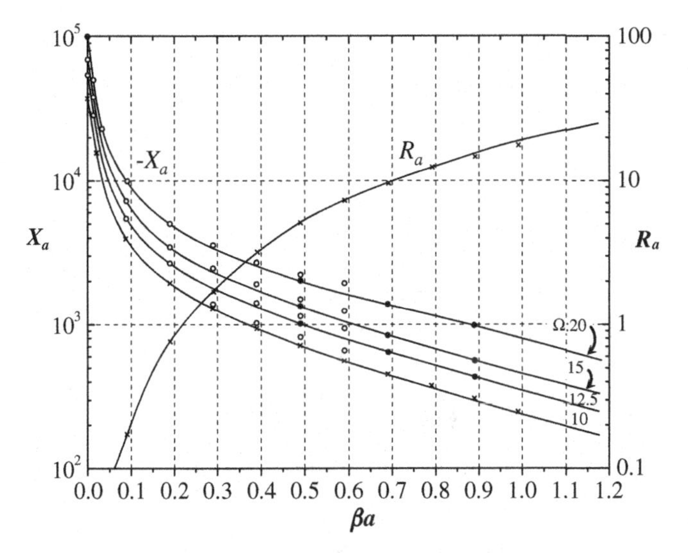


**Figure 3.2** Impedance characteristics of a dipole [1].

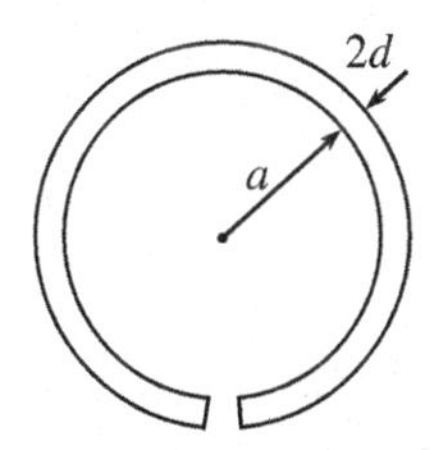


**Figure 3.3** A small circular loop antenna [1]–[3].

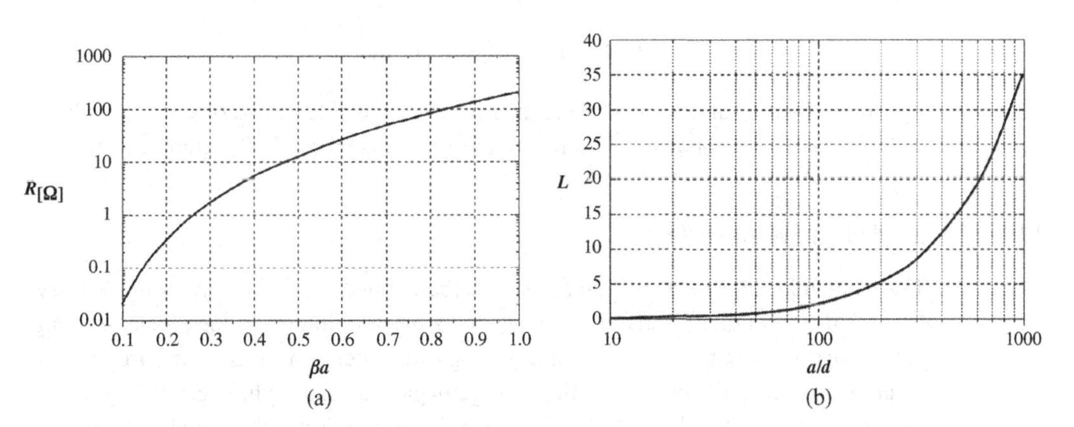


**Figure 3.4** Impedance characteristics of a loop antenna; (a) resistance component and (b) inductance.

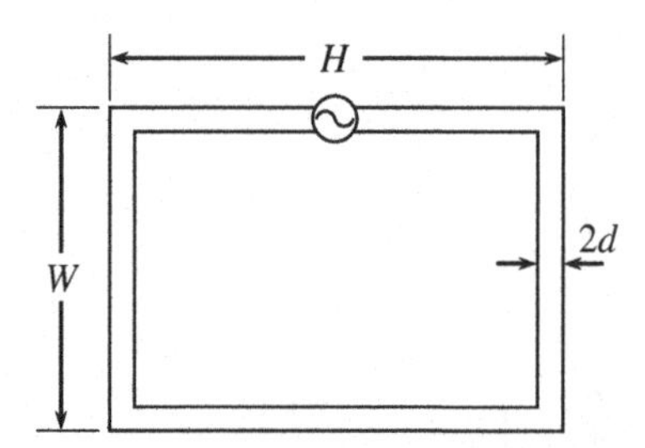


Figure 3.5 A small square loop antenna [3].

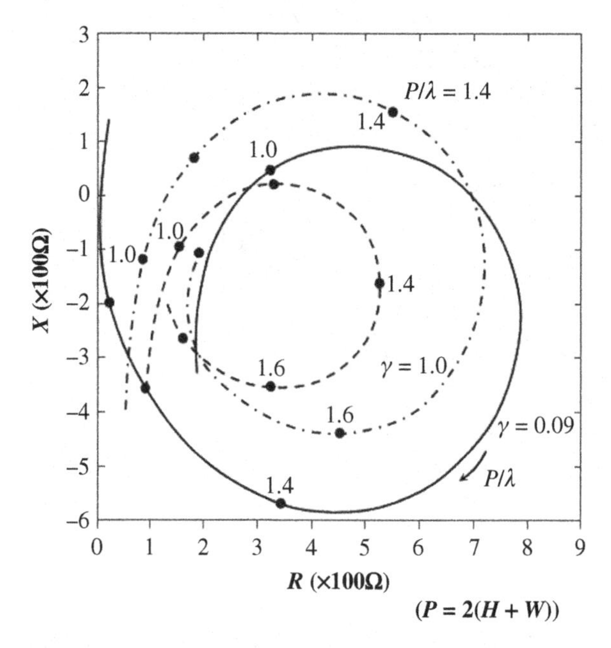


Figure 3.6 Impedance characteristics of a rectangular loop antenna [4].

For the square loop with sides $a$ and wire radius $d$ [3] (Figure 3.5),

$$L = 2\mu(a/\pi)[\ln(a/d) - 0.774]. \tag{3.5}$$

The impedance $R_a$ and $X_a$ of a rectangular wire loop of side $P = 2(H + W)$, $\gamma = W/H$ and wire radius $d$, is shown in Figure 3.6, where $P/\lambda$ is taken as the parameter [4].

### 3.1.2 Bandwidth BW and antenna *Q*

The bandwidth of an antenna cannot simply be defined the same way as in circuitry, because the functional bandwidth of an antenna may be limited by factors not existing in circuitry performance, for example, change of pattern shape or pattern direction, variation in impedance characteristics, change in gain, and so forth. In general, however, the bandwidth can be defined as a frequency band within which the antenna meets a

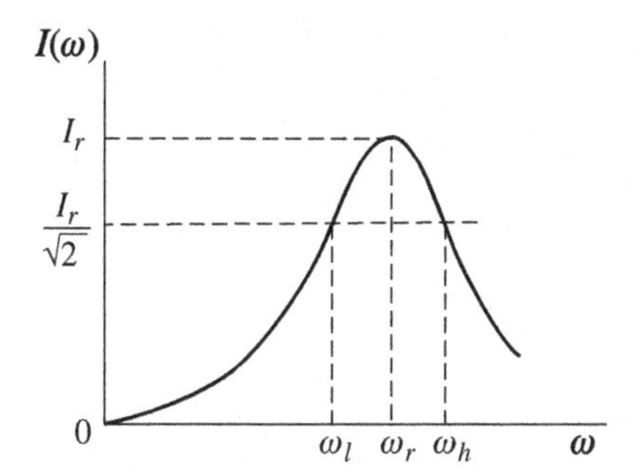


**Figure 3.7** Frequency characteristics of a LCR circuit near resonance.

given set of specifications, typically based on the impedance characteristics. In small antennas, either the antenna input impedance or the power spectra can be taken as the parameter to specify the bandwidth. The bandwidth so defined is the frequency difference $\Delta f$ between the highest frequency $f_h$ and lowest frequency $f_l$, both specified at a certain value of parameters such as VSWR $S$ or the return loss $R_L$, based on the input impedance $Z_a$, and by some level of power reduced from the maximum. These parameters, $S$ and $R_L$, are given by

$$S = (1 + |\Gamma|)/(1 - |\Gamma|) \tag{3.6}$$

$$\text{and } R_L = 20\log|\Gamma| \tag{3.7}$$

where $\Gamma$ (the reflection coefficient) $= (Z_a - Z_L)/(Z_a + Z_L)$.

Here $Z_L$ is the load impedance.

The load impedance $Z_L$ is usually taken to be a pure resistance of 50 ohms. Usually $S = 2$ is taken, which corresponds to $R_L \approx -10$ dB. $\Gamma$ for the impedance at the half-power point is 0.707 and $S$ for this $\Gamma$ is 5.828. The half-power point $\mathrm{BW}_{1/2}$ is usually used in circuitry and not generally used in the antenna case.

For convenience, the relative bandwidth RBW, which is a ratio of BW to the center frequency $f_0 = (f_h - f_l)/2$ or the resonance frequency $f_r$, is generally used, instead of BW. Sometimes, a ratio of $f_h/f_l$ is used as a measure of wideband capability.

Antenna $Q$ is defined from the physical implication as

$$Q = \omega P_{react}/P_{rad} \tag{3.8}$$

where $P_{react}$ denotes the reactive power stored in the antenna structure and $P_{rad}$ is the radiated power from the antenna.

Originally, $Q$ was used as a circuit parameter showing a measure of selectivity of a tuning circuit, generally when the BW is relatively narrow. Taking an LCR series circuit as an example, the impedance $Z = R + \mathrm{j}(\omega L - 1/\omega C)$. When a voltage $V$ is applied to the circuit, current $I$ flows in the circuit, that is, $I = V/|Z| = V/\sqrt{R^2 + (\omega L - 1/\omega C)^2}$. The current $I(\omega)$ with respect to frequency $\omega$ is depicted in Figure 3.7. At resonance $\omega = \omega_r$, the current $I_r = V_r/R$, because $\omega L = 1/\omega C$. Taking a ratio of $I/I_r = G$ in dB, $G$ is given by

$$G = -10\log[1 + (\omega L - 1/\omega C)^2/R^2](\mathrm{dB}). \tag{3.9}$$

At the half-power point, $I/I_r = 1/\sqrt{2}$, and $G = -3$ dB. The frequency bandwidth $\Delta\omega$ for the half-power point is $(\omega_h - \omega_l)$, where $\omega_h$ and $\omega_l$, respectively, are the higher and lower half-power point frequencies (Figure 3.7), and given by

$$\omega_h = \sqrt{\omega_r^2 + (R/2L)^2} + (R/2L) \text{ and}$$

$$\omega_l = \sqrt{\omega_r^2 + (R/2L)^2} - (R/2L) \tag{3.10}$$

$$\text{since } \Delta\omega = 2\pi\Delta f, \Delta f = f_h - f_l = R/2\pi L. \tag{3.11}$$

This $\Delta f$ is used when $f_h$ and $f_l$ respectively, are not far removed from the resonance frequency $f_r$.

At the resonance, voltage $V_c$ of the capacitance and $V_L$ of the inductance, respectively, are

$$V_c = (\omega_r L/R)V \quad \text{and} \quad V_L = (1/\omega_r CR)V. \tag{3.12}$$

Since at the resonance $\omega_r L = 1/\omega_r C$, dividing this by $R$ will give $Q$ by the definition.

$$\omega_r L/R = 1/(\omega_r CR) = Q. \tag{3.13}$$

Then, from (3.11),

$$\text{BW}/f_r = \text{RBW} = 1/Q. \tag{3.14}$$

This provides the relation of a RBW to an inverse of $Q$, when the BW is not very wide. In the antenna case, this is applicable only when $Q$ is much greater than unity.

The term $Q$ was originated to mean "quality" of the circuit component or the selectivity of the circuit and is abbreviated as "$Q$." In order to attain a highly selective circuit, a coil of low loss $R$, thus high $Q$ is desired. From this, $Q$ is used to represent a measure of high quality of the coil that also stands for high selectivity of the circuit. However, the $Q$ is generally understood from the physical implication rather than the quality of the circuit as described by (3.9).

The bandwidth of a small antenna can be improved by inserting a network to the matching circuit, whereas the size of antenna limits the bandwidth. Fano demonstrated bandwidth increase by means of matching with a network [5][6]. The Fano concept can be applied to matching of an antenna so as to achieve wider bandwidth. Matthaei *et al.* treated this problem and gave design parameters along with the improvement factors as follows [7]:

$$\begin{aligned} \tanh a/\cosh a &= \tanh (nb)/\cosh b \\ \cosh (nb) &= \Gamma \cosh (na) \\ \text{and } \sinh b &= \sinh a - 2(\kappa)\sin(\pi/2n). \end{aligned} \tag{3.15}$$

Here $a$ and $b$ are parameters to be determined, $\Gamma$ is the reflection coefficient, $\kappa$ is $1/Q$ of the load at the band edge, and $n$ is the number of additional matching circuits. Matthaei solved equations (3.15) and obtained numerical values of these parameters as shown in Table 3.1, where values of $a$, $b$, and $\kappa$ are shown for VSWR $\leq 2$. Hansen gave the

**Table 3.1** Parameters related with matching circuit design [7]

| $n$ | $a$ | $b$ | $\kappa$ |
|---|---|---|---|
| 1 | 1.81845 | 0.32745 | 1.33333 |
| 2 | 1.03172 | 0.39768 | 0.57735 |
| 3 | 0.76474 | 0.36693 | 0.46627 |
| 4 | 0.62112 | 0.33112 | 0.42416 |
| 5 | 0.52868 | 0.30027 | 0.40264 |
| $\infty$ | 0 | 0 | 0.34970 |

**Table 3.2** Bandwidth improvement factor [8]

| $n$ | $IF$ |
|---|---|
| 1 | 2.3094 |
| 2 | 2.8596 |
| 3 | 3.1435 |
| 4 | 3.3115 |
| $\infty$ | 3.8128 |

bandwidth improvement factor, which is defined as a ratio ($\kappa$ for $n = 1$)/($\kappa$ for $n > 1$), as Table 3.2 shows. In the table, values are given for VSWR $\leq 2$ [8].

A novel matching method to increase bandwidth is application of Non-Foster circuits. The Foster Reactance Theorem says that the frequency derivative of the reactance $dX/d\omega > 0$ for the circuit which is composed with only lossless reactance $X$. The Non-Foster circuit violates this theorem and may include not only passive components, but also active components in the network. A typical Non-Foster circuit is an NIC (Negative Impedance Converter), which transforms positive impedance to negative impedance [9]. Sussman-Fort demonstrated bandwidth improvement by applying a negative capacitance circuit to matching for a short dipole and obtained a 10-dB improvement over more than an octave of frequency [10].

In small antennas, $Q$ or bandwidth that can be attained with a given size is limited. The fundamental limitation of small antennas has so far been discussed by many researchers, first by Wheeler and then Chu and others. These will be described in the next chapter.

### 3.1.3 Radiation efficiency

The radiation efficiency $\eta$ is given by

$$\eta = R_{rad}/(R_{rad} + R_{loss}) \tag{3.16}$$

where $R_{loss}$ denotes the loss resistance, which includes $R_{La}$ in the antenna structure and $R_{Lc}$ in the matching circuit. Input impedance of a small electric dipole antenna has a large capacitive component, so for matching, a large conjugate impedance – i.e. an inductance – is required. When a small dipole, for example, is a short stub of 0.16 $\lambda$ with radius $r$ of $3.3 \times 10^{-4}$ $\lambda$, $R_{rad}$ is 0.1Ω and $X_{rad}$ (input reactance component) is –2071 Ω. The matching condition requires an inductive component of +2071 Ω. When $Q_L$ of this inductance is 100, for example, $R_{Lc}$ of the inductive component alone would be 20.7Ω. The radiation efficiency $\eta$ in this case would be about 0.005 – that is 0.5%. With $Q_L$ of 400, $\eta$ increases to 0.019, but that's still only 1.9%. This is because of the much greater loss resistance $R_{Lc}$ in the inductive component over the radiation resistance $R_{rad}$. Practically, a transformer between the low radiation resistance and the load resistance 50Ω is necessary and the loss at that transformer should be taken into consideration.

Meanwhile, when a small magnetic dipole of the circumference $0.1\lambda$ with the same thickness as the above dipole is used, $\eta$ is 9.28% if $Q_C$ of the capacitive component is 1000. If $Q_C$ of that capacitance is 4000, $\eta$ becomes 21.1%. Because $Q_C$ of the capacitive component is usually higher than $Q_L$ of an inductive component, a small magnetic antenna element is advantageous to obtain higher radiation efficiency when about the same size of small antenna is considered.

As such, matching design will be very serious as the size of antenna becomes smaller.

### 3.1.4 Gain

Gain $G$ of an antenna is related to its directivity by accounting for efficiency and two different matching factors.

$$G = \eta D M p \tag{3.17}$$

where $D$ is directivity, $M$ is the impedance-matching factor, and $p$ is the polarization-matching factor. $M$ is given by using reflection factor $\Gamma$ at the input terminals and VSWR $S$ as

$$M = 1 - |\Gamma|^2 \leq 1 \tag{3.18}$$

$$= 4S/(1+S)^2. \tag{3.19}$$

The polarization factor $p$ is given by

$$p = |\rho_H \cdot \rho_E|^2 = |\cos\psi_p|^2 \tag{3.20}$$

where $\rho_H$ and $\rho_E$ are unit vectors of the wave with horizontal and vertical polarization, respectively, and $\psi_p$ is the angle between the two vectors. Here the electric fields $E_H$ and $E_E$ of incoming waves $E_i$, respectively, are described as

$$E_H = \rho_H E_i \tag{3.21}$$

$$E_E = \rho_E E_i. \tag{3.22}$$

Gain of a very small antenna is determined almost by $\eta$ and $M$, because $D$ is around 1.5 independently of the size and $p$ is usually taken as unity, since polarization matching can be good in almost all cases. In practical small antenna design, a condition $M \neq 1$ will be often encountered. The $M$ may take much lower values than unity, as the antenna size tends to be smaller. Gain reduction due to mismatching occurs in many cases with improper design of small antennas, and specific design techniques are needed to avoid degradation of antenna performance due to mismatching. Impedance matching problems will be discussed in the next section.

## 3.2 Importance of impedance matching in small antennas

In communication systems, signal transmission through the network under the well-matched condition is essential for no loss or no serious degradation in the signal quality

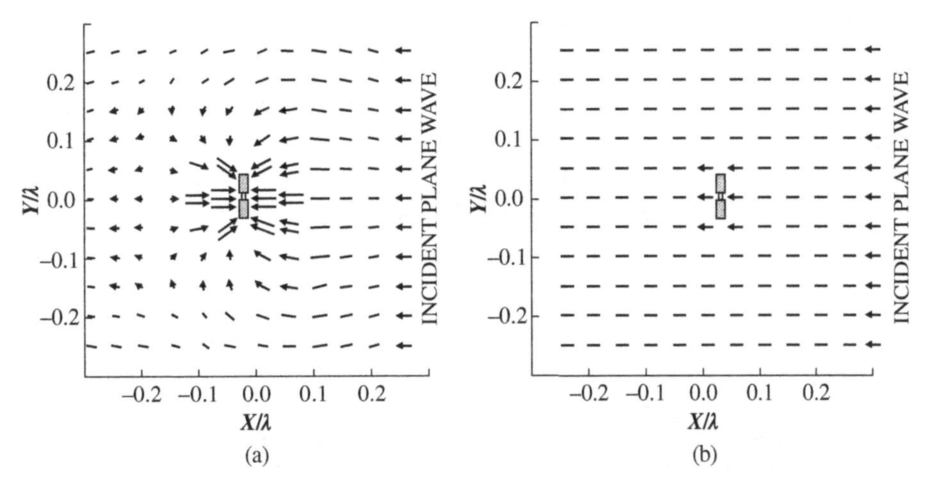


**Figure 3.8** Power flow of a plane wave incoming to a small dipole [11].

or quantity. The same is true for antennas. Achieving good matching condition is particularly a serious problem in small antennas, because of difficulty in proper matching of small antennas to the load of 50 ohms. There are often cases where proper matching is impossible.

The importance of matching in a small antenna can be visualized by a model shown in Figure 3.8, where a small dipole antenna is placed in an environment with a plane wave incoming [11]. In the figure, the plane wave energy flow is illustrated by arrows, showing amplitude with the length and the direction with the arrow. The plane wave energy is intercepted by the antenna perfectly when the antenna is in the perfectly matched condition as Figure 3.8(a) illustrates. On the other hand, when the antenna is in mismatched condition the plane wave energy flow passes through the antenna as Figure 3.8(b) illustrates. This demonstrates that the antenna does not function as a radiator when it is in mismatched condition, and indicates how crucial a problem matching is for small antennas.

In practice, the smaller the antenna will be, the harder the matching to the load of 50 ohms becomes. In addition, a transformer necessary for transferring the very low resistive component of the antenna to the load 50 ohms adds additional loss so that the radiation efficiency will be further lowered as a consequence.

When an antenna is constructed as a structure within which self-resonance is possible, no reactance component is necessary for the matching, and the loss problem – at least in the matching circuit – is relieved. Hence higher efficiency may be achieved. Examples of such antenna structures are NMHA (Normal Mode Helical Antenna), meander line, zigzag line, fractal shape, and composite antennas, in which resonance condition at lower frequencies can be achieved by combining antenna elements.

The latest technology offers the novel matching method, which applies the Non-Foster circuit and metamaterials to the matching circuits. Difficulty attributed to loss in the devices or the matching circuit can be overcome by applying the Non-Foster circuit to matching, and yet wideband matching may be expected.

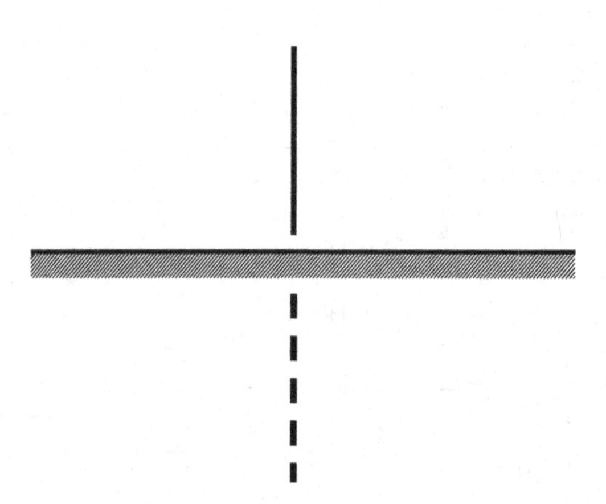

**Figure 3.9** Image of a vertical monopole on a ground plane.

Applications of metamaterials have been shown to be useful for matching in space. Stuart has introduced a new concept [3], which demonstrated an application of negative-epsilon material to matching in the near field of a very small monopole antenna as Figure 1.9 illustrated. The principle is to compensate the reactive (capacitive) fields outside the antenna with an inductive component due to a negative epsilon material. By this means, an ultra small antenna is possibly created.

## 3.3 Problems of environmental effect in small antennas

Antenna performance is affected by proximity effects due to materials existing near an antenna. A simple example is a monopole placed on a ground plane (GP) (Figure 3.9). When the size of the GP is infinite, a virtual monopole is produced beneath the GP by the image effect as shown with a dotted line below the GP in Figure 3.9. As a consequence the antenna is treated equivalently as a dipole. However, when the size of GP is finite, radiation current is induced on the GP by the excitation of the monopole, and the GP will act as a part of the antenna system. Thus the antenna together with the GP exhibits different performance from that of the dipole only.

Similar proximity effects are often observed in an antenna system located near some influential materials. The proximity effect is either advantageous or disadvantageous for an antenna system, depending on the dimensions of the antenna element and the materials. In general, one common situation is variation or degradation of the antenna performance due to the proximity effect. Contrarily, an advantage derived from the proximity effect is often observed for cases where the antenna performance is enhanced by the existence of the proximity effect. A practical example of such advantage is a PIFA installed inside a mobile terminal. The PIFA itself is an antenna of inherently low gain and narrow band, say about 1%–2% when the antenna is placed on an infinite or comparable size of GP; however, a PIFA installed on a finite size GP, for example that in a mobile phone terminal, exhibits higher gain and wider bandwidth than the antenna in free space [12]. This improvement is a result caused by the assistance of the GP, on

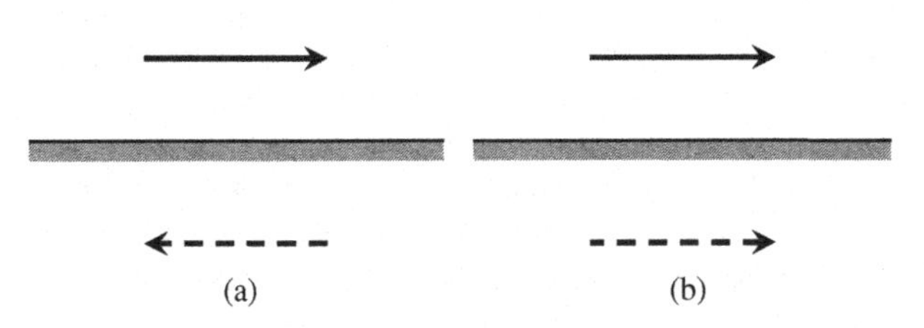


**Figure 3.10** Image of a horizontal monopole on a ground plane; (a) PEC and (b) N-PEC.

which the radiation current excited by the PIFA flows, and the performance of the entire antenna – that is, the PIFA plus the GP – is improved. In practice, the PIFA to be mounted on the mobile phone is designed intentionally to include the GP as a part of the antenna system so as to meet the system specification such as gain and bandwidth.

Another good example is a small ceramic-chip antenna encapsulated in a few-millimeter cube, which is a typical ESA and also PSA as well. When designing a small ceramic-chip antenna, dimensions of the mobile terminal unit, location to install the antenna, space to be occupied by the antenna, as well as the size of the antenna, should all be included in the antenna specifications.

It should be noted that a small antenna that is appropriately designed to satisfy the system requirements should not be placed near certain materials so that designed antenna performance is substantially harmed. It can be said generally that the smaller the antenna becomes, the larger the influence of such nearby materials is likely to be.

Recently, non-perfect electrically conducting (N-PEC) surface has drawn much interest worldwide to be used as a platform on which an antenna is placed. When a linear current source is placed near a PEC surface in parallel as shown in Figure 3.10(a), an image of the inverse phase is produced beneath the surface. However, when an N-PEC surface is used, the image is the same phase as the original source as shown in Figure 3.10(b). By using the N-PEC surface, a current source antenna can be placed very near to the surface, implying that such an antenna can have low-profile structure. Typical N-PEC surface can be composed by using EBG (Electronic Band-Gap) material, which is realized by either periodical structure or physically constructed material. Application of EBG surface has now become one of the popular topics in the field of small antennas. It is considered as a useful means to make an antenna low-profile [13], to reduce the coupling between two closely spaced antennas [14], and to miniaturize an antenna [15].

## References

[1] R. W. P. King, *The Theory of Linear Antennas*, Harvard University Press, 1956, pp. 184–192.

[2] R. W. P. King and C. W. Harrison, *Antennas and Waves: A Modern Approach*, The MIT Press, 1969, pp. 549–552.

[3] C. A. Balanis, *Antenna Theory*, 2nd edn., John Wiley and Sons, 1997, pp. 209–216.

[4] C. A. Balanis, *Antenna Theory*, 2nd edn., John Wiley and Sons, 1997, pp. 236–239.

[5] R. M. Fano, Theoretical Limitations on The Broadband Matching of Arbitrary Impedance, *MIT Research Lab of Electronics, Technical Report*, no. 41, 1948, pp. 56–83.

[6] R. M. Fano, A Note on the Solution of Current Approximation Problems in Network Synthesis, *Journal of The Franklin Institute*, vol. 249, March 1950, pp. 189–205.
[7] G. L. Matthaei, E. M. T. Jones, and L. Young, *Microwave Filters, Impedance-Matching Networks and Coupling Structure*, McGraw-Hill, 1964, sections 4.09 and 4.10.
[8] R. C. Hansen, *Phased Array Antennas*, John Wiley and Sons, 1998, section 5.4.3.
[9] L. G. Linvill, Transistor negative impedance converters, *Proceedings of IRE*, vol. 41, 1953, pp. 725–729.
[10] R. M. Rudish and S. Sussman-Fort, Non-Foster Impedance Matching improves S/N of Wideband Electrically-Small VHF Antennas and Arrays, *Proceedings of the Second IASTED International Conference* 19–21 July 2005, Banff, Alberta, Canada.
[11] K. Ishizone, Poynting Vector flow in the vicinity of a receiving dipole antenna, *IEICE Technical Report* AP-86-38, 1986, pp. 35–41.
[12] K. Satoh *et al.*, Characteristics of a Planar Inverted-F antenna on a Rectangular Conducting Body, *IEICE Transactions B*, vol. J-71-B, 1971, no. 11,71, pp. 1237–1243.
[13] F. Yang and Y. Rahmat-Sami, Reduction Phase Characteristics of the EBG Plane for Low Profile Antenna Applications, *IEEE Transactions on Antennas and Propagation*, vol. 51, 2003, pp. 2691–2703.
[14] E. Saenz *et al.*, Coupling Reduction Between Dipole Antenna Elements by Using a Planar Meta-Surface, *IEEE Transactions on Antennas and Propagation*, vol. 57, 2009, no. 2, pp. 383–392.
[15] J. L. Volakis *et al.*, Antenna Miniaturization Using Magnetic-Photonic and Degenerate Band-Edge Crystal, *IEEE Antennas and Propagation Magazine*, vol. 48, October 2006, no. 5, pp. 12–28.

# 4 Fundamental limitations of small antennas

## 4.1 Fundamental limitations

It is commonly understood that as antennas reduce in size, antenna gain and efficiency will degrade and the bandwidth tends to be narrower. Then, a question arises how small a dimension an antenna can take for practical use? Or, what will happen when the antenna dimension is limitlessly reduced? One answer was given by J. D. Kraus, who showed that a small antenna could have effective aperture of as high as 98 percent of that of a half-wave dipole antenna, if the antenna could be perfectly matched to the load [1]. It suggests that however small an antenna is, the antenna can intercept almost the same power (only 8 percent less) as a half-wavelength dipole does. In other words, there seems to be no limitation in reducing the antenna size so long as the antenna could be perfectly matched. Unfortunately, the perfect matching is impossible when an antenna becomes extremely small, because the smaller the antenna size tends to be, the harder the antenna matching will become, as was mentioned before. In addition, losses existing in the antenna structure and the matching circuit will exceed the radiation resistance, resulting in significant reduction of the effective aperture that corresponds to reduction of the radiation power and the degradation of the radiation efficiency. Regarding the antenna impedance, increase in reactive component and decrease in the resistive component results in high $Q$, and as a consequence bandwidth will be narrowed. Thus, the size reduction of an antenna also affects $Q$ and the bandwidth as well. Then it is rather natural to say that there is a fundamental limitation applying to the size reduction of antenna dimensions. This implies in turn that antenna gain, efficiency, $Q$, and the bandwidth are bounded by the antenna dimensions. Parameters that concern fundamental limitations of small antennas are antenna $Q$, relative bandwidth (RBW), radiation efficiency $\eta$, and the antenna size $ka$, where $a$ is the radius of a sphere circumscribing the antenna. In summary, relationships between these parameters and the antenna size are roughly approximated, when $ka$ is much smaller than unity, as follows:

(a) $Q$ is proportional to $(ka)^{-3}$
(b) RBW is proportional to $(ka)^3$, since RBW nearly equals $1/Q$
(c) $\eta$ is proportional to $(ka)^4$

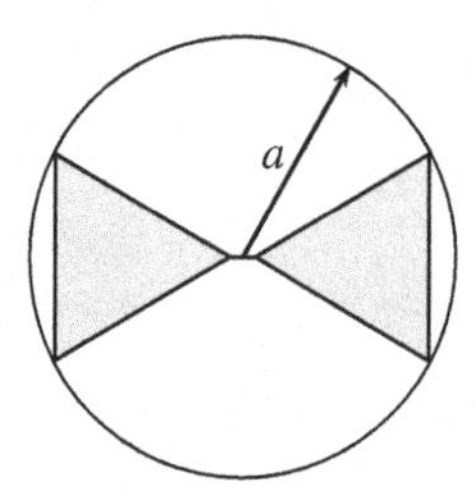


**Figure 4.1** A sphere model enclosing an antenna of radius $a$.

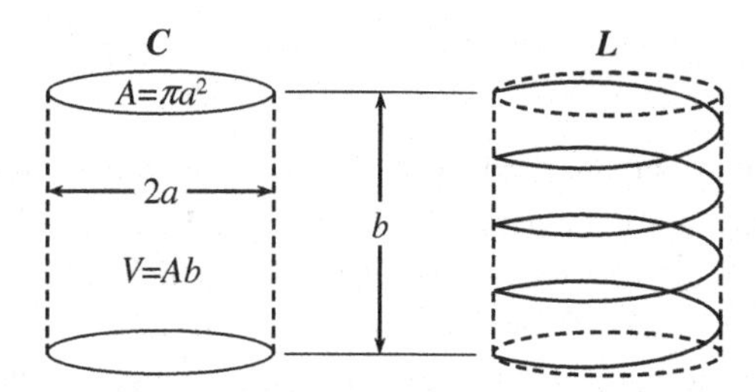


**Figure 4.2** A capacitor and an inductor with a cylindrical shape [2].

## 4.2 Brief review of some typical work on small antennas

Systematic analysis of small antennas was probably first made by H. A. Wheeler, who extensively treated small antennas practically as well as theoretically [2–5]. One of his notable works is on the limitation of small antennas. He showed that the size reduction of an antenna directly limits the radiation power factor (RPF), a ratio of the radiated power to the reactive power, implying that the radiation efficiency is constrained by the antenna size. It was also shown that the RPF equals the reciprocal of $Q$, thus the size reduction of antenna limiting $Q$ corresponds to a limitation of the RBW depending on the antenna size.

By assuming a sphere model of radius $a$, which encloses a small antenna as shown in Figure 4.1, Wheeler derived $Q$ for the most limiting case as

$$Q = 1/(ka)^3. \tag{4.1}$$

This equation indicates that $Q$ is inversely (cubic) proportional to the size of an antenna, so the antenna size imposes the fundamental limitation of the bandwidth.

Wheeler didn't work only on the theory of small antennas, but also on design and development of practical antennas. From the early to mid twentieth century, almost all antennas were "small antennas" because of their lower operating frequencies and electrically small size. Wheeler's development was large-sized antennas; however they were electrically small. His antennas were designed based on the concept introduced by himself.

Wheeler considered small antennas to behave as a capacitor $C$ or an inductor $L$, which was assumed to be circumscribed by a cylinder-shaped volume, as shown in Figure 4.2

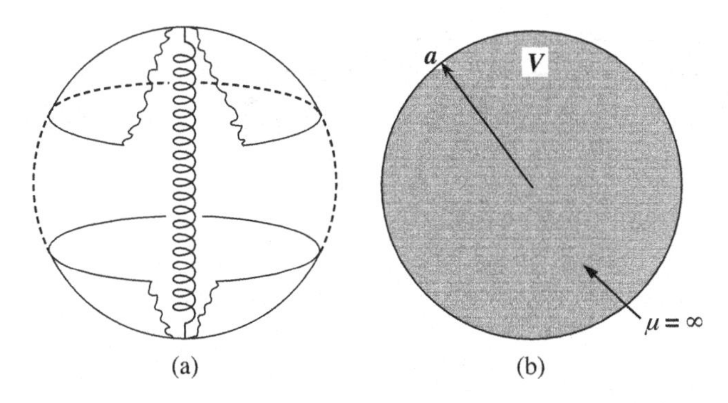


**Figure 4.3** A spherical coil (a) without and (b) with a magnetic core filled [2, 5].

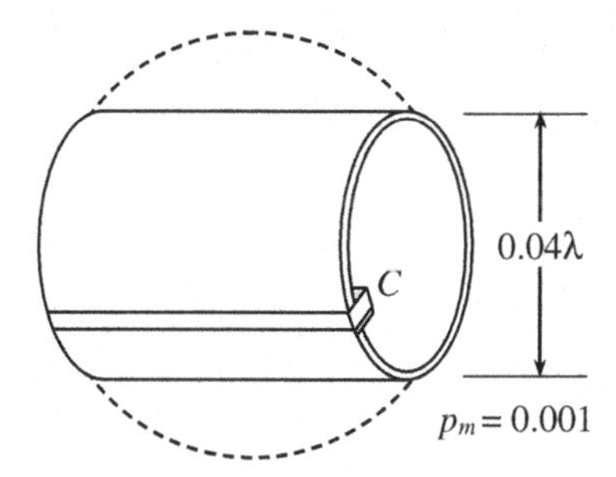


**Figure 4.4** A loop coil ferrite-loaded antenna [5].

[2]. He derived the radiation power factor (RPF), which is a ratio of the radiated power to the reactive power, for the $C$ or the $L$ types of small antennas [2, 5] and showed that it is directly proportional to the volume. His idea of the cylindrical radiator is extended to create a self-resonant helical coil that would radiate equal power in both TE and TM modes in phase quadrature so that circular polarization is yielded [6]. This type of antenna was later named as the Normal Mode Helical Antenna (NMHA) by J. D. Kraus [7]. Wheeler designed the NMHA for the first US communications satellite Telstar, on which the antenna handled the command and telemetry systems in the VHF band. He also invented a spherical coil, in which a magnetic core was filled (Figure 4.3) [5, 8]. Another practical antenna he developed was a one-turn loop of wide strip (Figure 4.4) [9], which was used for the oscillator of the proximity fuse in the nose cone on a small rocket [10]. A long coil on a ferrite core was commonly used for the radio-broadcasting receiver in the frequency range of 0.55–1.5 MHz and VLF. Some types of long coil ferrite-loaded antennas were found to be useful for horizontal mount on vehicles (Figure 4.5) [11]. Also, a VLF loop was applied to submarines [12].

Chu followed Wheeler's work on small antennas. He also analyzed the limitations of small antennas, but more exactly than Wheeler. In his analysis [13], he assumed a vertically polarized dipole antenna circumscribed by a sphere of radius $a$, and expressed the radiation field produced outside the sphere with a spherical function expansion, which can describe the fields with a sum of the spherical modes. By introducing the

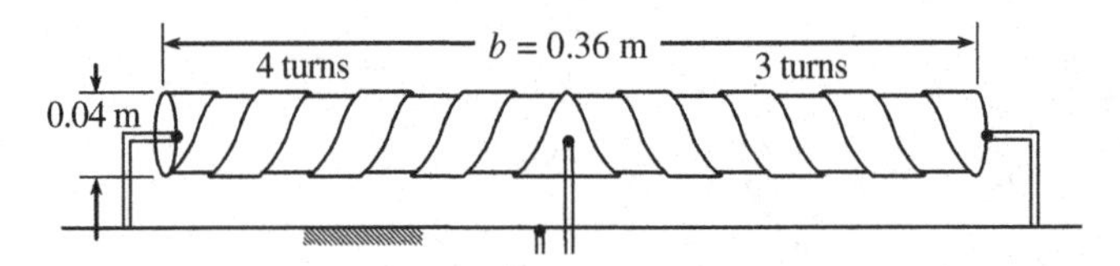


**Figure 4.5** Over turn loop of a wide strip [5].

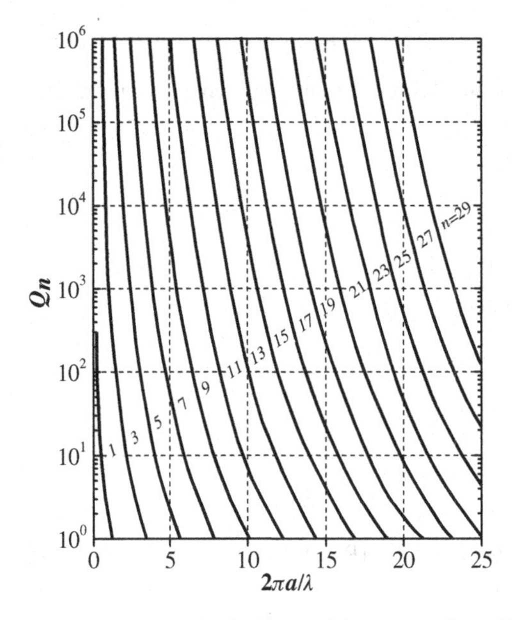


**Figure 4.6** Lowest obtainable $Q$ with antenna dimension maximum [13].

equivalent circuits for each mode, the modal impedance and the antenna $Q$ for each mode are derived. He obtained the lowest possible $Q$, highest obtainable Gain $G$ and the highest obtainable $G/Q$, and gave graphs as shown in Figures 4.6–4.8.

Chu's results have been understood as the important derivation that provides the lowest obtainable $Q$ that concerns the highest obtainable bandwidth, and the highest achievable $G/Q$, with maximum dimensions of antenna, although his analysis is accurate only for the lowest mode, whereas it includes rough approximations for the higher modes.

King worked extensively on linear antennas both theoretically and numerically [14–15], including small antennas. He showed fundamentals of linear antennas, demonstrating current distributions on antenna elements, numerical values of impedance, and $Q$ for various types and dimensions of linear antennas with theoretical and numerical data, and figures of these. He defined "Electrically Short Antenna," instead of "Electrically Small Antenna," as one that has the maximum length $ka$ smaller than 0.5. The impedance of a small dipole antenna was given previously in Figure 3.2. Table 4.1 shows resistance

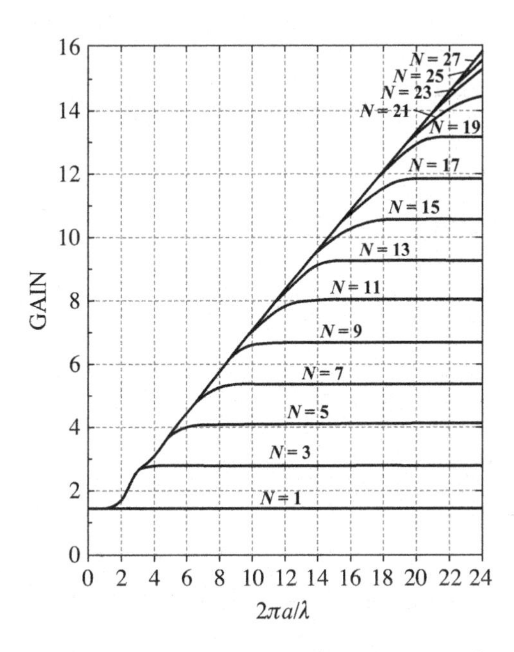


**Figure 4.7** Highest obtainable gain with antenna dimension maximum [13].

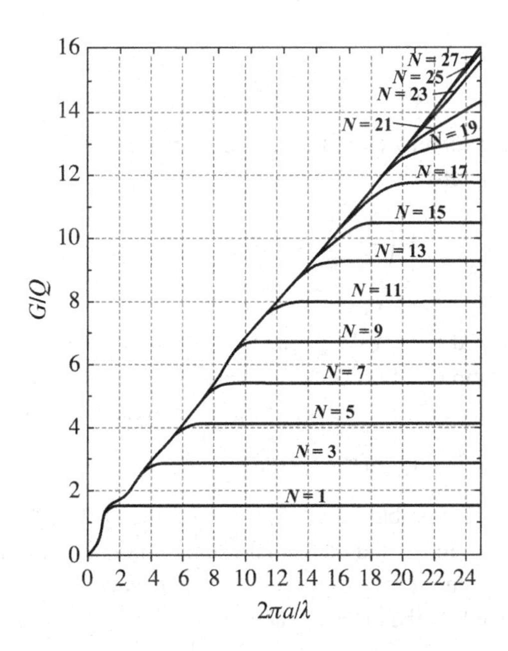


**Figure 4.8** Highest obtainable $G/Q$ with antenna dimension maximum [13].

**Table 4.1** Resistance and reactance for electrically short antennas [14, 15]

| $a$ | $Xa$ | (ohm.) $Ra$ |
|---|---|---|
| 0 | $-\infty$ | 0 |
| 0.1 | –3945 | 0.183 |
| 0.2 | –1950 | 0.732 |
| 0.3 | –1274 | 1.66 |
| 0.4 | –929 | 2.97 |
| 0.5 | –716 | 4.67 |
| 0.6 | –569 | 6.80 |
| 0.7 | –460 | 9.35 |
| 0.8 | –373.7 | 12.38 |
| 0.9 | –303.5 | 15.82 |
| 1.0 | –238.2 | 19.90 |

$R_a$ and reactance $X_a$ for electrically short antennas of length $2h$ and radius $a$ calculated by

$$R_a = 18.3\beta_0^2 h^2 \left(1 + 0.086\beta_0^2 h^2\right) \tag{4.2}$$

$$\text{and } X_a = -\frac{60\,(\Omega - 3.39)}{\beta_0 h} \tag{4.3}$$

where $\Omega = 2\ \ln\,(2h/a)$ and

$$\beta_0 = 2\pi/\lambda_0.$$

King gave the expression for antenna $Q_r$ at the resonance in terms of the impedance [16] by analogy with a lumped-constant RLC circuit as

$$Q_r = \left(\frac{\omega}{2R}\frac{dX}{d\omega}\right)_{\text{res}}. \tag{4.4}$$

Here $X = \omega L - 1/\omega C$ and $\omega L = 1/\omega C$ at resonance.

Collin and Rothschild improved previously given theory for obtaining $Q$ based on the field rather than the equivalent circuit [17]. They calculated $Q$ for cases where both TE and TM modes are used for excitation, and derived it for the lowest mode as

$$Q = 1/(ka)^3 + 1/(ka). \tag{4.5}$$

This $Q$ represents the minimum possible value. Fante generalized their treatment to include both TE and TM modes, and found that equal excitation of TE and TM modes does not provide $Q$ of half the value for the excitation of either mode alone. The actual result was somewhat larger than half the original value.

Hansen simplified Chu's expression for calculating $Q$ [18]. He stated that when $ka$ is roughly less than unity and only the lowest TM mode propagates, $Q$ is expressed by

$$Q = [1 + 3(ka)^3]/(ka)^3[1 + (ka)^2]. \tag{4.6}$$

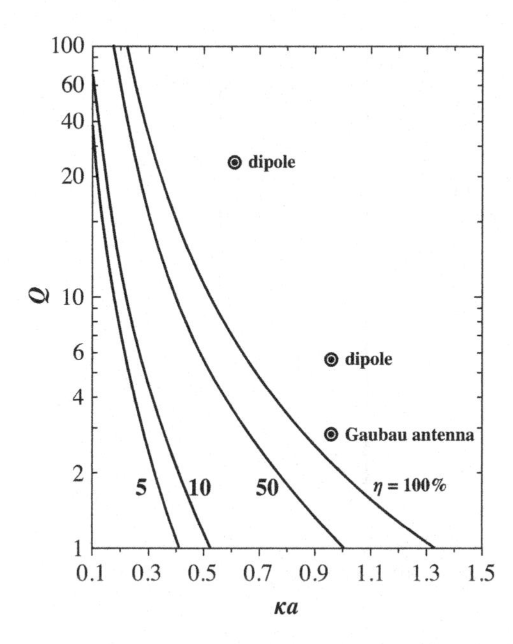


**Figure 4.9** Antenna $Q$ of a short dipole antenna [18].

Hansen concluded that the value of $Q$ would be halved when TM and TE modes are equally excited. He illustrated $Q$ with respect to the antenna size $ka$ along with the radiation efficiency $\eta$ as the parameter (Figure 4.9). This figure is useful and convenient for finding antenna $Q$, thus bandwidth, and the efficiency in relation with antenna size when designing a small antenna of a given size.

McLean corrected some previous derivations obtained by Wheeler, Chu, and Hansen, as he thought their results were too rough to establish fundamental limits of $Q$ because of the simple assumption or approximation [19]. He derived $Q$ as

$$Q = [1 + 2(ka)^3]/(ka)^3[1 + (ka)^2]. \tag{4.7}$$

He concluded that when $ka$ is very small, there is not much difference between the previous derivations and his own derivation, and equation (4.7) is an exact expression of the lower bound on $Q$ for a given antenna size.

Folts and McLean used a prolate spheroid that would better approximate an actual antenna like a dipole antenna (Figure 4.10) and evaluated $Q$ vs. $ka$ [20]. They considered that $Q$ previously obtained was not really close to the verifiable values for many antennas, because the volume of a sphere surrounding an antenna was not fully utilized, and that made the value of $Q$ larger than was expected. Figure 4.11 depicts $Q$, using the parameter $a/b$, where $a$ and $b$ respectively denote the major axis and the minor axis of the prolate spheroid.

Thiele obtained $Q$ by a different method, as he thought that previously obtained $Q$ was far from that of actual antennas, because the current distributions were not taken into consideration in the calculation [21]. He used the superdirective ratio SDR for

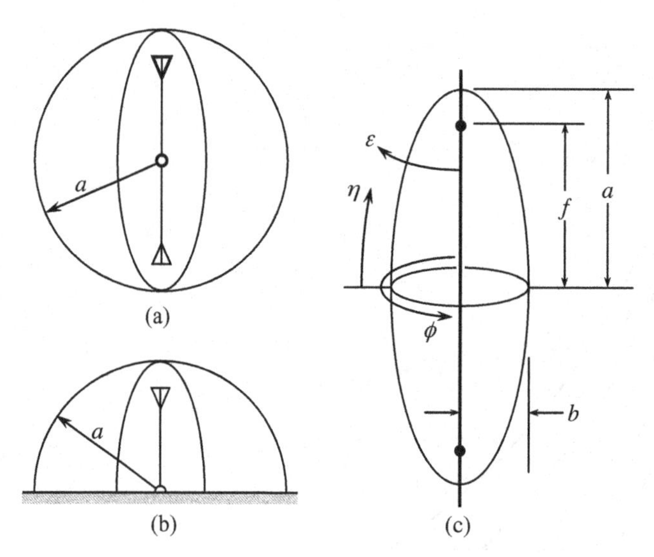


**Figure 4.10** A prolate spheroidal model approximating a small monopole and dipole antenna [20].

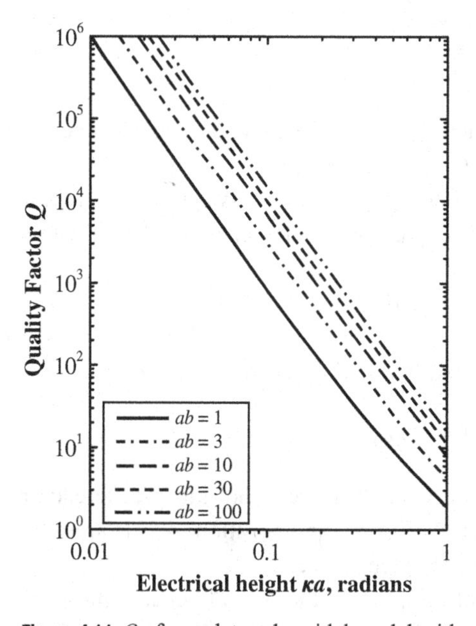


**Figure 4.11** $Q$ of a prolate spheroidal model with respect to electrical length $ka$ [20].

calculating $Q$, based on the concept that the radiation energy is concerned with the visible region in the radiation field, while the total stored energy is concerned with the invisible region. The calculated $Q$ for a thin dipole antenna in terms of $X/R$ McLean's result, lower bound for an ideal dipole with uniform current distribution, and far field $Q$ for a dipole with sinusoidal current distribution are shown in Figure 4.12. Figure 4.13 depicts $Q$ for a thin dipole of different radius, end-loaded dipole, and bow-tie antenna, respectively. In the figure, McLean's $Q$ was drawn as a comparison.

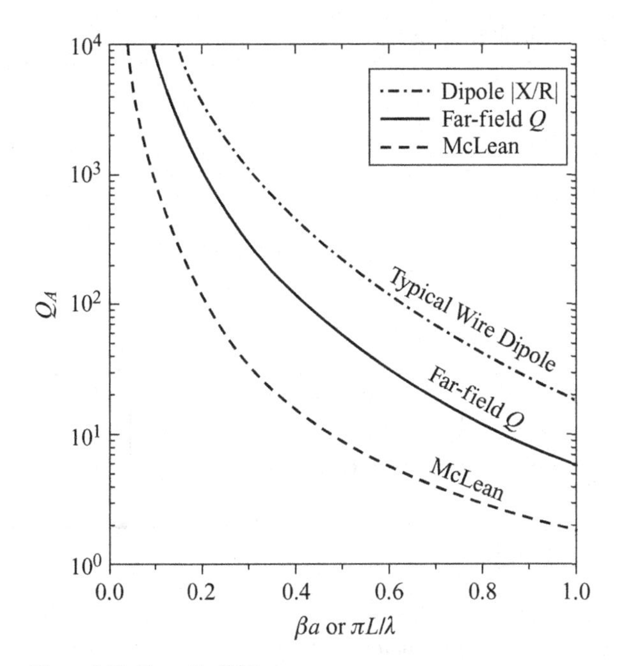


**Figure 4.12** $Q$ vs. $\beta$a [21].

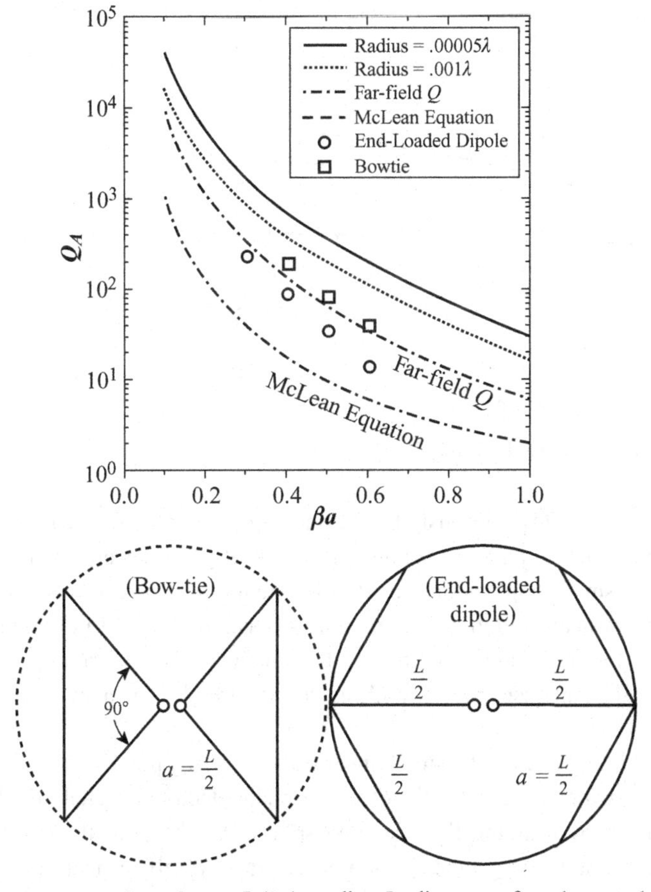


**Figure 4.13** $Q$ vs. $\beta$a or $\pi$L/$\lambda$ (a: radius, L: diameter of a sphere enclosing the radiator) [21].

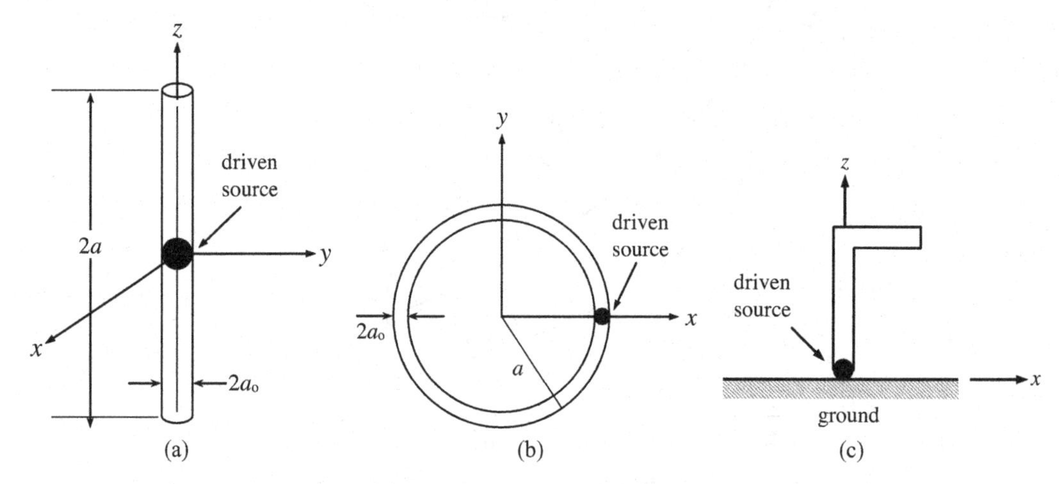


**Figure 4.14** Antenna models used for evaluation of $Q$: (a) a dipole, (b) a loop, and (c) inverted-L [23].

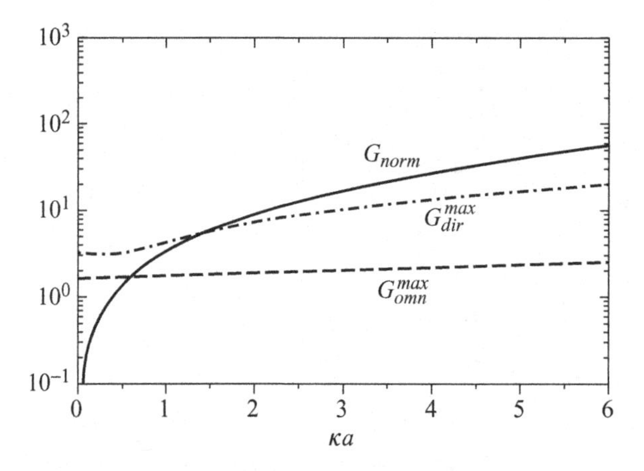


**Figure 4.15** Maximum obtainable gain [22].

Geyi studied $Q$ [22–24] rigorously by introducing complex power balance relations into an antenna structure and derived an expression of $Q$ for several typical antennas such as a dipole, small loop, and an inverted-L antenna as Figure 4.14 shows [23]. He also discussed physical limitations of small antennas, and derived minimum possible $Q$ for both TE and TM modes excited by the antenna, and the ratio of maximized $G/Q$ and normal gain, respectively, and his results are shown in Figures 4.15, 4.16 and 4.17 [22–24].

Best discussed $Q$ of electrically small linear and elliptically polarized spherical dipole antennas, based on the concept that the $Q$ of a resonant electrically small dipole antenna can be minimized by utilizing the occupied spherical volume to the greatest possible extent with the antenna geometry. He demonstrated a self-resonant (at 300 MHz)

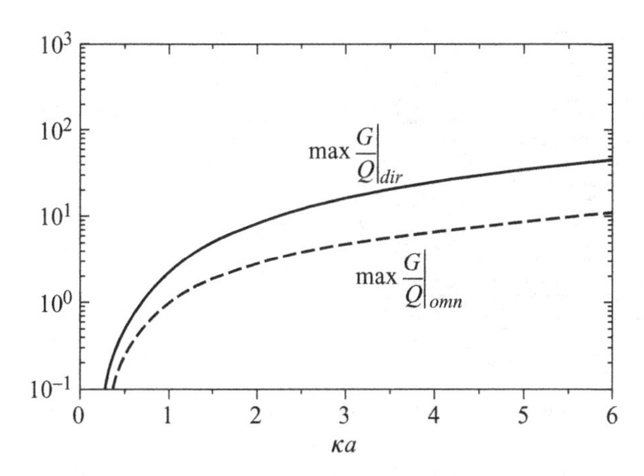


**Figure 4.16** Maximum obtainable $G/Q$ [22].

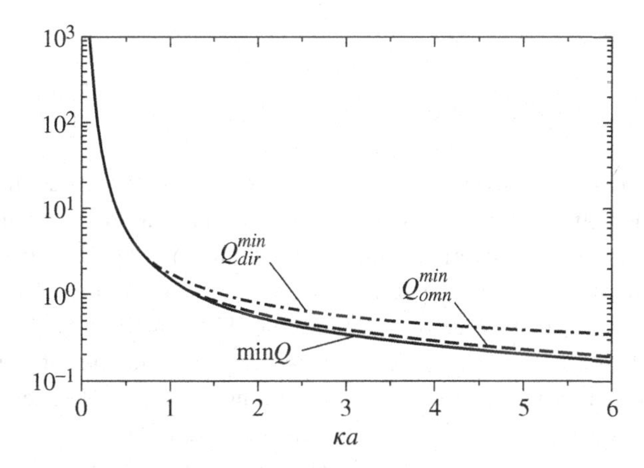


**Figure 4.17** Minimum possible $Q$ [22].

four-arm folded linearly polarized spherical dipole antenna that exhibits an impedance 47.5 Ω, an efficiency of 97.4%, and $Q$ of 87.3, which is within 1.5 times the fundamental lower bound at a value $ka$ of 0.263 [25]. This antenna may have the lowest $Q$ that can be realized with a practical antenna. He also showed that a self-resonant (at 300 MHz) elliptically polarized antenna exhibits $Q$ within two times the lower bound. The antenna has an efficiency of 95% and an impedance of 61.5 Ω.

Best also treated $Q$ and bandwidth in terms of antenna impedance and gave relationships between $Q$ and VSWR fractional bandwidth (FBW) [26–27]. He discussed the upper bound of VSWR FBW for loss-free antennas and showed it to be

$$FBW_{Vub} = \frac{2\sqrt{\beta}\,(ka)^3}{1+(ka)^2} = \frac{(ka)^3}{1+(ka)^2}\frac{s-1}{\sqrt{s}} \tag{4.8}$$

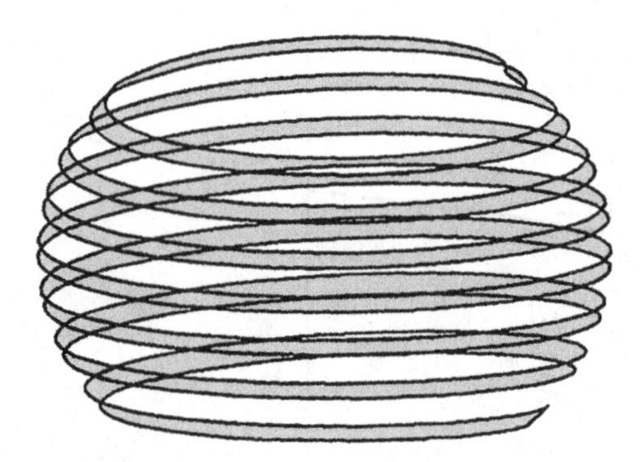

**Figure 4.18** An antenna model used for evaluating minimum $Q$ [29].

where $s$ denotes VSWR. The maximum available FBW when a number of tuned circuits are applied to the matching circuit in order to increase the BW is given by

$$FBW_{V\,\max} = \frac{1}{Q}\frac{\pi}{\ln\left(\frac{s+1}{s-1}\right)} = \frac{(ka)^3}{1+(ka)^2}\frac{\pi}{\ln\left(\frac{s+1}{s-1}\right)} \approx \frac{\pi\,(ka)^3}{\ln\left(\frac{s+1}{s-1}\right)}. \tag{4.9}$$

Best further has given evaluation of $Q$ for various different dipole-type antennas [25, 28].

Thal commented that previous treatment of $Q$ ignored the energy inside the sphere, resulting in considerably higher $Q$ limit for realizable antennas than the minimum values predicted by the previous work. He then derived minimum $Q$ values that include the contributions of energy within the sphere by using an antenna model formed by conducting thin wires on the surface of a sphere [29]. The antenna model he used is depicted in Figure 4.18. He used the equivalent circuits representing the antenna system valid for both outside and inside the sphere, accounting for the mode energy stored inside the sphere. He defined stricter limits that apply to a class of antennas consisting of conductors arranged to conform to a spherical surface and showed that the new limit he derived should be helpful for estimating the minimum $Q$ of wire antenna configurations not necessarily exactly spherical but reasonably approximating the configuration of a sphere.

Thal, after his discussion on the limit of $Q$ [29], explored the relationships between phase, gain, bandwidth, and the $Q$ lower bound, particularly for electrically small antennas [30], and provided general theory and numerical examples giving a lower bound on the $Q$ of any antenna with coupled TM–TE modes as a function of the electrical radius $ka$ and the relative phases of the radiated mode fields, by which the gain is determined. He showed values of radiation $Q$ with respect to $ka$ from 0.05 to 0.65 and the minimum $Q$s for TM and TE mode, respectively, are 1.5 and 3 times $Q_{chu}$, which is given by Chu [17].

Hansen and Collin also calculated internal stored energy and provided exact formulas for total stored energy $Q$ for TM modes and an approximate formula for the $Q$ of the lowest mode [31]. The calculation was based on the previous work [17], by which formulas for the $Q$ values of TM and TE modes are derived for a case where these modes

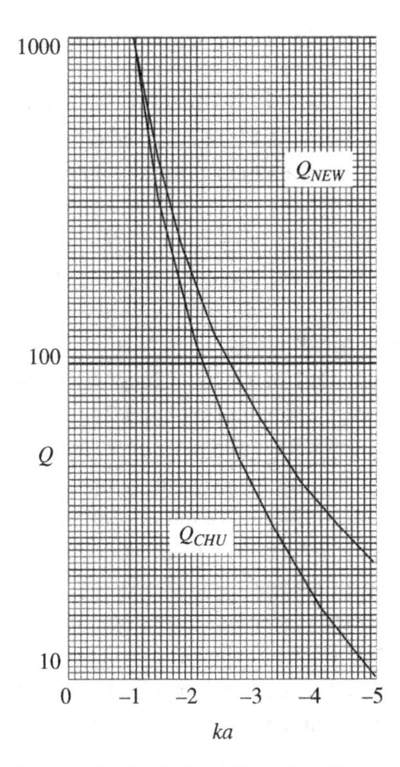


**Figure 4.19** $Q$ calculated by using Equation (4.10) and $Q_{chu}$ [31].

are excited by current sheets on a sphere and include the stored reactive energy within the sphere. Then by following Thal [29], new formulas to express $Q$ for TM modes are derived [31], by treating the reactive power inside the sphere circumscribing the antenna, whereas Chu's calculation did not include that. Thus the $Q$ is given by the $Q_{chu}$ plus an additional term which corresponds to the internal reactive energy. This formula is simplified to an approximate formula similar to (4.5), which can be expressed by

$$Q = 3/2(ka)^3 + 1/\sqrt{2}(ka). \tag{4.10}$$

The value of $Q$ with respect to the antenna dimension $ka$ is depicted in Figure 4.19 [31], where the $Q_{chu}$ is also given as a comparison.

Discussions on the $Q$ of radiating structures have continued. Gustafsson gave new physical bounds for antennas of arbitrary shape and illustrated numerical examples [32]. He presented physical bounds for antennas circumscribed by the rectangular parallelepiped, finite cylinders, and planar rectangles. The theory can directly be applied to analyze the potential performance of different antenna geometries using pattern and polarization diversity.

Vandenbosch derived general and rigorous expressions for calculating the reactive energy stored in the electromagnetic field around an arbitrary source device [33]. A straightforward application is to investigate the energies and $Q$ of radiating structures in terms of topology of the device. The expressions can be applied to much more general topology, particularly for small antennas.

Kim *et al.* studied $Q$ of electrically small current distributions and practical antenna designs radiating the $TE_{10}$-mode magnetic dipole field, and derived closed form expressions for the internal and external electric and magnetic stored energies and radiated power by a spherical, electric surface current density enclosing a magneto-dielectric core [34]. This result leads to determining the $Q$ for arbitrary values of $ka$. He demonstrated that for a given size of antenna and permittivity, there is an optimum permeability that ensures the lowest possible $Q$, and this optimum permeability is inversely proportional to the square of the antenna electrical radius. With permittivity of unity, the optimum permeability yields the lowest bound of $Q$ for a magnetic dipole antenna with a magneto-dielectric core. He obtained by simulation $Q$ of 1.24 times $Q_{chu}$ for the $TE_{10}$-mode multi-arm spherical helix antenna with a magnetic core, for $ka$ of 0.192.

Yaghjian obtained general expressions for the lower bounds on $Q$ of electrically small electric and magnetic-dipole antennas confined to an arbitrary shaped volume $V$ which are excited by general sources or by global electric current sources alone [35]. He showed that the lower bound expressions depend only on the direction of the dipole moment with respect to $V$, electrical size of $V$, and per unit volume static PEC (Perfect Electric Conductor) and PMC (Perfect Magnetic Conductor) electric and magnetic polarizabilities of $V$. He provided new expressions of the lower bound $Q$ for electrically small electric and magnetic dipole antennas, and sources restricted to global surface currents.

Stuart and Yaghjian investigated the effect of a thin shell of high permeability magnetic material on $Q$ for the electrically small, top-loaded electrical dipole antenna [36]. The magnetic polarization currents induced in the thin shell of magnetic material reduce the internal stored energy, leading to a lower $Q$. It was shown that for the case of spherical antennas, $Q$ approaches $Q_{chu}$.

Kim and Breinbjerg found that the lower bound for radiation $Q$ of spherical magnetic dipole antennas with pure solid magnetic core could approach $Q_{chu}$ as the antenna electrical size decreases [37]. With properly selected permeability, the antenna can exhibit a lower $Q$ than the air-core magnetic and electric dipole antennas of the same size in the range of $ka < 0.863$.

So far, Chu's limitation has been understood as the lowest achievable $Q$ for an antenna of given size, which, however, would be impossible in reality to achieve. Design of small antennas is an exercise in finding new techniques to approach Chu's limitations $Q_{chu}$ as closely as possible with small antennas. The main issues would be study on the antenna shape, current distributions on it, and relationships with radiation efficiency.

## References

[1] J. D. Kraus, *Antennas*, 3rd edn., McGraw-Hill, 2002, pp. 30–33.

[2] H. A. Wheeler, Fundamental Limitations of Small Antennas, *Proceedings of IRE*, vol. 35, Dec 1947, pp. 1479–1484.

[3] H. A. Wheeler, Small Antennas, in H. Jasik (ed.), *Antenna Engineering Handbook*, 2nd edn., McGraw-Hill, 1984, chapter 6.
[4] H. A. Wheeler, The Radian Sphere Around a Small Antenna, *Proceedings of IRE*, vol. 47, August 1959, pp. 1325–1331.
[5] H. A. Wheeler, Antenna Topics in My Experiences, *IEEE Transactions on Antennas and Propagation*, vol. 33, 1985, no. 2, pp. 144–151.
[6] H. A. Wheeler, A Helical Antenna for Circular Polarization, *Proceedings of IRE*, vol. 35, 1947, pp. 1484–1488.
[7] J. D. Kraus and R. J. Marhefka, *Antennas*, 3rd edn., 2002, McGraw-Hill, pp. 292–293.
[8] H. A. Wheeler, The Spherical Coil as an Inductor, Shield, and Antenna, *Proceedings of IRE*, vol. 46, 1958, pp. 1595–1602.
[9] pp. 147–148 in [5].
[10] p. 148 in [5] and H. A. Wheeler, Small Antennas, *IEEE Transactions on Antennas and Propagation*, vol. 23, 1975, pp. 462–469.
[11] pp. 146–147 in [5].
[12] H. A. Wheeler, Fundamental Limitations of a Small VLF Antenna for Submarines, *IEEE Transactions on Antennas and Propagation*, vol. 6, 1958, pp. 123–125.
[13] L. J. Chu, Physical Limitations of Omni Directional Antennas, *Research Laboratory of MIT, MIT Tech Report* no. 64, 1948.
[14] R. W. P. King, *Linear Antennas*, Harvard University Press, 1956.
[15] R. W. P. King, and C. W. Harrison, *Antennas and Waves*, The MIT Press, 1969.
[16] pp. 171–182 in [14].
[17] R. E. Collin and S. Rothschild, Evaluation of Antenna Q, *IEEE Transactions on Antennas and Propagation*, vol. 12, 1964, pp. 23–27.
[18] R. C. Hansen, Fundamental Limitations in Antennas, *Proceedings of IEEE*, vol. 69, 1981, no. 2, pp. 170–181.
[19] J. S. McLean, A Re-Examination of the Fundamental Limits on the Radiation Q of Electrically Small Antennas, *IEEE Transactions on Antennas and Propagation*, vol. 44, 1996, pp. 672–676.
[20] H. D. Folts and J. S. McLean, Limits on the Radiation Q of Electrically Small Antennas Re-Estimated to Oblong Bounding Regions, *IEEE Antennas and Propagation Society International Symposium*, July 1999, vol. 4, pp. 2702–2705.
[21] G. A. Thiele, P. L. Detweiler, and R. P. Peno, On the Lower Bound of the Radiation Q for Electrically Small Antennas, *IEEE Transactions on Antennas and Propagation*, vol. 51, 2003, pp. 1263–1269.
[22] W. Geyi, Physical Limitations of Antenna, *IEEE Transactions on Antennas and Propagation*, vol. 51, 2003, pp. 2116–2123.
[23] W. Geyi, A Method for the Evaluation of Small Antenna Q, *IEEE Transactions on Antennas and Propagation*, vol. 51, 2003, pp. 2124–2129.
[24] W. Geyi, P. Jarmauszewski, and Y. Qi, The Foster Reactance Theorem for Antennas and Radiation Q, *IEEE Transactions on Antennas and Propagation*, vol. 48, 2000, pp. 401–408.
[25] S. R. Best, Low Q Electrically Small Linear and Elliptical Polarized Spherical Dipole Antennas, *IEEE Transactions on Antennas and Propagation*, vol. 53, 2005, no. 3, pp. 1047–1053.
[26] A. D. Yaghjian and S. Best, Impedance, Bandwidth, and Q of Antennas, *IEEE Transactions on Antennas and Propagation*, vol. 53, 2005, no. 4, pp. 1298–1324.

[27] S. R. Best, Bandwidth and the Lower Bound on Q for Small Wideband Antennas, *IEEE APS International Symposium*, 2006, pp. 647–650.

[28] S. R. Best, A Low Q Electrically Small Magnetic (TE Mode) Dipole, *IEEE Antennas and Wireless Propagation Letters*, vol. 8, 2009, pp. 572–575.

[29] H. L. Thal, Gain and Q bounds for coupled TM–TE modes, *IEEE Transactions on Antennas and Propagation*, vol. 57, 2009, no. 7, pp. 1879–1885.

[30] H. L. Thal, New Radiation Q Limits for Spherical Wire Antennas, *IEEE Transactions on Antennas and Propagation*, vol. 54, 2006, no. 10, pp. 2757–2763.

[31] R. C. Hansen and R. E. Collin, A New Chu Formula for Q, *IEEE Antennas and Propagation Magazine*, vol. 51, 2009, no. 5. pp. 38–41.

[32] M. Gustafsson, C. Sohl, and G. Kristenssen, Illustrations of New Physical Bounds on Linearly Polarized Antennas, *IEEE Transactions on Antennas and Propagation*, vol. 57, 2009, no. 5, pp. 1319–1326.

[33] G. A. E. Vandenbosch, Reactive Energies, Impedance, and Q Factor of Radiating Structures, *IEEE Transactions on Antennas and Propagation*, vol. 58, 2010, no. 4, pp. 1112–1127.

[34] O. S. Kim, O. Breinbjerg, and A. D. Yaghjian, Electrically Small Magnetic Dipole Antennas With Quality Factors Approaching the Chu Lower Bound, *IEEE Transactions on Antennas and Propagation*, vol. 58, 2010, no. 6, pp. 1898–1905.

[35] A. D. Yaghjian and H R. Stuart, Lower Bounds on the Q of Electrically Small Dipole Antennas, *IEEE Transactions on Antennas and Propagation*, vol. 58, 2010, no. 10, pp. 3114–3121.

[36] H. R. Stuart and A. D. Yaghjian, Approaching the Lower Bounds on Q for Electrically Small Electric-Dipole Antennas Using High Permeability Shells, *IEEE Transactions on Antennas and Propagation*, vol. 58, 2010, no. 12, pp. 3865–3872.

[37] O. S. Kim and O. Breinbjerg, Lower Bound for the Radiation Q of Electrically Small Magnetic Dipole Antennas With Solid Magneto-dielectric Core, *IEEE Transactions on Antennas and Propagation*, vol. 59, 2011, no. 2, pp. 679–681.

# 5 Subjects related with small antennas

## 5.1 Major subjects and topics

The major subjects that have long been discussed on small antennas are investigation into the fundamental characteristics and performances of small antennas, and exploitation of methods or ideas to realize small antennas, particularly ESA (Electrically Small Antennas). Also, there have been practical matters discussed frequently as critical problems involved with designing small antennas. They are rather urgent problems that are concerned with design of antennas in unusual or uncommon environmental conditions often encountered in practical applications.

### 5.1.1 Investigation of fundamentals of small antennas

The essentials of small antennas have long been studied by many investigators; however, there are still many problems that must be discussed or solved. Among them, the limitations of small antennas is the most significant and interesting subject. It is important because knowledge of the limitations may give a clue to realizing novel small antennas. The physical-size limitation was first discussed by Wheeler in 1947 and then by Chu in 1948, as was mentioned in Chapter 4. Since then small antenna problems have been continually discussed down to the present day. Chief problems are those concerned with major antenna parameters regarding antenna quality factor $Q$, gain, bandwidth, and efficiency, in conjunction with the limitations of small antennas. How to obtain antenna $Q$ precisely still remains a controversial problem, since $Q$ directly relates with the radiation power, and exact calculation of the reactive part of the radiation power still remains a controversial subject. The antenna $Q$ is significant in relation to the bandwidth, since when the antenna $Q$ is appreciably high it is very nearly the reciprocal of the bandwidth. Thus achieving required bandwidth with an antenna of given size is related to realizing the $Q$ of a corresponding value. When the $Q$ is small, the bandwidth is wide; however, with small-sized antennas, it is usually a hard problem to obtain a small $Q$.

### 5.1.2 Realization of small antennas

How to achieve a small antenna with reasonably high gain, efficiency, bandwidth, etc. is one of the most serious propositions in the field of antennas and communications.

Various efforts for realizing antennas having performance as close as possible to the fundamental limitation of small antennas have been ongoing for many years. Along with study and research on realization of small antennas, various attempts to develop practically useful small antennas have been made repeatedly and much progress has been observed these days. Demand for small antennas to be used in various wireless systems recently deployed is rather urgent. Recent progress of downsizing in electronic components, devices, and thus wireless equipment is remarkable, whereas small-sizing of antennas has not necessarily followed the trend in downsizing of electronic devices. The basic concept of downsizing antennas is essentially different from that of electronic devices or components, because problems in antennas are inherently defined by the physical nature of electromagnetic waves in open space, while those in circuitry and devices are concerned with electronics parameters within closed space.

## 5.2 Practical design problems

Design of small antennas, particularly ESA, is to attain an antenna of physically small dimensions, yet having appreciably high performance, even though the dimensions become small. It is rather a challenging issue to realize an ESA, but the practical problem is to achieve a compromise between small-sizing and attaining higher performance in an antenna system. In addition to the physical dimensions, when reducing the size of an antenna there is need to consider various other important parameters such as antenna structure, including type, shape, materials of element, feed system, and environmental conditions, including the platform to install the antenna element, and nearby materials, to list a few.

When designing or developing small antennas, one often encounters complicated problems due to environmental conditions that must be met in order to have a practically useful small antenna. When an antenna is to be placed or installed in a complicated or unusual location, one should consider either how to avoid influence of such undesired conditions that may cause degradation of antenna performance, or how to include such conditions in the antenna design to utilize them for enhancing the antenna performance. It may be of almost no use to apply ordinary theory for designing an antenna located in such complicated situations.

One of the practical examples is a mobile antenna where mobile terminals are so small that the built-in antenna cannot be isolated electrically from nearby materials, because the antenna is forced to reside in a restrictive area inside the small terminal. A similar example is a case where the antenna must be placed in very limited space, such as in a small, thin plastic card or on a tiny dielectric chip. Design of small antennas in these conditions is not very easy and certainly not simple, because matching problems and environmental effects must be taken into account in the antenna design. It is rather usual, in these situations, to aim for either antennas having the maximum obtainable performance, or antennas having the performance attained as a compromise in the complicated situations.

Meanwhile, electromagnetic simulation techniques can be applied to the antenna design, even though antennas may be located in difficult environmental conditions. Simulation is a powerful means that allows calculation of antenna performance fairly precisely, and can obtain good approximation that is useful for the antenna design.

An example is an antenna installed inside a small mobile terminal, where two antenna elements are located in confined space with very close separation so that coupling between the elements cannot be avoided. Space to locate antenna elements inside the small unit is so narrow that the size of the antennas must be as small as possible and the separation between them is forced to be small. Then reduction of coupling between antenna elements is significant for avoiding degradation of antenna performance.

This type of problem is often observed in antennas used for MIMO systems; a typical example is antennas in small mobile WiMAX (Worldwide Interoperability for Microwave Access) terminals.

Materials existing near an antenna element give rise to effects on antenna performance that can be either advantageous or disadvantageous, depending on whether they can enhance or improve the antenna performance as a result of acting as a part of the antenna system. In small mobile terminals, the antenna element is usually located in so narrow a space that isolation of the antenna from nearby electronic components, devices, and hardware, including parts like speaker, filter, case, switch, etc. is almost impossible. The design problem is how to integrate these materials into the antenna system. A ground plane, on which an antenna element is installed, also acts as a part of the antenna system. In this case, the size of the ground plane and the location of the antenna on it should be specified so that integrated design of the antenna and the ground plane can be carried out effectively. The unit's metal case affects radiation too, and it may sometimes be utilized positively as a radiator, as the antenna dimension is essentially extended to the size of the unit. There is a circumstance where the ground plane or the unit case is utilized as one part of a diversity element. When the ground plane or the unit case is used as a part of the radiator, care must be taken for the effect of an operator's hand which can degrade the antenna performance.

Since materials existing near an antenna usually yield complicated environmental conditions for the antenna, the antenna design procedure must properly include these materials as part of the antenna system. Of course ordinary design procedure is useless in this regard. On the contrary, use of computer software is helpful, since modeling of the antenna system, in which even considerably complicated environmental conditions can be included, is made feasible by the appropriately selected software, namely electromagnetic simulators such as MoM (Method of Moments), FDTD, HFSS (High Frequency Simulation Software), IE3D, SEMCAD, FEM (Finite Element Method) and so forth. The latest simulation techniques have made remarkable progress and so one can rely on them to obtain reasonably precise results, providing very good approximations that assist in making design of antennas easier and yet more reliable than before. Nonetheless, however when one may try to approximate the antenna behavior, some errors are likely to remain, as the modeling cannot be completely perfect.

## 5.3 General topics

There have been increasing demands for small antennas, paralleling the rapid progress in modern wireless mobile systems, where the numbers of small terminals have explosively increased. The antennas for such mobile terminals are required to be small, compact, and built-in, and yet they must be capable of multiband, wideband, and variable-frequency operation, and in some cases they must exhibit signal processing performance. In practice, gain and bandwidth should be maintained at least the same as the prototypical antennas even before their size is reduced. Other growing demands have also arisen in various wireless systems, which are applied to not only communication, but also to transmission of data, video signals including moving images, and control signals. Typical of these applications are the WLAN (Wireless Local Area Network) systems, and WiMAX systems. Other important wireless systems are NFC (Near Field Communications) systems, including RFID (Radio Frequency Identification) systems, and UWB (Ultra Wideband) systems. Small-sized antennas are demanded for all of these systems. Apart from conventional mobile systems, small antennas are also required for the reception of the Digital Television Broadcasting (DTVB) in vehicles and mobile phones in Japan. Since the DTVB is received with small receivers not only outdoors but also in indoor environments, the antennas should be as small and light as possible. In the vehicle use, equipping with functions such as diversity and adaptive control is preferred.

Recent development of modern wireless systems is notable, being expanded widely into systems other than communications such as near-field systems, identification, data and image transmission, and other systems where use of the wireless systems is considered beneficial. Among these, RFID systems have been applied widely in the fields of business, industry, science, agriculture, medicine, and so forth. In practical use, RFID equipment takes the form of cards, pencils, key-holders, small boxes, notes, and so forth. Antennas to be used for such RFID systems must be simple, lightweight, and compact, and often they must be placed in a very limited space inside the small or tiny container. This makes design of these antennas difficult, and as a consequence many of these antennas may not exhibit the fully satisfactory performance that would be expected from their larger prototypes. Thus it is rather common to employ antennas which are understood to be only practically usable enough. In a way, design of these antennas will include how to compromise the performance with the specified size of antenna, the size of the platform, nearby materials, and other system requirements.

With knowledge of the limitations, antenna design turns out to be an effort on how to approach the limitation as closely as possible; in other words, how to obtain wider bandwidth with a given small size.

# 6 Principles and techniques for making antennas small

## 6.1 Principles for making antennas small

The general concept of making an antenna small is creation of an antenna having dimensions much smaller than the wavelength, keeping the performance the same as before the downsizing. From a slightly different aspect, attaining improved or enhanced antenna performance while keeping the physical dimensions of an antenna unchanged, should also be recognized as a concept similar to making an antenna small. The former is generally realization of an ESA (Electrically Small Antenna), while the latter is mostly concerned with creation of FSA (Functionally Small Antenna). When only some of the antenna dimensions are made much smaller than the wavelength, it is classified as a PCSA (Physically Constrained Small Antenna). If an antenna is simply physically small, it is a PSA (Physically Small Antenna). Typical examples of PSA are antennas used in higher frequency regions like microwave (MW) and millimeter wave (MMW) regions. These antennas have inherently small dimensions because of their operating frequencies. Antennas simply made small so that they can be placed on, for instance, a human palm, can be classified into the category of PSA. Principles and techniques to produce these small antennas differ depending on the types such as ESA, FSA, PCSA, and PSA.

Major principles for realizing an ESA are;

(1) lowering the resonance frequency for an antenna of given dimensions
(2) full use of space or volume circumscribing the maximum dimension of an antenna
(3) arranging for uniform current distributions on the antenna element, and
(4) increasing the number of radiation modes in an antenna structure.

Items (2) and (3) relate to conditions that improve the bandwidth of a given size of antenna. There are various techniques and methods to realize the above conditions.

The principles of producing an FSA are based on integration of either electronic components or functions into an antenna structure. Principles of producing PCSA and PSA are rather simpler than those used to produce other types of antennas. A PCSA can be constructed by making an antenna either low profile or with planar structures, whereas a PSA is simply an antenna made small enough to have dimensions perhaps comparable to the size of a human palm. Each of these will be discussed in the sections to follow.

**Figure 6.1** A zigzag line.

## 6.2 Techniques and methods for producing ESA

There are various techniques and methods for producing ESA based on the principles described in the previous section. What follows are the typical methods.

### 6.2.1 Lowering the antenna resonance frequency

Lowering the resonance frequency of an antenna of given dimensions is the most basic principle for making an antenna electrically small. When an antenna with physically very small dimensions, which ordinarily makes the resonance frequency higher, is made to have a lower resonance frequency, the antenna can be recognized as electrically small and classified into ESA. It is usually a very hard problem to attain lower resonance frequencies with physically small antennas; however, whatever dimensions of an antenna are physically small, achieving resonance at lower frequencies with smaller dimensions is one of the effective ways to produce an ESA, since matching, thus resonance at the operating frequency is inevitably necessary.

The most efficient way of lowering the resonance frequency of an antenna while keeping the dimensions unchanged is to compose the antenna with a slow wave (SW) structure. The SW is defined as the electromagnetic wave that travels a transmission path with smaller phase velocity $v_p$ than that of light in free space, $c$ [1, 2]. Then an antenna composed of the SW structure carries its electromagnetic wave travelling with phase velocity $v_p$ smaller than $c$.

$$v_p/c = (\omega'_p/\beta_p)/(\omega_0/\beta_0) < 1,$$

Here $\beta$ is phase constant, subscript $p$ denotes slow wave structure and 0 is the normal structure and prime (′) is taken for the slow wave structure. When an antenna (length $L_0$) and an antenna with slow wave structure (length $L_p < L_0$) are resonant at $\omega'_p = \omega_0$, $(\omega'_p/\beta'_p)/(\omega_0/\beta_0) = \beta_0/\beta'_p < 1$. From this, $\beta_0 < \beta'_p$, meaning $\lambda_0 < \lambda'_p$ and $f'_p < f_0$. This implies that with slow wave structure, even a shorter antenna can be made resonant at a lower frequency $f'_p$ than $f_0$, of the longer one.

#### 6.2.1.1 SW structure

The SW structures are composed using periodic structures, transmission lines, waveguides, modification of antenna geometry, and material loading using dielectric or magnetic materials.

##### *6.2.1.1.1 Periodic structures*

The simplest periodic structures are wire lines formed into helix (Figure 1.6 (a)), meandered shapes (Figure 1.6(b)), zigzag (Figure 6.1), and fractal shape (Figure 6.2) [3], all of which are modified from a linear line. Since a wave travels along the wire line of

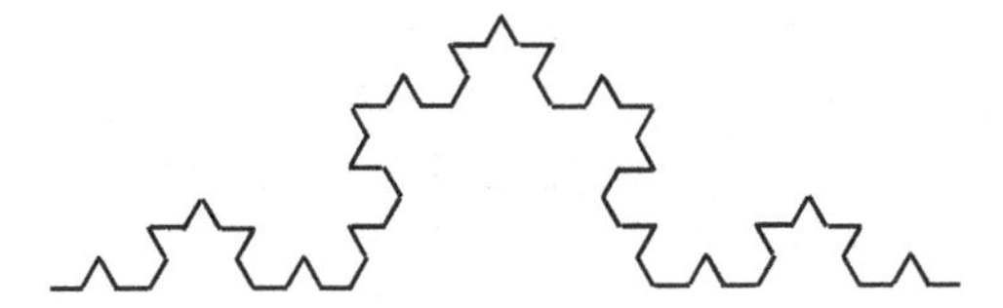

**Figure 6.2** A fractal line.

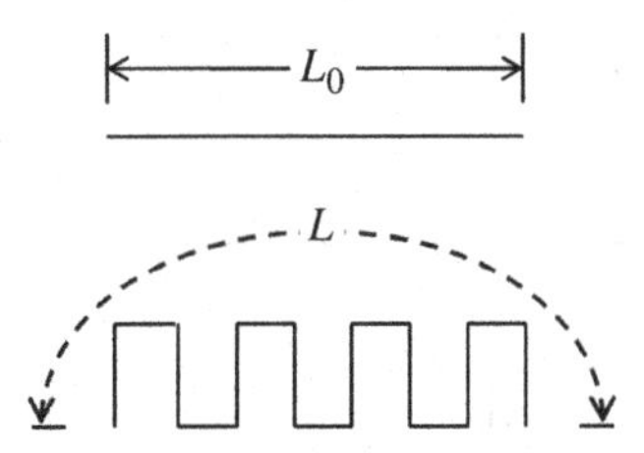


**Figure 6.3** Linear length and the length of the meander line.

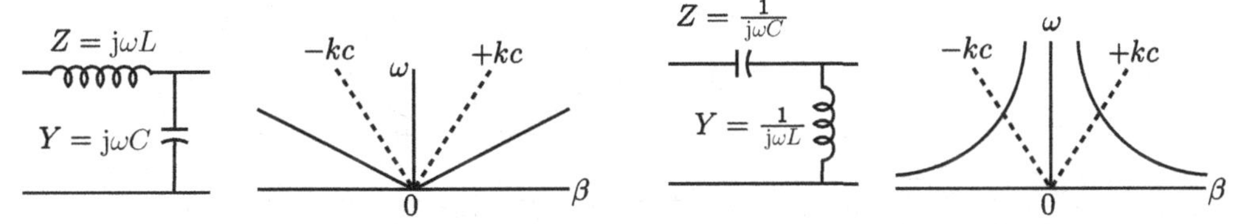


**Figure 6.4** A SW structures composed with $C$ and $L$, with $\beta$–$\omega$ diagram [after 4].

the line length $L$, it takes a longer path than that of the antenna (the linear length $L_0$ (Figure 6.3)). Consequently, the wave travel time on the antenna structure of the length $L_0$ becomes equivalently $t_p$ which is excess time beyond the travelling time $t_0$ on the length $L_0$. Thus, the phase velocity $v_p = L_0/t_p$ of the wave on the antenna structure is smaller than $c = L_0/t_0$, implying that the wave propagates more slowly on the antenna structure corresponding equivalently to the SW.

These wired SW structures are advantageous in designing a small antenna, since a self-resonance condition is easily attained, even though the antenna dimensions are small. In small antennas, self-resonance is significant, because it helps mitigate the matching difficulty in a small antenna arising from its impedance, which is a very low resistive component and a very large reactive component that may combine to produce a large loss and degrade the antenna performance. The wired SW structures have additional advantages; they are fabricated rather simply, with low cost, and can be constructed in planar form, which is very useful for small equipment such as small handsets, handy mobile terminals, and so forth.

The SW structures can be composed with transmission lines, on which impedance components or equivalent boundaries are loaded periodically. The loaded impedances are treated as ordinary lumped components. Examples are shown in Figure 6.4, in which

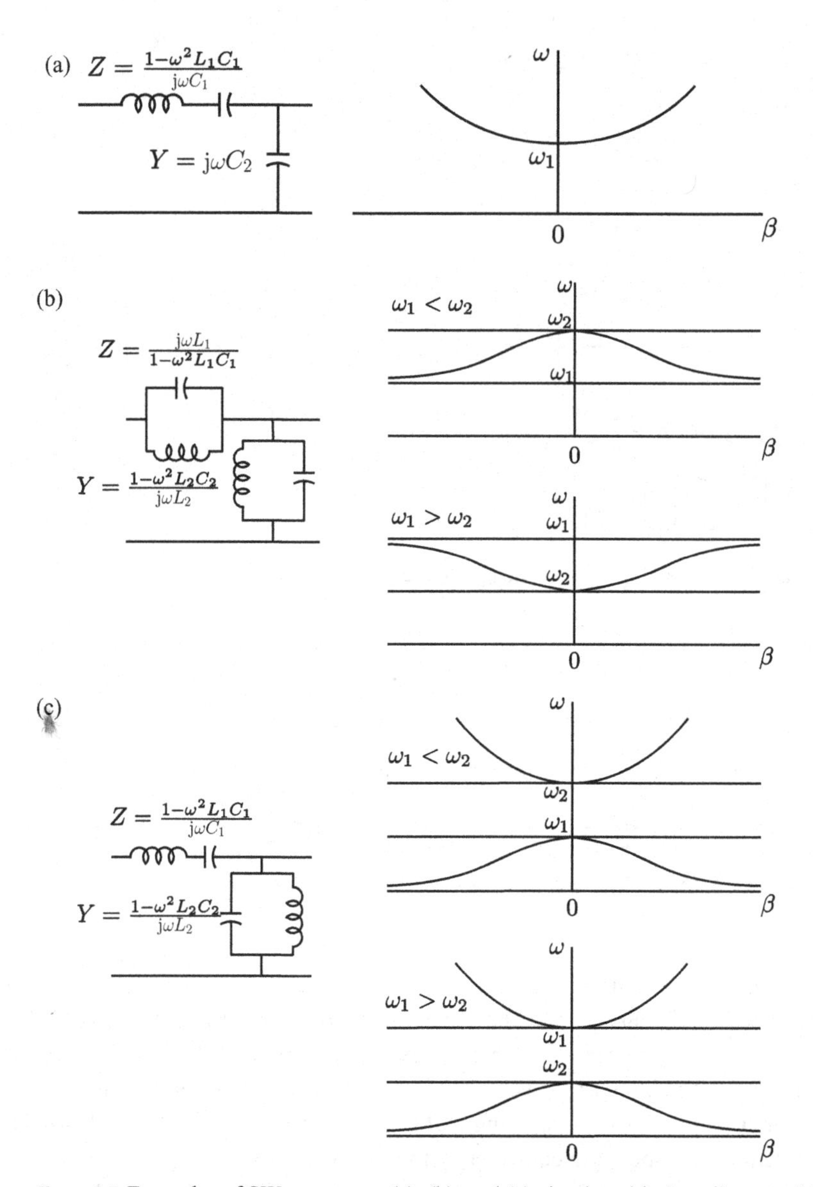


**Figure 6.5** Examples of SW structures: (a), (b), and (c) circuits with $\beta$–$\omega$ diagrams [4].

one segment of the unit length with $\beta$–$\omega$ diagrams for each circuit is illustrated [4]. Here the propagation constant $\gamma = \alpha + j\beta = \sqrt{ZY}$, where $\alpha$ is the attenuation constant, $\beta$ is the phase constant, $Z$ is the series impedance per unit length, and $Y$ is the shunt admittance per unit length. In the figure, the dotted line indicates $\pm kc$, below which $\beta$ is greater than $k$ (the wave number in free space), meaning that $v_p$ is smaller than $c$ and so the media supports the SW. The SW structure can be composed by setting appropriate impedances on the transmission line to make $\beta$ greater than $k$. Other examples of LC circuits that possibly realize SW structure are illustrated in Figure 6.5 (a), (b) and (c) with

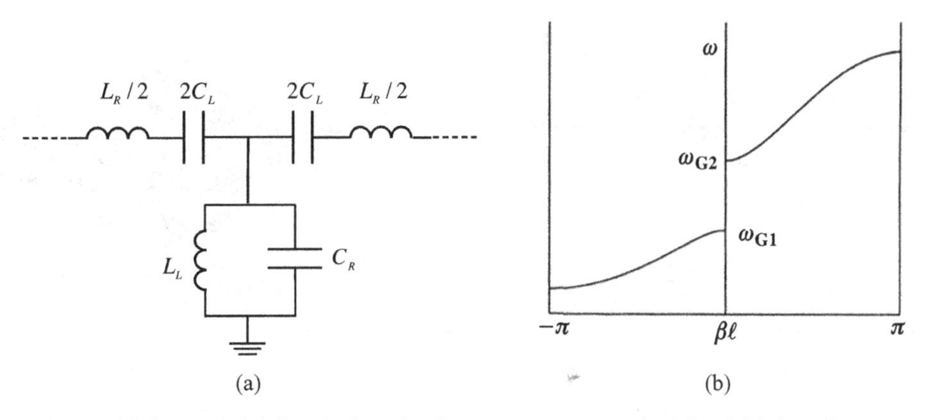


**Figure 6.6** (a) An example of equivalent circuit to represent metamaterial and (b) $\beta$–$\omega$ diagram [6].

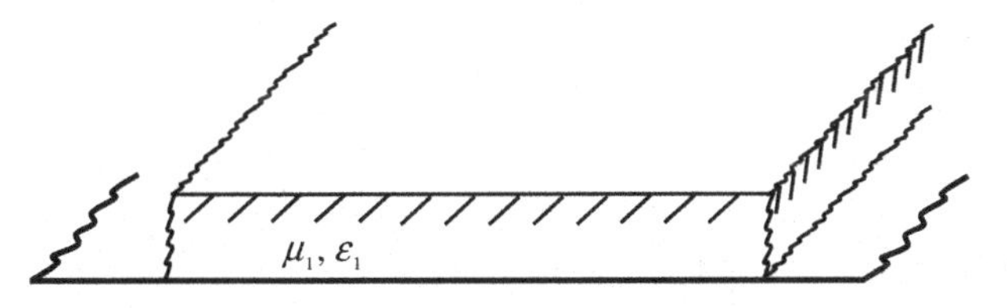


**Figure 6.7** A SW structure with dielectric sheet.

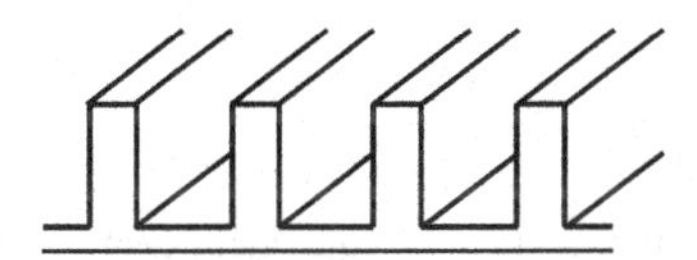

**Figure 6.8** A SW structure with corrugated surface.

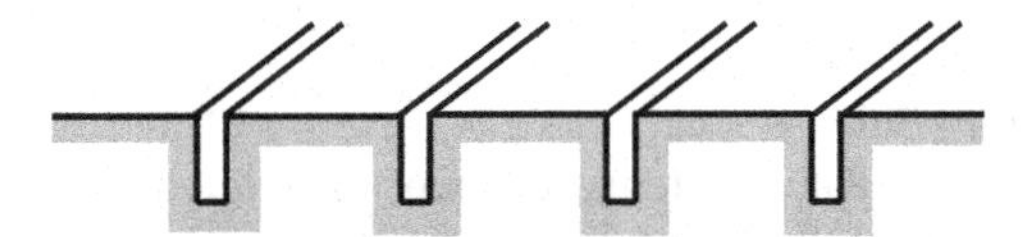

**Figure 6.9** A SW structure with trough on the surface.

the $\beta$–$\omega$ diagrams [5]. The circuits shown in Figure 6.5 can be utilized as metamaterials exhibiting negative phase constant $-\beta$ by referring to the $\beta - w$ diagram.

Transmission lines can also be applied to implement metamaterials performance exhibiting either negative $\varepsilon$ or negative $\mu$, or both by setting up impedance components appropriately [6]. Examples are shown in Figure 6.6. Applications of this technique to small antennas will be described in later sections.

Two-dimensional SW structures can be made with a dielectric sheet (Figure 6.7), corrugation (Figure 6.8), holey plate, and troughs on the surface (Figure 6.9) [7]. The SW structures can be formed in two-dimensional (2D) structures with transmission lines

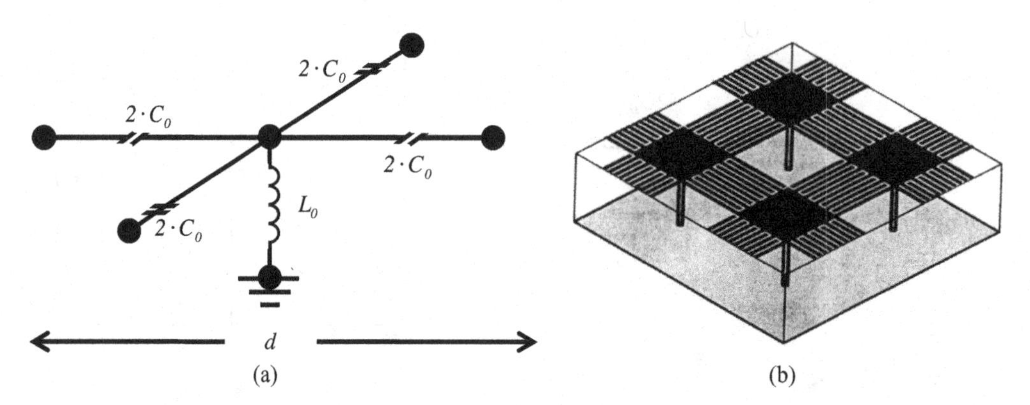


**Figure 6.10** A 2D metamaterial structure composed with transmission line; (a) equivalent circuit [17], (b) practical model [8].

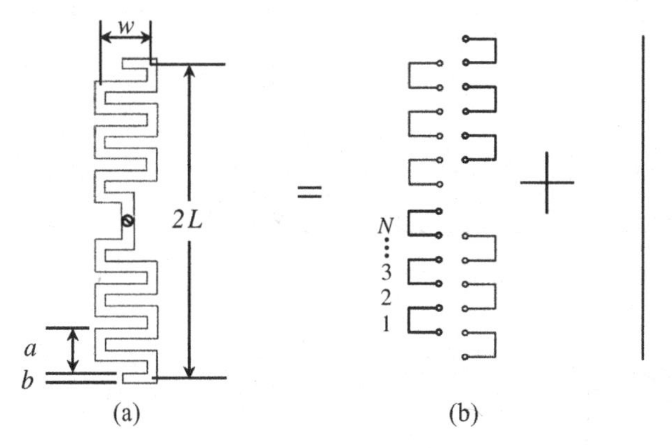


**Figure 6.11** Meander line antenna: (a) antenna structure and (b) an equivalent expression [10].

(Figure 6.10) [8]. Three-dimensional (3D) structure is also composed with transmission lines or bulk materials [9].

*6.2.1.1.1.1 Meander line antennas*

A meander line may be used as a typical antenna element for reducing the size, as it has the SW structure [10–12]. Applicable antenna types are not only dipoles, but also monopoles over a ground plane, and arrays. The antenna element can be made of thin wire and thin tape printed on the surface of dielectric structure. The performance of a meander line antenna (Figure 6.11(a)) can be found by treating the antenna structure as a series connection of a shorted two-wire transmission line plus a linear element as shown in Figure 6.11(b) [10]. However, the antenna is practically designed by using an electromagnetic simulator, for instance, MoM (Method of Moment) and FDTD (Finite-Difference Time-Domain) [11]. The simulation is simple and yet provides better results as compared with the calculation that uses the equivalent transmission line.

The phase velocity $v_p$ of an EM wave on the meander line is approximately given by

$$(L_p/L_0)c \tag{6.1}$$

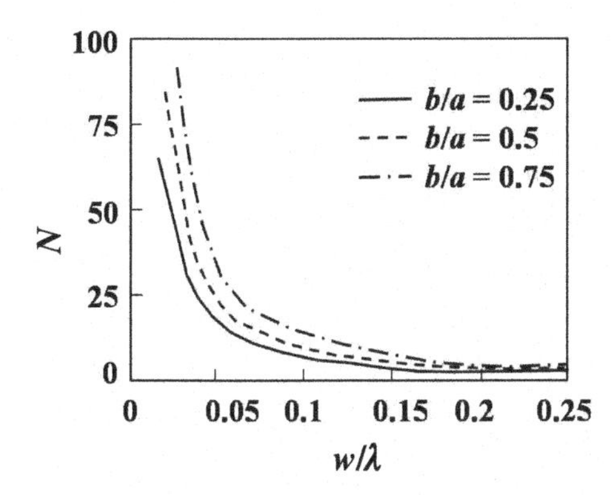


**Figure 6.12** Relationship between the number of segment $N$ and the width $w/\lambda$ [after 10].

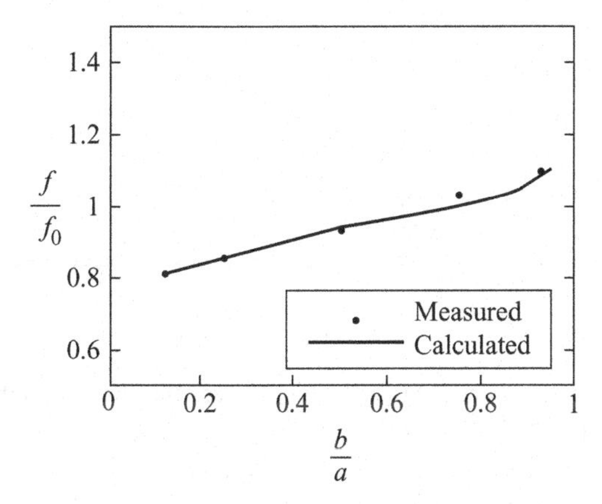


**Figure 6.13** Resonance frequency ($f/f_0$) dependence on the ratio of the pitch $b/a$ [10].

where $L_p$ denotes the length of the meandered line and $L_0$ is the length of the linear part of the line.

The resonance frequency $f_0$ is determined mainly by the total length $L_0$ of the line, but in a strict sense it depends on the structural parameters of antennas (the straight length $L$, width $w$, pitch $a$, element width $b$, and number of equivalent transmission lines $N$). With the equivalent transmission line model, the resonance wavelength $\lambda_0 = c/f_0$ is given in conjunction with $N$, which is given by

$$N = [\lambda/4 \log(2\lambda/b) - L \log(8L/b)]/\{w \log(2L/Nb)[1 + 1/3(\beta w/2)^2]\}. \quad (6.2)$$

This is a transcendental equation which can't be solved analytically but only numerically and given graphically as shown by Figure 6.12 and Figure 6.13, respectively, where relationships between $N$ and $w/\lambda$, and $b/a$ against $f/f_0$ are illustrated. The resonance frequency $f_0$ can be calculated from $\lambda_0$ and also read from Figure 6.13.

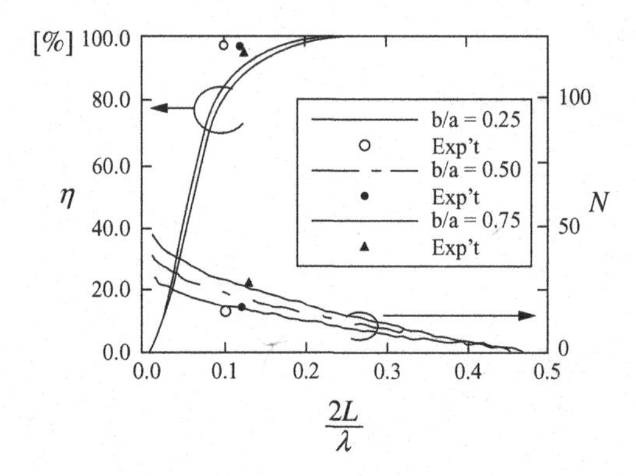


**Figure 6.14** Relationship between radiation efficiency $\eta$ and the length $2L/\lambda$ $(w/\lambda = 0.03)$ [10].

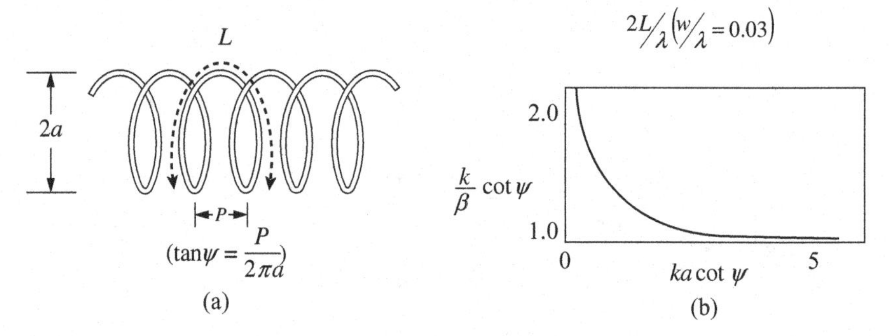


**Figure 6.15** NMHA; (a) antenna structure and (b) relationship between the phase velocity $v_p$ and the pitch angle $\psi$ in NMHA [14].

Efficiency $\eta$ of a meander line antenna is given by using antenna parameters as follows:

$$\eta = 1/[1 + 1/4\pi(Rs/Rd)(\lambda/b)(Nw/L)]. \tag{6.3}$$

The radiation efficiency $\eta$ is drawn with respect to $b/a$ as Figure 6.14 shows.

The efficiency of a meander line antenna is lower than that of a linear antenna of length $L_0$ because it has greater loss owing to longer length $L_p$ than the linear antenna of the length $L$.

Design of this type of antenna will be introduced in Chapter 7.

### *6.2.1.1.1.2 Helical antennas*

Here the helical antenna is taken only for the case where radiation is in the direction normal to the antenna axis; that is, Normal Mode Helical Antenna (NMHA) [13]. In the case of NMHA, the phase velocity $v_p$ is given approximately by [14]

$$v_p = c \sin \psi \tag{6.4}$$

where $\psi$ is the pitch angle (Figure 6.15).

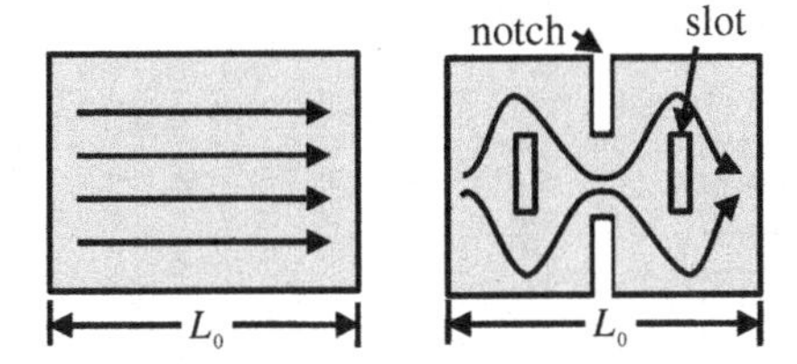


**Figure 6.16** An example of extension of the current path on an antenna structure with a slot and notch.

The resonance frequency depends on the antenna parameters such as the diameter $d$, pitch angle $\psi$, and the total length $L$; however, since it is hardly given analytically, practical design depends on either simulation or approximation.

Practical design methods will be discussed in Chapter 7.

### *6.2.1.1.1.3 Waveguide structures*

Planar surface waveguides propagate either fast or slow waves, depending on the structural parameters of the guides. Usually, these guides are used as relatively large, two-dimensional, flush mounted antennas; however, here only SW structures and those applicable to construct small antennas will be taken into consideration. Typical surface guides are planar dielectric sheet (Figure 6.7), holey plate structure, inductive guide structure, corrugated surface (Figure 6.8), capacitive guided structure, and so forth [1, 4].

These guides are not necessarily used as a radiating element, but applied to lowering resonance frequency by placement on a planar antenna surface.

This concept is adopted when equivalent metamaterial performance is required.

### *6.2.1.1.2 Modification of antenna geometry to extend the current path*

The SW structures can be produced by modifying an antenna geometry such that the current flow path is lengthened, resulting in the phase velocity $v_p$ equivalently smaller than $c$ in free space, and hence lowering the resonance frequency. A planar antenna, on which a slot or a notch is placed to lengthen the current path so that the electrical size is shortened as shown in Figure 6.16, is an example. In the figure, the direction of the current flow is illustrated with arrows.

The phase velocity $v_p$ of the wave travelling on this surface is modified by lengthening the current path $L_0$ to $L_p$ and is approximately given by

$$v_p = (L_0/L_p)c \tag{6.5}$$

where $L_p$ is the length of the current path with the slot and $L_0$ is the length without the slot. The resonant frequency is determined approximately by the length $L_p$.

Multiband operation can be realized by using several slots on a planar structure, by which the current flows on several paths, as shown in Figure 6.17. In this case, each resonance frequency corresponds to one length of the current path.

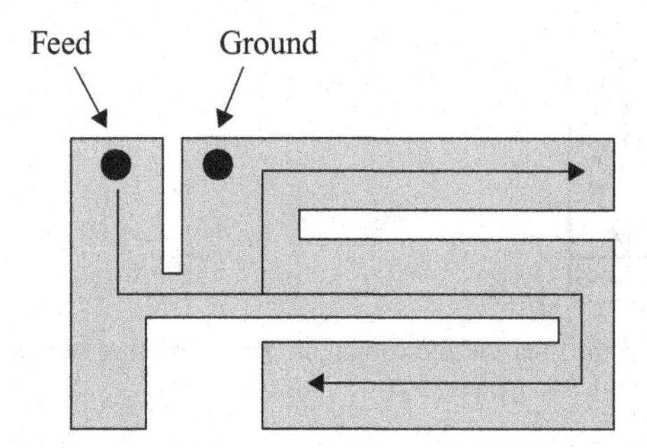


**Figure 6.17** Extended current paths with a notch and a slot producing multiband operation.

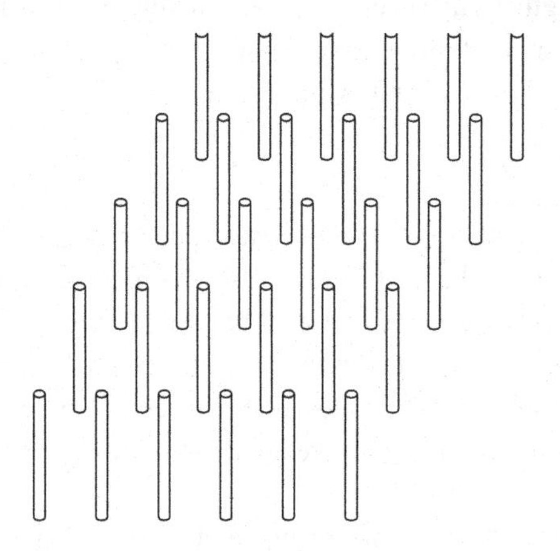

**Figure 6.18** Array of thin wires to compose negative permittivity surface [15, 16].

### *6.2.1.1.3 Material loading on an antenna structure*

Loading of materials such as dielectric, magnetic, or metamaterials on an antenna structure is a simple way to produce the SW structure. Since the phase velocity $v_p$ in such materials is expressed by using the permittivity $\varepsilon$ of the dielectric material and the permeability $\mu$ of the magnetic material, as

$$v_p = \omega_0/\beta = 1/(\varepsilon\mu)^{1/2}. \tag{6.6}$$

From this, $v_p$ is shown to be smaller than $c$, because

$$v_p/c = (\varepsilon_0\mu_0)^{1/2}/(\varepsilon\mu)^{1/2} = 1/(\varepsilon_r\mu_r)^{1/2} < 1 \tag{6.7}$$

since $\varepsilon = \varepsilon_r\varepsilon_0$ and $\mu = \mu_r\mu_0$, and $\varepsilon_r$ and $\mu_r$ are usually greater than unity.

Making use of metamaterials is another useful way to produce SW structure. The metamaterials are known as materials that exhibit uncommon electromagnetic properties not available in nature [15–17]. In reality, metamaterials do not exist; however, equivalent materials have been realized first by using a dense array of thin wires (Figure 6.18), and an array of split ring resonators (SRR) (Figure 6.19) [18–20]. The former exhibits a property of negative permittivity $\varepsilon$ and the latter shows negative permeability $\mu$ and

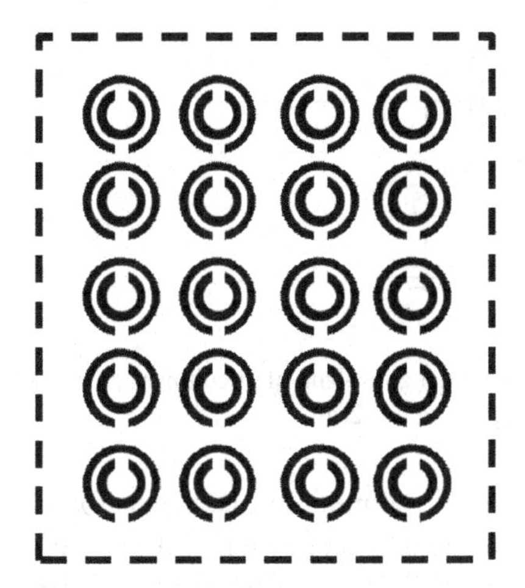

**Figure 6.19** Array of sprit ring resonators (SRR) to compose negative permeability surface [15, 16].

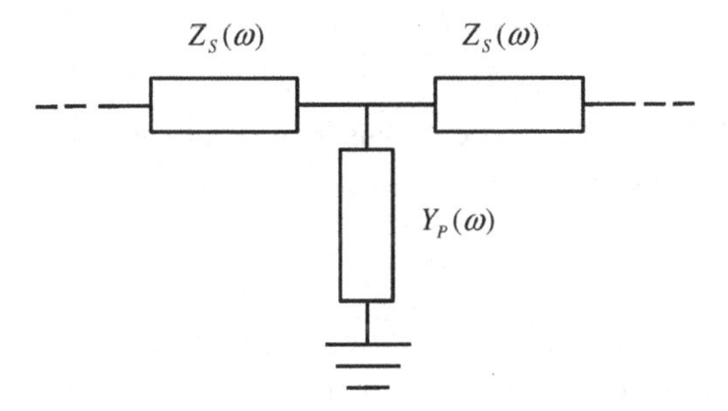


**Figure 6.20** A T-circuit model representing transmission line (one segment of the periodic connection) [26].

applications of them to antennas have been described in many papers [16–17, 21–23]. Among them, Itoh demonstrated an artificial metamaterial using transmission lines, which achieved equivalent metamaterial performances that exhibit a left-handed (LH) property that resulted from negative permeability and negative permittivity at the same time [24, 25].

Plane-wave propagation in isotropic and homogeneous dielectric can be treated by the concept of TEM propagation in a transmission line, which is comprised of periodic connections of a T-shaped circuit consisting of a series per-unit-length impedance $Z_s$, and a shunt per-unit-length admittance $Y_p$ as shown in Figure 6.20 [26]. Here, characteristic impedance $Z = \sqrt{Z_s/Y_p}$ and propagation constant $\gamma = \sqrt{Z_s Y_p}$, and $Z_s$ and $Y_p$, respectively, are described as

$$\begin{aligned} Z_s(\omega) &= \mathrm{j}\omega\mu \\ Y_p(\omega) &= \mathrm{j}\omega\varepsilon \end{aligned} \tag{6.8}$$

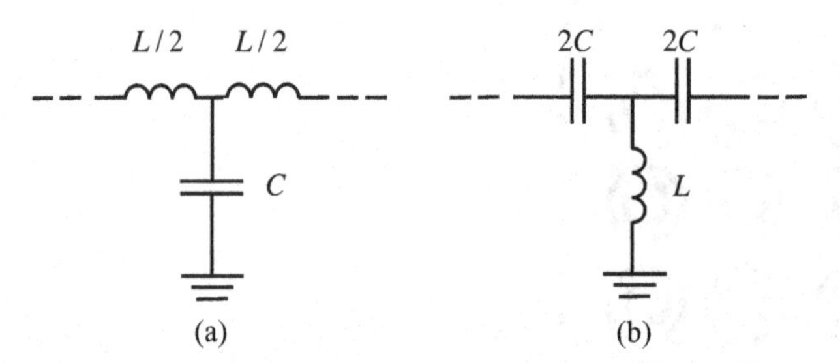


**Figure 6.21** The equivalent transmission line model consisting of circuit parameters *L-C*; (a) *L* in series and *C* in shunt and (b) *L-C* interchanged circuit [26].

where $\varepsilon$ and $\mu$, respectively are the dielectric permittivity and the magnetic permeability of the propagation medium [27]. The transmission line model consisting of circuit parameters *C* and *L*, is illustrated in Figure 6.21(a). From Eq. (6.8), the effective constitutive parameters $\varepsilon_{eff}$ and $\mu_{eff}$, respectively, are described as

$$\begin{aligned} \varepsilon_{eff} &= C/l \\ \mu_{eff} &= L/l \end{aligned} \tag{6.9}$$

where $l$ is the unit length.

Interchange of $C$ and $L$, results in the dual transmission line as shown in Figure 6.21(b). By using (6.8), $\varepsilon_{eff}$ and $\mu_{eff}$ can be related with the circuit parameters as

$$\begin{aligned} \varepsilon_{eff} &= -1/(\omega^2 Ll) \\ \mu_{eff} &= -1/(\omega^2 Cl) \end{aligned} \tag{6.10}$$

The dispersion characteristics of the equivalent circuit are the same as that shown previously in Figure 6.5(a) and (b). In the SW (Slow Wave) structure, a wave propagates with the phase constant $\beta$ greater than that in free space $k$. The transmission line with the constitutive parameters $-\varepsilon$ and $-\mu$ that are given by (6.10) is considered as the LH (Left Handed) media where the phase velocity $v_p$ and the group velocity $v_g$ are anti-parallel, and the wave propagates backward [28]. The LH media is characterized by the property of negative propagation constant $-k$, the attribute of $-\varepsilon$ and $-\mu$. The name is originated from the term Right-Hand (RH) Rule that is derived from use of a right hand to indicate the direction of the wave vector **k**; for example, $E \times H = S$, by the thumb of the right hand as Figure 6.22 shows. In the LH media, the direction of the wave vector **k** is the opposite to that of the cross product vector **S**, while in the RH case, directions of both **k** and **S** are the same as Figure 6.22(b) illustrates.

In reality, however, this transmission line cannot be implemented perfectly, because there exists unavoidable parasitic series inductance $L_p$ and shunt capacitance $C_p$. Then a CRLH (Composite Right/Left-Handed) model shown in Figure 6.23 is considered as the most general form of a structure with the LH attributes [26]. The dispersion characteristics of the CRLH transmission line model is depicted in Figure 6.24, which shows a case for $\omega_2 > \omega_1$ [26]. The region between two frequencies $\omega_1$ and $\omega_2$ denotes a bandgap, where waves do not propagate. This bandgap is generally called the EBG (Electromagnetic Band Gap).

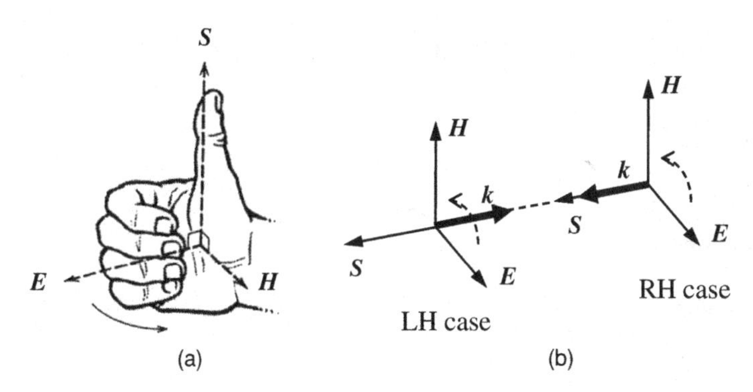


**Figure 6.22** The Right-handed Rule; (a) demonstration for vector cross product and (b) the Right-handed (RH) and Left-handed (LH) cases.

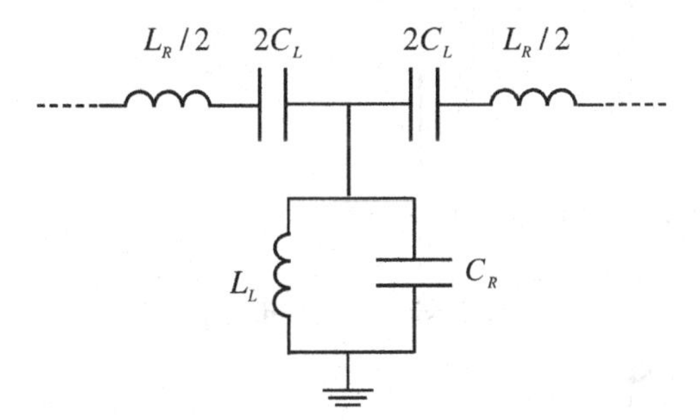


**Figure 6.23** CRLH model (one segment of equivalent transmission line) [26].

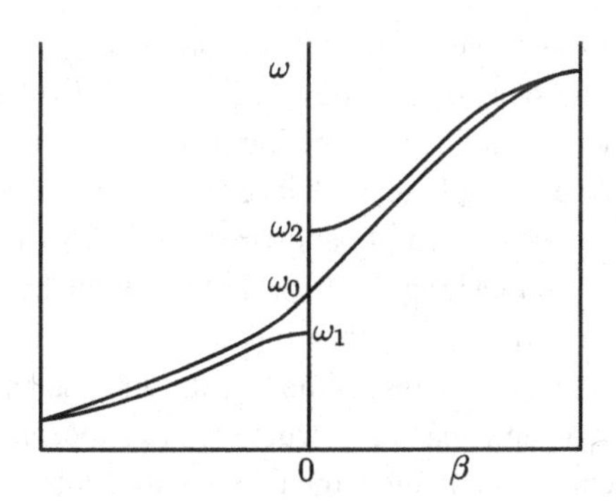


**Figure 6.24** Dispersion characteristics of the CRLH transmission line model [26].

The effective constitutive parameters $\varepsilon_{eff}$ and $\mu_{eff}$ of this CRLH model are given by

$$\begin{aligned} \varepsilon_{eff} &= (C_R - 1/\omega^2 L_L)/l \\ \mu_{eff} &= (L_R - 1/\omega^2 C_L)/l \end{aligned} \qquad (6.11)$$

This indicates the nature of CRLH of the artificial transmission line. An SNG transmission line can be realized by loading a host line with series capacitors (corresponding to $-\mu$ lines) or shunt inductors (corresponding to $-\varepsilon$ lines).

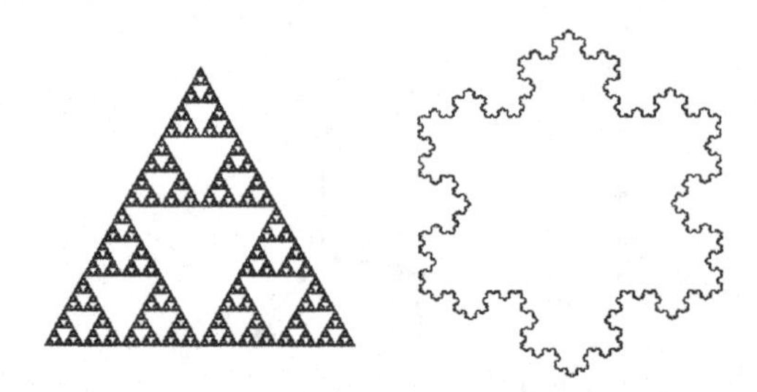

**Figure 6.25** Examples of fractal shape [31].

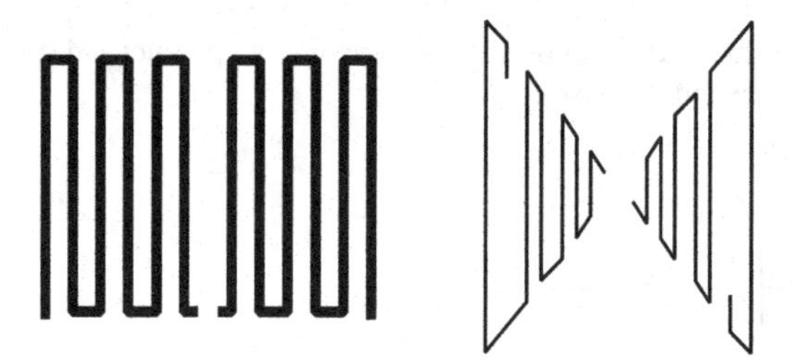

**Figure 6.26** Examples of meander line (dipole structure) [32].

Metamaterials composed by this transmission line model will be discussed in Section 6.2.5.

### 6.2.2 Full use of volume/space circumscribing antenna

There have been some discussions on improvement of bandwidth or maximization of gain in an antenna with a given antenna size. Balanis described in [29] that bandwidth of an antenna circumscribed by a sphere of radius $r$ can be improved only if the antenna utilizes efficiently the available volume of the sphere, with its geometrical configuration. Also Hansen commented in [30] that improvement of bandwidth for an ESA is possible only by fully utilizing the volume in establishing a TE or TM mode or by reducing efficiency. These comments can be interpreted as that full use of a volume which circumscribes an antenna structure is the only means to improve the antenna bandwidth. In other words, even with small size, if an antenna is constructed so as to occupy a whole volume of a sphere in which the antenna is contained, the bandwidth can be improved. Increase in bandwidth with a given size of an antenna corresponds to creation of a small antenna. However, in practice, this is an idealized concept, because an antenna structure can never occupy fully the space or volume that circumscribes the antenna. The concept may be followed by filling such space or volume efficiently; that is, to use the space with the antenna geometry as much as possible. The space is not necessarily three dimensional, but perhaps two dimensional depending on an antenna structure. Fractal shape is a typical example, which fills a space efficiently with the antenna geometry as Figure 6.25 shows [31]. There are some other examples of two-dimensional antennas; meander lines (Figure 6.26) [32], Peano geometry (Figure 6.27) [33], and Hilbert curve

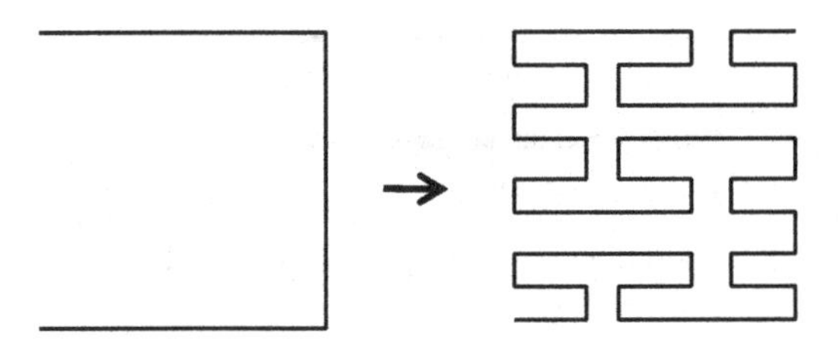

**Figure 6.27** Examples of Peano curve geometry (the original to the higher mode) [33].

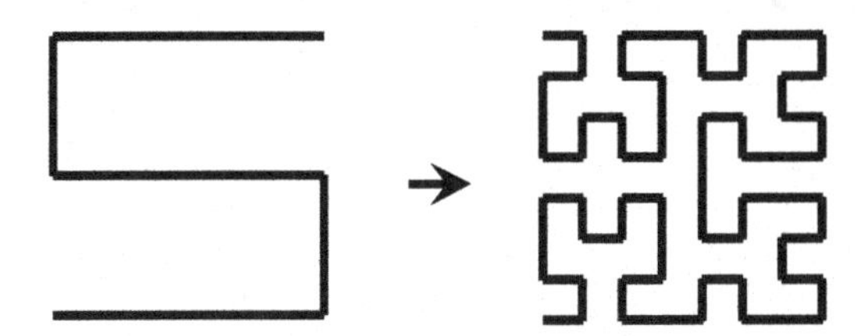

**Figure 6.28** Examples of Hilbert curve geometry (the original to the higher mode) [34].

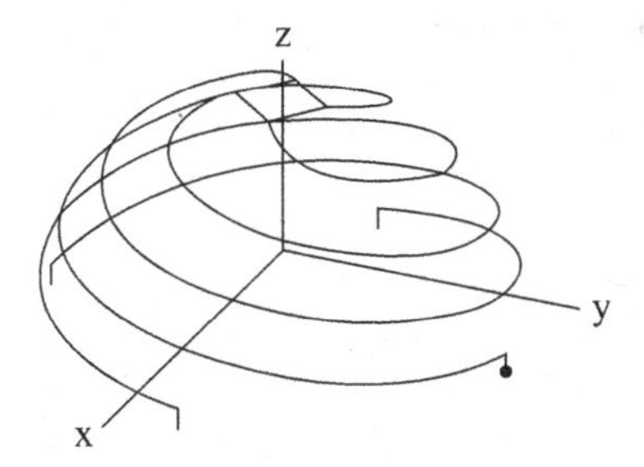


**Figure 6.29** A four-arm helical antenna [35].

geometry (Figure 6.28) [34]. With a helical structure, a four-arm helical antenna, which is designed to occupy a volume of a sphere as much as possible as shown in Figure 6.29, is one of the three-dimensional applications [35].

### 6.2.3 Arrangement of current distributions uniformly

A similar concept to the full use of space, but slightly different method, by which the maximum gain with a given size of an antenna is obtained, is to arrange the current distribution on an antenna element to be uniform. Chu discussed that the ideal current distribution to attain the maximum gain with a given size of an antenna is uniform [37] and the gain at this condition is $\pi r/4\lambda$, where $r$ is the radius of a sphere enclosing the antenna. However, uniform distribution can never be realized by a small antenna. On a small dipole, for instance, the current distribution tends to zero toward the end of the element, taking the maximum value at the center, that is, the feed point. Usually it is assumed to be a triangular shape with the maximum at the feed point and zero at the end as Figure 6.30(a) depicts, where a uniform case is also shown (Figure 6.30(b)). One way

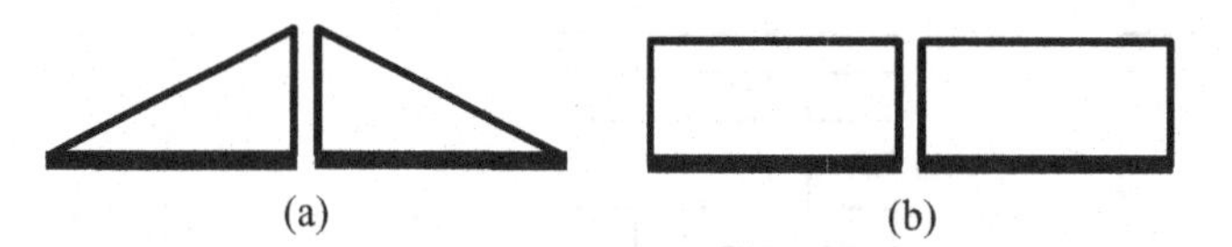


**Figure 6.30** Current distribution on a short dipole; (a) triangular shape and (b) uniform (ideal case).

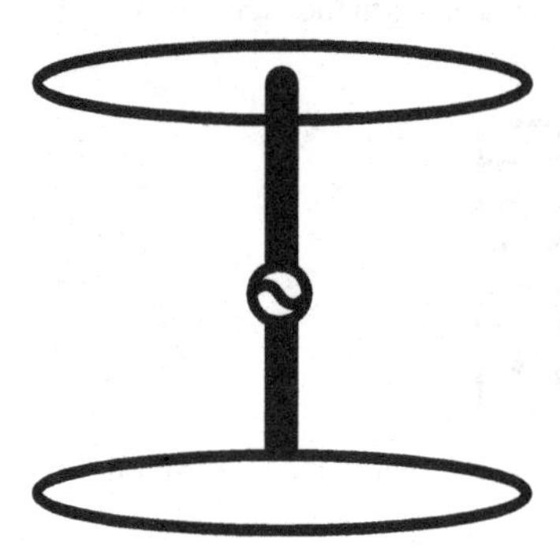

**Figure 6.31** Capacitance plate loaded on the top of a short dipole.

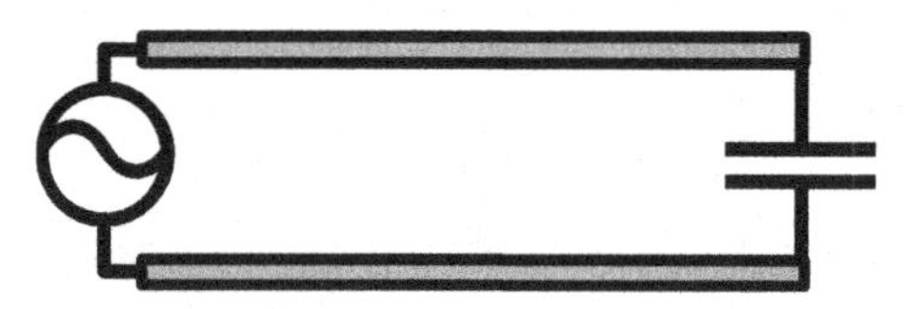

**Figure 6.32** A capacitance loaded transmission line.

to obtain uniform current distribution on a short dipole is to place a capacitance plate loading on its top as shown in Figure 6.31 [38]. By this means, the antenna gain can be the highest that is possible with that length of the dipole antenna. The longer the linear part of the antenna becomes, the nearer the current distribution approaches to uniform, as the distribution is gradually smoothed from the triangle. By loading a capacitor at the end terminal, the current distribution on the transmission line may be made nearly uniform (Figure 6.32).

### 6.2.4 Increase of radiation modes

Increasing radiation modes is another important method to create an ESA. Hansen discussed in [30] that antenna $Q$ will be reduced by $1/2$ with simultaneous excitation of TE and TM modes. Reduction of antenna $Q$ implies increase in the bandwidth. If this can be achieved in an antenna with a given size, it corresponds to creating an ESA. Practical methods to increase radiation modes are composing an antenna with complementary structure, or conjugate structure, combining different types of antenna, and so forth. These will be described in the following section.

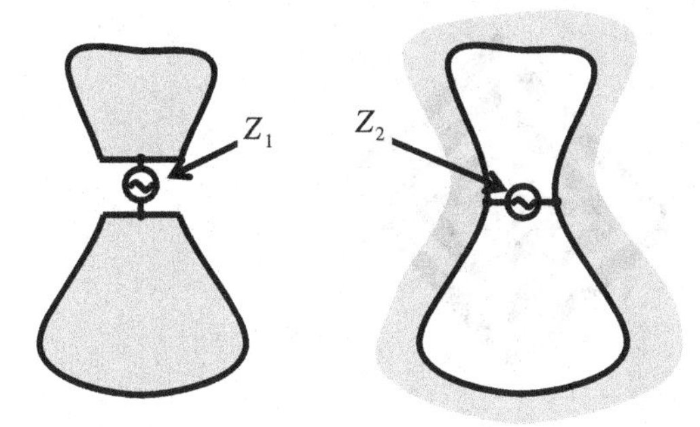


**Figure 6.33** A demonstration of the principle of self-complementary concept combining two (E and H) mode structure [39].

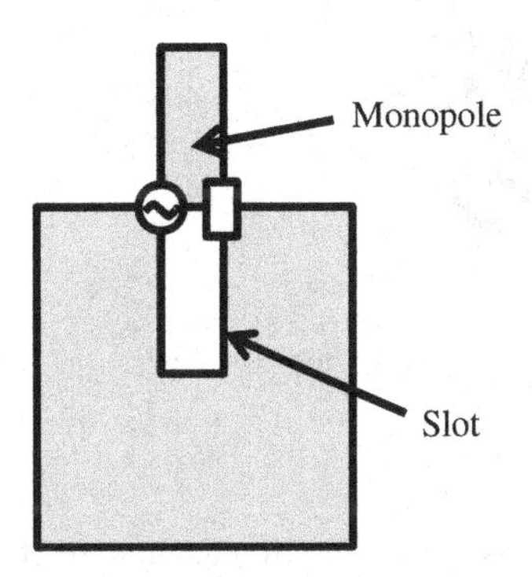


**Figure 6.34** An example of self-complementary antenna; composed with a monopole and a complementary slot on the ground plane [40].

#### 6.2.4.1 Use of self-complementary structure

Self-complementary structure is constructed by combining two-mode ($E$ and $H$) structures [39], as Figure 6.33 illustrates. An example is a monopole combined with a self-complementary shape of a slot on the ground plane as shown in Figure 6.34 [40, 41]. A self-complementary antenna has inherently frequency independent properties, as the impedance $Z_0$ of a complementary antenna is constant over infinite frequency range as shown by Mushiake in [40] with

$$Z^2 = Z_1 Z_2 = (Z_0/2)^2 = \text{const} \tag{6.12}$$

where $Z_1$ is the impedance of an E-mode antenna, $Z_2$ is that of a complementary H-mode antenna, and $Z_0$ is the intrinsic impedance of the medium.

The perfect frequency independent performance can be achieved only when the size of the ground plane is infinite. With a finite ground plane, the property of frequency independence is limited to some extent, and the frequency bandwidth is no longer infinite, although the bandwidth obtained will be still appreciably wide.

Some examples of planar self-complementary antennas are shown in Figure 6.35 [39].

**Figure 6.35** Examples of planar self-complementary structures [39].

### 6.2.4.2 Use of conjugate structure

An antenna composed of conjugate components consisting of a capacitive element and its conjugate inductive element so that the reactive component in the antenna structure can be compensated, resulting in self-resonance, can be designed to have appreciably wide bandwidth, although the size is fairly small. This is also a useful method to produce an ESA. Combination of an electric source with a magnetic source may become a pair to compensate the reactive component in the antenna structure; thus self-resonance is attained.

### 6.2.4.3 Compose with different types of antennas

It has been shown in the previous sections that radiation modes of an antenna can be increased by increasing radiation sources, realized by combining different types of antennas, which contribute to producing different radiation modes. In addition to previously described methods such as complementary and conjugate structures, other examples are introduced below.

An example is an ESA composed with a small loop and a ground plane, on which the receiver circuit is mounted [42]. The antenna is designed based on the concept for producing ESA; increasing radiation modes, and accomplishing self-resonance. Figure 6.36 illustrates an antenna system as an example, where a loop antenna is located inside a small unit and fed with a coaxial cable. The antenna system is expressed as a combination of a loop element and the ground plane (printed circuit board) as Figure 6.37(a) shows. At the feed terminals of this antenna system, the current $I_0$ flows, that can be divided

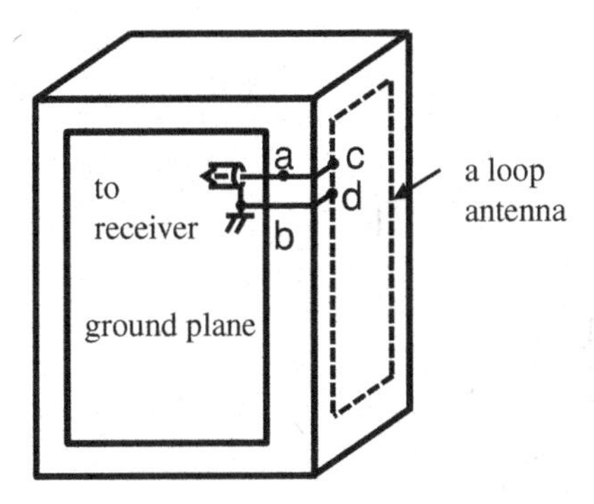


**Figure 6.36** An example of a composite antenna system composed of a loop and a dipole [41b].

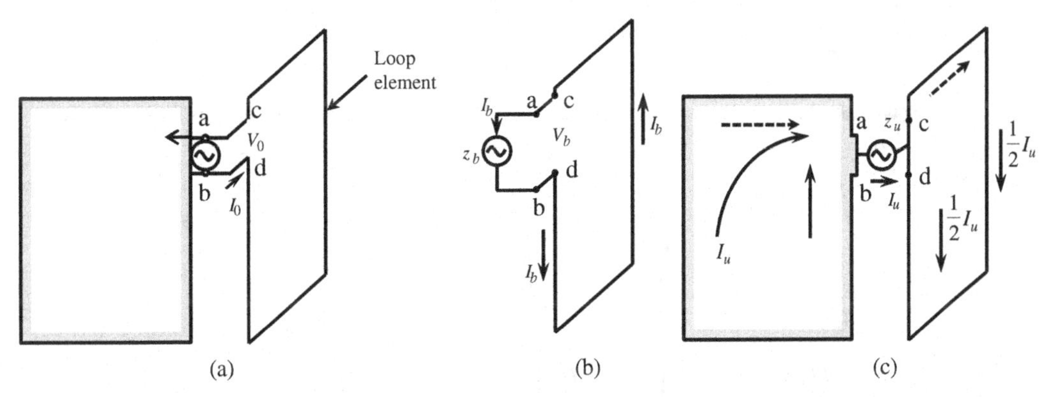


**Figure 6.37** Equivalent expression of the antenna system shown in Figure 6.36; (a) the antenna system, (b) the loop and (c) an equivalent dipole composed with the loop and the ground [41b].

into two modes; one part is the unbalanced current $I_u$ and another is the current $I_b$. These two modes are originated from feeding the connection of the loop with an unbalanced cable. The balanced current $I_b$ flows into the balanced terminals of the loop as shown in Figure 6.37(b), while the unbalanced current $I_u$, flows into both the ground plane and the loop element as Figure 6.37(c) shows. Because of the unbalanced current flow on the loop element, the loop element can be assumed to be an equivalent flat plate that yields a virtual planar dipole along with the ground plane as Figure 6.38 depicts.

This antenna was previously used in a VHF pager and brought several significant outcomes that were: (1) about a 3 dB increase in the receiver sensitivity when the pager was put in the operator's pocket due to the image effect of the loop, and (2) a change in the receiving pattern (sensitivity) to almost non-directional as a result of combination of a figure-eight pattern of the loop and another out-of-phase figure-eight pattern of the equivalent dipole. This antenna system simply appears to be only a small loop, but actually works with enhanced performance as a consequence of two-mode combination of a loop and a virtual dipole.

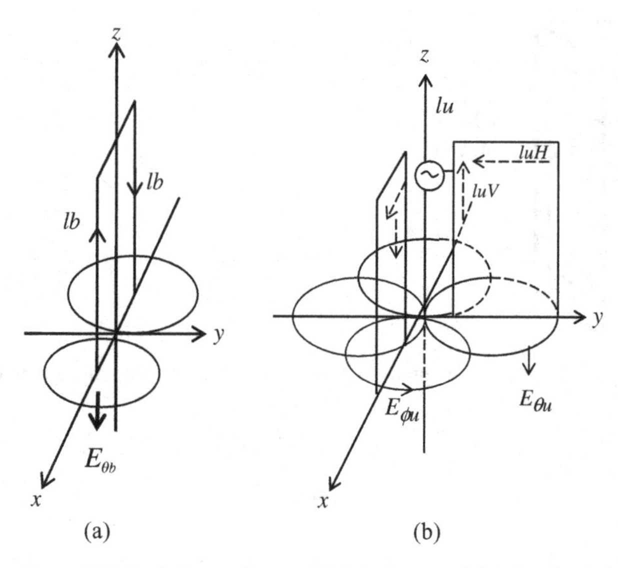


**Figure 6.38** Radiation patterns of (a) the loop and (b) the dipole having elements at right-angle [41b].

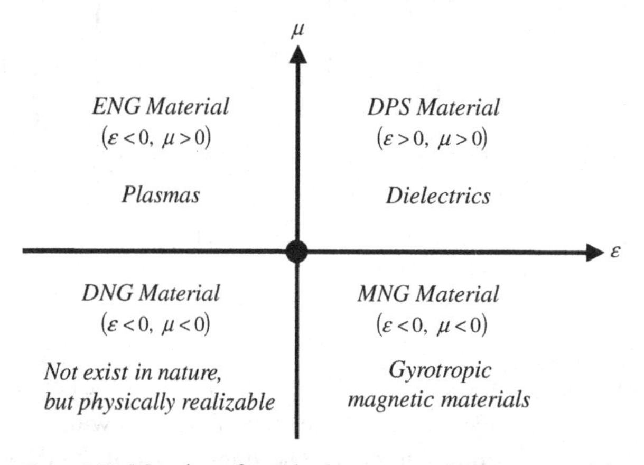


**Figure 6.39** Mapping of terminology expressing metamaterial properties.

### 6.2.5 Applications of metamaterials to make antennas small

There have been increasing trends to employ specific properties of metamaterials to create novel small antennas in the recent decade. Of course, metamaterials do not exist in reality. Artificial metamaterials, for example, transmission lines and waveguides of periodical structure, have been used for creating small antennas. Many papers treating metamaterial applications to small antennas have been published and some practically useful antennas have been reported already.

Terminologies to express metamaterial properties generally used are DNG (Double Negative), SNG (Single Negative), ENG (Epsilon Negative), and MNG (Mu Negative) and they are mapped on a permittivity–permeability chart as Figure 6.39 shows. DNG

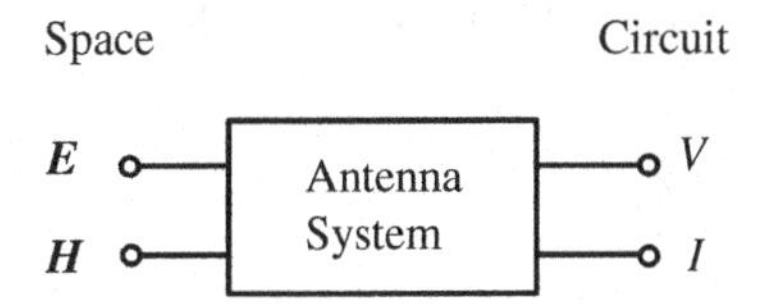


**Figure 6.40** A transducer representing an antenna system [42].

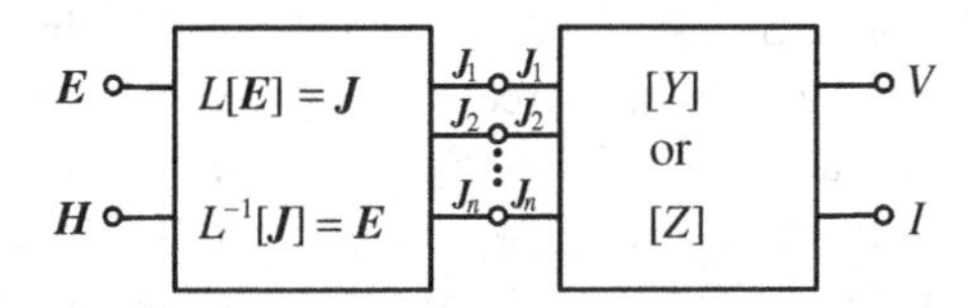


**Figure 6.41** Antenna system model divided into two parts: Space Domain and Immittance Domain [42].

materials are those exhibiting both $\varepsilon$ and $\mu$ negative, SNG materials are those having the property of either $\varepsilon$ or $\mu$ negative, and ENG and MNG materials respectively represent negative $\varepsilon$ and negative $\mu$ materials. There are some other terms, ENZ and MNZ, expressing materials of $\varepsilon$ near zero or $\mu$ near zero, respectively. The DNG property, as was described in the previous section, corresponds to properties such as LH (Left-Handed), and the NRI (Negative Refraction Index), both the attributes of negative constitutive parameters, $-\varepsilon$ and $-\mu$. The term LH is derived from the RH rule (Figure 6.22) as was described in the previous section. It is also well known that in the DNG media, the phase velocity $v_p$ and group velocity $v_g$ are anti-parallel.

Applications of metamaterials to antennas can be found in various ways: making use of SW structure realized by DNG materials for small sizing, use of composite transmission lines which exhibit both LH (Left-Handed) and RH (Right-Handed) property for small sizing and for radiation pattern control, employing SNG or DNG materials for matching either in space or at the load terminal, and so forth. For matching at the load terminals, a Non-Foster circuit realized by metamaterials is used, and an antenna of small size yet with high efficiency is obtained.

An antenna is a transducer which transforms the field parameters $E$ and $H$ (electric and magnetic field respectively) to the circuit parameters, voltage $V$ and current $I$ as shown in Figure 6.40 [42]. The antenna model represented by a transducer is divided into two parts: Space Domain and Immittance Domain as Figure 6.41 illustrates. The Space Domain defines relationships between the EM fields ($E$ and $H$) and the current distributions $J$ on the antenna system, which is given by

$$E = \text{Ł}[J] \tag{6.13}$$

$$\text{and } J = \text{Ł}^{-1}[E] \tag{6.14}$$

where Ł denotes the operator and $\text{Ł}^{-1}$ is the inverse operator. The Immittance Domain defines the relationship between the current distributions $J$ and the circuit parameters $V$

and $I$. These are expressed by using either the impedance matrix $[Z]$, or the admittance matrix $[Y]$ as

$$[V] = [Z][I] \tag{6.15}$$

$$[J] = [Y][V]. \tag{6.16}$$

Ordinary matching at the load terminals is made in the Immittance Domain, where the input impedance $Z_i$ of an antenna is matched to the load impedance, usually 50 ohms. The space matching is done in the near field of an antenna as follows. Consider the complex Poynting vector $S$, where $S = E \times H^* = \mathrm{Re}\ S + \mathrm{Im}\ S$. If $\mathrm{Im}\ S$ can be compensated by some vector $P$ produced by a source $F$ so that $\mathrm{Im}\ S + P = 0$, only real part $\mathrm{Re}\ S$ remains, and $\mathrm{Re}\ S$ is varied to make $\mathrm{Re}\ Z_i = R_i$ to be 50 ohms, then the matching process is completed, without regarding the size of antenna. The additional vector $F$ should produce such a field that $\nabla \times F$, will be zero by taking divergence so that

$$\nabla \cdot (S + \nabla \times F) = \nabla \cdot S \tag{6.17}$$

This implies that $\nabla \times F$ is a near-field component, corresponding to a reactive power. This $\nabla \times F$ component can be replaced by a metamaterial that provides the corresponding quantity to the $\nabla \times F$ component which is equal to $-\mathrm{Im}\ S$ so that the resonance in space can be attained. In a short dipole case, $\mathrm{Im}\ S$ is a reactive (capacitive) power, so the $\nabla \times F$ component should be a reactive (inductive) power, which can equivalently be represented by an ENG material. The ENG material also affects $\mathrm{Re}\ S$, so $Z_i$, (now $R_i$ since $X_i = 0$), will be adjusted to 50 ohms, by varying the material parameters such as size, location, geometry, and so forth. Even when the resistive component $R_i$ could not be made equal to 50 ohms, at least the resonance condition $X_i = 0$ in the near field can be achieved and the matching at the load terminal with low loss can be achieved very easily. In the mean time, MNG materials can be employed for the space matching when a small magnetic source like a small loop, which produces inductive field in the near field, is used. DNG materials can also be adapted for small antennas to enhance the radiation [46, 47].

#### 6.2.5.1 Application of SNG to small antennas

##### *6.2.5.1.1 Matching in space*

The concept of matching in space is to realize compensation of the reactive components in the near field of an antenna. The conjugate component is produced by an extra field in the near field of the antenna so that resonance condition is obtained, and at the same time the real component in the near field is varied so that the total antenna impedance is made equal to the complex conjugate of the load impedance. By this means, a very high efficiency small antenna would be realized. The bandwidth depends on the range of frequency over which the material can compensate the reactive component in the near field. The extra field is produced by an additional radiation source to the antenna; however, it can also be represented by a metamaterial located in the proximity to the antenna, which produces the conjugate field equivalent to that of the near field so that

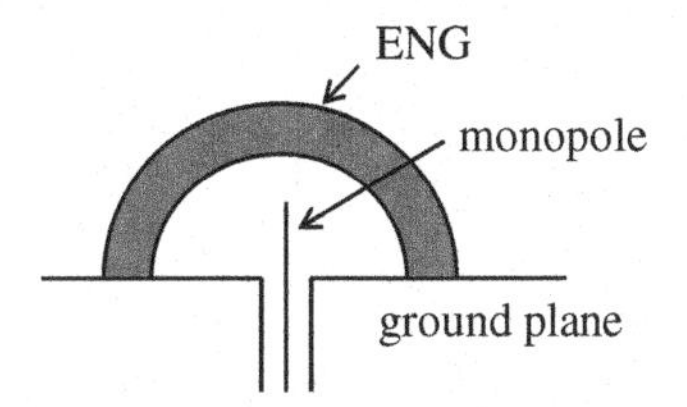


**Figure 6.42** An example of an ENG application to a short monopole antenna.

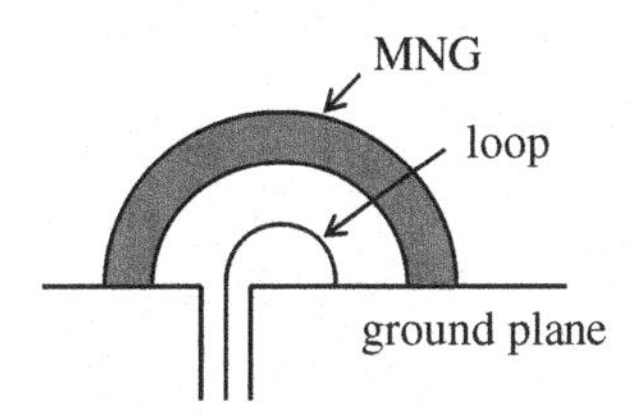


**Figure 6.43** An example of an MNG application to a loop antenna.

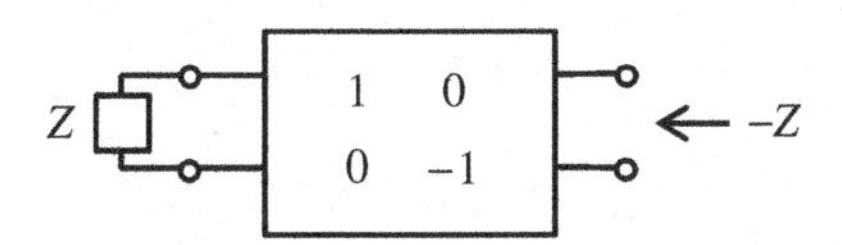


**Figure 6.44** An NIC network.

resonance and matching can be achieved in the near field of the radiation source. In practice, an epsilon-negative (ENG) material is placed near an electrical radiator like a small monopole (Figure 6.42), while a mu-negative (MNG) material is used near a magnetic radiator, for example a small loop (Figure 6.43). Stuart reported application of a negative permittivity material to a short monopole, by which the size reduction of an antenna was achieved (Figure 6.42) [44]. Bilotti *et al.* published a paper [45], in which they discussed application of a negative permeability material to a patch antenna to realize downsizing of an antenna (Figure 6.43).

The space matching can be achieved by using not only materials but also hardware that represents metamaterials. Some examples are shown in [46–48], where both ENG and MNG are realized by using a meander line, and an inter-digital capacitor circuit, respectively.

### *6.2.5.1.2 Matching at the load terminals*

It is taken for granted that matching occurs at the load terminal of an antenna; however, it is also well known that the smaller the size an antenna becomes, the harder the matching at the load terminal becomes, because the impedance tends rapidly to high reactive impedance and low resistive impedance. In order to overcome this problem, making use of an NIC (Negative Impedance Converter) at the matching circuit is considered very useful. The NIC is represented by a two-terminal network as Figure 6.44 shows [49].

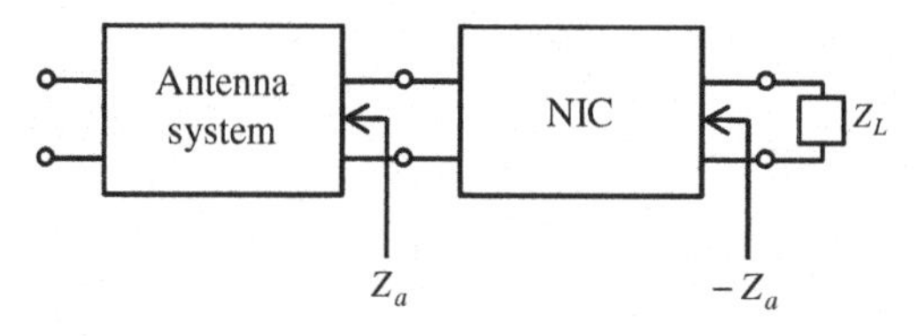


**Figure 6.45** An NIC application to antenna matching.

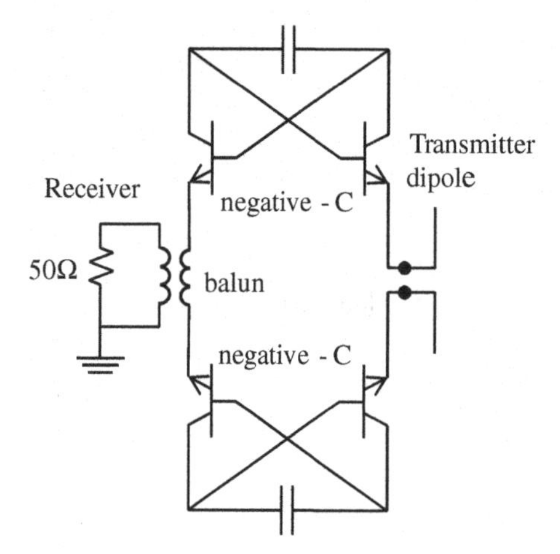


**Figure 6.46** A practical NIC implemented by a transistor circuit [49a].

The NIC network transforms an impedance $Z$ to its negative $-Z$, as shown in Figure 6.44, where the network parameters are given. By inserting an NIC network between the antenna output terminals and the matching circuit as shown in Figure 6.45 [50], the antenna impedance $Z_a$ of, for instance, a very short dipole, which has very high capacitive impedance, is converted into the negative impedance $-Z_a$, resulting in high inductive impedance at the output terminals of the NIC. The high inductive impedance can be compensated by high capacitive impedance, which has low loss. This is advantageous for matching a short dipole antenna, because it does not require high inductive impedance that has big loss, thereby reducing efficiency in an ordinary matching process. Practical NIC circuits can be implemented by a transistor circuit and some excellent results have been reported by Sussman (Figure 6.46) [49a]. This is an application of Non-Foster circuitry to the antenna matching circuit [51]. However, use of a transistor circuit has disadvantages, because of its uni-directionality against bi-directionality of the antenna. Recently real (not artificial) metamaterial, which exhibits negative permeability, has been developed by using composite ferrite material [52]. This material is doubly advantageous to be used in the antenna matching circuit, because the circuit is made bi-directional, and the material is made in a very small piece, so it does not take space as compared with other metamaterials like transmission lines.

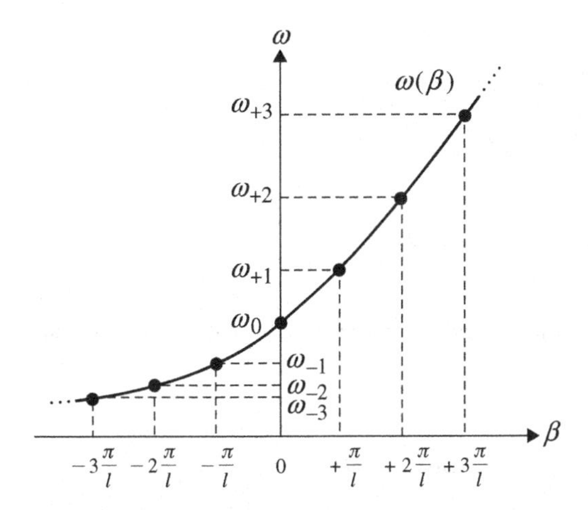


**Figure 6.47** CRLH transmission line media dispersion relation and resonance frequencies [16].

#### 6.2.5.2 DNG applications

The DNG property, as was described previously, is characterized by other terms such as LH (Left-Handed), $-k$ (the negative propagation constant), NRI (Negative Refractive Index), anti-parallel nature of $v_p$ (phase velocity) and $v_g$ (group velocity), as the attribute of the properties based on the negative constitutive constants ($-\varepsilon$ and $-\mu$). These properties can be implemented by combination of an array of thin wires and an array of SRRs, and periodically reactance-loaded transmission lines, in which the typical one is the CRLH transmission line. The combination of thin wires and SRR was first introduced as the metamaterial that exhibits the property of DNG, and discussions ensued about its application in various ways such as to planar lenses, filters, waveguides, and so forth, but not to radiators. Practical implementation of such metamaterials has been so far found in use of transmission lines, which are composed with lumped components periodically loaded into the structure. The typical one is the CRLH transmission line.

The CRLH transmission line constituted with a periodical connection of a series $L_R$ and $C_R$ circuit, and a shunt $C_L$ and $L_L$ circuit, as was shown in Figure 6.23, exhibits the dispersion characteristics having resonant modes as illustrated in Figure 6.24. As a conventional transmission line, a CRLH transmission line may also be open-ended or short-ended to produce a standing wave or resonance state. The resonance modes of the transmission line structure constituted with $N$ units of the length $l$ (= $Nd$; $d$ is the unit length) are given in relation with the length $l$ by

$$l_m = |m|\lambda/2 \quad (m = 0, \pm 1, \pm 2, \pm 3, \ldots) \tag{6.18}$$

and the phase constant for a mode $m$ is

$$\beta_m = \pm m\pi/l. \tag{6.19}$$

The resonance modes $m$ exist both positive and negative as shown in Figure 6.47. Each positive resonant mode ($m > 0$) at $\omega_{+m}$ corresponding to $\beta_{+m}$ (RH region) has twin-negative resonant modes ($m < 0$) at $\omega_{-m}$ corresponding to $\beta_{-m}$ (LH region) [53].

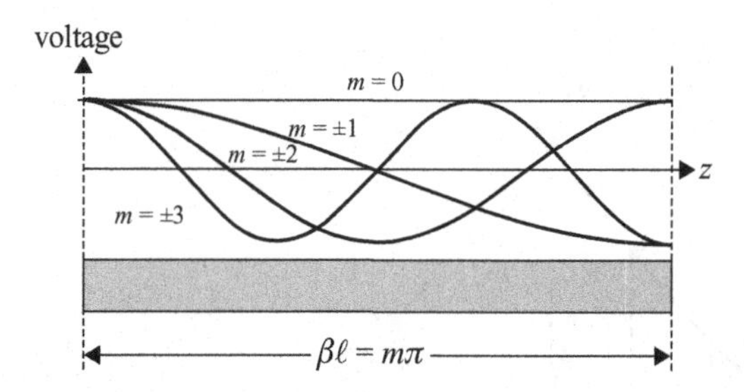


Figure 6.48 Typical field distribution of resonance modes on the CRLH transmission line [16].

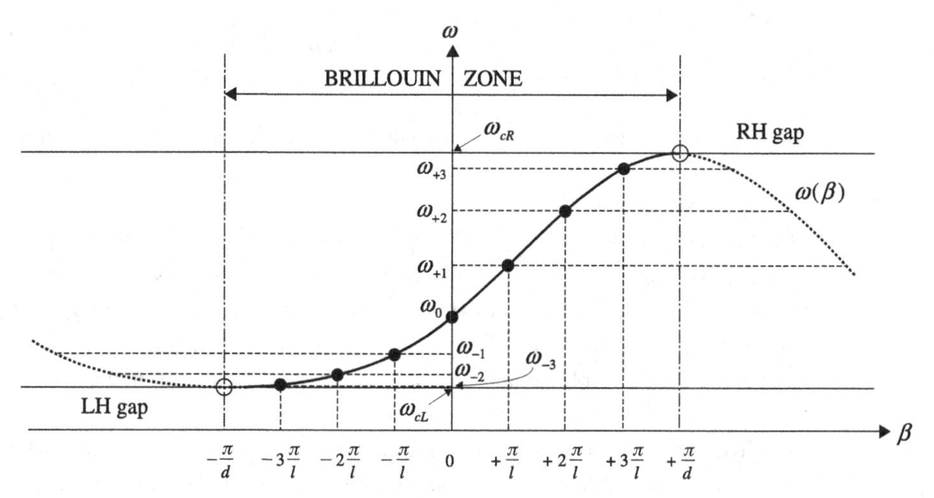


Figure 6.49 Resonance of CRLH transmission line resonator constituted of $N = 4$ units (balanced case) [16] ($l$: length of resonator, $d$: period).

The typical field distributions of the resonant modes on the CRLH transmission line are depicted as in Figure 6.48. It should be noted that there is a zeroth-order (ZOR) mode ($m = 0$), which exhibits at the transition frequency $\omega_0$ where $\lambda_g = \infty$ or $\beta = 0$, implying infinite wavelength or no phase variation [54, 55]. In the practical $N$-unit CRLH transmission line, there exist passbands as shown in Figure 6.49, where a case for $N = 4$ is illustrated. With $N = 4$, resonance modes $m$ are 0, $\pm 1$, $\pm 2$, $\pm 3$, and when $m = \pm 4$, where $l_m = \pm\, \pi/d$, it is not resonance, but the edge of the Brillouin zone. In general, the dispersion diagram of the $N$-unit CRLH transmission line is limited in frequencies at the limits $\beta = \pm\pi/d$ of the Brillouin zone, and resonance occurs at $m = \pm(N - 1)$ plus $m = 0$. For a balanced transmission case, there exist $(2N - 1)$ resonances, while an unbalanced case has $2N$ resonances. The resonance frequency $\omega_m$ is calculated from

$$\begin{aligned} 2\{1 - \cos(m\pi/N)\} &= (\omega_L/\omega_m)^2 + (\omega_m/\omega_R)^2 - (\omega_{sh}/\omega_{se})^2/(\omega_R)^2 \\ \text{where } \omega_{cR} &= 1/\sqrt{L_R C_R},\ \omega_{cL} = 1/\sqrt{L_L C_L},\ \omega_{se} = 1/\sqrt{L_R C_L}, \\ \text{and } \omega_{sh} &= 1/\sqrt{L_L C_R}. \end{aligned} \tag{6.20}$$

When $m = 0$ (ZOR mode), the resonance frequency is independent of physical length. The resonance frequency $\omega_0$ is either one of $\omega_{se}$ or $\omega_{sh}$. When the transmission line is a balanced structure, $\omega_0 = \omega_{se} = \omega_{sh} = 1/\sqrt{L_R C_L}$. The property of ZOR structure, i.e., size-independence of the resonance frequency, can be applied to reduce the antenna size; i.e., design of ESAs, as the physical size does not depend on the frequency, but on the circuit parameters $C$ and $L$ [56, 57]. Examples are introduced in Chapters 7 and 8.

## 6.3 Techniques and methods to produce FSA

Since an FSA does not necessarily have its size much smaller than the wavelength as in an ESA, techniques and methods to produce FSA differ from that of ESA. Generally, an FSA is constructed by integration techniques, by which either antenna performance can be improved or enhanced, or some function is integrated into an antenna structure so that the antenna can perform additional functions as a consequence of integration.

### 6.3.1 FSA composed by integration of components

Integration of components such as device, circuitry, or materials, including metamaterials, with antenna structure has been commented upon in the previous sections; for the purpose of generating uniform current distributions, composing SW structure, or self-resonance structure, and modifying antenna geometry so that various types of small antennas, not only FSA, but also ESA, can be created. The major objective of integration is generally to change current distributions on an antenna structure so that modification of antenna performance, change in antenna characteristics, and improvement or enhancement of antenna performances, are achieved. These techniques enable us to produce various interesting FSAs such as those having enhanced gain or bandwidth or capability of varying or controlling radiation patterns.

### 6.3.2 FSA composed by integration of functions

By integrating some function into an antenna, either enhancement of the antenna performance or improvement of the antenna characteristics can be expected. The functions to be integrated include amplification, oscillation, frequency conversion, power combining, and pattern control. By integration of amplification, gain increase can be expected, although extreme care about internal noise in the amplifier should be taken into account. The concept of smallness in the FSA should be understood in such a way that when, for instance, an enhanced gain or wider bandwidth that can be attained only by increasing the antenna size is realized, then the antenna of unchanged size can be said to be a small antenna. Integration of oscillator circuit, frequency converter, etc. will contribute to produce varieties of interesting FSAs such as a power oscillating antenna, an antenna having frequency mixing function, power combining in space, and so forth.

The integration technique may bring evolutional antenna systems that cannot be realized by conventional methods. Either passive or active integration, or a combination

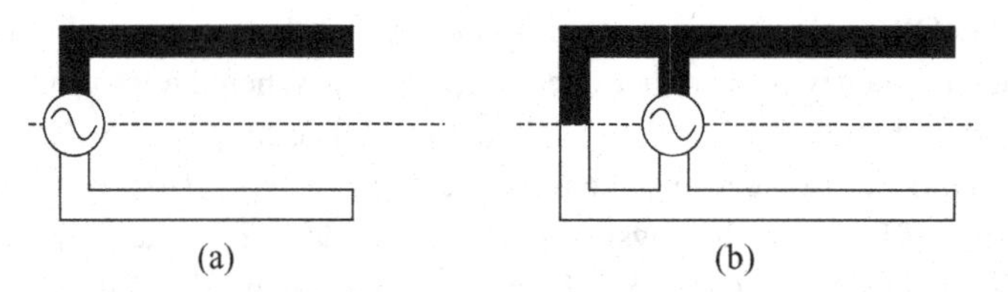


**Figure 6.50** Equivalent expression of (a) ILA and (b) IFA on the ground plane.

of these are possible. Enhancement of antenna performance or improvement of antenna characteristics may create novel antennas with small size that cannot practically be realized by conventional antennas, although it might not be as easy as expected.

Although integration techniques are very attractive and useful means to create novel FSAs, there are some serious problems to be taken into consideration. They involve noise and non-linear characteristics in devices and components. Noise produced from an active device deteriorates the FSA performance, while the non-linear characteristics inherently existing in devices, even in passive ones, may generate interference due to its intermodulation performance.

### 6.3.3 FSA of composite structure

Some composite structures have been described in previous sections, some of which are concerned with ESA. They are self-complement structure, conjugate structure, and combination of different types of antennas. This type of FSA can be composed by similar techniques as in ESA: (1) by loading impedance components, which includes conjugate components, (2) with complementary elements, and (3) by using travelling wave structure. These techniques are essentially based on application of IAS (Integrated Antenna Systems [58]) techniques.

## 6.4 Techniques and methods for producing PCSA

PCSA is a Physically Constrained Small Antenna, which has a part of its structure very small as compared with the wavelength as in ESA. PCSA is generally constructed with either low-profile or planar structures. In usual low-profile structures, a low-height antenna is placed on a ground plane and the antenna is treated equivalently as a symmetrical antenna constituted with its image, as shown in Figure 6.50, where (a) depicts an Inverted-L antenna, and (b) an Inverted-F antenna as the examples. Planar structures are treated in the same way as linear antennas. Images of these antennas must be included in the antenna design. The PCSA is an antenna which has very low height that is much smaller than the wavelength; however, other parts of the antenna are not necessarily small, but may have dimensions comparable to the wavelength; for instance, the periphery of a PIFA is about a half wavelength. No particular methods are used for the design of these antennas, excepting cases where proximity effects must be included.

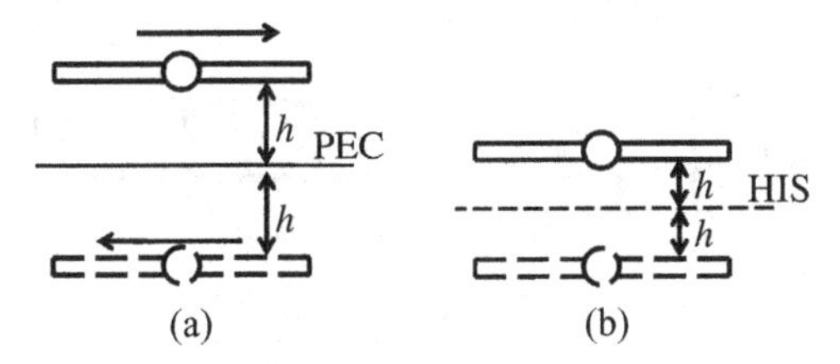


**Figure 6.51** A dipole placed on (a) PEC surface and (b) on HIS.

### 6.4.1 PCSA of low-profile structure

Generally low-profile antennas are mounted on the ground plane and in the evaluation of the antenna performance the image antenna is taken into consideration. When the ground plane is not infinite, but has appreciable size, some currents driven by the antenna flow on it, and the ground plane will act as a part of the antenna system. When the size of the ground plane is larger than several wavelengths, it can be assumed as an almost infinite ground plane. Design of low-profile antennas is not always simple, depending on the size of the ground plane and the environmental conditions. Well-known low-profile linear antennas are ILA, IFA, and a half-loop of either circle or square, while planar ones are parallel plate, MSA, PIFA, printed antenna, and tiny ceramic-chip antennas. Design methods of these antennas differ depending on the type of antenna, but are not necessarily specific, excepting cases where complexity surrounding the antenna system needs serious considerations. Use of a simulation technique may become useful for such cases where ordinary methods are not applicable, as very complicated problems such as inclusion of proximity effects, are involved. However, simulation methods are not a cure-all means, but are only an approximation, although they can provide fairly accurate and precise results as compared with experiments performed in extraordinary environmental conditions. However, it would be recommended that the results obtained by the simulation should better be assured by comparison with experimental results, although for smaller antenna sizes, the implementation of the experiment can become difficult.

### 6.4.2 PCSA employing a high impedance surface

When an ordinary dipole antenna is placed in parallel and very close to the PEC (Perfect Electric Conductor) surface, the performance of the antenna degrades significantly because of the negative influence caused by the image of the dipole, which has opposite phase to that of the primary dipole (Figure 6.51(a)). When, however, the PEC surface is replaced with a High Impedance Surface (HIS) [59–61], the performance of the dipole does not degrade, but may be modified to allow the dipole to be placed very close to the surface (Figure 6.51(b)), and have improved impedance characteristics, efficient radiation, and so forth.

The input impedance $Z_d$ of the dipole antenna placed on the HIS is given by [68]

$$Z_d = Z_{11} + Z_{12}e^{j\theta} \tag{6.20}$$

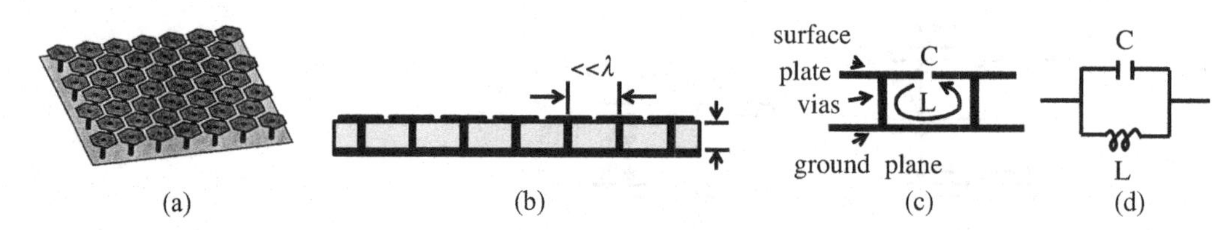


**Figure 6.52** Mushroom structure to represent EBG surface; (a) Mushroom type EBG implementation, (b) a cross section of several segments, (c) a cross section of one segment, and (d) equivalent circuit of one segment.

where $Z_{11}$ is the impedance of the primary dipole, $Z_{12}$ is the mutual impedance between the primary and the image dipole, and $\theta$ is the reflection phase angle, which is 180 degrees for the PEC surface. The $\theta$ can be determined by the surface media parameters. One of the typical HISs is EBG (Electromagnetic Band Gap) media [57], in which the Mushroom type [58] is the most popular one. Figure 6.52 illustrates (a) implementation of the Mushroom type EBG, (b) a cross section of the Mushroom structure, (c) a cross section of the one segment, and (d) an equivalent representation of the one segment with an L-C circuit. The capacitance C is due to the gap between the two surface planes and the inductance L is produced by the current flowing through the surface plane, vias, and the ground plane. Applications of the Mushroom type media are found in various situations; one is for lowering antenna height on the ground plane [64], another is to reduce mutual coupling between two antennas [65–66]. The first one contributes to creating a very low profile antenna, and the second one allows antennas to be located closely in a limited area. Thus, in a small wireless unit, in which more than one antenna is required to be installed in a small limited space, mounting several antenna systems simultaneously, for instance, placing such antennas as multiband antennas, MIMO antennas, and multi-purpose antennas in a limited area, becomes feasible. The HIS media can be applied to other significant purposes; reduction of surface wave, enhancement of radiation [67, 68], beam scanning [69, 70], and bandwidth enhancement [71], are some examples. The design and practices will be described in Chapters 7 and 8.

## 6.5 Techniques and methods for making PSA

A PSA is a Physically Small Antenna which has such small general physical dimensions that the antenna can be put on a human palm. High-frequency antennas in microwave (MW) and millimeter wave (MMW) regions are typical examples, as they have inherently small dimensions according to their frequency. Lower-frequency antennas, which have simply small dimensions, are also included in the PSA. Lately, various types of PSAs have been developed and employed in small terminals or units used in recently deployed wireless systems such as NFC (Near Field Communication) systems, including RFID (Radio Frequency Identification) systems, UWB (Ultra Wide Band) systems, and so forth.

Matching problems are often encountered in these systems, as antennas used for these systems are too small to satisfy appropriate matching conditions. Serious issues are found particularly in application to the RFID systems, where equipment is so small that the space to install an antenna is strictly limited to a very narrow area, and thus the size of the antenna is forced to be very small, and satisfying matching conditions becomes extremely difficult. Antennas used unfittingly in such situations without proper matching conditions often suffer insufficient system performance; for example, low sensitivity so that the service is limited to small areas.

### 6.5.1 PSA in microwave (MW) and millimeter-wave (MMW) regions

Antennas for MW and MMW regions are inherently small size because they are designed for high operating frequencies. Since they are physically very small, many types of them can be classified into ESA in addition to PSA, but they are not necessarily restrained by the limitations imposed on small antennas as the usual ESA. The basic structure of these antennas is the same as those used in lower-frequency regions; however, since aperture antennas such as horn, reflector types, and waveguide require generally some physical volume, they are not usually used in lower-frequency applications. But, thanks to their small size in MW and MMW regions, they are useful for application in various systems, especially in integrated RF devices with small antennas. Particularly, planar antennas are employed in such systems. Antennas integrated with RF circuits play very important roles in small mobile wireless systems, where small, compact, and yet functional antenna systems are required. Some of these antennas are thus classified into FSA.

Recently demands on antennas in MW or MMW regions have gradually increased, especially for signal transmission of high data rate, video, and multimedia applications. One example is an application to mobile phones, to which a small 60-GHz wireless module is attached, that is used for the purpose of transmission of high-rate data information or TV images at home. Small chip antennas or planar antennas integrated into RF parts constitute the front end of the wireless systems.

Design of these antennas differs depending on where to apply, how to use, the type of antenna, and so forth. Use of simulation is particularly helpful for design of antenna systems having complicated or extraordinary structure.

### 6.5.2 Simple PSA

There are wide varieties of PSAs. Some of them are ordinary types, while others are extraordinary types constituted with complicated or composite structure. Typical ordinary antennas are small wire antennas such as dipole, monopole, loop, IFA, meander line, helices, fractal structure, and so forth, whereas planar antennas are PIFA, MSA, printed antennas, and related types. The wire elements can be printed on the surface of a planar substrate to comprise a planar antenna. They can be implanted into a small bulk substructure to form a tiny chip structure. Meanwhile, many of the extraordinary types of antennas are composed by integrating impedance components, materials, or other types of antenna elements into an antenna structure.

Realization of antennas with physically small size is in other aspects to achieve either wider bandwidth or higher efficiency with the antenna size being kept unchanged.

Practical wireless systems sometimes require a very small antenna to fit to the very small electronic unit. Antenna designers thus should attempt to realize an antenna having the practically obtainable performances for the required size, although there could be still some difficulty in achieving it.

It should not be neglected to have impedance matching conditions in any antenna systems in addition to the above conditions. Matching is the inevitable condition, as being understood commonly in ordinary antenna systems; however, in small antennas, especially with physically small size, proper matching will become difficult and final results may be left without perfect matching status. Nevertheless, the lower frequency resonance should be satisfied either within the antenna structure or by network at the antenna load terminal. The matching is significant particularly in PSA, and the matching technique needs skill and experience. The latest advances are an application of NF (Non-Foster) circuits to the matching circuit, and matching in space by using metamaterial, which can compensate the reactive component in near field so that resonance in space is attained.

## 6.6 Optimization techniques

Generally, antenna miniaturization is at the expense of efficiency, bandwidth, and gain. As previously noted, small antenna design is a compromise between performance and the dimension as well as manufacturability, materials, and the operating environment. Experience and intuition are essential in the antenna design process as well. However as a certain level of design complexity is reached, design optimization tools become extremely valuable if not a necessity. In this section, some optimization methods adopted for antenna design are discussed. Specifically, Genetic Algorithm (GA), Particle Swarm Optimization (PSO), Topology Optimization, and Volumetric Material Optimization schemes are considered for antenna design improvement.

### 6.6.1 Genetic algorithm

Genetic Algorithms (GAs) are search methods based on the principles and concepts of natural selection and evolution. These optimization methods consider a set of trial solutions (in parallel), based on a parametric variation of a set of coded geometric and material features. GA employs already known concepts, such as chromosomes, genes, alleles, mating, and mutation, to the coding and best design selection. Figure 6.53 outlines the GA optimization process. More details can be found in papers and books covering the application of GA in electromagnetics [69–73]. The most studied application of GAs in electromagnetics is antenna design. Specifically, GAs have been applied in reducing side lobes [74, 75], aperture amplitude or phase tapering [76, 77], and for adaptive array optimization [78, 79]. Several uses of GA in single element designs have also been reported [80–82] in uses of circuit elements for antenna loading. GAs have also been used to optimize the performances of standard antennas such as reflectors [83], and Yagis [84]. As is the case with all optimization algorithms, a search for an optimal

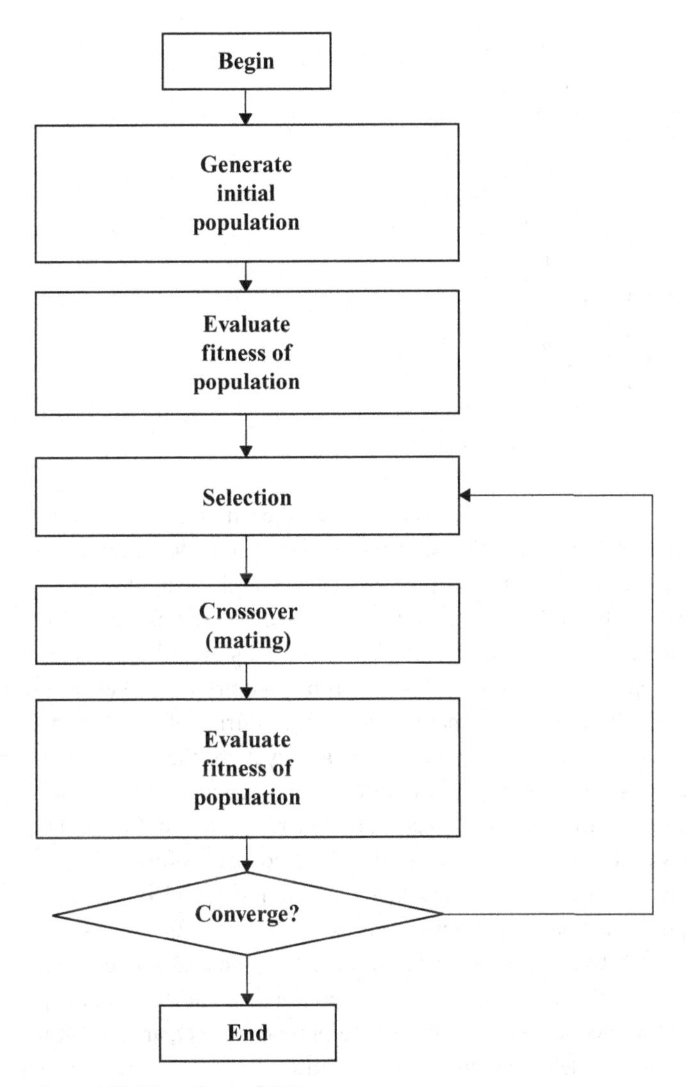


**Figure 6.53** Flow chart of GA.

solution is carried out subject to pre-specified performance criteria (such as bandwidth, gain, return loss, etc.). Figure 6.53 displays a typical flow chart of the GA optimization process.

Antenna optimizations via GA have been demonstrated. The examples are a miniaturization of a microstrip patch antenna [85], and design of a biocompatible antenna for the 402–405 MHz band [86], where the metal patch is divided into numbers of rectangular cells, and some of them are removed to provide reduction in the patch resonance frequency. Prior to these applications, similar approaches have already been introduced in [87], in which improvement of the quality factor ($Q$) of a resonator is achieved, and another [88] had shown use of GA to obtain wideband patch antennas.

| 1 | 1 | 1 | 0 | 1 | 1 | 0 | 0 | 0 |
|---|---|---|---|---|---|---|---|---|
| 1 | 1 | 0 | 0 | 1 | 1 | 1 | 1 | 0 |
| 0 | 1 | 0 | 0 | 0 | 1 | 0 | 0 | 0 |
| 0 | 0 | 0 | 0 | 1 | 1 | 1 | 1 | 0 |
| 1 | 0 | 0 | 0 | 1 | 1 | 0 | 0 | 0 |
| 1 | 1 | 1 | 0 | 1 | 1 | 1 | 1 | 1 |
| 0 | 0 | 0 | 1 | 1 | 0 | 0 | 1 | 1 |
| 1 | 0 | 1 | 0 | 1 | 1 | 0 | 0 | 1 |
| 1 | 1 | 1 | 1 | 1 | 1 | 0 | 1 | 0 |

**Figure 6.54** Generic microstrip patch antenna.

In the GA based optimization design procedure, the variables are the number of metal cells and their location, while all the other antenna parameters remain constant. The initial common rectangular microstrip patch as an example is divided into $N \times M$ equal rectangular cells. The coding is straightforward as a "1" represents a cell with metal and a "0" represents a cell with no metal. All the cells are allowed to be metallic or nonmetallic except the cell where the coaxial probe is connected, which necessarily must be metallic. Each chromosome is, therefore, coded as a string of $N \times M$ bits (genes). The initial population is generated randomly. For each cycle of the optimization procedure a start frequency, stop frequency, and number of frequency points are chosen. At each frequency point and for each individual the input reflection coefficient is obtained. If the return loss is less than or equal to 10 dB the cost function is set to zero. If it is greater than 10 dB, the resonance frequency of the individual is set as the frequency where the input reflection coefficient is minimum. The cost function is then calculated as the difference between the resonance frequencies, the higher the cost function the higher the fitness of the individual. A new generation is obtained from the previous one using tournament selection with elitism, single-point crossover, and mutation. The cycle continues until the predefined number of generations limit is reached. Since there is no information about an "absolute" or "unique" solution, a new cycle can be started. In this case, a new start frequency and stop frequency are defined to guarantee the optimization procedure efficiency.

In an example introduction in [85], a patch having dimensions of 32.94 mm in both the width $w$ and the length $L$, is printed on the substrate with a thickness of 62 mils, a relative dielectric constant of 2.20, and a loss tangent of 0.0007. The conventional parameters have been chosen to provide a resonance frequency around 3 GHz in GA optimization, in which $N \times M = 9$ and populations of ten individuals are used. For the optimized patch shown in Figure 6.54, a resonance frequency of 1.738 GHz has been obtained compared with 3 GHz of the conventional patch.

The space-filling curves are chosen to design small antennas because they utilize the space more efficiently to improve the antenna performance with a given small size. Some representative antenna geometries and structures using space-filling curves like meander,

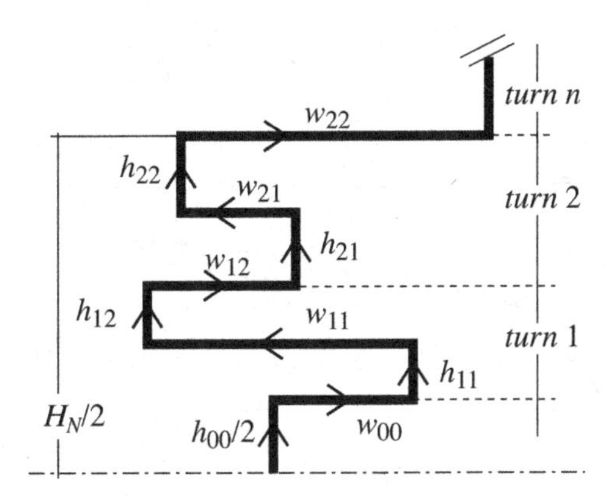


**Figure 6.55** Scheme for non-uniform MLA with indication of parameters to be optimized only half antenna is visible.

spiral/helix, and fractals are described later in Section 7.1 in conjunction with design of small antennas. However, the resonant frequency and the antenna performances do not depend only on the total wire length of space-filling curves, but also on their geometrical configuration. Thus the GAs are introduced to design these kinds of small antenna to obtain better performance while maintaining miniaturization.

Meandering structure is widely used in small antennas for wireless applications. However, for a fixed height, an ordinary meander line antenna (MLA) does not exhibit the optimum gain [89], especially when the conductor losses due to lengthening the meandered line cannot be neglected. It has already been shown that the inherent geometry of the meander line will affect the resonant frequency and the performance, such as $Q$, input impedance, and gain. Thus a GA-based MLA is designed to obtain better gain while taking into account the wire conductivity [90]. An MLA can be described by only three parameters: height, width, and number of turns. Whereas in the GA design, more parameters will be considered as shown in Figure 6.55: the number of turns $N$, the length of the horizontal ($w_{n1}$ and $w_{n2}$) and vertical ($h_{n1}$ and $h_{n2}$) segments of the $n$th turn, and the length of the central segments ($w_{00}$ and $h_{00}$). The above optimization process, involving a trade-off between miniaturization with self-resonance (long wire length) and minimization of loss (short wire length), requires all the vertical and horizontal segments to be independently designed and can be efficiently handled by the GA approach. For each $p$th antenna of the GA population at the $k$th generation, the following fitness function is then evaluated:

$$f_p^{(k)}(G_p, H_p, X_p) = r_1 \frac{G_p}{G_0} + r_2 \frac{H_p}{H_{\max}} + r_3 \frac{X_0}{|X_p| + X_0} \tag{6.21}$$

where $H_p, G_p, X_p$ are the $p$th antenna height, maximum gain, and input reactance, respectively. Parameters have been chosen as: $r_1 + r_2 + r_3 = 10$, $G_0 = 1.63$ (maximum gain of half-wavelength perfect conductor dipole), $X_0 = 1\Omega$. The fitness function converges to $f_p = 10$ as antenna gain equals $G_0$, the height equals $H_{max}$ and the antenna is at resonance.

To investigate the GA optimization to enhance MLA gain, several trials of designs have been performed for different maximum sizes ($H_{max} \times W_{max}$) ranging from $3 \times 3$ cm$^2$

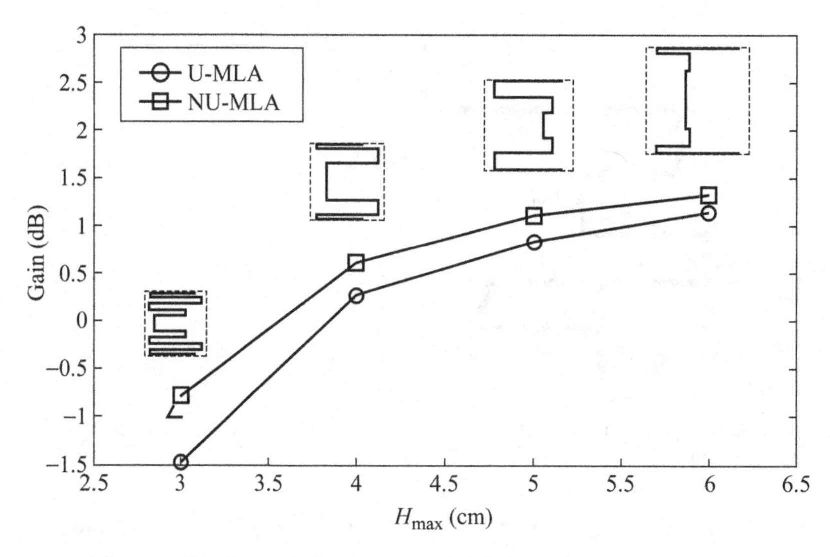


**Fig 6.56** Copper-wire GA-optimized MLA antennas for different maximum available areas (dashed shapes).

to $6 \times 6$ cm$^2$. The antenna elements are made of copper. As depicted in Figure 6.56, the GA-optimization remains effective over the uniform MLA as the maximum available area decreases. For sizes as small as $4 \times 4$ cm$^2$, an $N = 2$ antenna has been considered to obtain self-resonance. As the available space increases, the antenna tuning requires shorter horizontal segments, mainly localized at the wire's ends to minimize losses.

GA techniques have also been applied to the design of pre-fractal antenna elements. For example a GA-engineered second-order Koch-like dual antenna, having a compact size and low voltage standing wave ratio (VSWR) is presented in [91]. Later Pantoja *et al.* extended the work to seek an optimum set of solutions in terms of resonance frequency, bandwidth, efficiency, and the design, using a multi-objective GA of wire pre-fractal Koch-like antennas [92]. Moreover, the GA code is employed to search for nonfractal structures, namely zigzag and meander type antennas. It is shown that for a given overall wire length and antenna size, GAs find Euclidean geometry designs that perform better than do their pre-fractal counterparts.

Altshuler used GA to optimize wire antennas filling the volumetric space. In [93] various wire antennas were designed. One of them consists of seven wires, with the locations and lengths determined by the GA alone, that radiates waves for GPS/IRIDIUM applications [94]. Also two Yagis were designed for different goals, one is for a broad frequency band and low sidelobes, and the other is for high gain at a single frequency. All of these antennas have unusual shapes compared to conventional ones. In [95] a Self-Resonant Wire Antenna (S-RWA) was designed. First, Altshuler chose a target frequency and then constrained the algorithm to create a configuration of linear wire segments, connected in series within the volume of a cube, such that the resulting antenna was self-resonant. As a consequence of cancellation of the inductance and capacitance of the wire configuration at the target frequency, Altshuler found that as the size of the cube was reduced, more wire segments were required. Altshuler used GA to optimize the S-RWA with a minimum $Q$ and maximum bandwidth. In [96], an antenna consisting of a set of wires connected in series and with impedance loads is investigated. The shape of the antenna, the location of the loads, and their impedance are optimized using a GA. The resultant antenna mounted over a ground plane radiates elliptically polarized waves in almost near hemispherical coverage. It has a VSWR less than about 4.5 over the 50 to 1 frequency band ranging from 300 to 15 000 MHz.

It is well known that antenna miniaturization impacts antenna efficiency as well as bandwidth. Therefore multi-objective optimization should be performed to design small antennas with improved efficiency. An optimal set of designs using Pareto GA approach is implemented for designing electrically small wire antennas taking into consideration both bandwidth and efficiency [97]. For the antenna configuration, multi-segment wire structure is employed. The resulting GA designs followed the trend of the fundamental limit, but were about a factor of two below the limit. To further improve the performance of the GA-designed antennas, other design freedoms such as variable characteristic impedance, multi-arm wires, and multiple wire radii could be considered.

### 6.6.2 Particle swarm optimization

As a novel evolutionary algorithm proposed in the mid 1990s [98, 99], particle swarm optimization (PSO) has been introduced into the EM community [100, 101], and its applications have received enormous attention in recent years. Unlike GAs, the swarm intelligence in nature is modeled by fundamental Newtonian mechanics in PSO for optimization purposes. This corporative scheme manifests in PSO the concise formulation, the ease in implementation, and many distinct features in different types of optimizations. PSO is based on the principle that each solution can be represented as a particle (agent) in a swarm. PSO has been widely used in antenna optimal design in recent years, such as antenna arrays [102–104], and wideband printed antennas [105–107].

Figure 6.57 shows the working procedure of the PSO algorithm using a simple flow chart. The detailed description of PSO can be found in [101, 102], where, as an example, a microstrip array and a corrugated horn have been optimized.

PSO has also been applied to small antenna design in recent years, but the design procedure is not much different from GA. For example, the hybrid real-binary PSO algorithm is used to design a dual-band handset patch antenna operating at 1.8 GHz and 2.4 GHz [108]. The unique hybrid representation of candidate antenna designs using real and binary variables enables the optimized benefit from the advantages of both continuous and discrete optimization techniques. With the fitness function evaluated by an MoM-based full-wave simulator, the design is accomplished in quite a limited space with a dimension of $0.23\lambda_g \times 0.13\lambda_g$ at 1.8 GHz ($\lambda_g$ is a wavelength).

### 6.6.3 Topology optimization

A full three-dimensional (3D) antenna design methodology is presented using concurrent shape, size, metallization as well as dielectric and magnetic material volume optimization

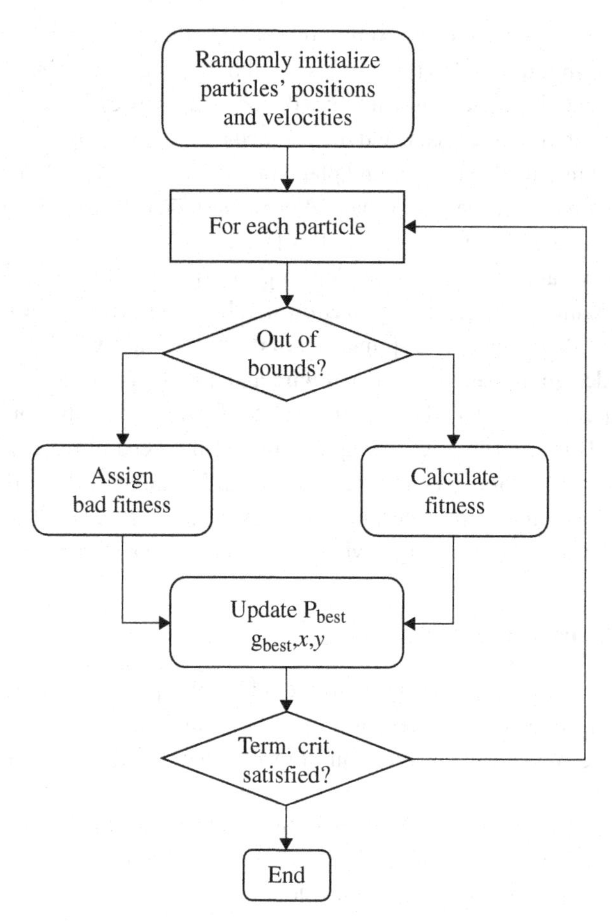


**Figure 6.57** Flow chart depicting the PSO algorithm.

by Volakis *et al.* [109, 110]. It is reasonable to expect that designs resulting from the optimal selection of materials, metallization, and matching circuits would lead to novel configurations with much higher performance as compared to simpler approaches employing a subset of these parameters. Nevertheless, volumetric optimization would provide for greater design flexibility.

In the early works [109, 110], optimum topology/material design of dielectric substrates was applied to a patch antenna and obtained remarkable improvement in the bandwidth. The topology algorithm has generally been used to optimize size/shape of objective and it provides a possibility of volumetric design [111]. In applying it to RF design, the optimization scheme will bring the best geometrical and topological configuration from its volumetric variation, along with consideration of geometry and physical dimensions as well as material composite. It performs iteration of evaluation to reach the optimum (maximum) value initially set to obtain the objective shape/configuration. It was demonstrated that the volumetric material design led to significant improvement of bandwidth in a simple patch antenna as a consequence of the application of the topology optimization method.

### 6.6.4 Volumetric material optimization

To take advantage of volumetric variations in antenna design, various optimization schemes can be pursued for material and topological selections subject to various electrical and size/volume constraints while considering geometry, physical dimensions as well as material composition. Several approaches have already been employed to facilitate antenna design by integrating analysis tools with formal optimization. In general, much emphasis has been placed on the shape optimization either by changing geometric parameters or by concentrating on metallization shape. Material optimization has not been considered except for a few recent cases.

Volumetric Material Optimization [112] is novel optimization algorithm, which permits full volumetric antenna design by combining GA with finite element-boundary integral method. This is the first time that a full 3D antenna design is pursued using concurrent shape, size, metallization as well as dielectric and magnetic material volume optimization.

Later, a full 3D antenna design is pursued using concurrent shape, size metallization as well as dielectric and magnetic material volume optimization. Multi-objective optimization is used for consideration of higher gain, increased bandwidth, and size reduction simultaneously. An aggressive optimization scheme was employed using GAs in conjunction with a fast finite element-boundary integral code. The approach employs a wide-frequency sweep using a single geometry model, thus, enhancing speed, along with several discrete material choices for realizable optimized designs. The presented color-coded graphs allow for quick and efficient evaluation of the best/optimized cases. Design examples for circularly and linearly polarized antennas were presented. The final designs are associated with very thin ($0.01\lambda$) material substrates and yield as much as 15% bandwidth using a $0.1{\sim}0.4\lambda$ aperture subject to various gain and bandwidth requirements. In particular, optimization cycles using multilayer magneto-dielectric stacked patches produced a small ($0.11\lambda$ aperture) antenna design exhibiting 4.7 dBi gain with 3.4% bandwidth, as shown in Figure 6.58.

### 6.6.5 Practice of optimization

In this section, Particle Swarm Optimization (PSO) to optimize the method of cutting the ground plane (GP) under a U-shaped Folded Dipole Antenna is described. The UFDA [114–116] element is designed for WiMAX (using frequency of band 2.3 GHz~2.7 GHz and 3.4 GHz~3.8 GHz).

#### 6.6.5.1 Outline of particle swarm optimization

PSO was first introduced by James Kennedy and Russell Elberhart in 1995 [117]. It has the algorithm based on the simplified social model of swarming theory. The algorithm

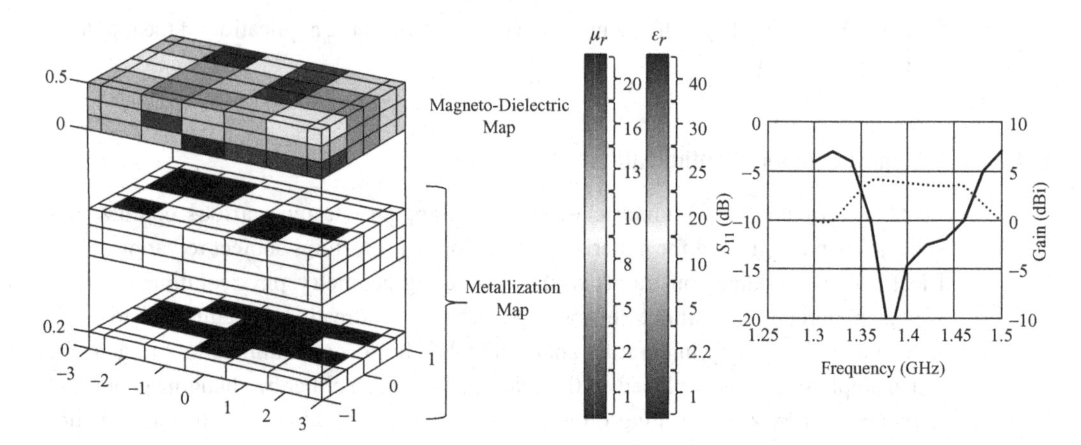


**Figure 6.58** Examples of optimized metallomagnetodielectric LP designs; left: resulting material and metallic geometries and right: corresponding return loss and gain.

of PSO imitates the social behavior of birds and insects, etc., that keep living by maintaining swarm actions. The algorithm of PSO is regarded as simpler than other methods and moreover, it can lead to a convergence quickly. The algorithm is described by renewing two standard quantities of each particle in the swarm: coordinate and velocity. At first, the coordinate and velocity of each particle in the swarm are set with initial values randomly inside the searching space. It is supposed that, $x_i(k)$ and $v_i(k)$ are the coordinate and velocity of particle number $i$ at step $k$, respectively, and they are updated into their new values at step $k+1$ as $x_i(k+1)$ and $v_i(k+1)$ by applying the two equations below:

$$v_i(k+1) = \omega v_i(k) + c_1 r_1 [p_i(k) - x_1(k)] + c_2 r_2 [p_g(k) - x_i(k)] \tag{6.22}$$

$$x_i(k+1) = x_i(k) + v_i(k+1)\Delta t \tag{6.23}$$

where $p_i(k)$ and $p_g(k)$ are the best coordinates of particle number $i$ and the whole swarm until step $k$, respectively. Both $r_1$ and $r_2$ get independent random values between 0 and 1. Cognitive parameter $c_1$ indicates how much a particle trusts in itself and social parameter $c_2$ indicates how much it trusts in the swarm while searching the optimum solution of the optimization problem. According to reference [117], in order to obtain a good convergence for the PSO algorithm, $c_1$ is set to1.5 and $c_2$ is set to 2.5. The unit time step $\Delta t$ is used throughout the optimization process with a constant value as 0.7. The inertia weight $\omega$ plays a role as controlling the exploration properties of the algorithm, and it is adjusted dynamically from 1.4 to 0.35 during the optimization. Movements of the whole particles of the swarm are controlled and repeated by the two equations above, and finally one optimal solution will be derived when the algorithm results a convergence.

### 6.6.5.2 PSO application method and result

Figure 6.59 shows the model of U-shaped Folded Dipole Antenna (UFDA) on the ground plane (GP). The previous study [116] shows that in the low-profile case, when the space between UFDA and GP becomes smaller, then UFDA gets more effect from GP that makes its VSWR characteristic worsen (GP w/o cutting in Figure 6.63). But when a part

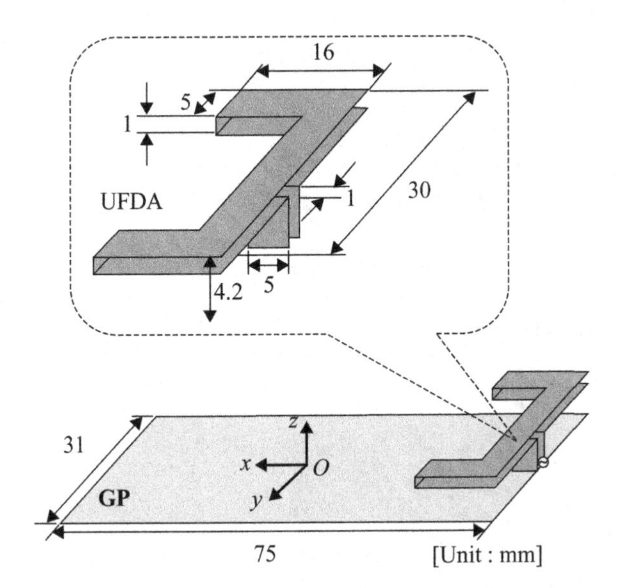


**Figure 6.59** Model of UFDA on GP.

of GP is cut practically such as shown in Figure 6.60(b) with the cutting area $S_{\text{cut}} = 88\ \text{mm}^2$, it has been shown that VSWR can be improved with a wider-band characteristic (GP cut w/o using PSO in Figure 6.63). However, it is desired to find the best way to cut GP most effectively to get the wider broadband feature, and simultaneously find the minimum value for the cutting area $S_{\text{cut}}$. Therefore, PSO is applied, to this problem.

How to cut the GP using PSO is decided by defining three design variables $x_1, x_2$ and $x_3$ as shown in the Figure 6.60(a) where $x_1$ indicates the cutting position, $x_2$ represents the deletion distance along the $O$-$x$ axis and $x_3$ represents the deletion distance along the $O$-$y$ axis. These three variables can change freely within GP's limit considering the position of the feed strip and short strip. Thus, the constraint conditions for them are as follows:

$$0 < x_1, x_2, x_3 \tag{6.24}$$

$$x_1 + x_2 \leq 75 \tag{6.25}$$

$$x_3 \leq 14 \tag{6.26}$$

The objective function $F$, the function expressing the purpose we target in the optimization problem is set by the following equations:

$$F = wV + (1 - w)S \tag{6.27}$$

$$V = w_1 V_1 + w_2 V_2 + w_3 V_3 \tag{6.28}$$

$$S = S_{\text{cut}}/S_0 = 2x_2 x_3/S_0 \tag{6.29}$$

where $w = 0.5$, $w_1 = w_2 = w_3 = 1/3$ are inertia weight coefficients. $V_1$, $V_2$, $V_3$ are VSWR values at three frequencies $f_1 = 2.4$ GHz, $f_2 = 3.2$ GHz, $f_3 = 3.95$ GHz where VSWR is apt to increase easily. The smaller $V$ (composed of $V_1$, $V_2$, $V_3$) is, the wider covered

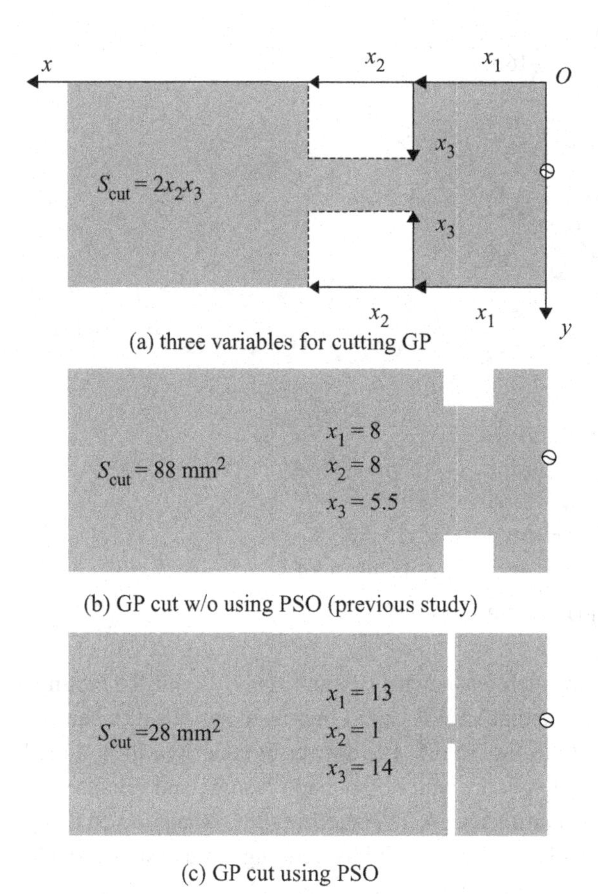


**Figure 6.60** Shape of GP (a) three variables for cutting GP, (b) GP cut w/o using PSO (previous study), and (c) GP cut using PSO.

frequency band is achieved. The cutting area $S_{cut}$ is transformed to a normalization value as $S$ with $S_0 = 100$ mm$^2$ to fit to $V$ in Eq. (6.27) because $V$ is dimensionless. The value of function $F$ is minimized by the algorithm of PSO. This process will give us the minimum of $V$, which means that it is possible to obtain a wider broadband characteristic and the minimum of $S$ simultaneously.

A swarm with 50 particles to search for the optimum solution of this problem through 40 steps is used. The convergence result of objective function $F$ is shown in Figure 6.61 which demonstrates that $F$ converges to the minimum value $F_{min} = 0.8686$ from the 18th step. Naturally, design variables ($x_1$, $x_2$, $x_3$) also converge to the optimum solution (13, 1, 14) at the same step as in Figure 6.62. This means that the most appropriate shape of GP is the shape shown in Figure 6.60(c). In this case, cutting area $S_{cut}$ is minimized to 28 mm$^2$ which is much smaller than the result $S_{cut} = 88$ mm$^2$ of a previous study when PSO was not used. Comparisons of VSWR characteristics between this case and the previous result are shown in Figure 6.63. Covered frequency band (VSWR $\leq 3$) of the optimized result (GP cut using PSO) is obviously wider than the case GP cut w/o using PSO, and the difference between them is about 600 MHz.

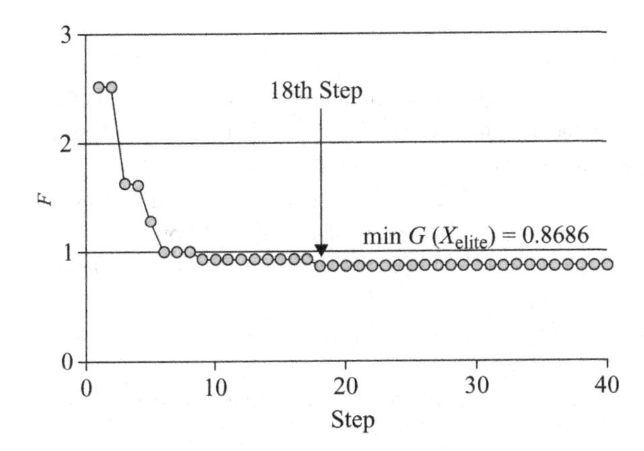


**Figure 6.61** Convergence of objective function $F$.

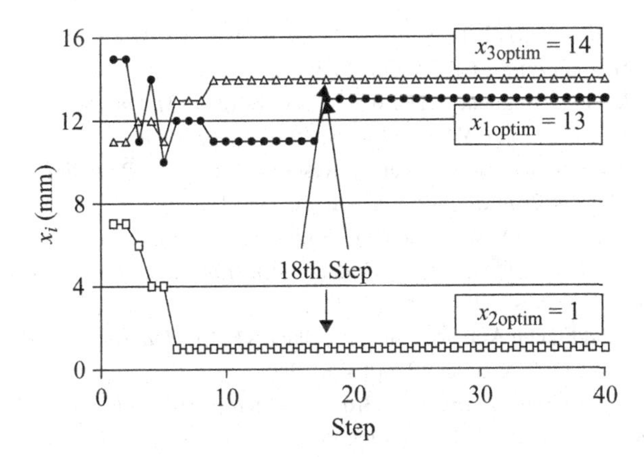


**Figure 6.62** Convergence of design variables $x_i$ ($i = 1, 2, 3$).

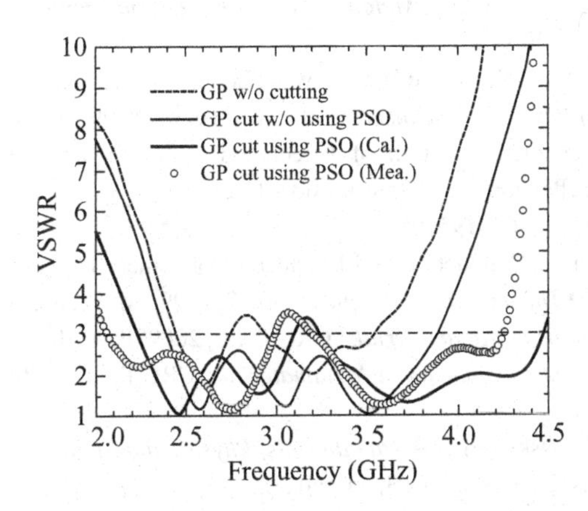


**Figure 6.63** VSWR characteristics.

## References


[1] C. H. Walter, *Travelling Wave Antennas*, Dover 1965, p. 14 and chapters 5 and 6.

[2] K. Wu, Slow Wave Structures, in K. Chang (ed.) *Encyclopedia of RF Microwave Engineering*, John Wiley and Sons, 2005, vol. 3, pp. 4744–4760.

[3] D. H. Werner and S. Gargiel, An Overview of Fractal Antenna Engineering Research, *IEEE Antennas and Propagation Magazine*, vol. 45, February 2003, pp. 38–57.

[4] J. Ramo, J. R. Whinnery, and T. V. Duzer, *Fields and Waves in Communications*, John Wiley and Sons, 1965, Appendix IV, p. 737.

[5] J. Ramo, J. R. Whinnery, and T. V. Duzer, *Fields and Waves in Communications*, John Wiley and Sons, 1965, pp. 738–739.

[6] E. Martin and B. Marquis, 9.1, in F. Capolino (ed.), *Applications of Metamaterials*, CRC Press, 2009, pp. 9.2–9.3.

[7] J. Ramo, J. R. Whinnery, and T. V. Duzer, *Fields and Waves in Communications*, John Wiley and Sons, 1965, p. 475 and C. H. Walter, *Travelling Wave Antennas*, Dover 1965, pp. 237–269.

[8] C. Carlos and T. Itoh, CRLH Transmission Line Metamaterials, in F. Capolino (ed.), *Applications of Metamaterials*, CRC Press, 2009.

[9] G. E. Eleftheriades, in N. Engheta and R. W. Ziolkowsky (eds.) *Metamaterials, Physics and Engineering Exploration*, John Wiley and Sons, 2006, chapter 5, p. 5.2.

[10] T. Endo, Y. Sunahara, S. Sato, and K. Katagi, Resonant Frequency Radiation Efficiency of Meander Line Antennas, *Transactions of IEICE*, vol. J80-BE, 2000 no. 12, pp. 1044–1049.

[11] A. Z. Elsherbeni, C-W. Paul Huang, and C. E. Smith, 12.2 Meander Line Antenna for Personal Wireless Communications, in L. C. Godara (ed.), *Handbook of Antennas in Wireless Communications*, CRC Press, 2002.

[12] S. R. Best, Small and Fractal Antennas, in C. A. Balanis (ed.), *Modern Antenna Handbook*, John Wiley and Sons, 2008, chapter 10, pp. 488–498.

[13] J. D. Kraus and R. J. Marhefka, *Antennas*, 3rd edn. McGraw Hill, 2002, pp. 292–297.

[14] J. Ramo, J. R. Whinnery, and T. V. Duzer, *Fields and Waves in Communications*, John Wiley and Sons, 1965, pp. 467–468.

[15] F. Capolino (ed.), *Theory and Phenomena of Metamaterials*, CRC Press, Part I 1.1.

[16] N. Engheta and R. W. Ziolkowsky (eds.), *Metamaterials, Physics and Engineering Exploration*, John Wiley and Sons, 2006.

[17a] J. B. Pendry, A. J. Holden, D. J. Rollins, and W. J. Stewart, Low Frequency Plasmons in Thin Wire Structures, *Journal of Physics: Condensed Matter*, vol. 10, June 1998, pp. 4785–4809.

[17b] P. R. Smith and S. Shultz, Determination of Effective Permittivity and Permeability of Metamaterials from Reflection and Transmission Coefficients, *Physical Review B*, vol. B65, 2002, pp. 65–195104–65–195105.

[17c] J. B. Pendry, Magnetism from Conduction and Enhanced Non-Linear Phenomena, *IEEE Transactions on Microwave Theory and Techniques*, vol. 47, 1999, no. 11, pp. 2075–2084.

[18] F. Capolino (ed.), *Applications of Metamaterials*, CRC Press, 2009, Part III 9.

[19] F. Capolino (ed.), *Theory and Phenomena of Metamaterials*, CRC Press 2009, 1.2.2 and 1.2.3.

[20] N. Engheta and R. W. Ziolkowsky (eds.), *Metamaterials, Physics and Engineering Exploration*, John Wiley and Sons, 2006, sections 3.2.1 and 3.2.2.

[21] A. Lai and T. Itoh, Composite Right/Left Handed Transmission Metamaterials, IEEE Microwave Magazine, September 2004, pp. 34–50.

[22] C. Caloz, I. Itoh, and A. Rennings, CRLH Metamaterial Leaky–Wave and Resonant Antennas, *IEEE Antennas and Propagation Magazine*, vol. 50, 2008, no. 5, pp. 25–39.

[23] D. Guha and Y. M. M. Antar, *Microstrip and Printed Antennas*, John Wiley and Sons, 2011, Chapter 11.

[24] A. Lai and T. Itoh, Composite Right/Left Handed Transmission Metamaterials, IEEE Microwave Magazine, September 2004, pp. 36–40.

[25] C. Caloz, T. Itoh, and A. Rennings, CRLH Metamaterial Leaky–Wave and Resonant Antennas, *IEEE Antennas and Propagation Magazine*, vol. 50, 2008, no. 5, pp. 25–28 and F. Capolino (ed.), *Applications of Metamaterials*, CRC Press, 2009, Part IV 22.

[26] F. Capolini (ed.), *Applications of Metamaterials*, CRC Press, 2009 Part III 9.1, pp. 9.2–9.5.

[27] N. Engheta and R. W. Ziolkowsky (eds.), *Metamaterials, Physics and Engineering Exploration*, John Wiley and Sons, 2006, and A. Lai and T. Itoh, Composite Right/Left Handed Transmission Metamaterials, IEEE Microwave Magazine, September 2004, p. 37.

[28] N. Engheta and R. W. Ziolkowsky (eds.), *Metamaterials, Physics and Engineering Exploration*, John Wiley and Sons, 2006, section 7.2.

[29] C. A. Balanis, *Antenna Theory and Design*, 2nd edn, John Wiley and Sons, 1997, p. 570.

[30] R. C. Hansen, Fundamental Limitations in Antennas, *Proceedings of IEEE*, vol. 69, 1981, no. 2, February pp. 172–173.

[31] C. P. Baliarda, J. Romeu, and A. Curdama, The Koch Monopole: A Small Fractal Antenna, *IEEE Transactions on Antennas and Propagation*, vol. 48, November 2000, pp. 1773–1781.

[32] J. Rashed-Mohassed, A. Mehdipour, and H. Alinkbarian, New Scheme of Size Reduction in Space Filling Resonant Dipole Antennas, *3rd EuCAP Conference*, 23–27 March 2009, pp. 2430–2432.

[33] J. Zhu, A. Hoorfer, and N. Engheta, Peano Antennas, *IEEE Transactions on Antennas and Propagation Letters*, vol. 3, 2004, pp. 71–74.

[34] X. Chen, S. Safavi, and Y. Liu, A Down-sized Printed Hilbert Antenna for VHF band, *IEEE APS International Symposium*, 2003, vol. 2, 22–27 June 2003, pp. 581–584.

[35] S. R. Best, The Radiation Properties of Electrically Small Folded Dipole Antennas, *IEEE Transactions on Antennas and Propagation*, vol. 52, April 2004, no. 4.

[36] L. J. Chu, Physical Limitations of Omni-Directional Antennas, *Journal of Applied Physics*, vol. 19, 1948, no. 28, pp. 1113–1175.

[37] W. L. Weeks, *Antenna Engineering*, McGraw Hill, 1968, 21, p. 28.

[38] W. L. Stutzman and G. A. Thiele, Antenna Theory and Design, 2nd edn, John Wiley & Sons, 1998, p. 58.

[39] Y. Mushiake, *Self-Complementary Antennas*, Springer Verlag, 1996.

[40] P. Xu and K. Fujimoto, L-Shaped Self-Complementary Antenna, *IEEE APS International Symposium 2003*, no. 3, pp. 95–98.

[41a] K. Fujimoto *et al.*, *Small Antennas*, Research Studies Press, UK, 1987.

[41b] K. Fujimoto, Small Antennas in K. Chang (ed.), *Encyclopedia of RF and Microwave Engineering*, vol. 5, John Wiley and Sons, 2005, p. 4777.

[42] H. Morishita, Y. Kim, and K. Fujimoto, Design Concept of Antennas for Small Mobile Terminals and the Future Perspective, *IEEE Antennas and Propagation Magazine*, vol. 44, October 2002, no. 5, pp. 30–43.

[43] H. R. Stuart and A. Pievertsky, Electrically Small Antenna Elements Using Negative Permittivity resonator, *IEEE Transactions on Antennas and Propagation*, vol. 54, 2006, no. 6, pp. 1644–1653.

[44] F. Bilotti, A. Alu, and L. Vegri, Design of Miniaturized Metamaterial Patch Antennas, *IEEE Transactions on Antennas and Propagation*, vol. 56, 2008, no. 6, pp. 1640–1647.
[45] R. W. Ziolkowsky and A. Erentock, Metamaterial-Based Efficient Electrically Small Antennas, *IEEE Transactions on Antennas and Propagation*, vol. 54, 2006, no. 7, pp. 2113–2130.
[46] A. Erentock and R. W. Ziolkowsky, Metamaterial-Inspired Efficient Electrically Small Antennas, *IEEE Transactions on Antennas and Propagation*, vol. 56, 2008, no. 3, pp. 691–707.
[47] R. W. Ziolkowsky and A. D. Kipple, Application of Double Negative Materials to Increase the Power Radiated by Electrically Small Antennas, *IEEE Transactions on Antennas and Propagation*, vol. 51, October 2003, no. 10, pp. 2626–2640.
[48] J. L. Linvill, Transistor Negative-Impedance Converters, *Proceedings of the IRE*, June 1953, pp. 725–729.
[49a] R. M. Rudish and S. E. Sussman-Fort, Non-Foster Impedance Matching Improves S/N of Wideband Electrically-Small VHF Antennas and Arrays, *Proceedings of the Second IASTED International Conference on Antennas, Radar, and Wave Propagation*, July 19–21 2005, Alberta, Canada.
[49b] S. E. Sussman-Fort, Matching Network Design Using Non-Foster Impedance, *International Journal of RF and Microwave-Computer Engineering*, March 2006, pp. 135–142.
[50] J. T. Aberle, Two-Port Representation of an Antenna with Applications to Non-Foster Networks, *IEEE Transactions on Antennas and Propagation*, vol. 56, May 2008, no. 5, pp. 1218–1222.
[51] T. Tsutaoka *et al.*, Negative Permeability Spectra of Magnetic Materials, *IEEE iWAT 2008*, Chiba, Japan, p. 202, pp. 279–281.
[52] N. Engheta and R. W. Ziolkowsky (eds.), *Metamaterials, Physics and Engineering Exploration*, John Wiley and Sons, 2006, Part IV, pp. 15.2–15.6 and pp. 16.12–16.13.
[53] N. Engheta and R. W. Ziolkowsky (eds.), *Metamaterials, Physics and Engineering Exploration*, John Wiley and Sons, 2006, pp. 205–209.
[54] N. Engheta and R. W. Ziolkowsky (eds.), *Metamaterials, Physics and Engineering Exploration*, John Wiley and Sons, 2006, Part IV, pp. 16.12–16.19.
[55] N. Engheta and R. W. Ziolkowsky (eds.), *Metamaterials, Physics and Engineering Exploration*, John Wiley and Sons, 2006, Part IV, p. 16.13 and p. 16.20.
[56] N. Engheta and R. W. Ziolkowsky (eds.), *Metamaterials, Physics and Engineering Exploration*, John Wiley and Sons, 2006, pp. 207–209.
[57] K. Fujimoto, Integrated Antenna Systems, in K. Chang (ed.) *Encyclopedia of RF and Microwave Engineering*, John Wiley and Sons, 2005.
[58] C. A Balanis, *Modern Antenna Handbook*, John Wiley and Sons, 2008, Part III, ch. 15.
[59a] A. Lai and T. Itoh, Composite Right/Left Handed Transmission Metamaterials, IEEE Microwave Magazine, September 2004, Part VI 31.
[59b] [17a] or [17b] or [17c] chs. 8–9, and chs. 11–12.
[60] D. Sievenpiper *et al.*, High-Impedance Electromagnetic Surfaces with a Forbidden Frequency Band, *IEEE Transactions on Microwave Theory and Techniques*, vol. 47, November 1989, no. 11, pp. 2059–2074.
[61] F. Young and Y. Rahmat-Samii, Reflection Phase Characteristics of the EBG Ground Plane for Low Profile Antenna Applications, *IEEE Transactions on Antennas and Propagation*, vol. 51, October 2003, no. 10, pp. 2691–2703.
[62] F. Young and Y. Rahmat-Samii, Microstrip Antenna Integrated with Electromagnetic Bandgap (EBG) Structure; A Low Mutual Coupling Design for Array Applications, *IEEE Transactions on Antennas and Propagation*, vol. 51, no. 10, October 2003, pp. 2936–2946.

[63] N. Engheta and R. W. Ziolkowsky (eds.), *Metamaterials, Physics and Engineering Exploration*, John Wiley and Sons, 2006, pp. 276–281.

[64] R. Gonzalo, P. de Yang, K. P. Ma, and T. Itoh, Enhanced Patch Antenna Performance by Suppressing Surface Wave using Photonic-Bandgap Substances, *IEEE Transactions on Microwave Theory and Techniques*, vol. 47, 1999, pp. 2131–2138.

[65] S. Enoch, B. Glalak, and G. Tayeh, Enhanced Emission with Angular Confinement from Photonic Crystals, *Applied Physics Letters*, vol. 81, 2002, no. 9, pp. 1588–1590.

[66] F. Young and Y. Rahmat-Samii, Bent Monopole Antenna on EBG Ground Plane with Reconfigurable Radiation Patterns, *IEEE APS International Symposium 2004*, vol. 2, July 2004, pp. 1819–1820.

[67] N. Engheta and R. W. Ziolkowsky (eds.), *Metamaterials, Physics and Engineering Exploration*, John Wiley and Sons, 2006, 11.11 p. 305.

[68] K. L. Virga and Y. Rahmat-Samii, Low Profile Enhanced-Bandwidth PIFA Antenna for Wireless Communications Packaging, *IEEE Transactions on Microwave Theory and Techniques*, vol. 45, 1997, no. 10, pp. 1879–1888.

[69] R. L. Haupt, An Introduction to Genetic Algorithms for Electromagnetics, *IEEE Antennas and Propagation Magazine*, vol. 37, April 1995, no. 2.

[70] D. S. Whiele, E. Michielssen, Genetic Algorithm Optimization Applied to Electromagnetics: A Review, *IEEE Transactions on Antennas and Propagation*, vol. 45, March 1997, no. 3.

[71] J. M. Johnson, Y. Rahmat-Samii, Genetic Algorithms in Engineering Electromagnetics, *IEEE Antennas and Propagation Magazine*, vol. 39, August 1997, no. 4.

[72] R. L. Haupt, D. H. Werner, *Genetic Algorithms in Electromagnetics*, Wiley-IEEE Press, 2007.

[73] Y. Rahmat-Samii, E. Michielssen (eds.), *Electromagnetic Optimization by Genetic Algorithms*, Wiley Series in Microwave and Optical Engineering, 1999.

[74] R. L. Haupt, Thinned Arrays Using Genetic Algorithms, *IEEE Transactions on Antennas and Propagation*, vol. 42, 1994, no. 7, pp. 993–999.

[75] M. J. Buckley, Linear Array Synthesis Using a Hybrid Genetic Algorithm, *IEEE Antennas and Propagation Society International Symposium Proceedings*, Baltimore, MD, July 1996, pp. 584–587.

[76] R. L. Haupt, Optimum quantised low sidelobe phase tapers for arrays, *Electronics Letters*, vol. 31, 1995, no. 14, pp. 1117–1118.

[77] M. Shimizu, Determining the excitation coefficients of an array using genetic algorithms, *IEEE Antennas and Propagation International Symposium Proceedings*, Seattle, WA, June 1995, pp. 530–533.

[78] A. Tennant, M. M. Dawoud, and A. P. Anderson, Array pattern nulling by element position perturbations using a genetic algorithm, *Electronics Letters*, vol. 30, 1994, no. 3, pp. 174–176.

[79] D. S. Weile and E. Michielssen, The control of adaptive antenna arrays with genetic algorithms using dominance and diploidy, *IEEE Transactions on Antennas and Propagation*, vol. 49, October 2011, pp. 1424–1433.

[80] Z. Altman, R. Mittra, J. Philo, and S. Dey, New designs of ultrabroadband antennas using genetic algorithms, *IEEE Antennas and Propagation International Symposium Proceedings*, Baltimore, MD, July 1996, pp. 2054–2057.

[81] M. Bahr, A. Boag, E. Michielssen, and R. Mittra, Design of Ultrabroadband Loaded Monopoles, *Proceedings of IEEE Antennas and Propagation Society International Symposium*, Seattle, WA, June 1994, pp. 1290–1293.

[82] A. Boag, A. Boag, E. Michielssen, and R. Mittra, Design of Electrically Loaded Wire Antennas using Genetic Algorithms, *IEEE Transactions on Antennas and Propagation*, vol. 44, May 1996, pp. 687–695.

[83] S. L. Avila, W. P. Carpes, Jr., and J. A. Vasconcelos, Optimization of an Offset Reflector Antenna Using Genetic Algorithms, *IEEE Transactions on Magnetics*, vol. 40, March 2004, no. 2, pp. 1256–1259.

[84] D. S. Linden and E. E. Altschuler, The Design of Yagi Antennas Using a Genetic Algorithm, *Proceedings USNC/URSI Radio Science Meeting*, Baltimore, MD, July 1996, p. 283.

[85] N. Herscovici, M. F. Osorio, and C. Peixeiro, Miniaturization of Rectangular Microstrip Patches Using Genetic Algorithms, *IEEE Antennas and Wireless Propagation Letters*, vol. 1, 2002, pp. 94–97.

[86] P. Soontornpipit, C. M. Furse, and Y. C. Chung, Miniaturized Biocompatible Microstrip Antenna Using Genetic Algorithm, *IEEE Transactions on Antennas and Propagation*, vol. 53, June 2005, no. 6, pp. 1939–1945.

[87] C. Delabie, M. Villegas, and O. Picon, Creation of New Shapes for Resonant Microstrip Structures By Means of Genetic Algorithms, *Electronics Letters*, vol. 33, August 1997, pp. 1509–1510.

[88] H. Choo, A. Hutani, L. C. Trintinalia, and H. Ling, Shape Optimization of Broadband Microstrip Antennas Using Genetic Algorithm, *Electronics Letters*, vol. 36, December 2000, pp. 2057–2058.

[89] G. Marrocco, A. Fonte, and F. Bardati, Evolutionary Design of Miniaturized Meander-line Antennas for RFID Applications, *Proceedings of the IEEE AP-S International Symposium*, vol. 2, San Antonio, TX, 2002, pp. 362–365.

[90] G. Marrocco, Gain-Optimized Self-Resonant Meander Line Antennas for RFID Applications, *IEEE Antennas and Wireless Propagation Letters*, vol. 2, 2003, pp. 302–305.

[91] D. H. Werner, P. L. Werner, and K. H. Church, Genetically engineered multiband fractal antennas, *Electronics Letters*, vol. 37, 2001, no. 19, pp. 1150–1151.

[92] M. Fernández Pantoja, F. García Ruiz, A. Rubio Bretones, *et al.*, GA Design of Wire Pre-Fractical Antennas and Comparison With Other Euclidean Geometries, *IEEE Antennas and Wireless Propagation Letters*, vol. 2, 2003.

[93] E. E. Altshuler and D. S. Linden, Wire-antenna Designs Using Genetic Algorithms, *IEEE Antennas and Propagation Magazine*, vol. 39, April 1997, pp. 33–43.

[94] E. E. Altshuler, Design of a Vehicular Antenna for GPS/Iridium Using a Genetic Algorithm, *IEEE Antennas and Propagation*, vol. 48, June 2000, pp. 968–972.

[95] E. E. Altshuler, Electrically Small Self-resonant Wire Antennas Optimized Using a Genetic Algorithm, *Antennas and Propagation, IEEE Transactions on Antennas and Propagation*, vol. 50, March 2002, no. 3, pp. 297–300.

[96] E. E. Altshuler and D. S. Linden, An ultra-wideband impedance-loaded genetic antenna, *IEEE Transactions on Antennas and Propagation*, vol. 52, November 2004, pp. 3147–3151,

[97] Hosung Choo, Robert L. Rogers, and Hao Ling, Design of Electrically Small Wire Antennas Using a Pareto Genetic Algorithm, *IEEE Transactions on Antennas and Propagation*, vol. 53, March 2005, pp. 1038–1046.

[98] J. Kennedy and R. Eberhart, Particle swarm optimization, *Proceedings of IEEE International Conference on Neural Networks*, vol. 4 Perth, Australia, 1995 November, pp. 1942–1948.

[99] R. Poli, J. Kennedy, and T. Blackwell, Particle Swarm Optimization: An Overview, *Swarm Intelligence*, vol. 1, 2007, no. 1, pp. 33–57.

[100] J. Robinson and Y. Rahmat-Samii, Particle Swarm Optimization in Electromagnetics, *IEEE Transactions on Antennas and Propagation*, vol. 52, 2004, no. 2, pp. 397–407.

[101] Nanbo Jin and Y. Rahmat-Samii, Advances in Particle Swarm Optimization for Antenna Designs: Real-Number, Binary, Single-Objective and Multiobjective Implementations, *IEEE Transactions on Antennas and Propagation*, vol. 55, March 2007, no. 3, pp. 556–567.

[102] N. Jin and Y. Rahmat-Samii, Analysis and Particle Swarm Optimization of Correlator Antenna Arrays for Radio Astronomy Applications, *IEEE Transactions on Antennas and Propagation*, vol. 56, May 2008, pp. 1260–1279.

[103] D. W. Boeringer and D. H. Werner, Particle Swarm Optimization Versus Genetic Algorithms for Phased Array Synthesis, *IEEE Transactions on Antennas and Propagation*, vol. 52, March 2004, pp. 771–779.

[104] P. J. Bevelacqua and C. A. Balanis, Minimum Sidelobe Levels for Linear Arrays, *IEEE Transactions on Antennas and Propagation*, vol. 55, December 2007 pp. 3442–3449.

[105] F. Afshinmanesh, A. Marandi, and M. Shahabadi, Design of a Single-Feed Dual-Band Dual-Polarized Microstrip Antenna Using a Boolean Particle Swarm Optimization, *IEEE Transactions on Antennas and Propagation*, vol. 56, July 2008, pp. 1845–1852.

[106] N. Jin and Y. Rahmat-Samii, Parallel particle swarm optimization and finite-difference time-domain (PSO/FDTD) algorithm for multiband and wide-band patch antenna designs, *IEEE Transactions on Antennas and Propagation*, vol. 53, November 2005, pp. 3459–3468.

[107] L. Lizzi, F. Viani, R. Azaro, and A. Massa, A PSO-Driven Spline-Based Shaping Approach for Ultrawideband (UWB) Antenna Synthesis, *IEEE Transactions on Antennas and Propagation*, vol. 56, August 2008, pp. 2613–2621.

[108] N. Jin and Y. Rahmat-Samii, Particle swarm optimization for multi-band handset antenna designs: A hybrid real-binary implementation, *IEEE Antennas and Propagation Society International Symposium, 2008*, AP-S 2008, 5–11 July 2008, pp. 1–4.

[109] G. Kiziltas and D. Psychoudakis, J. L. Volakis, and N. Kikuchi, Topology design optimization of dielectric substrates for bandwidth improvement of a patch antenna, *IEEE Transactions on Antennas and Propagation*, vol. 51, October 2003, pp. 2732–2743.

[110] G. Kiziltas, and J. L. Volakis, Shape and Material Optimization for Bandwidth Improvement of Printed Antennas on High Contrast Substrates, *IEEE International Symposium on Electromagnetic Compatibility*, 2003. EMC'03, vol. 2, 11–16 May 2003, pp. 1081–1084.

[111] T. Nomura, K. Sato, S. Nishiwaki, and M. Yoshimura, Topology Optimization of Multiband Dielectric Resonator Antennas Using Finite-Difference Time-Domain Method, *IEEE International Workshop on Antenna Technology*, March 2007, pp. 147–150.

[112] S. Koulouridis, E. Psychoudakis, and J. L. Volakis, Multiobjective Optimal Antenna Design Based on Volumetric Material Optimization, *IEEE Transactions on Antennas and Propagation*, vol. 55, March 2007, no. 3, pp. 594–603.

[113] S. Ohtsu, S. Endo, N. Michishita, and H. Morishita, A Study on Configurations of Magnetic Material Loaded Planar Inverted-F Antenna Using Topology Optimization, *IEEE Antennas and Propagation Society International Symposium*, July 2008.

[114] A. Kajitani, Y. Kim, H. Morishita, Y. Koyanagi, Wideband Characteristics of Built-in Folded Dipole Antenna for Handsets, *IEEE AP-S Conference Proceedings*, June 2007, pp. 3548–3551.

[115] A. Kajitani, Y. Kim, H. Morishita, and Y. Koyanagi, Self-balanced and Wideband Characteristics of Improved Built-in Folded Dipole Antenna for Handsets, *ISAP AP-S Conference Proceedings*, August 2007, pp. 1178–1181.

[116] N. T. Hung, S. Watanabe, and H. Morishita, Fundamental Investigation of U-shaped Folded Dipole Antenna for WiMAX, *IEICE Society Conference*, Sep. 2010, B-1–128.

[117] G. Venter, Particle Swarm Optimization, *AIAA Journal* vol. 41, August 2003, no. 8.

# 7 Design and practice of small antennas I

## 7.1 Design and practice

This chapter describes practical design and examples of small antennas. The chapter is composed of four sections, which deal with four types of small antenna: ESA (Electrically Small Antennas), FSA (Functionally Small Antennas), PCSA (Physically Constrained Small Antennas), and PSA (Physically Small Antennas). Since this grouping is rather flexible according to the antenna structure, types of applications, and so forth, some antennas may be classified into more than a single group.

## 7.2 Design and practice of ESA

### 7.2.1 Lowering the resonance frequency

#### 7.2.1.1 Use of slow wave structure

*7.2.1.1.1 Periodic structure*

*7.2.1.1.1.1 Meander line antennas (MLA)*

Meander line antennas have been applied to various small wireless systems such as mobile phones, digital TV receivers, RFID, and so forth, as the antennas can be easily and more flexibly constituted in planar structure than simple wire types. The meander line structure has practically been employed with modified versions such as an asymmetric dipole for two-frequency bands, a composite with a line or a slot for multiband operation, a horizontal element of an IFA (Inverted-F antenna), and so forth. These antennas used as a part of another type of antenna like the horizontal element of a PIFA are designed partly as an MLA and integrated into the other type of antenna.

*7.2.1.1.1.1.1 Dipole-type meander line antenna*

The principle of application of a meander line structure to antennas is as mentioned in Section 6.2.1.1, application of SW (Slow Wave) performances, the main feature of which is the smaller phase velocity as compared with that of free space so that the resonance frequency is lowered, resulting in the antenna size reduction. As one way to evaluate the size reduction effect, the length-shortening ratio *SR* is proposed [1], and demonstrated

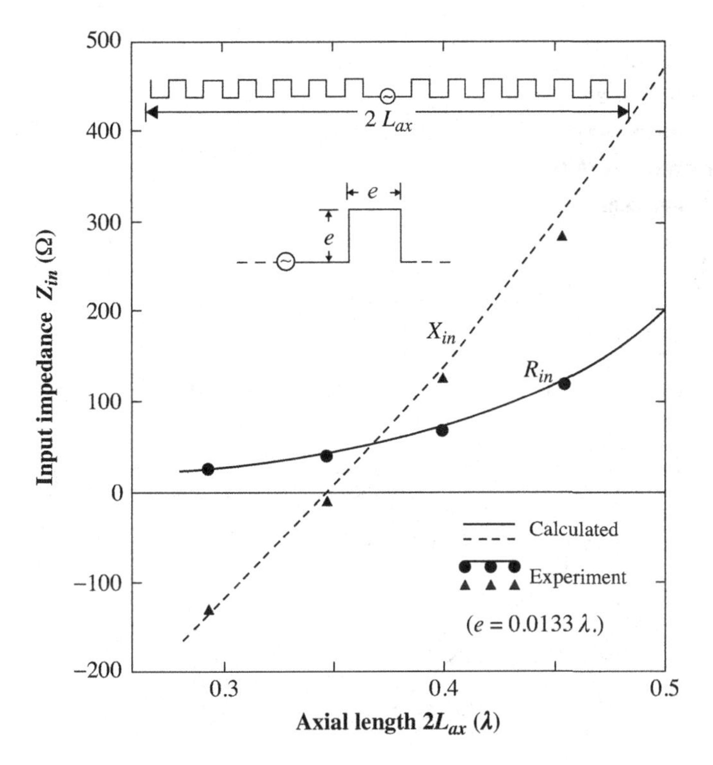


**Figure 7.1** Input impedance ($Z_{in} = R_{in} + \mathrm{j}\, X_{in}$) of MLA (from [1], copyright ©1984 IEEE).

by considering two models; a meander line antenna and a zigzag antenna. The *SR* is defined as

$$SR = (\lambda/2 - 2L_{ax})/(\lambda/2) \tag{7.1}$$

where $L_{ax}$ is the axial length of the antenna (the length taken directly from one end to another end of the antenna structure) and λ is the resonance wavelength. The phase velocity $V_p$ in this case is expressed by using *SR* as

$$V_p = (1 - SR)c. \tag{7.2}$$

Here *c* is the velocity of light in free space.

The input impedance $Z_{in}$ ($= R_{in} + \mathrm{j}X_{in}$) of the antenna is calculated and depicted in Figure 7.1, where the inset illustrates the antenna model (a dipole type), and $Z_{in}$ is drawn with respect to the axial length $2L_{ax}$. This figure indicates that the resonance occurs with the length of $2L_{ax} = 0.35\ \lambda$ (the length $L_0$ of the meandered wire extended into a straight line $= 0.70\ \lambda$ with $e = 0.0133\ \lambda$), and $R_{res}$ ($R_{in}$ at the resonance) $= 43\ \Omega$. In this case, *SR* is 0.3. The radiation pattern is similar to that of a half-wave dipole with gain of 1.95 dB and half-power beam width of ±47 degrees.

The dipole-type MLA is also described in [2], providing data on monopole type, dipole type, tapered type, folded type, and other types, including examples of practical application to mobile phones.

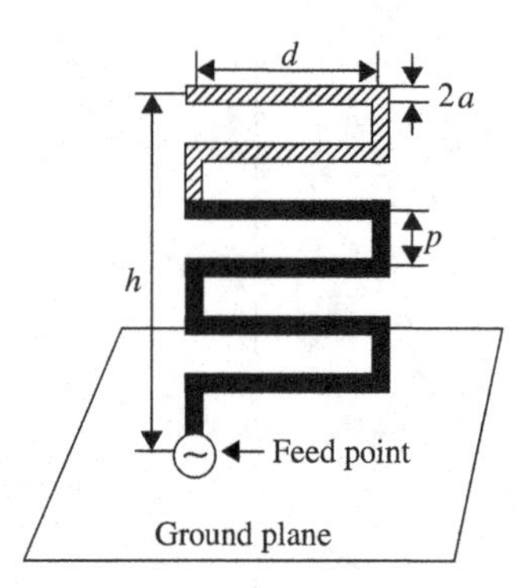


**Figure 7.2** Monopole-type MLA placed on a ground plane (from [3]).

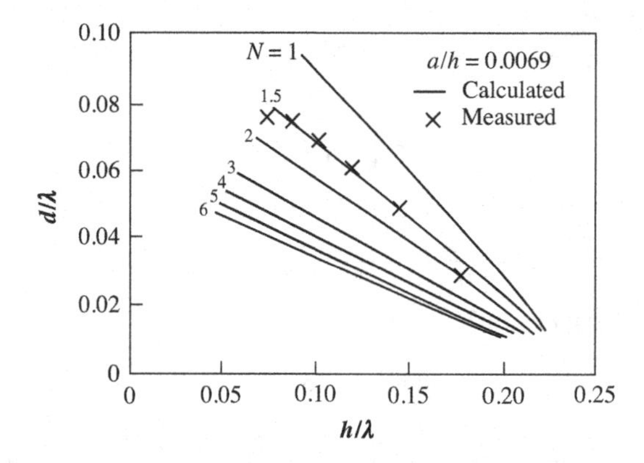


**Figure 7.3** Relationship between antenna height $h/\lambda$ and the width $d$ (from [3] copyright ©1998 IEICE).

### *7.2.1.1.1.1.2 Monopole-type meander line antenna*

An MLA can be constituted in a monopole type and placed on a ground plane as shown in Figure 7.2, where a hatched part shows one segment of the antenna and the dimensional parameters are given [3]. In practice, this type of antenna has been often seen in small mobile terminals. The relationship between the antenna height $h/\lambda$ and the width $d/\lambda$ for the antenna in the resonance state is drawn in Figure 7.3, where the number $N$ of segments is taken as the parameter and $\lambda$ is the wavelength at the resonance. The $h/\lambda$ varies linearly with $d/\lambda$ and all lines in the figure concentrate at a point (0, 0.25), where $d/\lambda = 0$, corresponding to the resonance point of an ordinary quarter-wave monopole. An empirical equation for $d/\lambda$ with respect to $h/\lambda$ is given by

$$d/\lambda = (0.25 - h/\lambda)/1.70N^{0.54}. \quad (7.3)$$

Radiation resistance $R_{in}$ in terms of $h/\lambda$ is shown in Figure 7.4, where both calculated and experimental results are shown with respect to $N$. $R_{in}$ does not vary substantially with $N$. When $N = 3$, $R_{in}$ is given approximately by

$$R_{in} = 545(h/\lambda)^{1.82}. \quad (7.4)$$

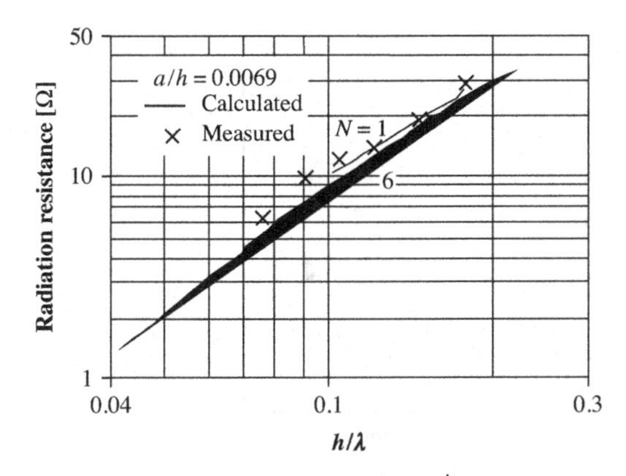


**Figure 7.4** Radiation resistance with respect to antenna height $h/\lambda$ (from [3], copyright ©1998 IEICE).

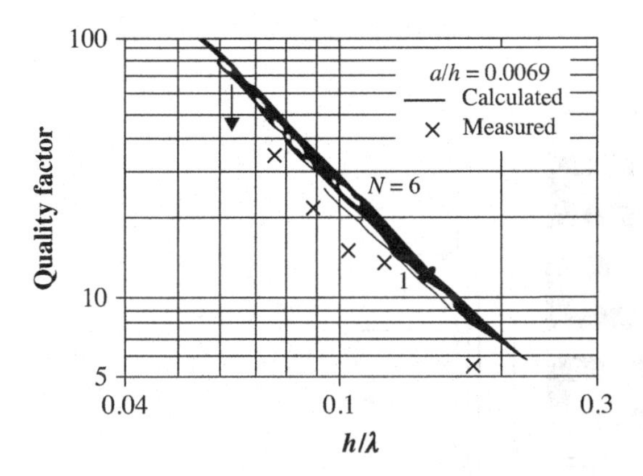


**Figure 7.5** Antenna $Q$ with respect to antenna height $h/\lambda$ (from [3], copyright ©1998 IEICE).

$Q$ is related with $h/\lambda$ and shown in Figure 7.5, where both calculated and experimental results are provided, and given empirically for $N = 3$ as

$$Q = 0.777/(h/\lambda). \tag{7.5}$$

Radius $a$ of the wire has almost no influence on the impedance characteristics.

Design parameters of monopole-type MLAs are also given in [4]. As a practical example, there has been a report on an MLA etched on both sides of a thin dielectric substrate [5]. The antenna model is depicted in Figure 7.6 with one side in (a) and the other side in (b). This antenna has wide enough bandwidth to receive the terrestrial TV broadcasting at UHF bands between 550 and 800 MHz, and gain of –3dB. The size is $80 \times 100 \times 1.6\ \text{mm}^3$.

Another example is a monopole-MLA driven by a small loop mounted on a ground plane [6]. The meander line winding is optimized by using NEC (Numerical Electromagnetic Code) in conjunction with a Pareto GA (Genetic Algorithm). A further example

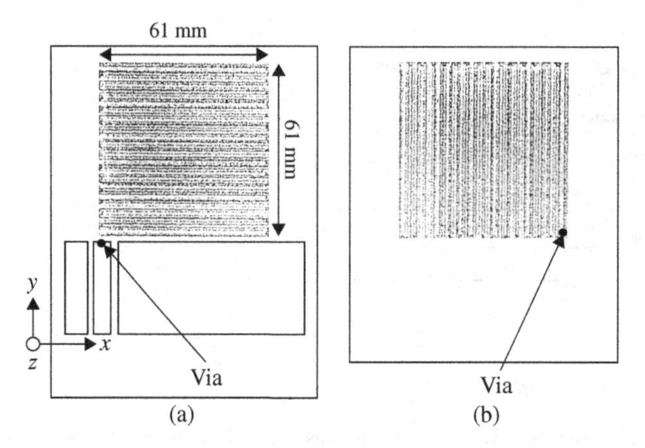


**Figure 7.6** MLA etched on both side of thin dielectric substrate (from [5], copyright ©2006 IEICE).

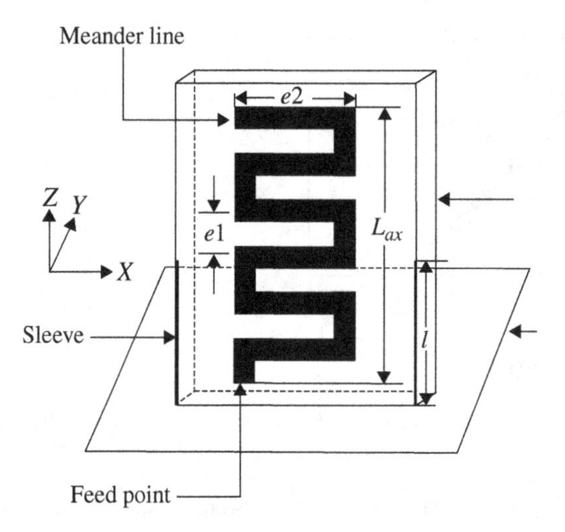


**Figure 7.7** A small MLA with two sleeves located besides the antenna element (from [7]).

is a printed MLA [7], which is mounted on a ground plane and has two sleeves besides the meander line element as shown in Figure 7.7. The antenna impedance is calculated by using FDTD simulation and is shown to have operating bandwidth in the 0.9 to 3.0 GHz band, although the ground plane is comparatively small. As one more example, Figure 7.8 illustrates an MLA constituted in a loop structure in order to attain multi-band operation [8]. In the figure (a) is the antenna mounted on the ground plane, (b) shows antenna structure (with bent part extended), and (c) illustrates the side view of the antenna mounted on the dielectric substrate. This antenna is designed as an internal antenna installed inside a mobile terminal used for GSM/DCS/PCS operations.

Tapered MLAs are described in [9a] and their applications to cellular units are introduced in [9b].

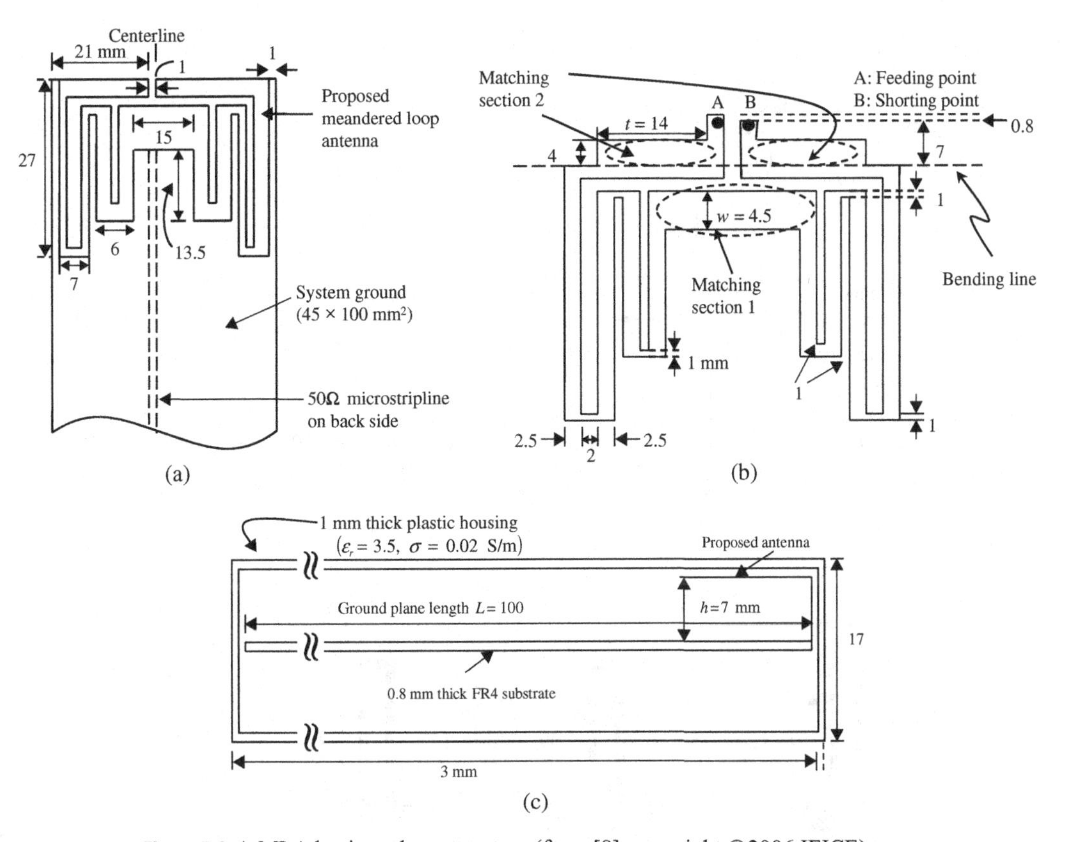


**Figure 7.8** A MLA having a loop structure (from [8], copyright ©2006 IEICE).

### *7.2.1.1.1.1.3 Folded-type meander line antenna*

As can be seen in Figure 7.5, $Q$ of a meander line antenna becomes higher as the size becomes smaller, meaning that the smaller the antenna size becomes, the narrower the bandwidth will be. In order to increase the bandwidth, a folded type can be useful [10], since a folded structure has substantially two modes, balanced and unbalanced modes, and by arranging the susceptances of these two modes appropriately, bandwidth can be increased. Figure 7.9 is an antenna model, constituting a folded structure with two meander lines, which are connected at the top. The folded structure is decomposed into two parts; balanced and unbalanced modes as shown in Figure 7.10 [11], in which (a) shows the folded structure, on which currents $I_1$ and $I_2$ flow on each element, and (b) illustrates decomposed modes; the unbalanced mode (the current $I_u$) plus balanced mode (the current $I_b$). The unbalanced mode current $I_u$ equals $(I_1 + I)_2/(1 + \gamma)$, and the balanced mode current $I_b$ equals $(\gamma I_1 - I_2)/(1 + \gamma)$, where $\gamma$ stands for the ratio of the unbalanced current on the element 1 and that on the element 2 [10]. This folded model is equivalently expressed by a circuit shown in Figure 7.11, in which the circuit parameters such as the source voltage $V$, currents $I_1$, the decomposed mode currents $I_u$ and $I_b$, and the impedances for both modes, $Z_u$ (unbalanced) and $Z_b$ (balanced), are provided.

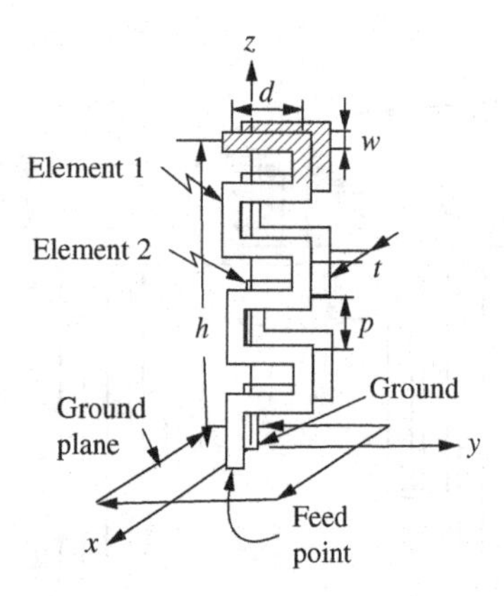


**Figure 7.9** A folded antenna model with two meander lines (from [10], copyright ©1999 IEICE).

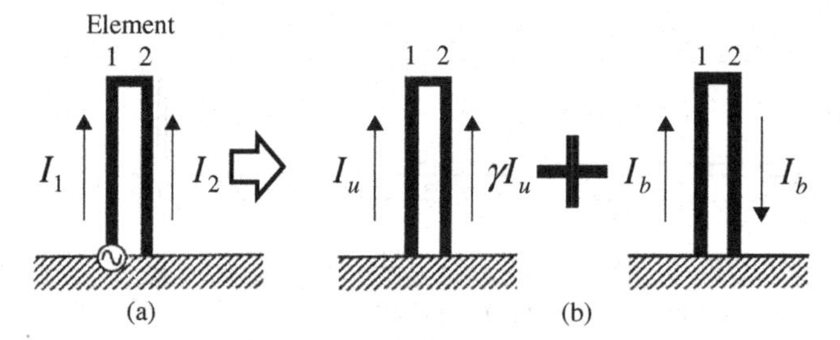


**Figure 7.10** Currents on a folded structure and on the decomposed balanced and unbalanced modes ([10], copyright ©1999 IEICE).

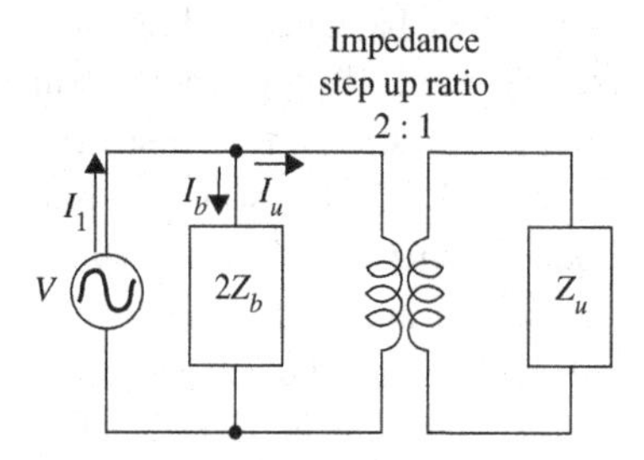


**Figure 7.11** Equivalent circuit corresponding to a folded structure ([10], copyright ©1999 IEICE).

By using these circuit parameters, $Z_u$ and $Z_b$ respectively, are given by

$$n^2 Z_u = V/I_u \tag{7.6}$$

and

$$Z_b = V/(2I_b) \tag{7.7}$$

where $I_u = (I_1 + I_2)/(1+\gamma)$, $I_b = (\gamma I_1 - I_2)/(1+\gamma)$, and $\gamma$ denotes a ratio of unbalanced currents on the element 1 and 2, and $n = 1 + \gamma$.

From (7.6), the unbalanced impedance $n^2\,Z_u = V/I_n{}^2$, and is given by using radiation resistance $R$, antenna $Q$, and resonance frequency $f_0$ as

$$n^2 Z_u = R + \mathrm{j}\,R_u \tag{7.8}$$

$$\text{where } u = Q(f/f_0 - f_0/f) = Q(2\Delta f)/f_0 \quad (\Delta f = f - f_0). \tag{7.9}$$

Admittance (unbalanced) $Y_u$ is expressed from (7.8) as

$$\begin{aligned} Y_u &= 1/(n^2 Z_u) = G_u + \mathrm{j}\,B_u \\ \text{and } G_u &= 1/\{R(1+u^2)\} \end{aligned} \tag{7.10}$$

$$B_u = -u/\{R(1+u^2)\} \tag{7.11}$$

In the same way, the balanced mode admittance $Y_b$ is

$$Y_b = 1/Z_b = \mathrm{j}B_b \tag{7.12}$$

$$\text{and } B_b = -B_0 u/Q. \tag{7.13}$$

Here $B_0$ is defined by using a parameter $K$ as

$$B_0 = KQ. \tag{7.14}$$

$$\text{Then } B_b = Ku. \tag{7.15}$$

Now the total susceptance $B_t = B_u + B_b$ will be

$$B_t = -\frac{1}{R}\frac{u}{(1+u^2)} + Ku. \tag{7.16}$$

From (7.11) and (7.14), when $u$ is very small, $B_u$ approaches $1/R$, and by selecting an appropriate value for $K$, $B_t$ can be made to be zero. The reflection coefficient $\Gamma$ of the line is given by

$$\Gamma(u) = \{Y_0 + Y_i(u)\}/\{Y_0 - Y_i(u)\}. \tag{7.17}$$

$\Gamma\,(u)$ is a function of $u$, and $Y_0$ is the characteristic admittance of the line. The voltage standing wave ratio (VSWR) $S$ as function of $(u)$ is

$$S(u) = \{1 + \Gamma(u)\}/\{1 - \Gamma(u)\}. \tag{7.18}$$

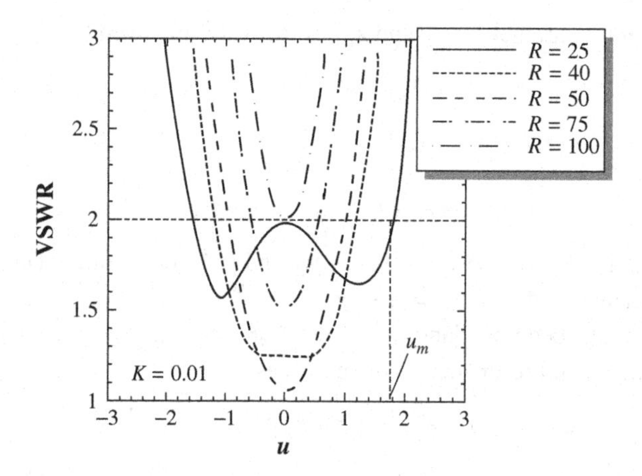


**Figure 7.12** VSWR characteristics in relation to parameter $u$ with variation of input resistance $R$ ([10], copyright ©1999 IEICE).

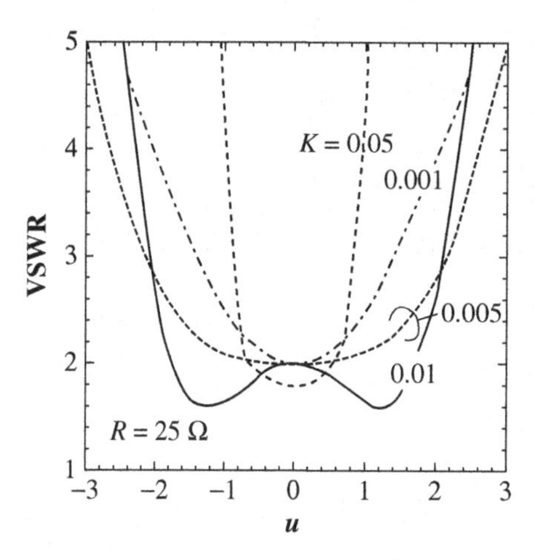


**Figure 7.13** VSWR characteristics in relation to parameter $u$ with variation of $K$ (from [10], copyright ©1999 IEICE).

Now desired relative bandwidth RBW is expressed by using bandwidth $\Delta f_m$ specified at a required value of $S_m$ and $u_m$ that correspond to $S_m$ as

$$\mathrm{RBW} = 2\Delta f_m / f_0 \tag{7.19a}$$

$$= u_m / Q. \tag{7.19b}$$

Then, (7.19b) shows that increase of RBW is possible by setting $u_m$ at its maximum value.

$S$ with respect to $u$ is illustrated in Figure 7.12, where $R$ is taken as a parameter, and in Figure 7.13, where $K$ is used as a parameter, respectively. Design parameters can be known from Figure 7.14 and Figure 7.15, in which relationships between $S_m$ and $K$, and

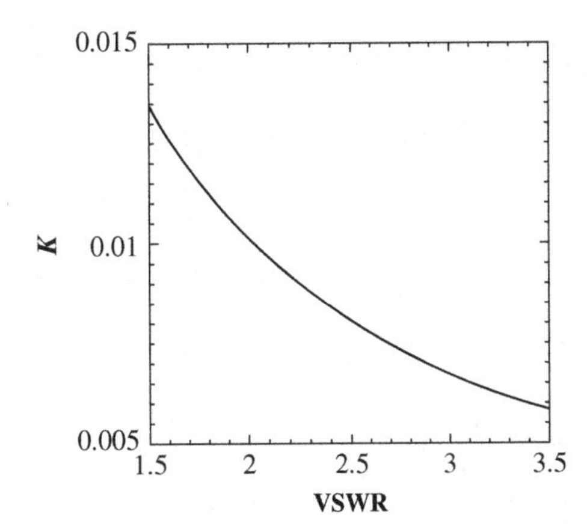


**Figure 7.14** Relationship between $K$ and $S_m$ (from [10], copyright ©1999 IEICE).

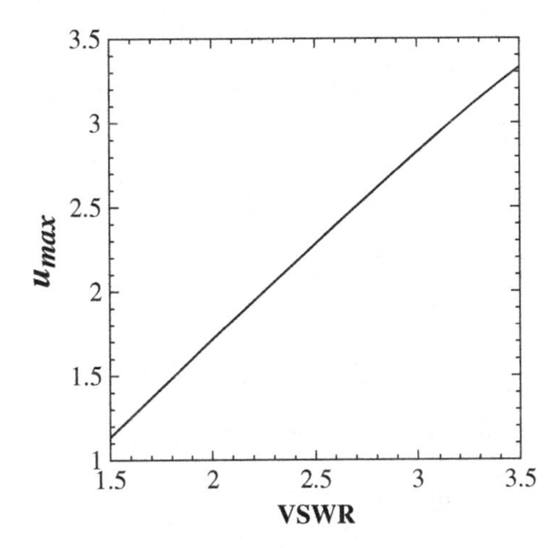


**Figure 7.15** Relationship between $u_{max}$ and $S_m$ (from [10], copyright ©1999 IEICE).

$S_m$ and $u_m$, respectively, are provided. RBW in relation to $S_m$ is given in Figure 7.16, where $Q$ is used as the parameter.

In practical designs, one of these antennas may be decomposed into two modal parts: balanced and unbalanced modes. The balanced mode can be treated equivalently as a two-wire transmission line of length $L$ with characteristic impedance $Z_0$. Here the length $L$ is that of the meander line from the feed point to its end. The input susceptance $B_b$ of this line is written as

$$B_b = -(1/Z_0)\cot(2\pi L/\lambda). \tag{7.20}$$

By using the wavelength $\lambda_0$ at the resonance,

$$B_b = -(1/Z_0)\cot\{(2\pi L/\lambda_0)(1 + \Delta f/f_0)\}. \tag{7.21}$$

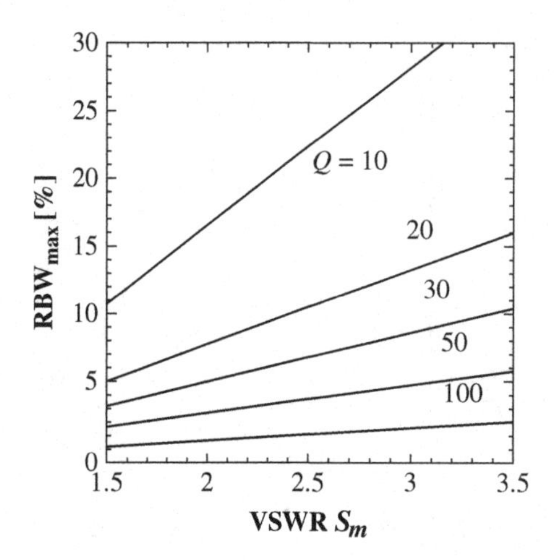


**Figure 7.16** Relationship between $Q_{max}$ and $S_m$ (from [10], copyright ©1999 IEICE).

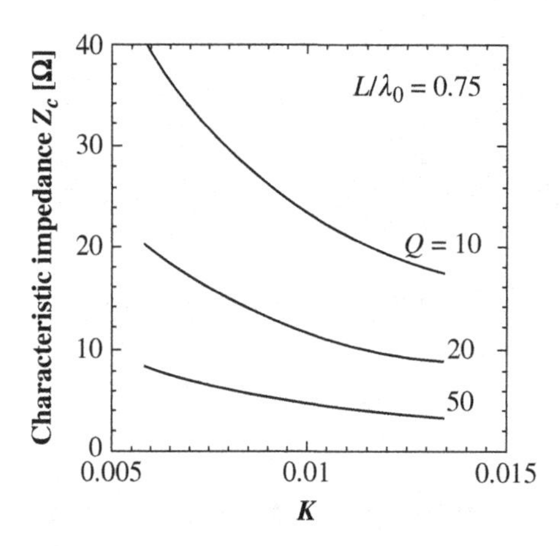


**Figure 7.17** Relationship between $Z_c$ and $K$ with variation of $Q$ (from [10], copyright ©1999 IEICE).

$L$ can be written by using number $n$ of the resonance mode as

$$L = (2n - 1)\lambda_0/4(n = 1, 2, 3, \ldots). \quad (7.22)$$

$$\text{Then, } B_b = (1/Z_t)\tan\{(2\pi L/\lambda_0)(\Delta f/f_0)\} \quad (7.23)$$

$$= Ku. \quad (7.24)$$

$$\text{Here } K = (1/Z_0)(L/\lambda_0)/Q \quad (7.25)$$

where the approximate expression taking only the first-order variable for tangent is used. From this, $K$ can be determined by selecting $Z_0$ and $L/\lambda$. $Z_0$ in relation to $K$ is expressed by Figure 7.17, where $Q$ is taken as the parameter. By using the meander line parameters $w$ (width) and $a$ (wire radius) in terms of $t$ (separation of two lines), which are given in

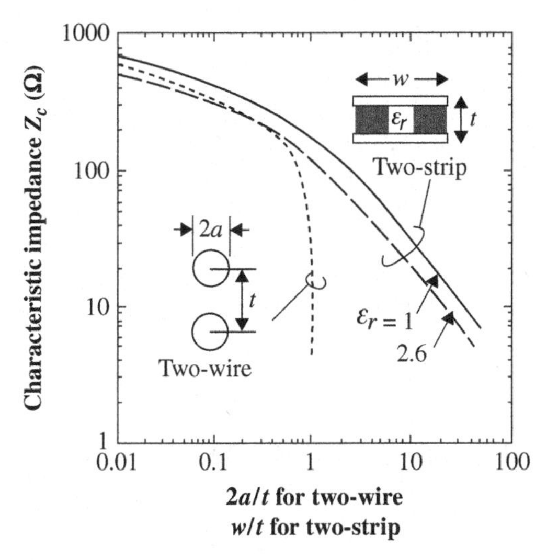


**Figure 7.18** Relationship between $Z_c$ and wire radius $a$ in terms of $t$ (from [10]).

Figure 7.9, $Z_0$ is expressed as shown in Figure 7.18. These are useful for designing an antenna consisting of two parallel strip-type meander lines.

A practical model having $h = 0.11\lambda$ (144.4 mm), $d = 0.057\lambda$ (74.8 mm), and $N$ (number of segments) $= 6$, is considered. In this case, $f_0$ is set to 228 MHz, $S_m = 1.5$, and the thickness of the substrate (Teflon, $\varepsilon = 2.6$) is $0.0006\lambda$ (0.8 mm). From Figure 7.14 and Figure 7.15, $u_m$ is 1.1 and RBW is 0.013. When the elements 1 and 2 are connected at the end to form the folded structure, $L/\lambda$ becomes 0.75, and $Z_0$ is found to be 9.58 Ω. From these, $w/t$ is 18, and thus $w$ becomes $0.011\lambda$ (14.4 mm), as $t = 0.8$ mm. RBW is obtained to be about 7.2%. Measured impedance (200 MHz step between 200 and 260 MHz) is shown in Figure 7.19, where a square ground plane of 0.76 λ (1 m) is used. Radiation efficiency obtained was about 90% to 95%. $S_m$ in terms of frequency $f$ is shown in Figure 7.20.

A tapered folded-type MLA is described in [9a], including folded type with a plate top-loaded MLA.

### *7.2.1.1.1.1.4 Meander line antenna mounted on a rectangular conducting box*

When an MLA is mounted on a rectangular conducting box [12], as can be often seen in practical applications (Figure 7.21), the resonance frequency $f_r$ varies with the distance $S$ of the antenna element from the surface of the substrate on which the antenna is printed. $R$ and $f_r$ with respect to $S$ are given in Figure 7.22. Figure 7.23 illustrates impedance loci on the Smith chart. The locus A is for the MLA mounted on the rectangular box, while the locus B is for the MLA placed on a circular ground plane with a diameter of 1 m. In both cases the impedance loci encircle near the center of the Smith chart, providing evidence of wideband characteristics. The frequency range used in the measurement is between 800 MHz and 900 MHz.

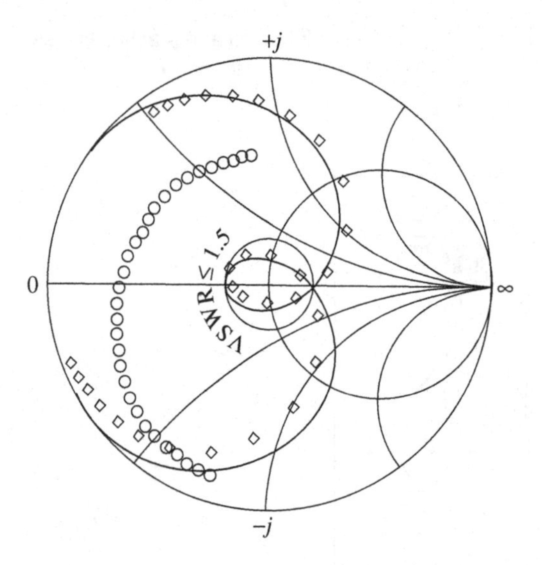

**Figure 7.19** Measured impedance ([10], copyright ©1999 IEICE).

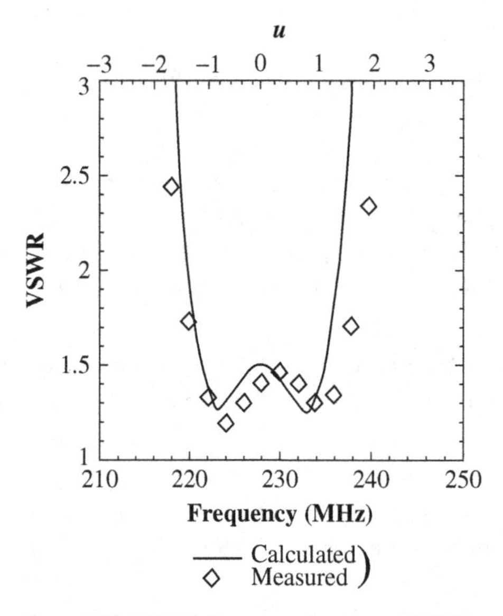

**Figure 7.20** VSWR $S_m$ versus frequency $f$ ([10], copyright ©1999 IEICE).

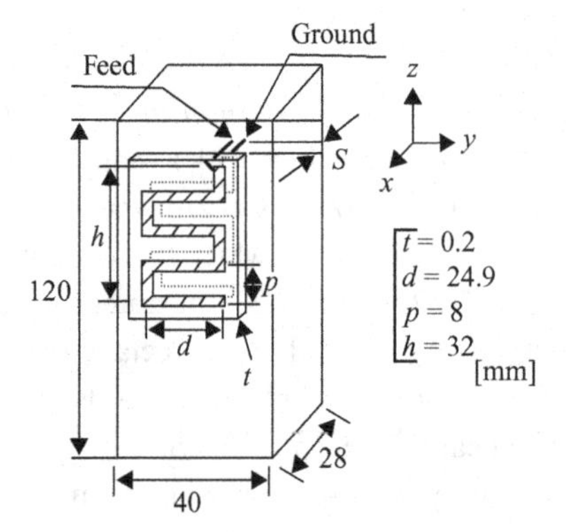

**Figure 7.21** An MLA mounted on a conducting box ([12], copyright ©1999 IEICE).

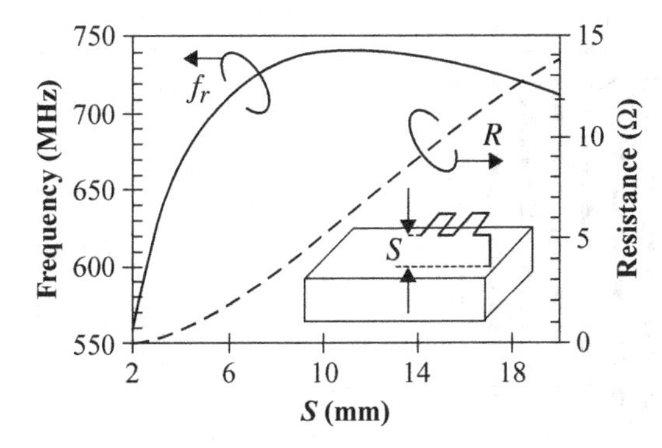


**Figure 7.22** Resonance frequency $f_r$ and radiation resistance $R$ with respect to distance $S$ ([12], copyright ©1999 IEICE).

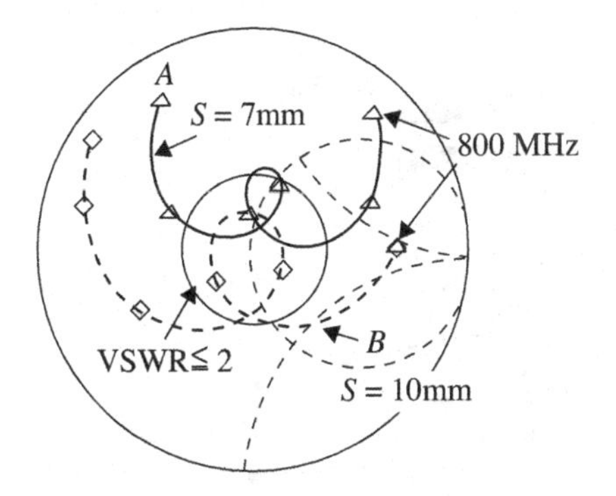


**Figure 7.23** Impedance loci of MLAs: a two-wire type mounted on a rectangular conducting plate and a one-wire type mounted on a circular conducting plane ([12], copyright ©2004 IEICE).

*7.2.1.1.1.1.5 Small meander line antennas of less than 0.1 wavelength [13]*

Small meander line dipole-type antennas having their lengths smaller than 0.1λ have been extensively studied theoretically and experimentally, and the design parameters have been provided [13]. The basic antenna structure is shown in Figure 7.24, where antenna dimensional parameters are given. The antenna performance was calculated by using an EM simulator IE3D and the operating frequency is taken to be 700 MHz. Since the wire width $d$ of the element is considered one of the significant parameters in designing the antenna, when the antenna length $L$ is first set to a desired value, the number $N$ of meandering turns is determined in conjunction with the width $d$. In this case, $d$ is the maximum value $d_{max}$ that is determined by the desired length $L$. Figure 7.25 shows $d_{max}$ in relation to the number $N$ for different $L$ (from 0.025 λ to 0.1 λ) of the antenna length where the separation $s$ of wire elements is 0.1 mm. This type of antenna can easily be set in self-resonance condition, as the wire length $L_a$ of the antenna can be made long enough to be nearly a quarter wavelength by selecting the number $N$ and the antenna width $W$ appropriately. Relationships between $W/\lambda$ and $L/\lambda$ in the resonance condition are given in Figure 7.26, where N is used as the parameter. In the figure, two cases when $d$ is 0.3mm ($2.3 \times 10^{-4}$ λ) and 0.1mm ($7.0 \times 10^{-4}$ λ), respectively, are

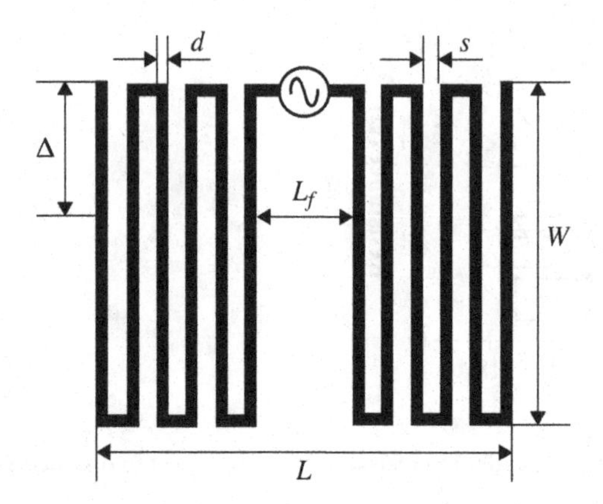


**Figure 7.24** MLA model with the length less than 0.1λ ([13], copyright ©2004 IEICE).

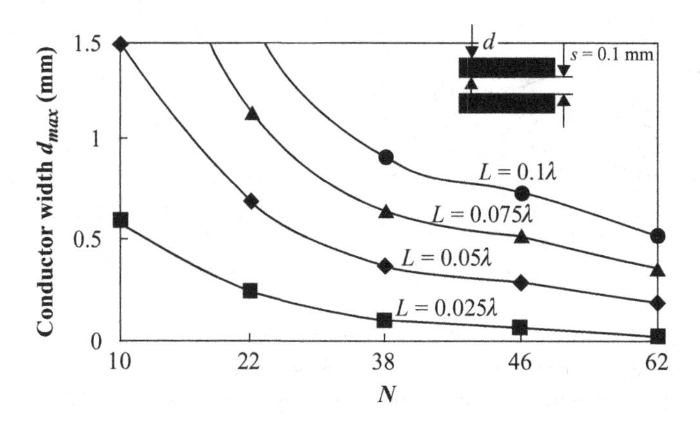


**Figure 7.25** Antenna width $d_{max}$ with respect to the segment number $N$ ([13], copyright ©2004 IEICE).

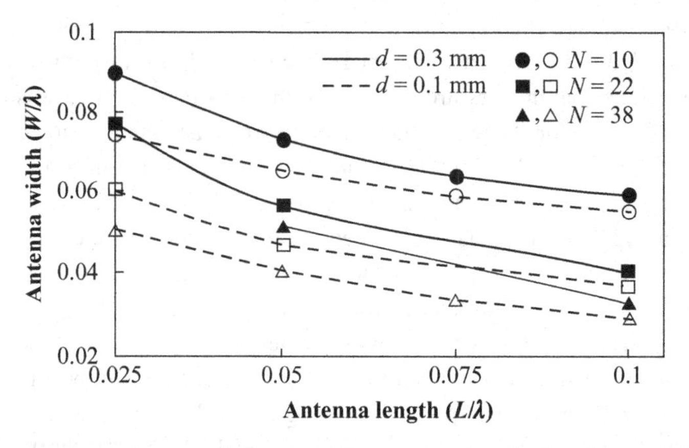


**Figure 7.26** Antenna width $w/\lambda$ with respect to the wire length $L/\lambda$ ([13], copyright ©2004 IEICE).

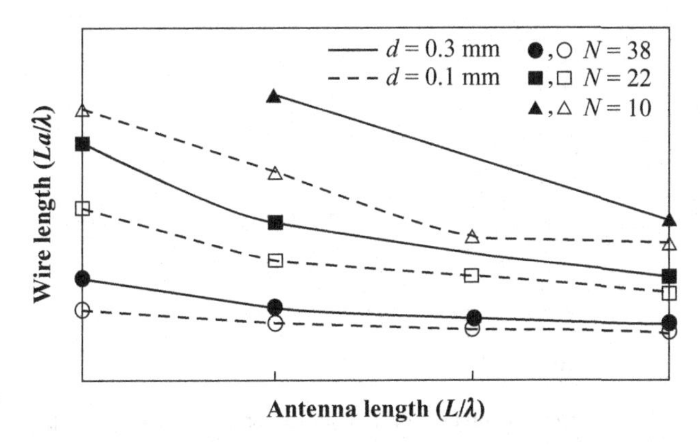


**Figure 7.27** Wire length $L_a$ versus antenna length $L$ at resonance state ([13], copyright ©2004 IEICE).

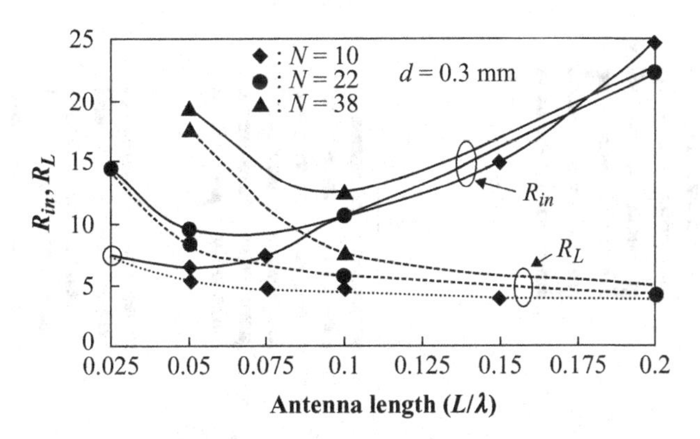


**Figure 7.28** Input resistance $R_{in}$ and loss resistance $R_l$ with respect to the antenna length $L/\lambda$ ([13], copyright ©2004 IEICE).

shown. The antenna wire length $L_a/\lambda$ in relation with $L/\lambda$ at the self-resonance state is illustrated in Figure 7.27 for cases $d = 0.1$ and 0.3, and $N = 10$, 22, and 38.

Input resistance $R_{in}$ and loss resistance $R_L$ of an antenna in relation to the antenna length $L/\lambda$ for cases $N = 10$, 22, and 38, and $d = 0.3$ are shown in Figure 7.28. The loss resistance $R_l$ of a conductor with the skin depth $\delta$ and the conductivity $\sigma$ is given by

$$R_l = 1/(\delta\sigma), \tag{7.26}$$

$$\text{where } \delta = \sqrt{1/(f\pi\mu\sigma)}. \tag{7.27}$$

When the conductor has planar structure with length $L$, width $w$, and thickness $t$, loss resistance $R_L$ of an antenna is expressed by

$$R_L = R_l L/2(w+t) = L\sqrt{\sigma}\sqrt{f\pi\mu}/2(w+t). \tag{7.28}$$

When the conductor is copper, $\sigma$ is $5.8 \times 10^7$ [S/m], then $\delta = 2.5$ mm at 700 MHz.

Since $R_L$ increases with the antenna wire length $L_a$ and $\sqrt{f}$, when $L$ becomes less than $0.1\lambda$, it increases notably with increase in $N$ that makes $L_a$ large.

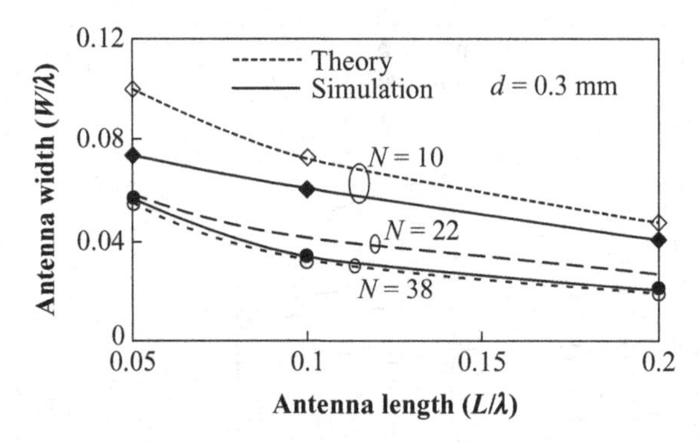


**Figure 7.29** Relationship between antenna length $L/\lambda$ and antenna width $w/\lambda$ ([13], copyright ©2004 IEICE).

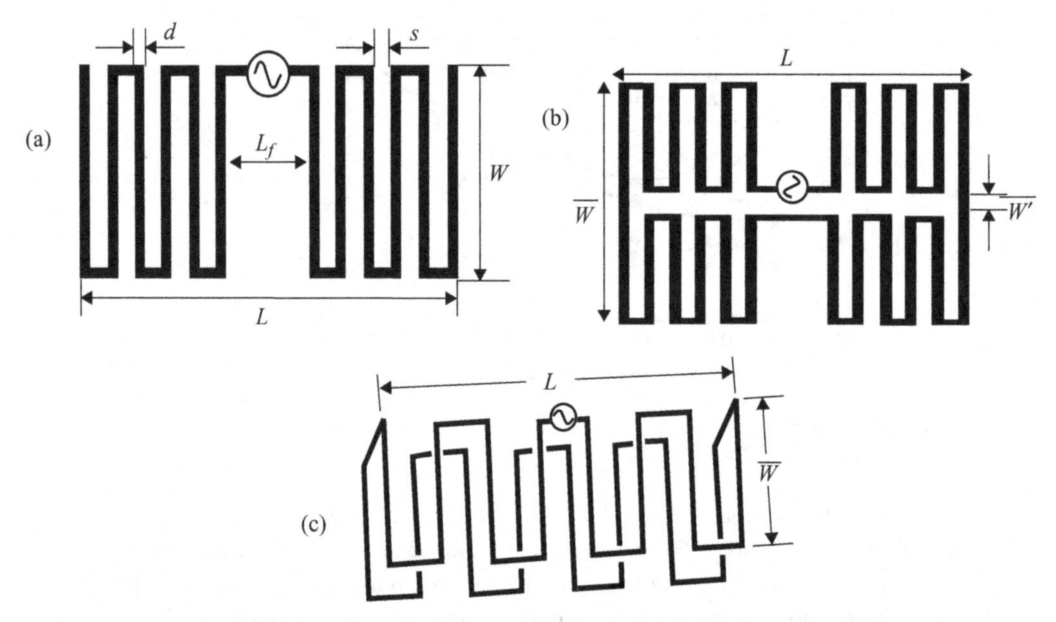


**Figure 7.30** Typical structure of folded-type MLA: (a) dipole type, (b) a folded-back type, and (c) folded-up type ([13], copyright ©2004 IEICE).

Relationships between the antenna length $L/\lambda$ and the antenna width $W$ for cases $N = 10$, 22, and 38 and $d = 0.3$ are shown in Figure 7.29.

When an antenna is a folded type, increase in the antenna impedance is expected. Figure 7.30 illustrates two types of folded MLAs along with an original dipole-type MLA (hereafter called "dipole-type" MLA); (a) is a dipole-type MLA, (b) is a folded-back type MLA, and (c) is a folded-up type MLA. The current distributions on an MLA are shown in Figure 7.31, in which arrows indicate directions of the current flows. Figure 7.31(a) depicts current flows on a dipole type and (b) that on a folded type. These figures suggest that substantial radiation is produced by the currents on the horizontal

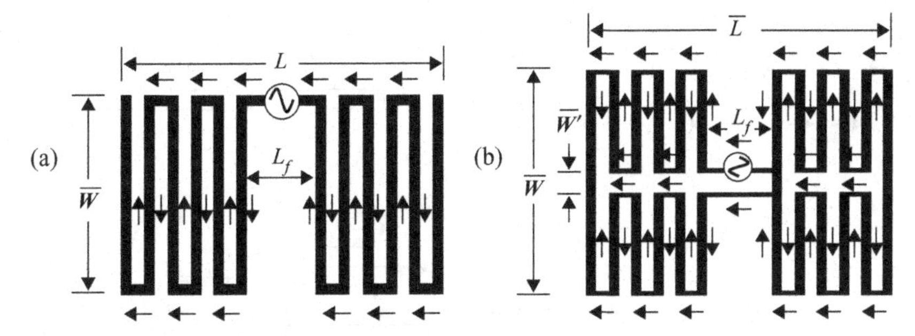


**Figure 7.31** Current distributions on the MLA element: (a) dipole type and (b) folded-back type ([13], copyright ©2004 IEICE).

elements (in the figure), because the directions of their current flows are the same. Contrastingly, the currents on the vertical elements (in the figure) do not contribute to the radiation, because the current flows on parallel elements are opposite in direction, so that their radiations cancel each other. Hence, radiation from a dipole-type MLA (the line diameter $d$ with the separation $w$ of the two current flows as shown in Figure 7.32(a)) is equivalently the same as that of a thicker dipole of the effective radius $d_{eff} = \sqrt{dw'}$ with the length $L$ as Figure 7.32(b) shows. This dipole is driven by a voltage $V$, and a current $I$ flows on the element. Meanwhile, a folded-type MLA shown in Figure 7.32(c), in which antenna parameters such as a driven voltage $V$ and currents $I_1$ and $I_2$ on each element separated with a distance $W'$ are provided. Current flows on these modes can be decomposed into two modes; unbalanced (radiation) and balanced (transmission line) modes as shown in Figure 7.32(d). On the unbalanced mode, current $I_u$ $(= (I_1 + I_2)/2)$ flows in the same direction on the two lines, while on the balanced mode, current $I_b$ $(= (I_1 - I_2)/2)$ flows in opposite directions on the two lines. The equivalent representation for these modes is further remodeled as a radiation mode and a transmission line mode as shown in Figure 7.32(e), which depicts a dipole driven by the voltage $V$, and the current $2I_u$ flowing on a single element with the thicker diameter $d_E$ $(= \sqrt{d_{eff}w'})$. From these, the impedance $Z_u$ of the unbalanced mode is found by $(V/2)/2I_u$, and that of the unbalanced mode $Z_b$ is $(V/2)/I_b$. The impedance $Z$ of the MLA is thus found by

$$Z = 4Z_u Z_b/(2Z_u + Z_b). \tag{7.29}$$

Meanwhile, the balanced mode is represented by two shorted two-wire transmission lines driven by a voltage $V/2$ with opposite-direction currents $I_b$ on each line as shown in Figure 7.32(e). The impedance $Z_b$ of the balanced mode is given by $2Z_{b0}$, which is the parallel combined impedance of the shorted two-wire transmission line of the length $L/2$. Here $Z_{b0}$ is given by $(V/2)/I_b$ and by $\mathrm{j}Z_0 \tan(\frac{\beta L}{2})$, where $Z_0$ is the characteristic impedance of the two-wire transmission line. From these relationships, the impedance $Z_f$ of the folded meander line antenna is derived as

$$Z_f = 4Z_u Z_{b0}/(2Z_u + Z_{b0}). \tag{7.30}$$

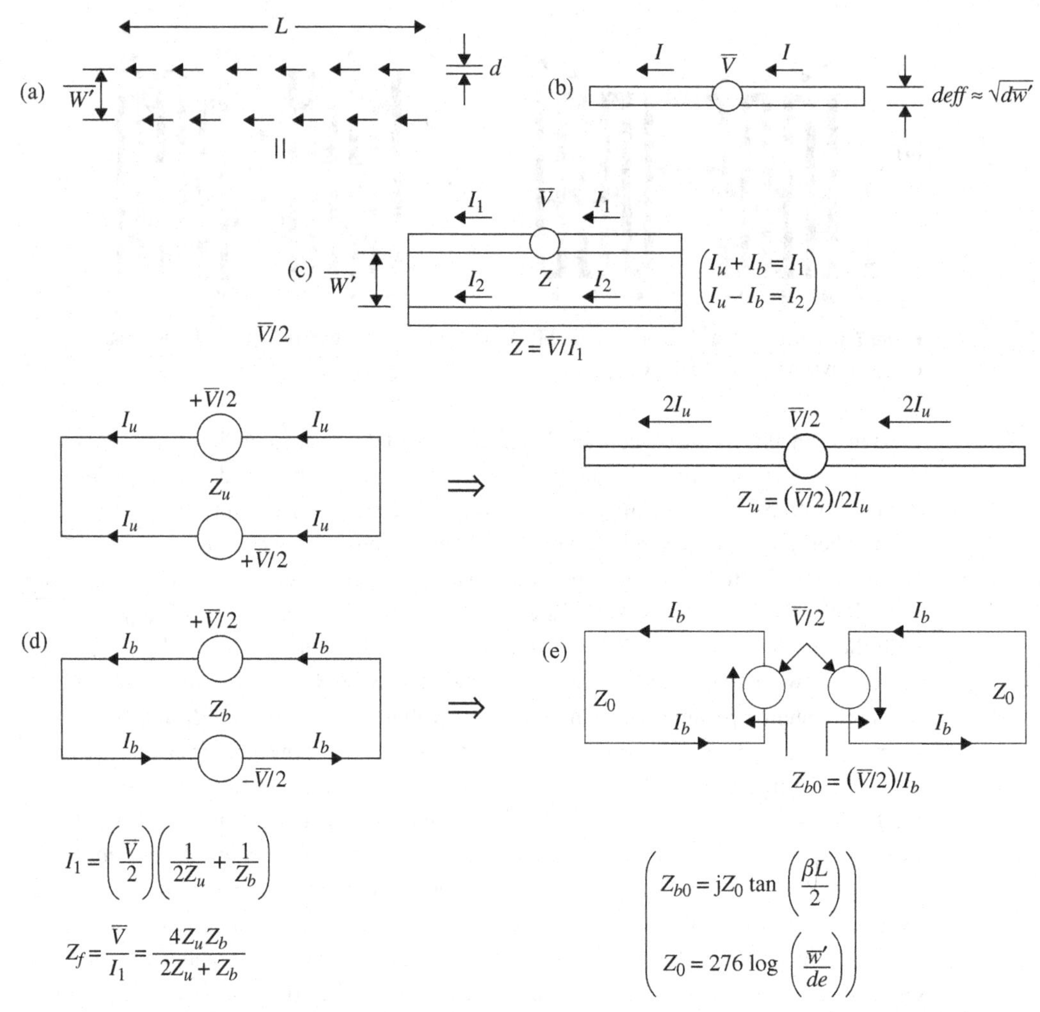


**Figure 7.32** Equivalent representation of current flows on MLA element; (a) currents, which contribute to radiation, on a meander line element, (b) an equivalent dipole for radiation mode, (c) an equivalent two-wire transmission line for the balanced mode, (d) current flows on a decomposed unbalanced ($I_u$) and balanced modes ($I_b$), and (e) an equivalent expressions for radiation and transmission line modes ([13], copyright ©2004 IEICE).

Since when an antenna is at the resonance condition, $Z_b$ becomes infinite, then

$$Z_f = 4Z_u. \tag{7.31}$$

This implies that the impedance of a folded antenna increases by a factor of four compared with that of its simple dipole equivalent. The factor is called the step-up ratio, by which the radiation resistance can be increased, regardless of small sizing. The radiation efficiency is particularly significant because of small size. The loss resistance $R_L$ of a meander line antenna with length $L$, planar line width $d$, and thickness $t$ is found by

$$R_L = R_l L/2(d + t), \tag{7.32}$$

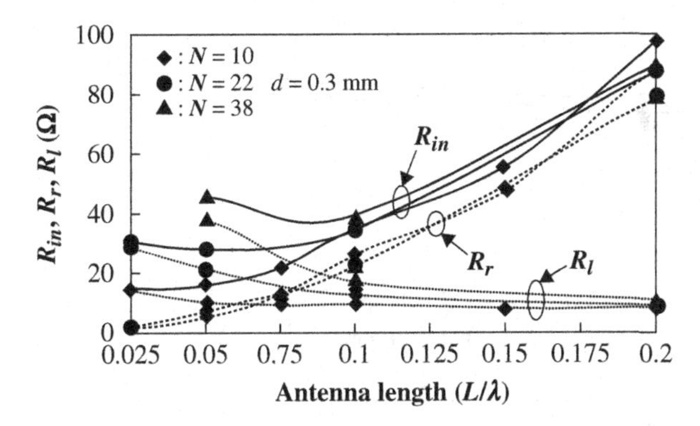


**Figure 7.33** Resistance components of a folded MLA ([13], copyright ©2004 IEICE).

where $R_l$ is the loss resistance of a conductor with skin depth $\delta$ and conductivity $\sigma$, and is given by

$$R_l = \sqrt{\delta\sigma}. \tag{7.33}$$

Here $\delta$ equals $\sqrt{f\pi\mu/\sigma}$.

Then, $R_L$ is expressed by

$$R_L = L_a\sqrt{f\pi\mu}/2(d+t)\sqrt{\sigma}. \tag{7.34}$$

Here, $L_a$ is the line length (when extended). The resistance components of a folded meander line antenna with respect to the antenna length $L/\lambda$ are shown in Figure 7.33, in which the radiation resistance $R_r$, the loss resistance $R_L$, and input resistance $R_{in}$, are depicted for cases $N = 10$, 22, and 38, and $d = 0.3$ mm. $R_L$ is calculated simply from

$$R_L = R_{in} - R_{in}(\sigma = \infty). \tag{7.35}$$

It is shown in the figure that, as the antenna length $L$ becomes shorter, the loss resistance $R_L$ increases, suggesting decrease in the radiation efficiency. It is also observed from the figure by comparing with the resistance components of a dipole type shown before in Figure 7.28 that a folded type exhibits greater resistance components than the dipole type. $R_{res}$ is about four times greater than that of the dipole type, showing the stepping-up effect due to the folded structure. The loss resistance $R_L$ is, however, not four times, but only about two times, and yet increases when the antenna length becomes shorter. This is because of $R_L$ dependence on the line length $L_a$, which increases with increase in $N$. $R_{in}$, on the contrary, decreases rapidly as the antenna length $L$ becomes shorter.

The radiation efficiency $\eta$ is calculated for both dipole-type and folded-type antennas with respect to the antenna length $L/\lambda$, and is depicted in Figure 7.34 for $N = 10$, 22, and 38. With greater $N$, the radiation efficiency $\eta$ rapidly decreases as $L$ becomes smaller, particularly in the dipole type, whereas less variation appears in the folded type.

Trial antennas of length 0.1 $\lambda$ of both dipole-type and folded-type MLAs are manufactured. The parameters of the folded-type MLA are: $N = 10$, $L = 43$ mm $(0.1\lambda)$,

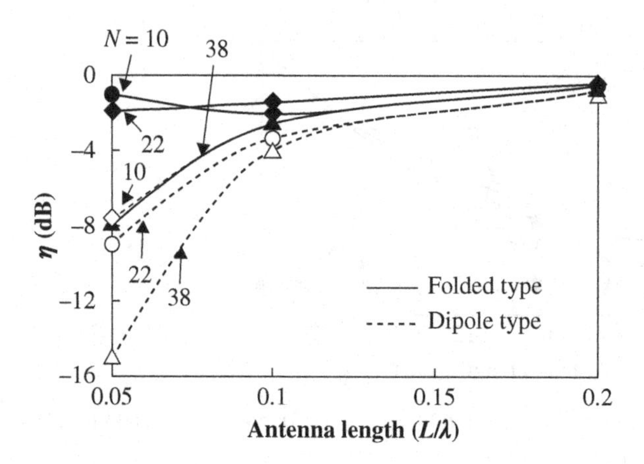


**Figure 7.34** Radiation efficiency of a folded MLA ([14], (copyright ©2005 IEEE).

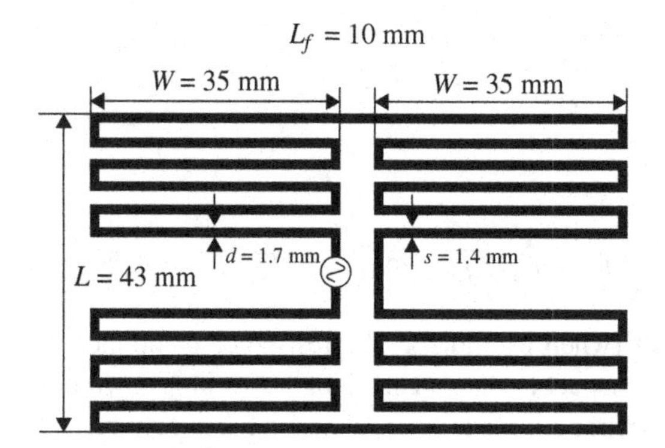


**Figure 7.35** A trial model of 0.1λ folded MLA ([13], copyright ©2004 IEICE).

$W = 35$ mm (0.08λ), $L_f = 10$ mm (0.02λ), $d = 1.7$ mm ($4.0 \times 10^{-3}$ λ), $s = 1.4$ mm ($3.3 \times 10^{-3}$ λ), and the thickness of the element made with thin copper plate $t = 0.1$ mm ($2.3 \times 10^{-4}$ λ) (Figure 7.35). Measured impedance characteristics of both dipole and folded types are shown in Figure 7.36. The folded type is shown to have higher impedance than that of the dipole type as was mentioned before. Radiation patterns of both dipole and folded types are given in Figure 7.37. Gain is evaluated as –5.99 dBd for the dipole type and –2.28 dBd for the folded type, respectively. The gain obtained by using the IE3D simulator was –5.2 dBd for the dipole type and –1.5 dBd for the folded type, showing higher gain of about 3.7 dB in the folded type [13].

### *7.2.1.1.1.1.6 MLAs of length* L = *0.05 λ [13, 14]*

Smaller antennas with length 0.05λ were studied and shown to be practically useful, with appreciably high gain of –12 dBd for the dipole type and –10 dBd for the folded type, respectively, even in this small size [14]. A type of MLA, which is sandwiched by two planar dielectric substrates of high dielectric constant $\varepsilon_r$ (Figure 7.38), was developed, and the antenna performances were studied. Figure 7.39 provides dimensional

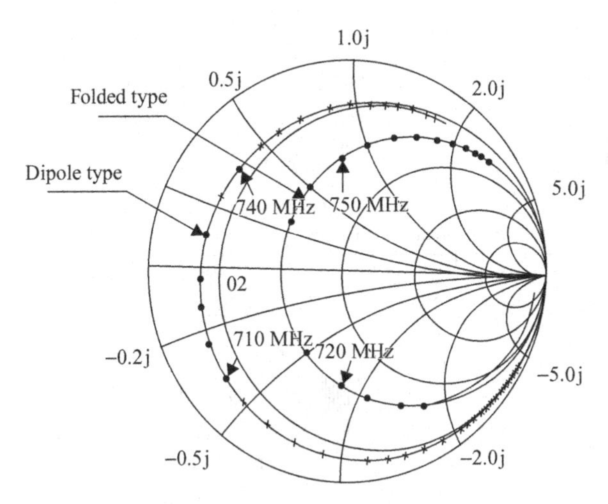


**Figure 7.36** Measured impedances of 0.1λ dipole-type and folded-type MLAs ([13], copyright ©2004 IEICE).

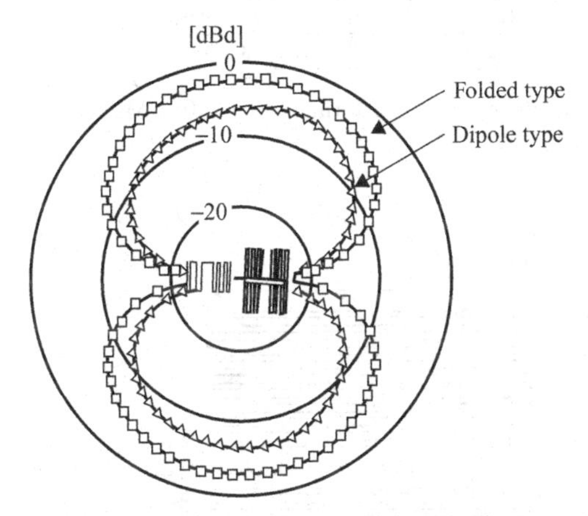


**Figure 7.37** Radiation patterns of 0.1λ dipole-type and folded-type MLAs ([13], copyright ©2004 IEICE).

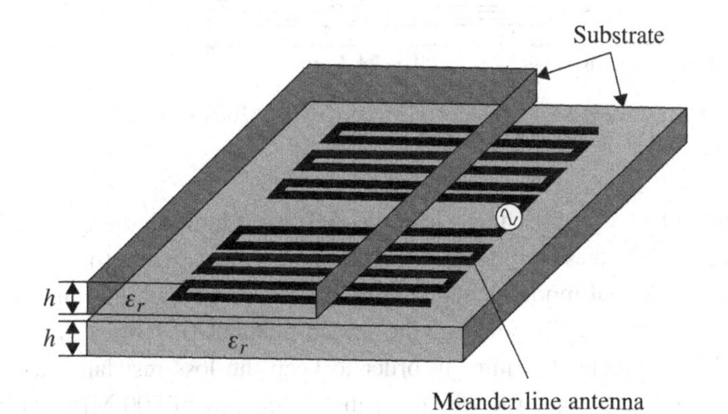


**Figure 7.38** A trial 0.05λ MLA model sandwiched by two high-$\varepsilon_r$ dielectric plates ([14], (copyright ©2005 IEEE).

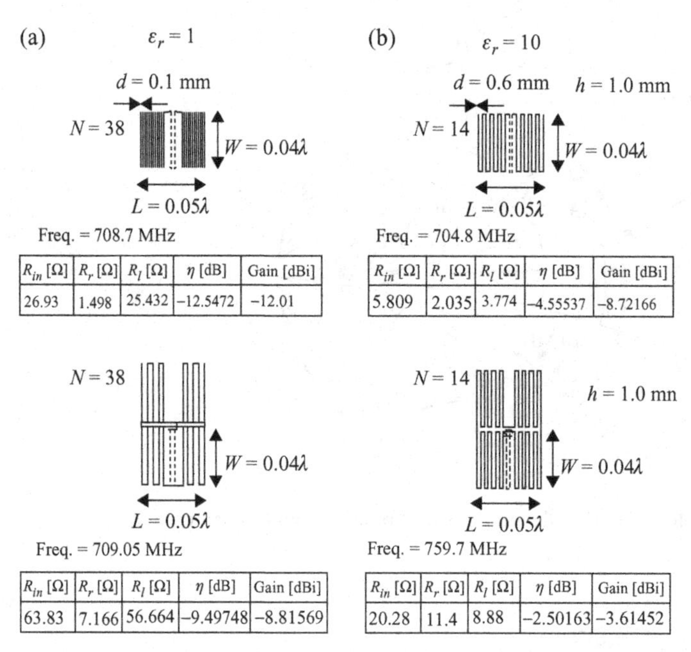


(a), Freq. = 708.7 MHz

| $R_{in}$ [Ω] | $R_r$ [Ω] | $R_l$ [Ω] | $\eta$ [dB] | Gain [dBi] |
|---|---|---|---|---|
| 26.93 | 1.498 | 25.432 | -12.5472 | -12.01 |

(b), Freq. = 704.8 MHz

| $R_{in}$ [Ω] | $R_r$ [Ω] | $R_l$ [Ω] | $\eta$ [dB] | Gain [dBi] |
|---|---|---|---|---|
| 5.809 | 2.035 | 3.774 | -4.55537 | -8.72166 |

(a), Freq. = 709.05 MHz

| $R_{in}$ [Ω] | $R_r$ [Ω] | $R_l$ [Ω] | $\eta$ [dB] | Gain [dBi] |
|---|---|---|---|---|
| 63.83 | 7.166 | 56.664 | -9.49748 | -8.81569 |

(b), Freq. = 759.7 MHz

| $R_{in}$ [Ω] | $R_r$ [Ω] | $R_l$ [Ω] | $\eta$ [dB] | Gain [dBi] |
|---|---|---|---|---|
| 20.28 | 11.4 | 8.88 | -2.50163 | -3.61452 |

**Figure 7.39** Dimensional parameters and radiation performances of 0.05λ MLAs for cases $\varepsilon_r = 1$ and 10 ([14], copyright ©2005 IEEE).

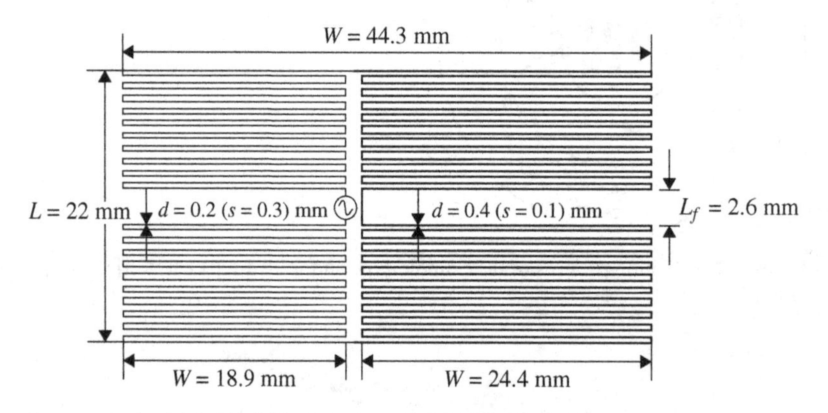


**Figure 7.40** A trial 0.05λ folded MLA model ([13], copyright ©2004 IEICE).

parameters and antenna performances of 0.05λ MLAs of both dipole type and folded type. In the figure, (a) gives results when the substrate $\varepsilon_r = 1$, while (b) represents the substrate $\varepsilon_r = 10$. A trial model is depicted in Figure 7.40, which has an asymmetric structure, with the element width $d$ of the driven element (0.2 mm) different from that of the folded element (0.4 mm) in order to keep the loss resistance as small as possible. Antennas are designed for the resonance frequency of 700 MHz for both the

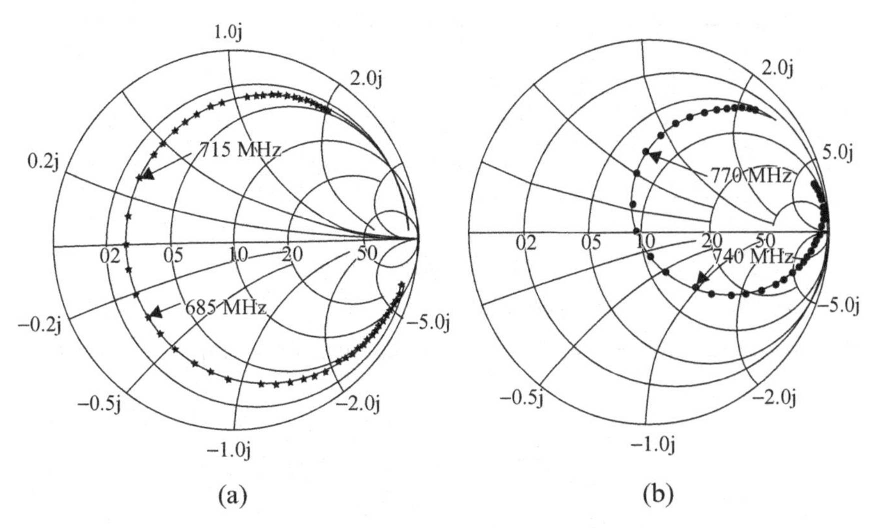


**Figure 7.41** Impedance characteristics of 0.05λ dipole-type and folded-type MLA ([13], copyright ©2004 IEICE).

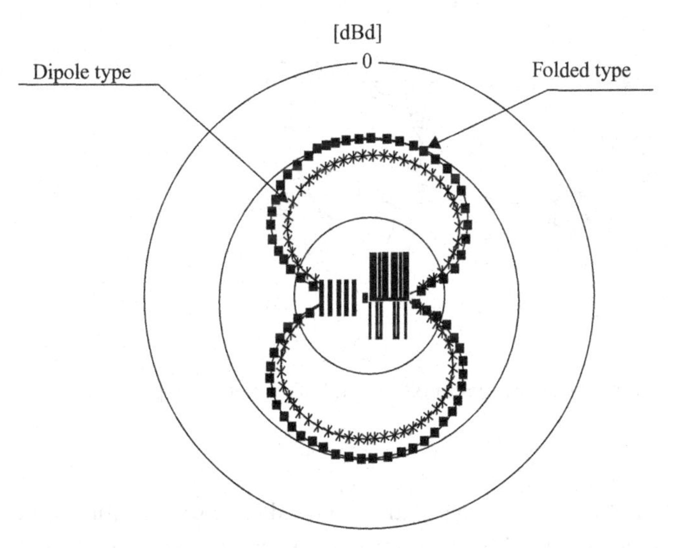


**Figure 7.42** Radiation patterns of 0.05λ dipole-type and folded-type MLA ([13], copyright ©2004 IEICE).

driven element and the folded element. Figure 7.41 illustrates impedance characteristics; (a) for the dipole type and (b) for the folded type. Figure 7.42 shows radiation patterns of both dipole type and folded type. These figures give evidence of advantageous performances of the folded type against the dipole type. Figure 7.43 depicts VSWR performance, showing the bandwidth of about 12 MHz for VSWR = 2, which

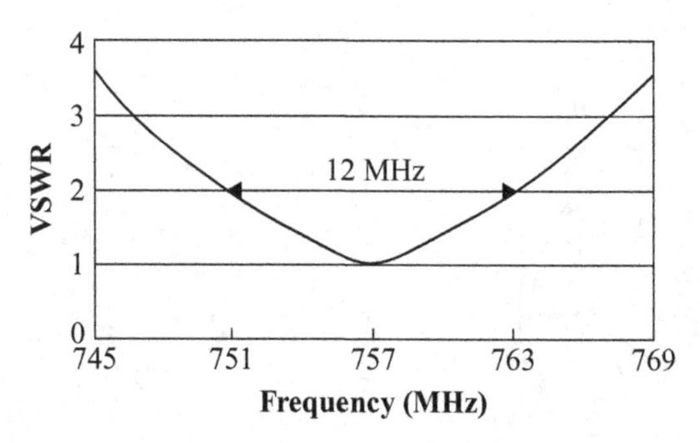


**Figure 7.43** VSWR performance of 0.05λ folded-type MLA ([13], copyright ©2004 IEICE).

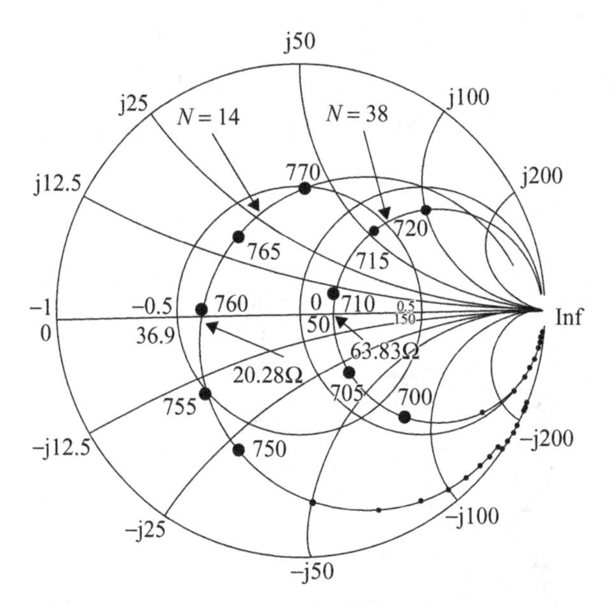


**Figure 7.44** Impedance characteristics of 0.05λ MLA for different $N$ ([14], copyright ©2005 IEEE).

corresponds to 1.6% in terms of the relative bandwidth. Figure 7.44 illustrates impedance characteristics for cases of $N = 14$ and 38, corresponding to $\varepsilon_r = 10$ and 1, respectively. Figure 7.45 shows radiation patterns, also for cases $N = 14$ and 38. Increase in gain for the case $N = 14$ was observed.

An example of a practically developed small MLA introduced previously (Figure 7.6) [5] is a monopole type etched on both sides of a dielectric substrate (glass epoxy) fed with coplanar waveguide. The antenna element occupies the substrate surface with an area of 61 mm × 61 mm and the meander line of 0.5 mm width is spaced periodically with 0.5 mm. The antenna pattern is designed to receive vertical polarization by the front side and horizontal polarization by the back side by arranging the meander line

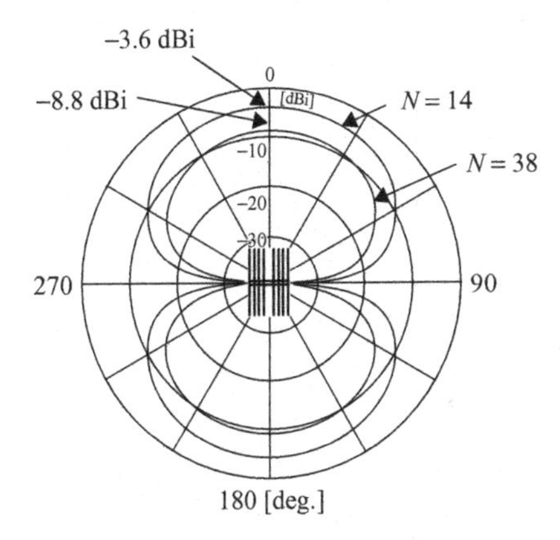


**Figure 7.45** Radiation patterns of 0.05λ MLA for different $N$ ([14], copyright ©2005 IEEE).

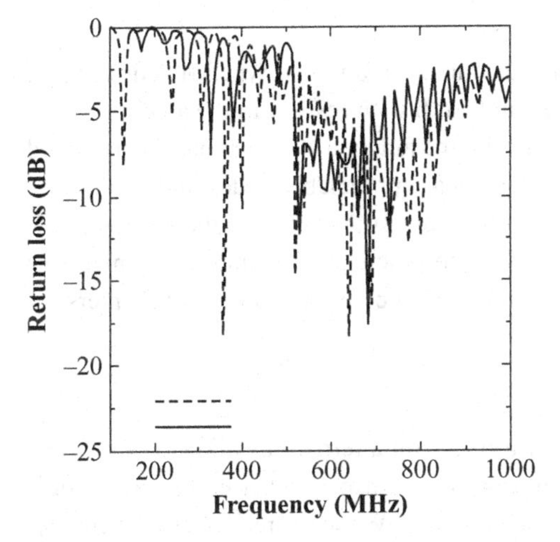


**Figure 7.46** Return-loss characteristics of a MLA shown in Figure 7.6 ([5], copyright ©2006 IEICE).

pattern parallel to the ground plane on the front side, while vertical on the back side. The return-loss characteristics, gain, and calculated radiation patterns, respectively, at 600 MHz are given in Figures 7.46–48. In Figure 7.48, (a) and (b) illustrate radiation patterns on the azimuth plane ($y$–$z$ plane) and the elevation plane ($x$–$z$ plane), respectively, for the antenna on the front side, and (c) and (d) show those on the azimuth plane and the elevation plane, respectively, for the antenna on the back side. The gain is evaluated to be greater than –3 dBd in the frequency range between 550 MHz and 800 MHz, and the return-loss characteristic is less than –5 dBd for the frequency range of 500 MHz to 700 MHz. This antenna is designed to receive UHF band terrestrial TV broadcasting.

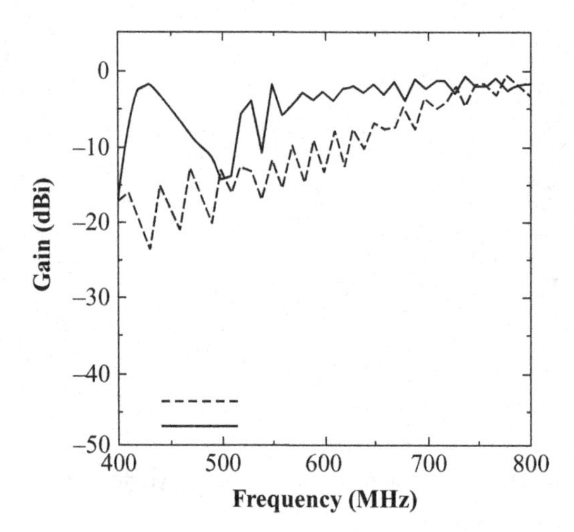


**Figure 7.47** Gain characteristics of a MLA shown in Figure 7.6 ([5], copyright ©2006 IEICE).

Figure 7.49 illustrates another example; a monopole-type MLA driven by a small loop. Figure 7.50 shows the relative bandwidth with respect to the antenna size $kr$. Here, $r$ is the radius of a sphere circumscribing the antenna structure and the bandwidth is shown for cases $1/Q$ and $2/Q$. This antenna has appreciably wide bandwidth; for instance, the relative bandwidth is evaluated to be about 10%, although the size $kr = 0.5$.

There are many other MLAs applied practically to small equipment such as mobile phones. Discussions on these antennas will be provided in later chapters.

### *7.2.1.1.1.2 Zigzag antennas*

A zigzag antenna composed of a line with alternating salient and re-entrant angles is depicted in Figure 7.51. The zigzag line is formed either uniformly or log-periodically. Since the zigzag line carries a slow wave (SW) with smaller phase velocity than that of free space, the antenna size can be made small. Study of zigzag antennas should go back to a paper published in 1960 [15], in which operation of typical logarithmically periodic zigzag antennas based on near-field measurement had been discussed. Later, other papers treating balanced backfire zigzag antennas [16], theory [17], input impedance [1], and analysis [18], which facilitated antenna design, were published. In [18] two types of zigzag antenna are described; one is a uniform balanced zigzag antenna over a ground plane, and another is a balanced log-periodic zigzag antenna as Figure 7.52 shows. Input impedance of a zigzag dipole antenna is depicted in Figure 7.53, where input impedance of a linear dipole antenna is shown as a comparison. Current distributions on the antenna element are calculated by solving integral equations for antennas, and the numerical results are illustrated in Figure 7.54, where (a) depicts that of the uniform zigzag antenna and (b) the log-periodic zigzag antenna. A $k$–$\beta$ diagram is shown in Figure 7.55, where the wire diameter $d$ of the antenna is $0.008\lambda$ and two cases in tooth angle, 20 degrees and 30 degrees, are shown.

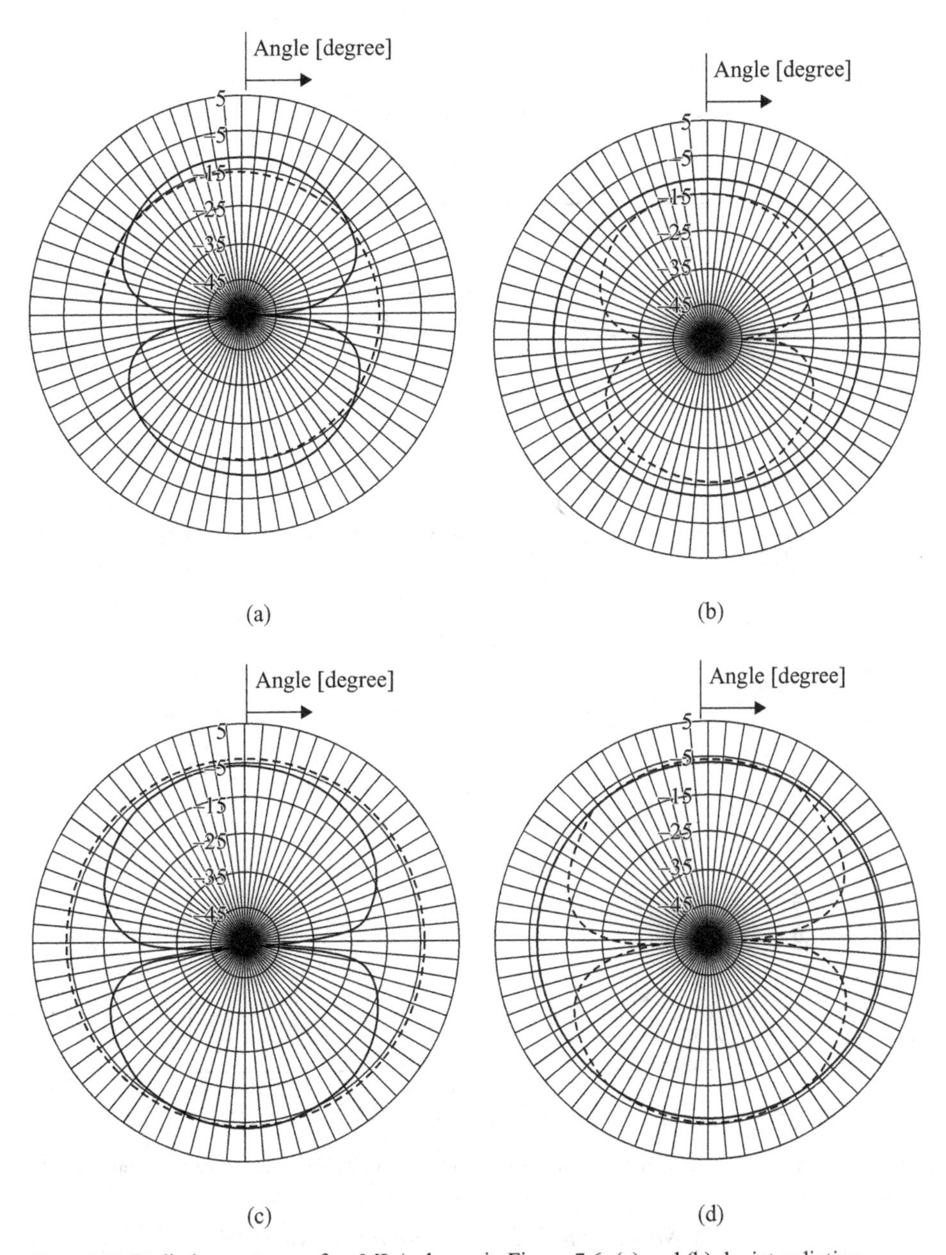


**Figure 7.48** Radiation patterns of an MLA shown in Figure 7.6: (a) and (b) depict radiation patterns on the azimuth plane and the elevation plane, respectively, for antenna on the front side, and (c) and (d) show those on the azimuth plane and the elevation plane, respectively, for antenna on the back side ([5], copyright ©2006 IEICE).

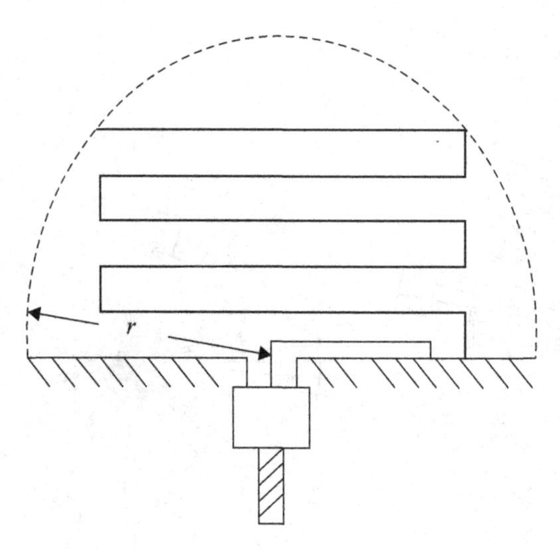


**Figure 7.49** A meander line monopole antenna driven by a small loop ([6], copyright ©2003 IEEE).

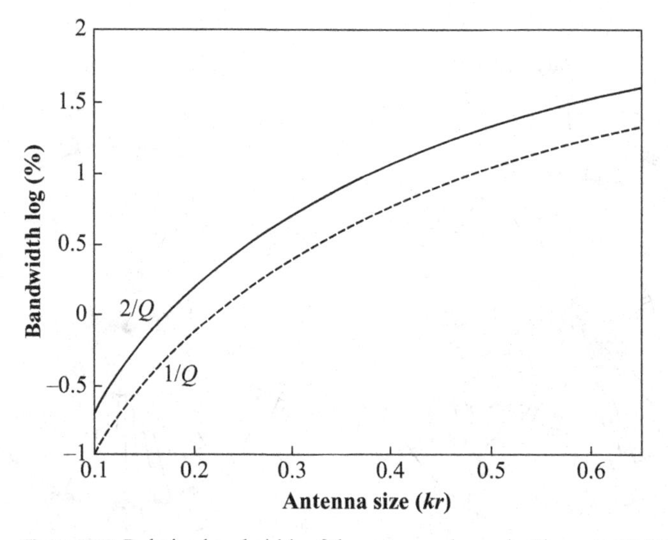


**Figure 7.50** Relative bandwidth of the antenna shown in Figure 7.49 ([6], copyright ©2003 IEEE).

*7.2.1.1.1.3 Normal mode helical antennas (NMHA)*

The NMHA (Figure 7.56) has been used practically in various small wireless terminals like mobile phones, as it can be produced in small size, and yet with higher radiation efficiency than other types of antennas having the same size. Radiation pattern is essentially the same as that of a linear dipole, and the most significant feature of the NMHA may be in the antenna structure, with which self-resonance can easily be attained, even though the size is small. Self-resonance is favorable for obtaining higher radiation efficiency, because a reactance component, which typically gives rise to a matching-network loss,

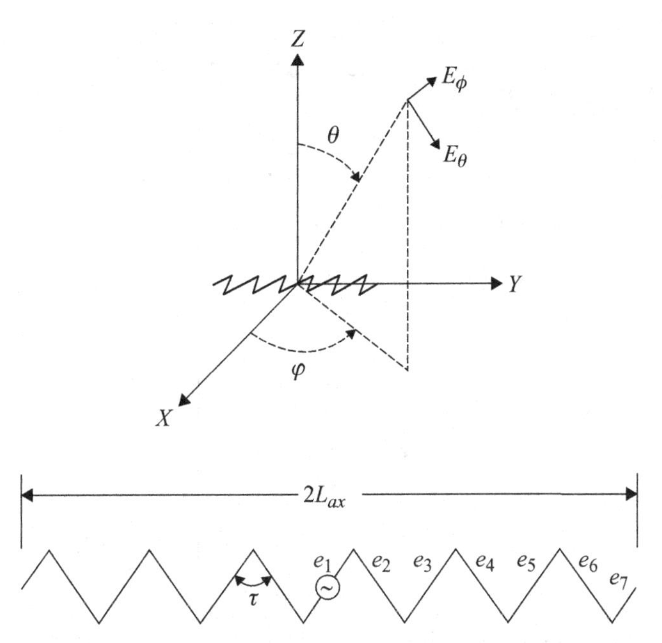


**Figure 7.51** Zigzag antenna model with the coordinate systems ([1], copyright ©1984 IEEE).

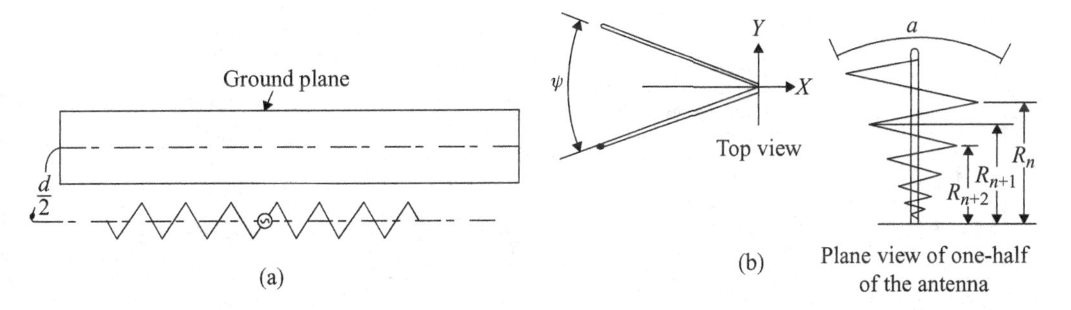


**Figure 7.52** Two types of zigzag antenna model ([18], copyright ©1970 IEEE).

is not necessary. It is a common understanding that the smaller the antenna becomes, the lower the antenna efficiency tends to be, but self-resonance in small antennas may resolve this problem.

Analysis of the NMHA by using a simulator is not simple, because its helical structure is constituted with very small helix radius and pitch values compared with the wavelength, and even though meshes as small as hundredths of a wavelength may be used to describe the antenna model, still accurate results are difficult to obtain. A dipole-type NMHA has been analyzed by using Method of Moments, in which unique techniques to solve the integral equations were contrived [19], and numerical data obtained by calculation are provided. Later, the design data based on the analysis described in [19] were

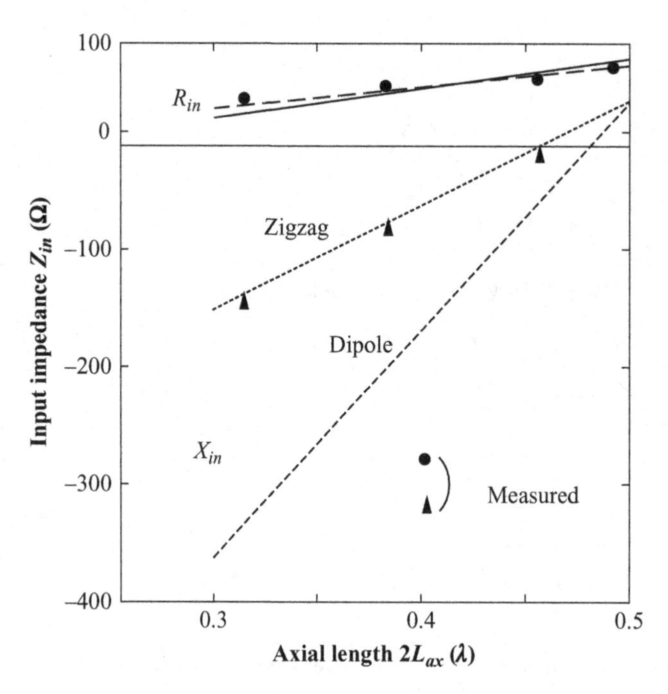


**Figure 7.53** Impedance characteristics of a zigzag dipole antenna ([1], copyright ©1984 IEEE).

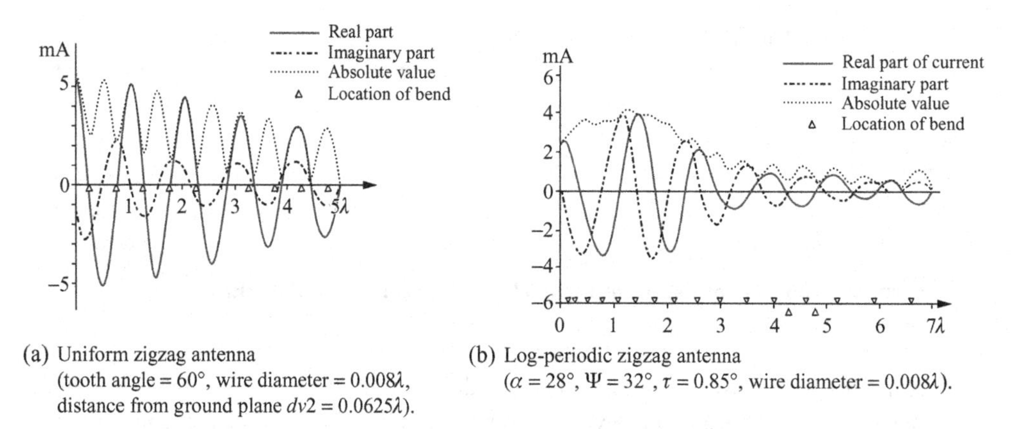


(a) Uniform zigzag antenna (tooth angle = 60°, wire diameter = 0.008λ, distance from ground plane $dv2$ = 0.0625λ).

(b) Log-periodic zigzag antenna ($\alpha$ = 28°, $\Psi$ = 32°, $\tau$ = 0.85°, wire diameter = 0.008λ).

**Figure 7.54** Current distributions on (a) the uniform zigzag antenna and (b) the log-periodic zigzag antenna ([18], copyright ©1970 IEEE).

provided in a book [20], where some practical examples, including NMHAs modified to improve bandwidth and gain and so forth, were described. A further study about the analysis of NMHA has been given in [21], in which the method of analysis shown in [19] has been extended to obtain more accurate numerical results efficiently by using improved expansion function and weighting function.

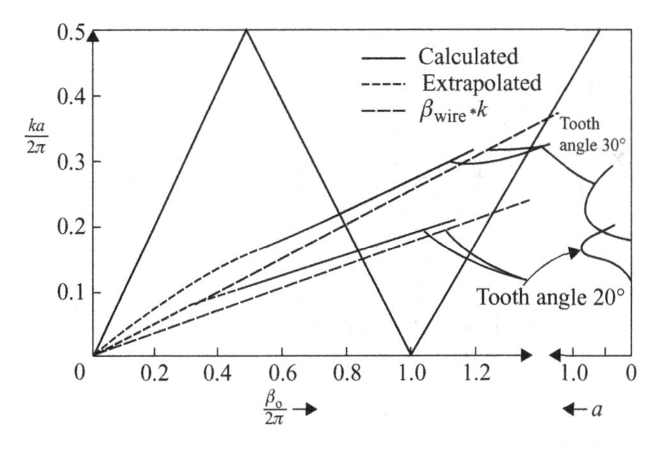


**Figure 7.55** $k$–$\beta$ diagram of a zigzag antenna ([18], copyright ©1970 IEEE).

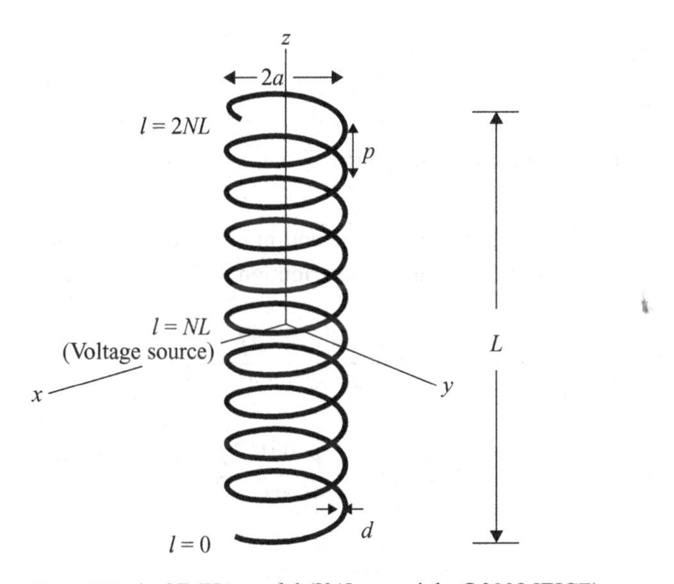


**Figure 7.56** An NMHA model ([21], copyright ©2008 IEICE).

Study on NMHA has progressed as requirements for small antennas such as the NMHA have increased. Various types of NMHAs such as dielectric-loaded NMHA [21], in which efficient numerical calculation is demonstrated, log-periodic NMHA [22], folded NMHA [23], and two-wire NMHA [24], have so far been used.

Here analysis and design of the NMHA will be briefly introduced by quoting the analysis described in [19–21], and providing design data based on the analysis. The antenna model used for the analysis is shown in Figure 7.57 with the coordinate systems and antenna parameters such as helix radius $a$, wire diameter $d$, pitch of helical winding

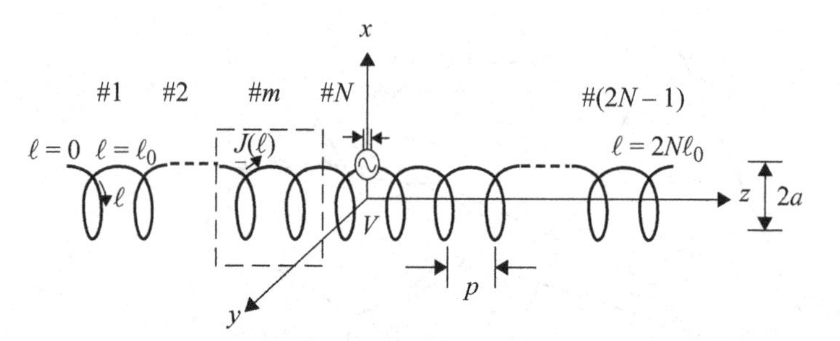


**Figure 7.57** NMHA model used for the analysis with the coordinate systems [19].

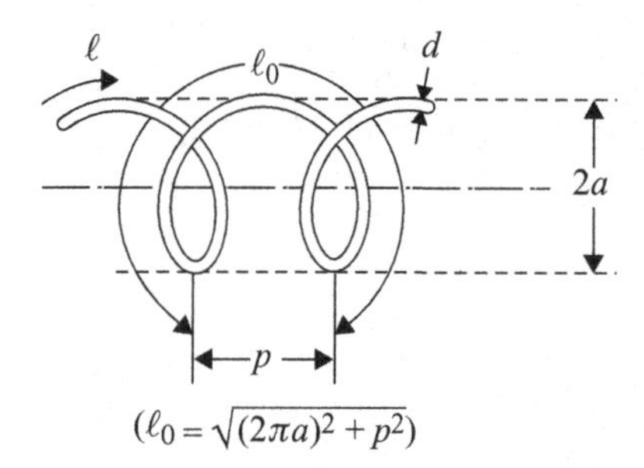


**Figure 7.58** A segment consisting of two whole turns of helical structure [19].

$p$, helix one-turn length $\ell_0 = \sqrt{(2a\pi)^2 + p^2}$, and number of helix windings $N$. The integral equation with respect to the current distribution $J(\ell)$ on the wire is

$$\frac{\mathrm{j}}{\omega\varepsilon_0}\int_0^{2NL}\left[k_0^2 J(\ell')G(\ell,\ell') + \frac{dJ(\ell')}{d\ell'}\cdot\frac{\partial G(\ell,\ell')}{\partial\ell'}\right]d\ell' = E^i(\ell) \tag{7.36}$$

$$\text{where } G(\ell,\ell') = \frac{e^{-\mathrm{j}k_0 r(\ell,\ell')}}{4\pi r(\ell,\ell')} \tag{7.37}$$

$$E^i(\ell) = -\frac{V}{\delta}rect\left(\frac{\ell - NL}{\delta}\right). \tag{7.38}$$

Here $V$ is the feed voltage, $\delta$ is a gap at the feed terminal, $k = \omega\sqrt{\mu_0\varepsilon_0} = 2\pi/\lambda$ stands for phase constant in free space, and $r$ is distance between two points on $\ell$, and $\ell'$. In order to transform the derivative of Green's function to that of current functions and to ease solving the integral equation, the Galerkin method will be adopted along with a weighting function, which is zero at the end of the segment. The antenna model is divided into $(2N-1)$ segments. A segment consists of two whole turns of helical structure taken at an arbitrary place on the antenna, which has length of $2\ell$ as shown in Figure 7.58. Then, $\ell$ at the $m$th segment is described as $(m-1)L \leq \ell \leq (m+1)L$, where $m$ takes the values 1, 2, …, $(2N-1)$, and the current $J_m(\ell)$ on the $m$th segment is expressed by $I_m$ times $f(\ell)$, which is a function having the maximum value unity at $\ell = mL$, and zero

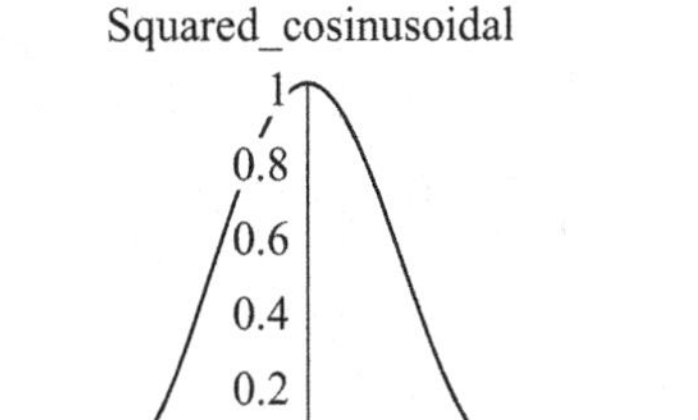


**Figure 7.59** A squared cosine function ([21], copyright ©2008 IEICE).

at $\ell = (m \pm 1)L$. Here, $I_m$ is an unknown coefficient to be determined. Then the current on the antenna $J(\ell)$ is given by

$$J(\ell) = \sum_{m=1}^{2N-1} I_m f(\ell - mL). \tag{7.39}$$

$I_m$ can be determined by using the circuit equation

$$\sum_{m=1}^{2N-1} Z_{n,m} I_m = V_n \tag{7.40}$$

$$Z_{n,m} = \mathrm{j}\frac{\eta_0}{4\pi} \int_{-k_0 L}^{k_0 L} \int_{-k_0 L}^{k_0 L} \frac{e^{-\mathrm{j}k_0 r_{n,m}}}{k_0 r_{n,m}} \times \left[ f(\ell) f(\ell') \vec{a}(\ell) \cdot \vec{a}(\ell') - \frac{1}{k_0^2} f'(\ell) f'(\ell') \right] d(k_0\ell) d(k_0\ell') \tag{7.41}$$

$$V_n = \int_{-\delta/2}^{\delta/2} f(\ell) \frac{V}{\delta} rect\left(\frac{l - NL}{\delta}\right) d\ell. \tag{7.42}$$

Now a function $f(\ell)$, a squared cosine function (Figure 7.59) is selected to give a good approximation as follows

$$f(\ell) = \cos^2(\pi\ell/2L). \tag{7.43}$$

Through some tedious manipulations, constitutional parameters of the NMHA at resonance are analyzed [21]. In the analysis, antenna parameters of relative helix radius $a/\lambda$ and relative pitch $p/\lambda$ with respect to the wavelength are taken into consideration and divided into two groups A and B depending on the size. In group A, both $a/\lambda$ and relative pitch $p/\lambda$ are relatively larger, lying between $10^{-3} \sim 10^{-4}$, and in group B, they are relatively smaller with $a/\lambda$ between $5 \times 10^{-4} \sim 10^{-3}$, and $p/\lambda$ between $10^{-4} \sim 5 \times 10^{-4}$. Number of turns $N$ at the resonance condition is expressed by illustrating the relationship between $a/\lambda$ and $p/\lambda$ in Figure 7.60 for the group A and in Figure 7.61 for the group B, respectively.

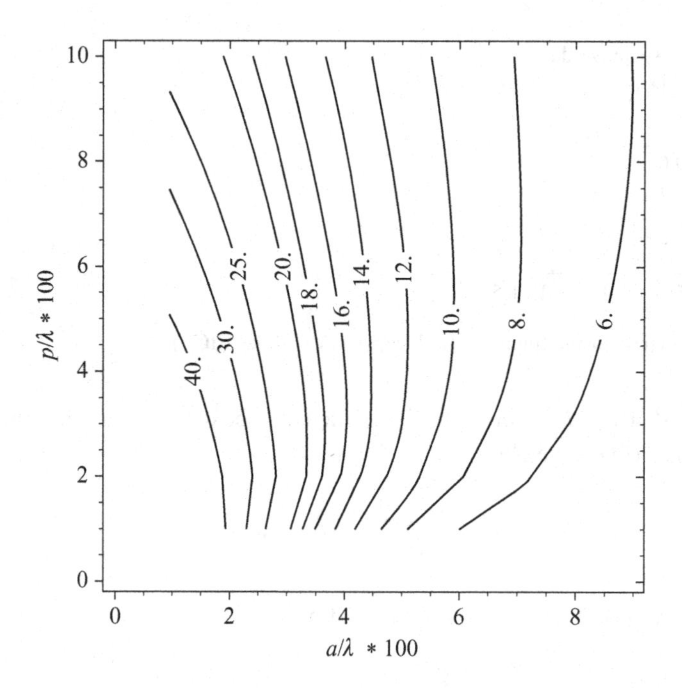


**Figure 7.60** Number of turns $N$ at resonance in relation to the helix radius $a/\lambda$ and the pitch $p/\lambda$ for the group A ([21], copyright ©2008 IEICE).

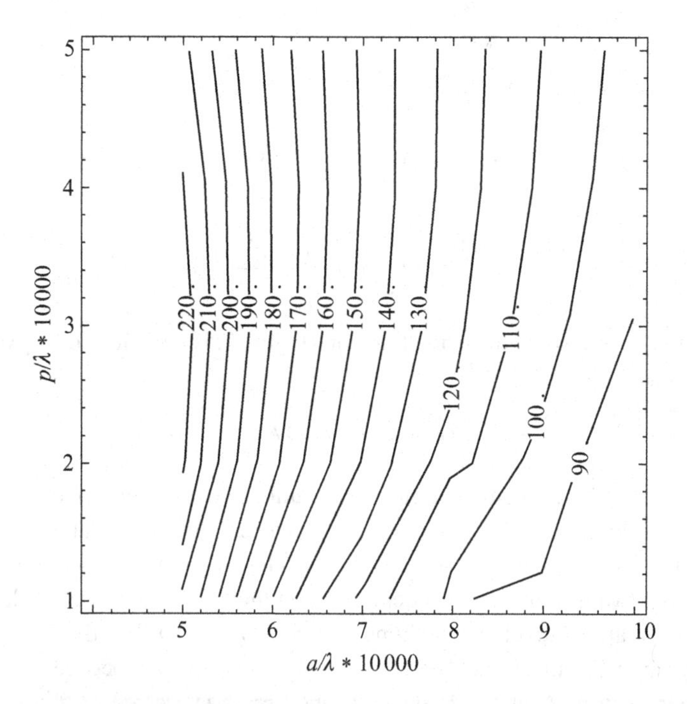


**Figure 7.61** Number of turns $N$ at resonance in relation to the helix radius $a/\lambda$ and the pitch $p/\lambda$ for the group B ([21], copyright ©2008 IEICE).

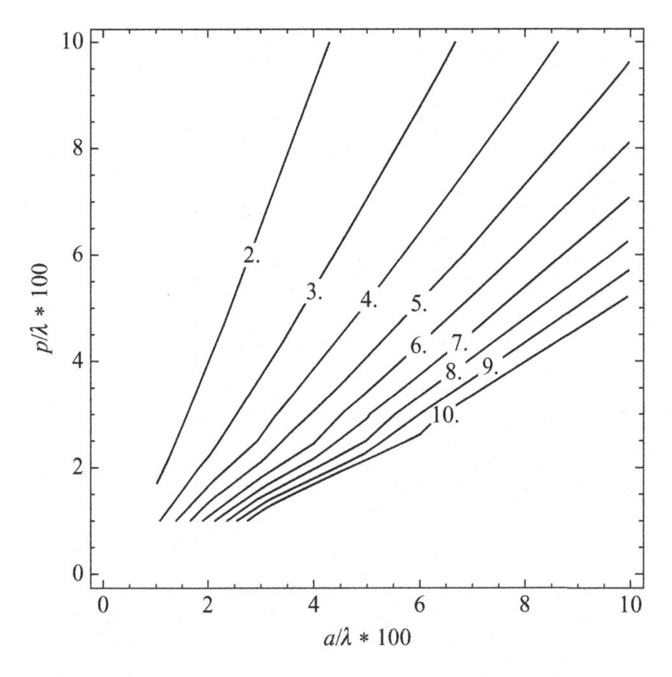


**Figure 7.62** Shortening factor $S_1$ of the antenna length $L$ at resonance in relation to the helix radius $a/\lambda$ and the pitch $p/\lambda$ for the group A ([21], copyright ©2008 IEICE).

In order to evaluate small-sizing, the shortening factor $S_1$ of the antenna length is introduced by taking the ratio of the NMHA length and a quarter wavelength of a linear antenna, which is

$$S_1 = (\lambda/4)/(Nrp). \tag{7.44}$$

Also the shortening factor $S_2$ of the wire length $L$ (extended length of helix), which is a ratio of $L$ and a quarter wavelength of a linear antenna, is introduced as

$$S_2 = (\lambda/4)/(NrL). \tag{7.45}$$

These $S_1$ and $S_2$, respectively, are illustrated by showing relationships between $a/\lambda$ and $p/\lambda$ in Figure 7.62 and Figure 7.64 for the group A, and Figure 7.63 and Figure 7.65 for the group B.

It is a general understanding that radiation resistance $R_{rad}$ becomes smaller when antenna resonance length is made shorter. The contour of this $R_{rad}$ is expressed by showing relationships between $a/\lambda$ and $p/\lambda$ in Figure 7.66 for the group A, and in Figure 7.67 for the group B.

These illustrations can be used for designing an NMHA.

#### *7.2.1.1.1.4 Discussions on small NMHA and meander line antennas pertaining to the antenna performances*

Antenna performances of small NMHA and meander line antennas are numerically obtained by using simulators NEC 4.1 of EZNEC Pro over a frequency range of 50 MHz

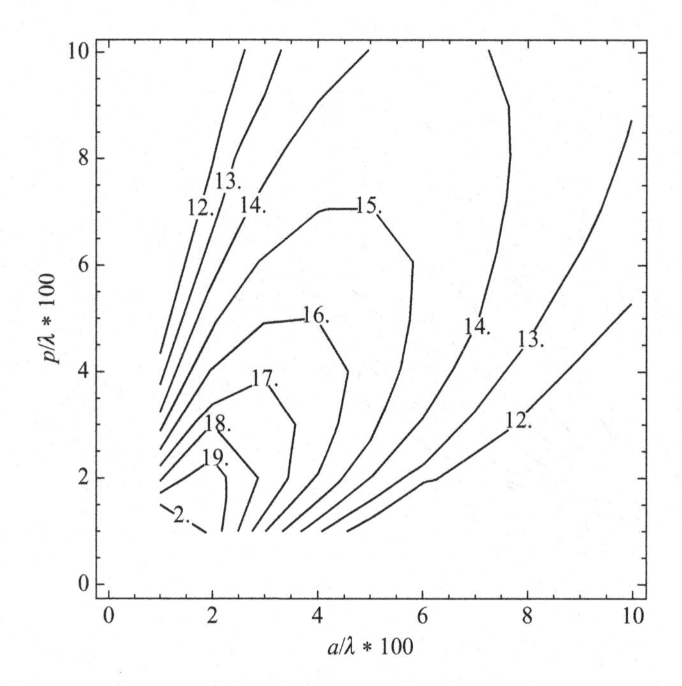


**Figure 7.63** Shortening factor $S_1$ of the antenna length $L$ at resonance in relation to the helix radius $a/\lambda$ and the pitch $p/\lambda$ for the group B ([21], copyright ©2008 IEICE).

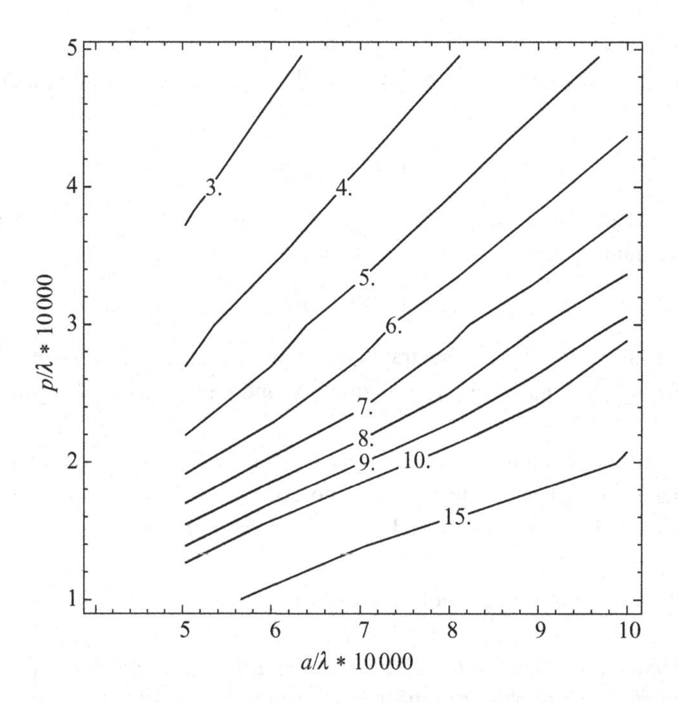


**Figure 7.64** Shortening factor $S_2$ of the antenna wire length $L_a$ (extended length) at resonance in relation to the helix radius $a/\lambda$ and the pitch $p/\lambda$ for the group A ([21], copyright ©2008 IEICE).

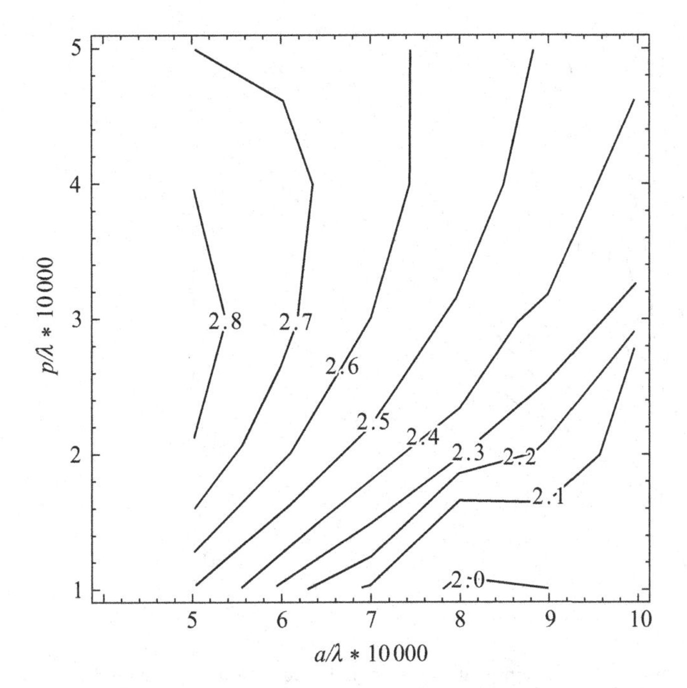


**Figure 7.65** Shortening factor $S_2$ of the antenna wire length $L_a$ (extended length) at resonance in relation to the helix radius $a/\lambda$ and the pitch $p/\lambda$ for the group B ([21], copyright ©2008 IEICE).

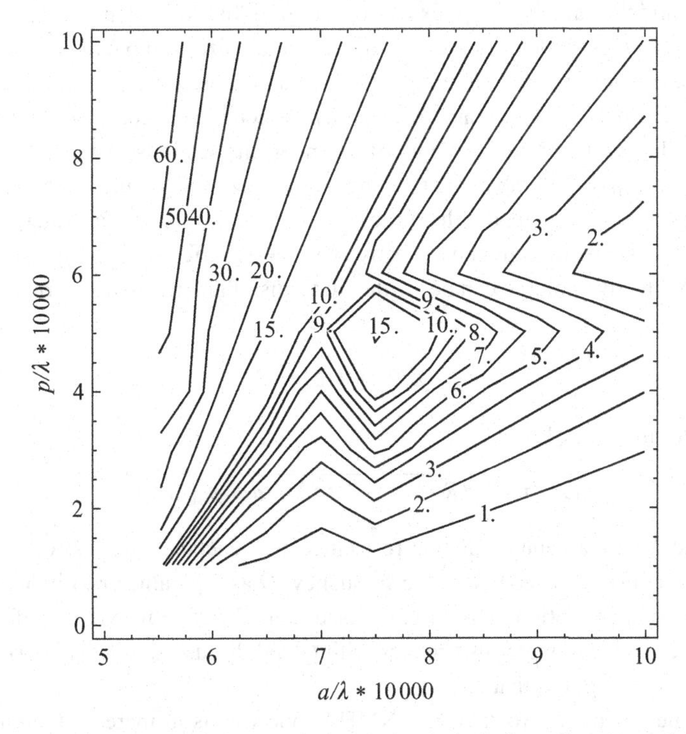


**Figure 7.66** Radiation resistance in relation to the helix radius $a/\lambda$ and the pitch $p/\lambda$ for the group A ([21], copyright ©2008 IEICE).

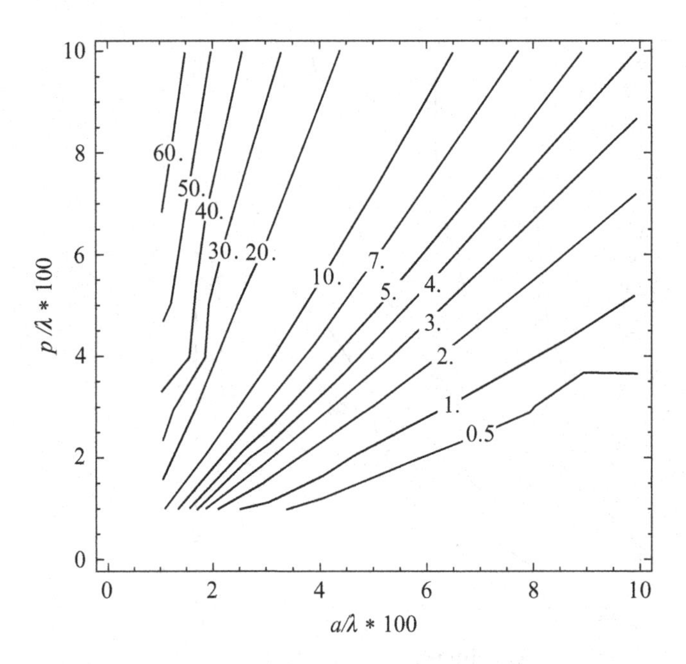


**Figure 7.67** Radiation resistance in relation to the helix radius $a/\lambda$ and the pitch $p/\lambda$ for the group B ([21], copyright ©2008 IEICE).

to 500 MHz, and they are discussed extensively in [26]. In [26], antenna length $kh = 0.5$ is taken according to the definition of electrically small size, and parameters to achieve self-resonance, which is desired in designing small antennas, are discussed. The antenna dimensional parameters used are antenna length $h$, which is 6 cm corresponding to $0.08\lambda$ at 400 MHz, diameter of NMHA, width of meander line sections, number $N$ of helical windings for a NMHA, number $N$ of segments for a Meander Line Antenna (MLA), the total length of wire (extended length of wire), wire radius, and antenna geometry. The parameters also include antenna effective volume $V_e$, effective height, and spherical radius $r$. The effective volume can be written in terms of an effective spherical radius $r$, which is given by [27]

$$r = (\lambda/2\pi)(9/2Q)^{1/3}. \tag{7.46}$$

Here $Q$ is determined by [28]

$$Q = (\omega/2\pi)\sqrt{(dR_A/d\omega)^2 + (dX_A/d\omega + \mid X_A \mid/\omega)^3} \tag{7.47}$$

where $R_A$ and $X_A$ are antenna radiation resistance and reactance, respectively. Antenna performances considered are resonance frequency, $Q$ as a parameter to indicate bandwidth, and resonant radiation resistance. In discussion, the effective volume of spherical radius $r$ is used as a parameter that relates with $Q$ and discussed by using normalized $r$ with the effective height $h$; that is, $r/h$.

To lower the resonance frequency in NMHA, one way is to increase the number of windings $N$ of the helix as shown in Figure 7.68, which depicts three models: NMHA

**Table 7.1** The resonant properties of the three NMHA models ([26], Copyright ©2004 IEEE)

| Antenna | Resonant frequency (MHz) | Radiation resistance (Ohms) | $Q$ |
|---|---|---|---|
| NMH1 | 408.2 | 5.1 | 66 |
| NMH2 | 331.4 | 3.5 | 115 |
| NMH3 | 278.4 | 2.6 | 184 |

**Table 7.2** The physical properties of the three NMHA models ([26], Copyright ©2004 IEEE)

| Antenna | Height (cm) | Total wire length (cm) | No. of turns |
|---|---|---|---|
| NMH1 | 6 | 29.25 | 9 |
| NMH2 | 6 | 38.42 | 12 |
| NMH3 | 6 | 47.65 | 15 |

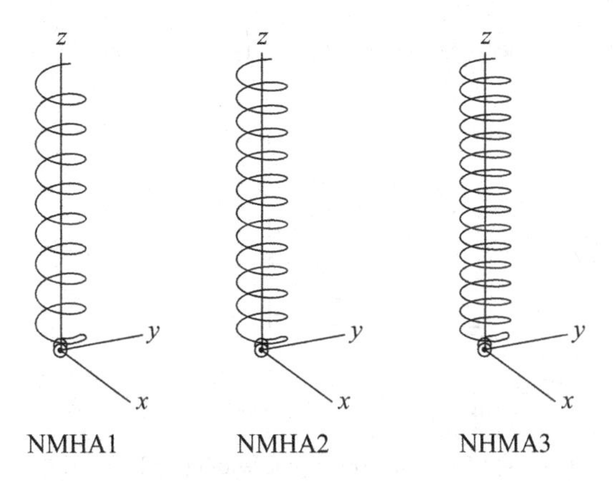


**Figure 7.68** Three models of NMHA with different number of winding $N$ ([26], copyright ©2004 IEEE).

1 to NMHA 3. Here the wire diameter is 0.5 mm. The total wire length also increases and the resonance frequency is lowered from 408.2 MHz to 278 MHz as Table 7.1 and Table 7.2 describe. Radiation resistance and $Q$ also decrease.

In the case of MLA, one way of lowering the resonance frequency is the same as in NMHA; that is, increase the number of meander line segments, thus increasing the total wire length as shown in Figure 7.69, and Table 7.3, which describes increase in the total wire length in models M1 to M2.

Another way is to increase the width of meander line segments as shown also in Figure 7.69, which depicts models M3 and M4 with wider width. In this case, the total length of M3 is decreased from that of M2, but the effective volume is increased so that

**Table 7.3** The physical properties of the six MLA models ([26], copyright ©2004 IEEE)

| Antenna | Height (cm) | Total wire length (cm) | Meander width (cm) |
|---|---|---|---|
| M1 | 6 | 29.25 | 0.97 |
| M2 | 6 | 38.42 | 0.95 |
| M3 | 6 | 29.25 | 1.66 |
| M4 | 6 | 29.25 | 3.87 |

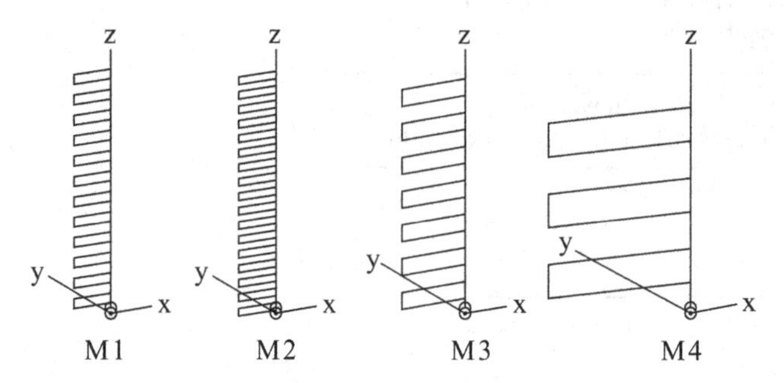


**Figure 7.69** MALs with different number of segment and width ([26], copyright ©2004 IEEE).

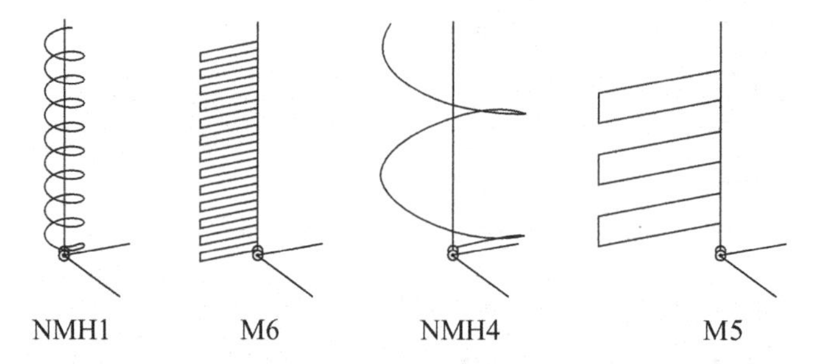


**Figure 7.70** Comparison of NMHAs and MLAs in terms of the width ([26], copyright ©2004 IEEE).

the effective capacitance against the ground is increased. In addition, increase in the effective diameter decreases capacitances between the meander line segments, which results in increase in the effective self-inductance of the wire. These effects contribute to lowering the resonance frequency. This is also the case with models M3 and M4, where the total wire length is kept the same, while the width is increased as is described in Table 7.3. As a consequence, the resonance is lowered.

An MLA is compared with an NMHA at the same resonance frequency of 408 MHz, taking models of NMHA 1, NMHA 4, M6, and M5 (Figure 7.70) as examples. Table 7.4 provides comparisons of the physical and radiation properties of these models. With increase in the width, the total wire length increases in M6 from that of NMHA 1, whereas keeping the same width, that of NMHA 4 increases as compared with that of

**Table 7.4** The radiation properties of the two NMHA models and two MLA models ([26], copyright ©2004 IEEE)

| Antenna | Height (cm) | Total wire length (cm) | Width (cm) | Resonant frequency (MHz) | Radiation resistance (Ohms) | $Q$ | $r/h$ |
|---|---|---|---|---|---|---|---|
| NMH1 | 6 | 29.25 | 1.0 | 408.2 | 5.1 | 66 | 0.80 |
| M6 | 6 | 49.0 | 1.66 | 408.12 | 4.71 | 65 | 0.80 |
| NMH4 | 6 | 21.99 | 3.7 | 408.2 | 4.2 | 61 | 0.82 |
| M5 | 6 | 28.16 | 3.7 | 408.2 | 4.4 | 61 | 0.82 |

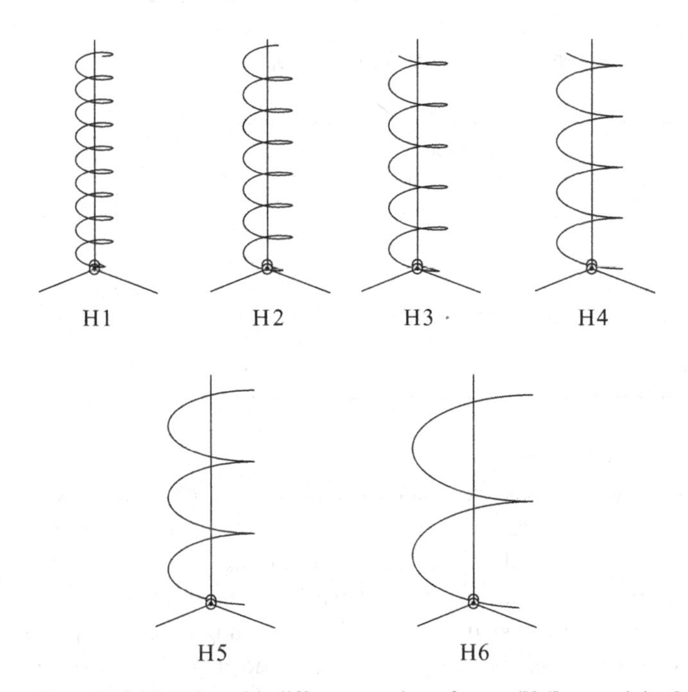


**Figure 7.71** NMHAs with different number of turns ([26], copyright ©2004 IEEE).

M5. Radiation resistance decreases in M6 as compared with that of NMHA 1, while a small increase is observed in M5 from NMHA 4. $Q$ and $r/h$ are equal or almost equal regardless of models, NMHA 1, NMHA 4, M6, and M5. The numerical values of these are also given in Table 7.4.

When diameter of the NMHA is increased as shown in Figure 7.71, the total wire length and also the number of turns becomes smaller for the same resonance frequency (fixed at 408 MHz). The physical properties and resonance properties for models H1 to H6 are given in Table 7.5 and Table 7.6, respectively.

When antennas have the same effective height and volume, their resonance properties are essentially identical, independently of any differences in their total wire length and their geometry. The above discussions are useful for designing MLAs and NMHAs.

**Table 7.5** The physical properties of the six NMHA models ([26], copyright ©2004 IEEE)

| Antenna | Height (cm) | Total wire length (cm) | Diameter (cm) | No. of turns |
|---|---|---|---|---|
| H1 | 6 | 29.25 | 1.0 | 9 |
| H2 | 6 | 27.45 | 1.24 | 6 ¾ |
| H3 | 6 | 26.05 | 1.50 | 5 ¾ |
| H4 | 6 | 25.09 | 1.77 | 4 ¾ |
| H5 | 6 | 23.63 | 2.31 | 3 |
| H6 | 6 | 22.36 | 3.18 | 2 |

**Table 7.6** The resonant properties of the six NMHA models ([26], copyright ©2004 IEEE)

| Antenna | Resonant frequency (MHz) | Radiation resistance (ohms) | $Q$ | $r/h$ |
|---|---|---|---|---|
| H1 | 408.2 | 5.13 | 66 | .80 |
| H2 | 408 | 5.18 | 63 | .81 |
| H3 | 408.3 | 5.11 | 61 | .82 |
| H4 | 408.2 | 5.03 | 59 | .82 |
| H5 | 408.3 | 4.85 | 57 | .83 |
| H6 | 408.2 | 4.55 | 57 | .84 |

#### 7.2.1.2 Extension of current path

This section describes antennas other than wire types introduced in the previous sections. Typical examples are antennas which have one or more slots or notches inserted in the current path on the surface of the antenna element. A term "slit" is often used instead of "notch." Slots embedded on the surface of a planar antenna element force the current into a roundabout geometry as shown in Figure 7.72, lengthening the current path so that the resonance frequency is lowered. Notches or slits embedded on the edge of a rectangular surface also lengthen the current path (Figure 7.73). Slits embedded alternately on both sides of a rectangular surface make the current path meandering and the resonance frequency lower (Figure 7.74). Another example is a pair of triangular (bow-tie) patches (Figure 7.75(a)), which is constituted by expanding the mouth of a notch to the end of the planar surface (Figure 7.75(b)). The current flows in this case spread toward the end of the triangular surface taking paths with different lengths so that the resonance occurs at not only one frequency, but also at two or more. Current flows on the surface of these antennas are shown in each of their corresponding figures with arrows. Operating at lower resonant frequency in a given antenna size implies an equivalent downsizing of the antenna, as normally an antenna of lower resonance frequency should have larger size.

Slots or notches are not only employed for the purpose of making antennas small, but also for attaining such operations as dual band, multiband, wideband, and circular polarization by arranging the size, shape, and the number of slots, and the places for embedding them. A slot may have an L-shape (Figure 7.76(a)), a folded shape (Figure 7.76(b)), or

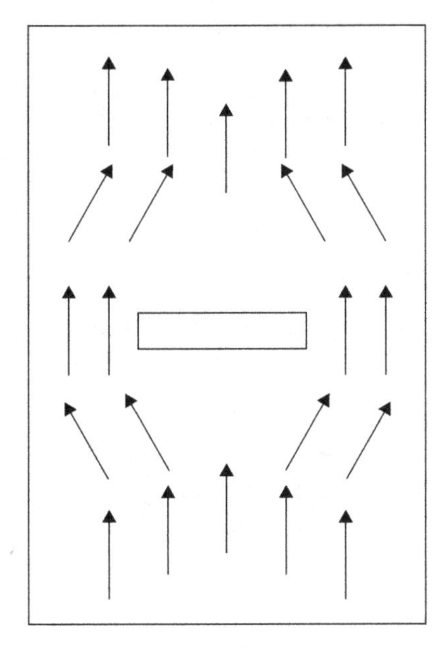

**Figure 7.72** Current flow on the surface of planar element with a slot.

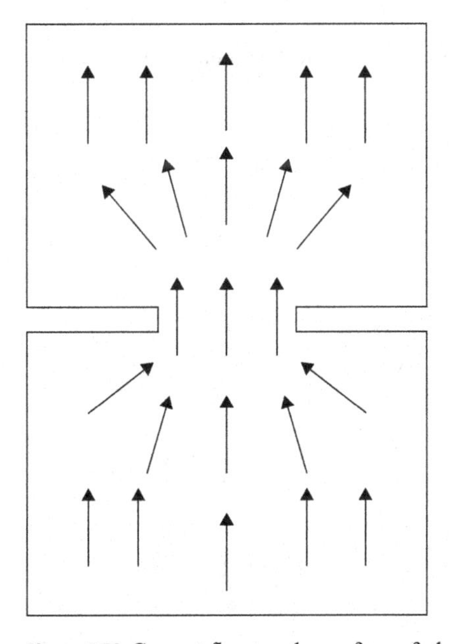

**Figure 7.73** Current flow on the surface of planar element with a notch.

be combined with a notch (Figure 7.77(a)) or with a bow tie (Figure 7.77(b)). Various other examples are illustrated in Figure 7.78, showing (a) cross slots, (b) two bent slots, (c) four bent slots, (d) four notches, (e) a square slot, (f) four square slots, (g) a circular slot, and (h) an offset circular slot [29].

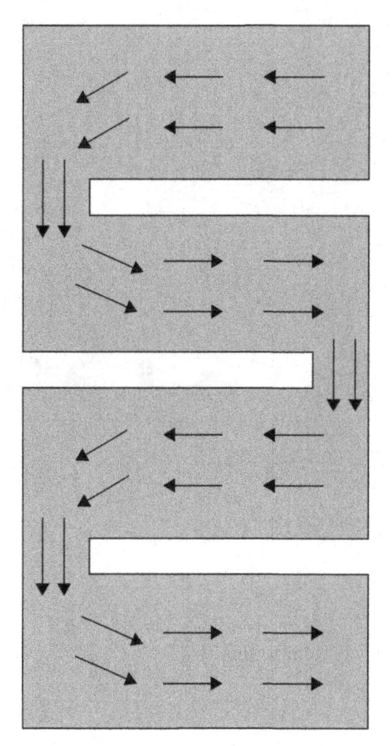

**Figure 7.74** Current flow on the surface of a planar element with meandered notches [29].

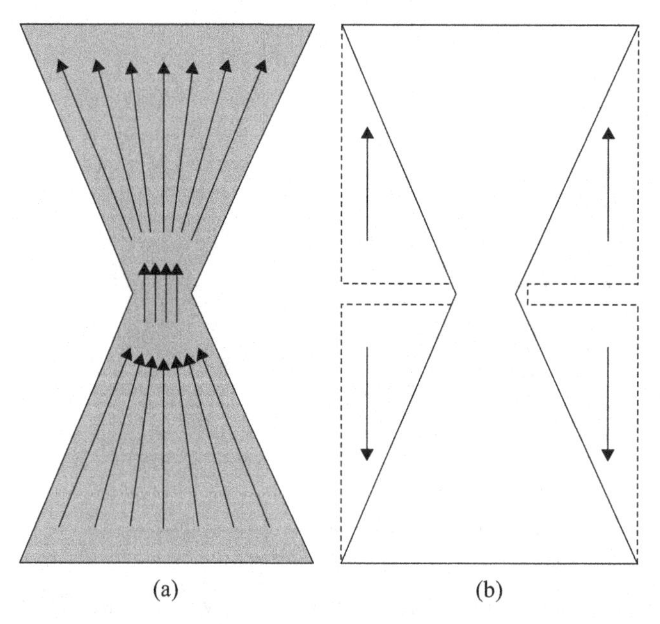


**Figure 7.75** Current flow on a pair of triangular patch (bow-tie) surfaces: (a) current flow and (b) end of notch expanded to form triangle [29].

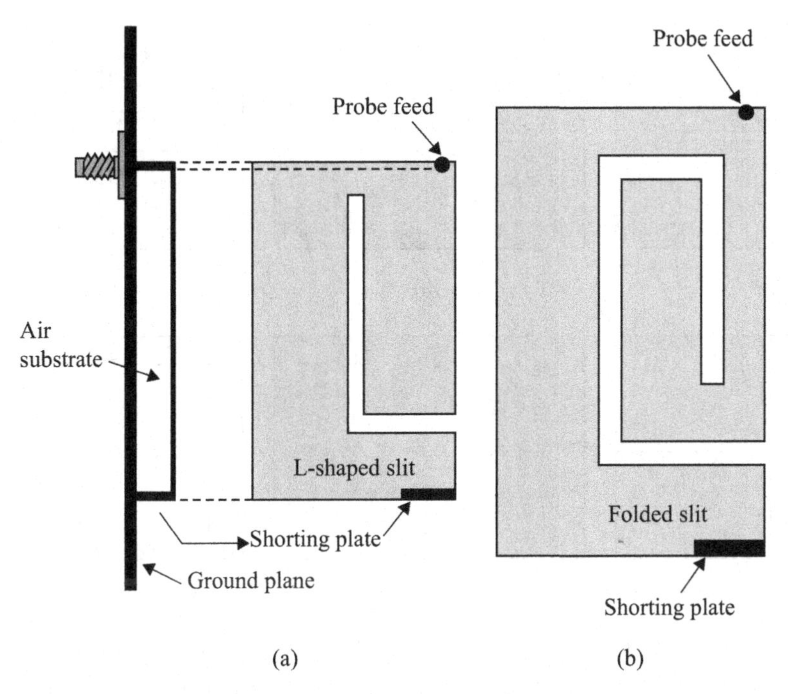


**Figure 7.76** Variation of slit on the surface of patch; (a) L-shape and (b) folded shape.

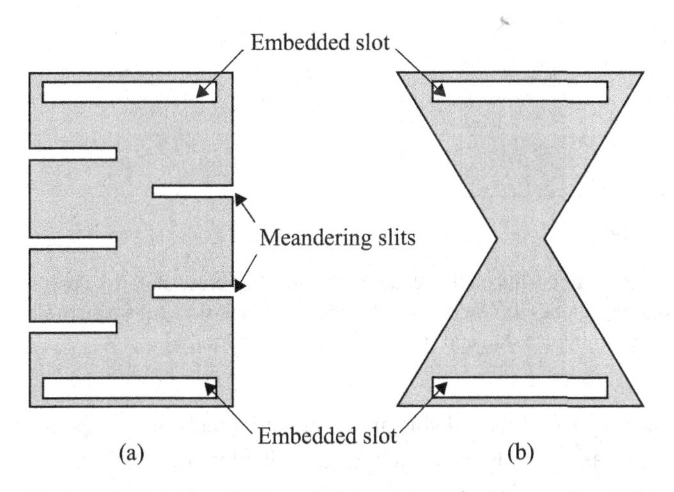


**Figure 7.77** Variation of slits and slots on a patch: (a) meandered slits and two slots on a rectangular patch and (b) two slots on triangle patch [29].

Mostly, these antennas have been developed not merely to create small antennas, but also to enhance antenna performance such as wide bandwidth, multiband operation, circular polarization, and so forth. Then, most such antennas are treated as FSA (Functionally Small Antenna) rather than ESA, and designed to satisfy demands for achieving small or compact antennas and yet having improved performance without increasing

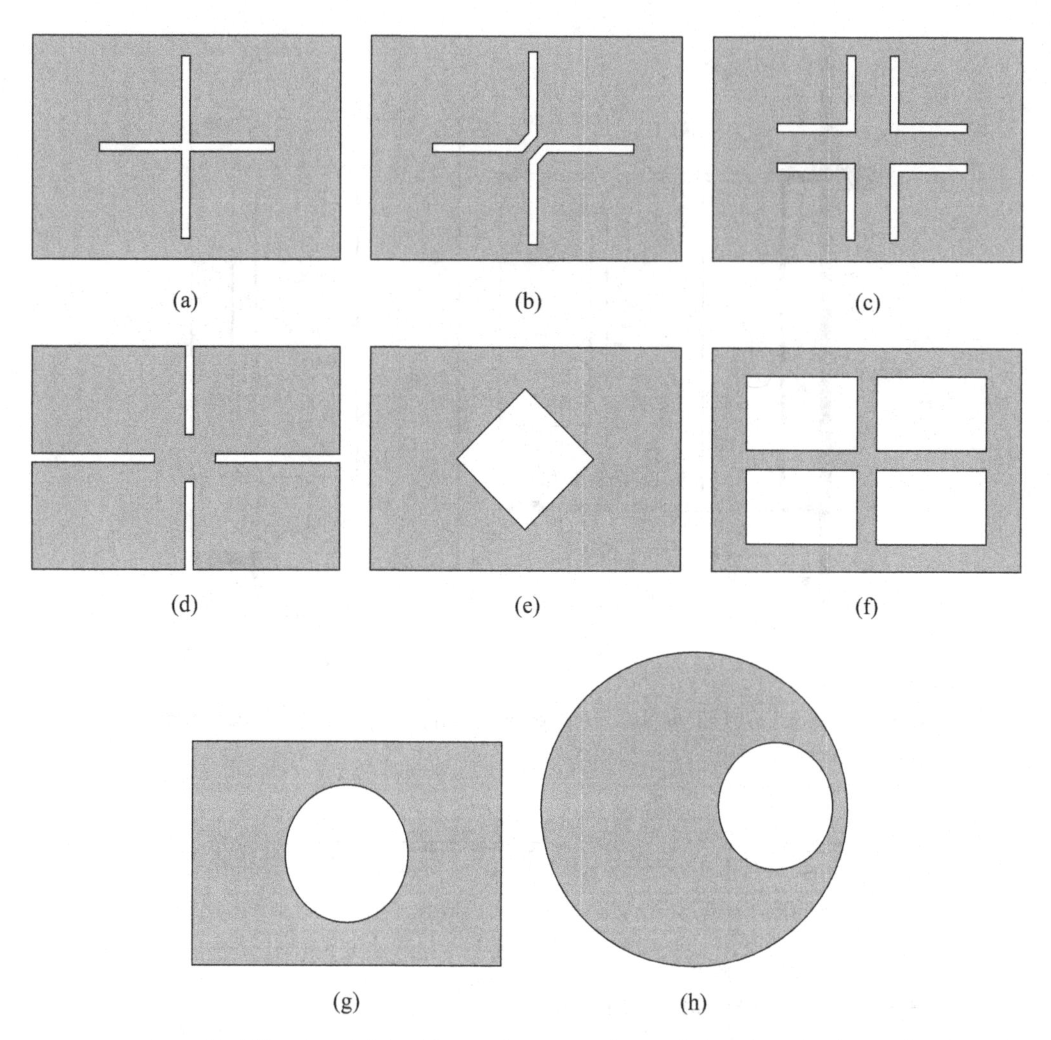


**Figure 7.78** Variations of slots/slits on the rectangular patch: (a) cross slot, (b) a pair of bent slots, (c) a group of four bent slots, (d) four slits on the center of each side, (e) a square slot, (f) four rectangular slots, (g) a circular slot, and (h) an offset circular slot [29].

the size. For that purpose, slots and slits are combined to enhance the performance and achieve downsizing as well. The FSAs will be described in the next chapter.

### 7.2.2 Full use of volume/space

The concept of filling volume/space by an antenna element is based on Wheeler's work showing that antennas fully utilizing the volume that circumscribes the maximum size of an antenna, will provide the lowest $Q$ as compared with other geometries within the same volume. He examined the antenna $Q$ in relation to the antenna size by taking a cylindrical shape of capacitor and inductor representing a C-type and an L-type dipole

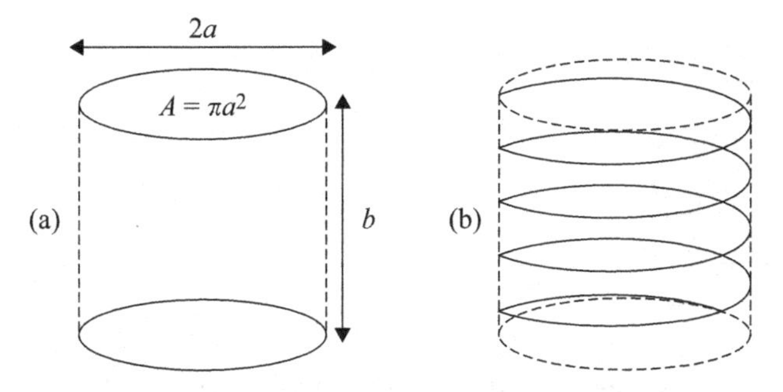


**Figure 7.79** (a) Small C-type antenna and (b) L-type antenna occupying the same volume ([30], copyright ©1947 IEEE).

antenna, respectively, as is depicted in Figure 7.79, and he showed that the $Q$ is inversely proportional to the volume of the antenna [30]. In his analysis, Wheeler used circuit parameters, capacitor $C$, inductance $L$, and radiation resistance $R_{rad}$ or conductance $G_{rad}$ of these antennas, and derived $Q$. The $C$ and $L$ became:

$$C = \varepsilon_0 k_C A/b \tag{7.48}$$

$$L = \mu_0 \pi^2 A/k_L b \tag{7.49}$$

where $k_C$ and $k_L$, respectively, are the shape factor of the C-type and the L-type antenna, $A$ denotes the cross section area, and $b$ is the height, as Figure 7.79 shows.

Radiation resistance $R_{rad}$ and conductance $G_{rad}$, respectively, are derived as

$$R_{Crad} = 20(kb)^2, \text{ and } G_{Crad} = (k^2 k_C A)/(6\pi Z_0) \tag{7.50}$$

for the C-type antenna, and

$$R_{Lrad} = 20(nk^2 A)^2, \text{ and } G_{Lrad} = (kk_L A)^2/(6\pi n^2 Z_0) \tag{7.51}$$

for the L-type antenna.

Here $k$ is the wave number $(2\pi/\lambda)$, $n$ is number of turns of the coil, and $Z_0$ is the free space impedance. Then by using Eqs. (7.48–7.50), $Q_C$ of the C-type antenna and $Q_L$ of the L-type antenna, respectively, are derived as

$$Q_C = \omega C/G_C = 6\pi k^3/(k_C Ab) = 9V_{RS}/2V_{eff} \tag{7.52}$$

and

$$Q_L = \omega L/R_L = 6\pi k^3/(k_L Ab) = 9V_{RS}/2V_{eff} \tag{7.53}$$

where $V_{RS}$ denotes the volume of a sphere having radius $\lambda/2\pi$ (the radian sphere), and $V_{eff}$ is the effective volume, which is defined by

$$V_{eff} = \sigma V_{phy}. \tag{7.54}$$

Here $\sigma = k_C, k_L$ and $V_{phy}$ is the physical volume of the antenna.

Equations (7.52) and (7.53) indicate that $Q$ is inversely proportional to $V_{eff}$, thus $V_{phy}$, as $V_{phy}$ is proportional to $V_{eff}$, and that by constituting an antenna in a sphere, which

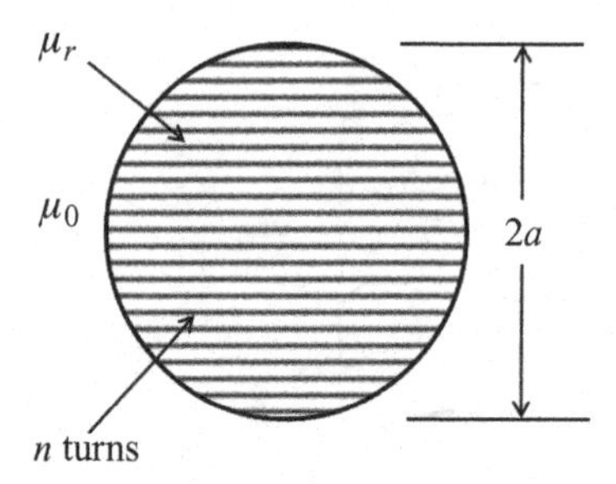


**Figure 7.80** Wheeler's spherical coil ([31], copyright ©1958 IEEE).

encloses its minimum, so as to occupy the entire volume of the sphere, the lowest $Q$ will be obtained.

Wheeler examined what type of small antenna can most effectively utilize the volume or has the highest volume and lowest $Q$. He took up a constant pitch spherical coil of $n$ turns with magnetic core as shown in Figure 7.80 [31], where the excitation voltage is across the poles of the coil, and he calculated the circuit parameters, from which $Q$ can be found as

$$Q = (1 + 2/\mu_r)/(ka)^3 = V_{RS}/(\sigma V_{phy}). \tag{7.55}$$

Where $\mu_r$ is the relative permeability of material filled in the sphere, and $a$ is the radius of the sphere. The lowest $Q_{min}$ is obtained when $\mu_r$ is infinite; that is,

$$Q_{min} = 1/(ka)^3. \tag{7.56}$$

It should be noted that this is a case for an antenna radiating only either $TE_{10}$ or $TM_{10}$ mode. When both $TE_{10}$ and $TM_{10}$ are radiated, $Q_{min}$ becomes

$$Q_{min} = 1/2(ka)^3. \tag{7.57}$$

Chu worked on a similar problem. There is a well-known bound for $Q$ that is often referred to as the Chu limit; the theoretical minimum $Q$ relates to the spherical volume, which embraces the maximum dimensions of the antenna. Here assuming a hypothetical sphere having radius $a$, the Chu limit [32] prescribes that the lower bound of $Q$ is a function of $ka$ ($k = 2\pi/\lambda$), the dimensional parameter of the sphere, and is given by

$$Q = (ka^3)^{-1} + (ka)^{-1}. \tag{7.58}$$

To obtain a $Q$ of the closest lower bound of Eq. (7.58), the antenna must effectively utilize the total sphere volume defined by the dimensional parameter $ka$. In case of a thin linear dipole having length $l$ and diameter $d$, a hypothetical sphere which circumscribes the dipole should have the radius $l/2$, and hence the dipole occupies only a small part of the volume of the sphere, implying very inefficient use of the volume of the sphere. Meanwhile, in case of a spherical dipole, which occupies the outermost regions of the sphere of radius $l/2$, the volume of the sphere is utilized very efficiently. The lowest $Q$ may be closely approached by this sort of spherical dipole.

The concept of filling volume can be applied to the two-dimensional case; that is, when a space is occupied fully with thin metallic conductors a lower obtainable $Q$

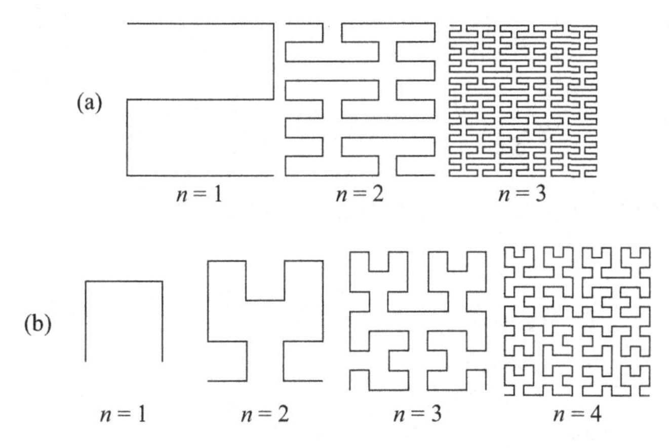


**Figure 7.81** (a) Peano curve; order one through three and (b) Hilbert curve; order one through four [33].

can be achieved within a limited space. For this purpose, the concept of space-filling curves, which are generally a continuous mapping of the normalized interval onto the normalized square, is used. The space-filling curve was first proposed by G. Peano in 1890 [33], and now it is called the Peano curve. D. Hilbert introduced his version of space-filling curve in 1891 [33], now called the Hilbert curve. These curves are composed through iterative generating procedures, by which contours (patterns) with infinitely complicated structure to fill a space are produced. The pattern is formed by infinitely repeating combinations of scaling and rotating an initial pattern. It is expressed in terms of the iteration numbers and the space is gradually filled as the iteration number approaches infinity. Figure 7.81(a) illustrates Peano curves of orders $n = 1$ through 3 and (b) shows Hilbert curves of orders $n = 1$ through 4.

In addition to these two curves, fractal patterns are also known as a class of space-filling curve, and have been used for lowering resonance frequency and thus small sizing of antennas. Fractal stands for "Fractional Dimension," and incorporates the meaning of "fragmented" or "broken." The term fractal has been applied to a species of cumulus clouds or stratus clouds with ragged, shred appearance. The fractal patterns are produced by similar iteration procedures, starting from a "generator," which replaces an "initiator," the initial geometry, and repeats this iteration process infinitely. Representative Fractals include Koch [34a], Minkowsky [34b], and Sierpinsky [34c], respectively, as shown in Figures 7.82, 7.83, and 7.84. In the case of a Koch Fractal, the initiator is a line, and onethird of the line is replaced by the generator, which is an equilateral triangle, at its center, and this procedure is repeated an infinite number of times to create a geometry that has intricate patterns on an ever-shrinking scale.

These fractal structures are expressed in terms of iteration number. In addition, as the iteration order of the pattern increases, the length of the element increases, while the footprint area is maintained the same. By this means, the resonance frequency is lowered without increasing the space, and hence it can be used for creating small antennas. Furthermore, when it is used for an antenna, it offers another advantage that input impedance can be increased as the length of the fractal pattern element increases,

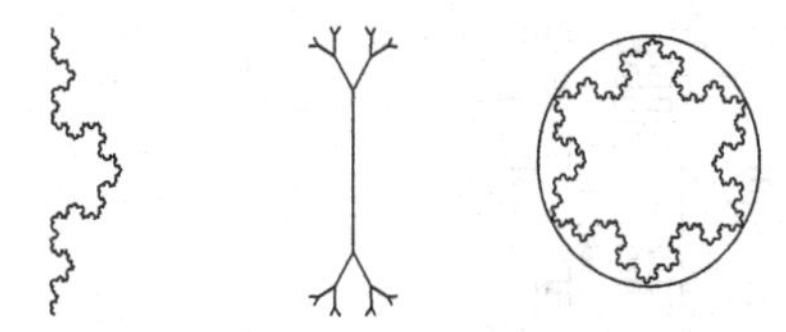

**Figure 7.82** Koch fractal models ([34a], copyright ©2000 IEEE).

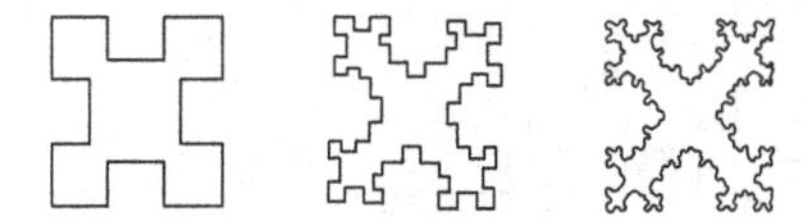

**Figure 7.83** Minkowsky fractal models ([34b], copyright ©2002 IEEE).

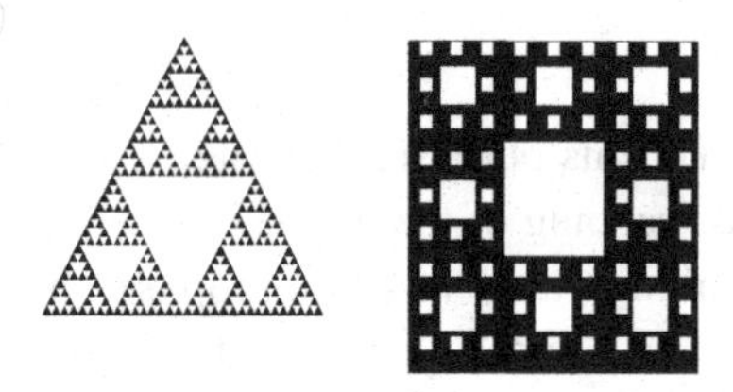

**Figure 7.84** Sierpinsky fractal models ([34c], copyright ©2003 IEEE).

whereas an ordinary small antenna suffers having very small input impedance that hinders good matching.

Features of space-filling curves, including fractal structures, when applied to antennas, result in not only lowering the resonance frequency, and increasing antenna impedance, but also in increasing the surface impedance as a result of lengthened space-filling elements within the limited space, that will produce a high impedance surface (HIS). The HIS can offer various salient antenna performances and novel antenna structure, and can decrease mutual coupling between antenna elements. Furthermore, enhancement of the antenna performance is attained by utilizing positive image effects, and a decrease of coupling between antenna elements allows close location of several antenna elements on a limited space, that will bring an increase in the transmission data-rate when it is applied to the MIMO system. Suppression of surface current contributes to decreasing undesired radiation; for instance, resulting in increase of the front-to-back ratio due to the reduction of backward radiation.

Spiral patterns are considered as a class of space-filling pattern as they occupy an area of given size without much vacancy and contribute to making bandwidth wide. This does not necessarily imply small-sizing of an antenna in terms of an entire frequency range, but at the lower resonance frequencies the antenna should be recognized as small-sized in terms of the operating wavelength. Then, spiral antennas should be partially classified into Functionally Small Antenna and partially into Electrically Small Antenna.

**Table 7.7** Parameters of basic meander lines of three types; N1, N2, and N3, shown in Figure 7.85 [35]

| Antennas | $f_0$(MHz) | $R_{rad}(\Omega)$ | $Q$ | $kr$ |
|---|---|---|---|---|
| N1 | 543.44 | 26.52 | 59.8 | 0.499 |
| N2 | 548 | 120 | 59 | 0.498 |
| N3 | 325.1 | 5.362 | 74.6 | 0.296 |

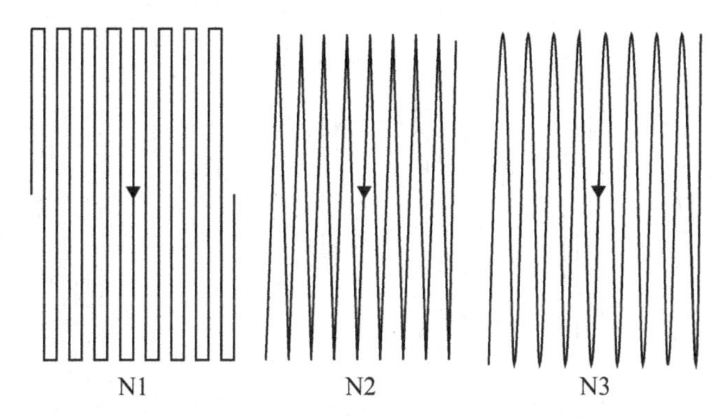


**Figure 7.85** Variation of meander line patterns: N1: original (rectangular), N2: triangular and N3: sinusoidal ([35], copyright ©2009 IEEE).

### 7.2.2.1 Two-dimensional (2D) structure

#### *7.2.2.1.1 Meander line*

A Meander line pattern is the simplest one to fill a space of given area, and has been applied in various small wireless systems. A representative pattern is that of a dipole shown in Figure 7.24, which is described in Section 7.2.1.1. Variations of the basic meander line profile from the original rectangular one are triangular profile and sinusoidal profile (Figure 7.85 N1, N2, and N3) [35]. The meander line with triangle profile raises radiation resistance from 26.5 Ω to 120 Ω, and the sinusoidal profile meander line significantly reduces resonance frequency from 543 MHz to 325 MHz, indicating a comparable reduction of antenna size. These parameters are listed in Table 7.7, where resonance frequency, radiation resistance, $Q$, and the size are provided.

In order to use space more efficiently and improve antenna performances, further modification of meander line profile are taken into consideration by taking various geometries shown in Figure 7.86, where four types, M1 through M4, are illustrated. Performances were studied and found lower resonance frequency with smaller size, moderate value of $Q$, and lower cross polarization level, although radiation resistance decreases (Table 7.8). Meandered wire itself can be a meandered structure to further lower the resonance frequency so that the antenna size can be reduced.

Some types of planar meandered antennas used in mobile phones will be introduced as examples of practical applications in later sections. Examples of other types of planar meandered antennas developed for RFID tags were a meandered rectangular loop

**Table 7.8** Parameters of meander line patterns of four different types; M1, M2, M3, and M4 [35]

| Antennas | Radiation resistance (Ω) | Resonant frequency (MHz) | $ka$ | $Q$ | $Q_{Chu}$ | Cross-Pol. (dB) |
|---|---|---|---|---|---|---|
| M1 | 11.68 | 512.92 | 0.645 | 114.2 | 5.277 | −115.9 |
| M2 | 18.32 | 474.5 | 0.67 | 66.25 | 4.871 | −118.6 |
| M3 | 15.7 | 423.9 | 0.6 | 102.5 | 6.296 | −117.1 |
| M4 | 5.89 | 414.95 | 0.588 | 54.39 | 6.62 | −122.7 |

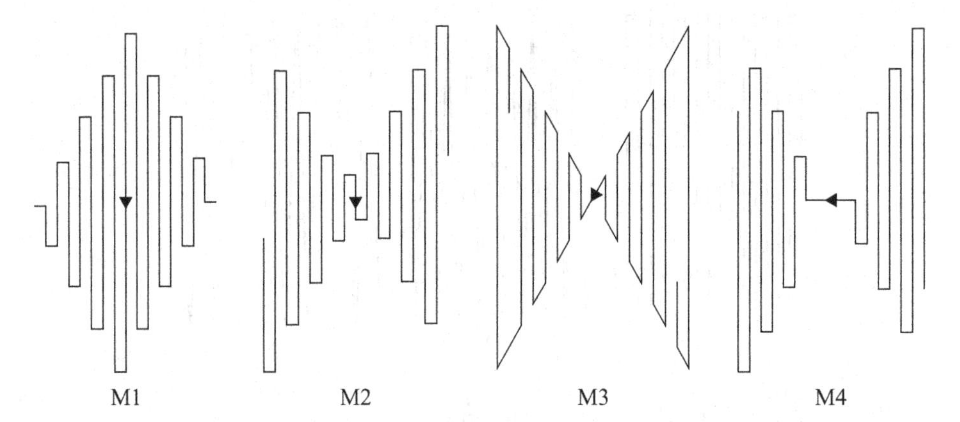


**Figure 7.86** Variation of meander line patterns: M1, M2, M3, and M4 ([35], copyright ©2009 IEEE).

[36] and a C-shaped folded monopole structure [37], operating at 900-MHz band and 400-MHz bands, respectively.

*7.2.2.1.2 Koch Fractal curve, Peano curve and Hilbert curve*

*7.2.2.1.2.1 Monopole using Koch, Peano, and Hilbert curves*

It is well known that use of space-filling geometry, such as Koch, Peano, and Hilbert curves [38–42] for antenna structure to lengthen the antenna in a given space so that the resonance frequency is lowered, results in small-sizing of an antenna. It is also useful for obtaining wideband or multiband operation with small antennas, as the wire length (stretched straight) increases with number of iterations of initial geometry.

A Koch monopole is generated by an iterative procedure that repeats replacing one third of the line at its center with an equilateral triangle, then scaling and rotating it infinitely as shown in Figure 7.87, which depicts the generator of the pattern and iteration process [38]. A Koch monopole having height $h = 6$ cm, being placed over the 80-cm ground plane is illustrated in Figure 7.88. The computed and measured input impedances with respect to number $n$ of the iteration are shown in Figure 7.89, where (a) is the resistance and (b) is the reactance. The stretched length $l$ of the antenna is expressed by $l = h\,(4/3)^n$ and an antenna constructed by iteration of $n$ times is denoted as $K_n$, where $n = 0$ means the Euclidean version of the antenna, that is an ordinary straight monopole. A line expressing the small-antenna limit $kh < 1$ is drawn as the

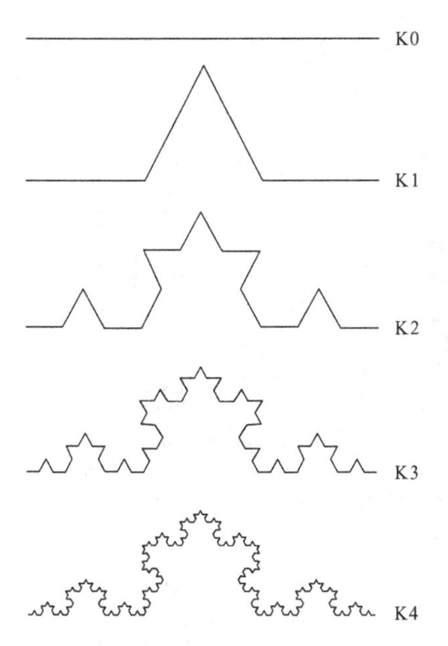


**Figure 7.87** Koch monopole structures; generator K0 through fourth stage K4, ([38], copyright ©2000 IEEE).

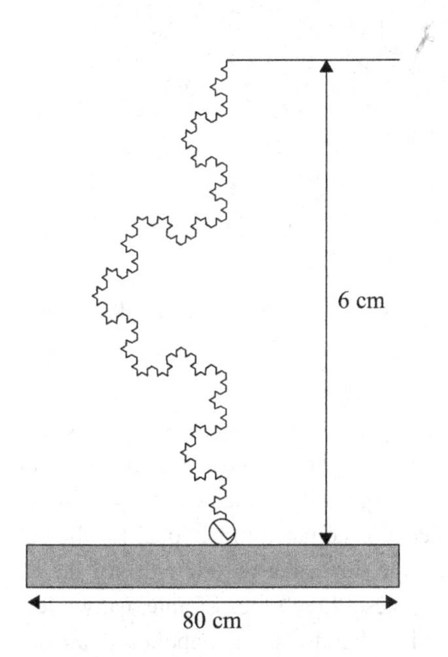


**Figure 7.88** Koch monopole antenna of the fourth stage ([38], copyright ©2000 IEEE).

reference. The figure indicates that the resonance frequency is lowered with increase in $n$, resulting in reduced sizing of the antenna corresponding to the lowered frequency.

The Koch Fractal monopole is considered to be useful for lowering resonance frequency and thus small-sizing an antenna. However, comparison with two other types

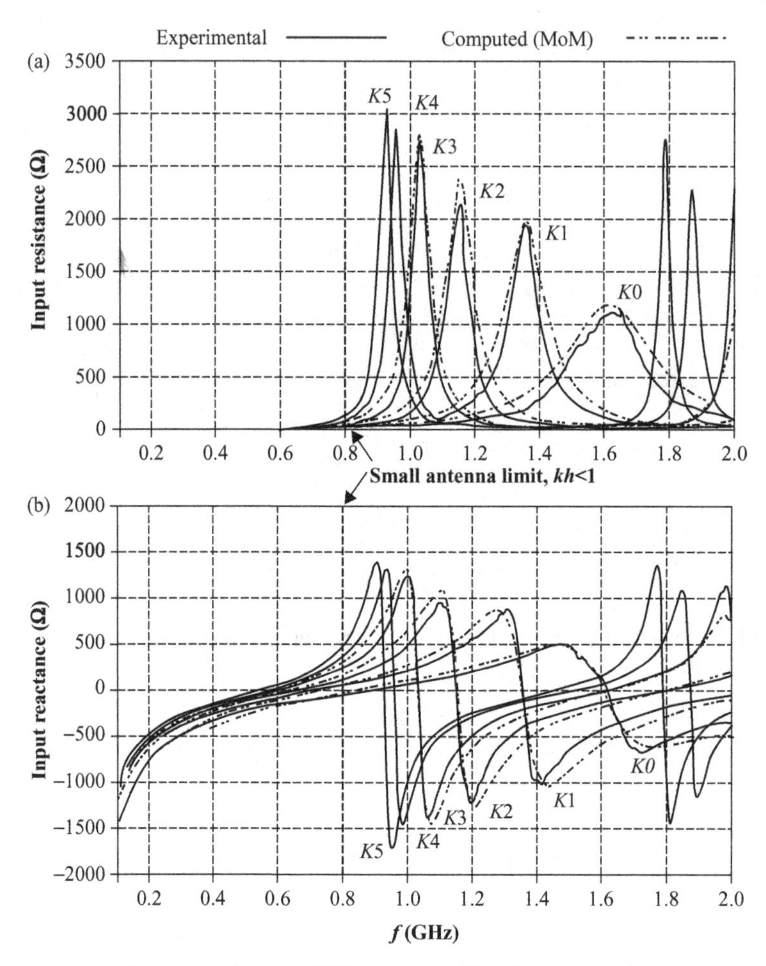


**Figure 7.89** Input impedance of different stage of Koch monopole: (a) resistance and (b) reactance ([38], copyright ©2000 IEEE).

of wire antennas, meander line antenna (MLA) and helical antenna (NMHA), reveals that the Koch Fractal antenna has relatively higher resonance frequency compared with that of MLA and NMHA (Figure 7.90) [39], for the same (stretched) wire length $l$. This means that when the resonance frequency is adjusted to be equal for the three types of antenna, keeping the antenna height $h$ the same, the wire length $l$ of the Koch Fractal monopole becomes the longest among these three types of antenna, while the antenna performances, radiation resistance, and the bandwidth, are about the same, as Table 7.9 shows [39].

This fact suggests that the antenna specific geometry is not necessarily significant for small-sizing of an antenna, but the electrical length, not physical length, is essential.

Instead of a single wire structure, a given space is filled more efficiently by single Peano or Hilbert curve elements. Consider these two curve structures, Peano and Hilbert, previously illustrated in Figure 7.81 (a) and (b), respectively, embedded in a square of 3 × 3 cm [41]. Induced currents due to normally incident plane waves for these

**Table 7.9** Comparison of antenna performances for three types at the same resonance frequency ([39], copyright ©2003 IEEE)

| Antennas | Resonant frequency (MHz) | Overall height (cm) | Total wire length (cm) | Resonant resistance (ohms) | 2:1 SWR bandwidth (%) |
|---|---|---|---|---|---|
| K0 | 1201 | 6 | 6 | 36.8 | 8.5 |
| K3 | 745.3 | 6 | 14.22 | 15.4 | 3.5 |
| M3 | 745.3 | 6 | 12.32 | 15.6 | 3.5 |
| H3 | 745.3 | 6 | 11.46 | 15.8 | 3.5 |

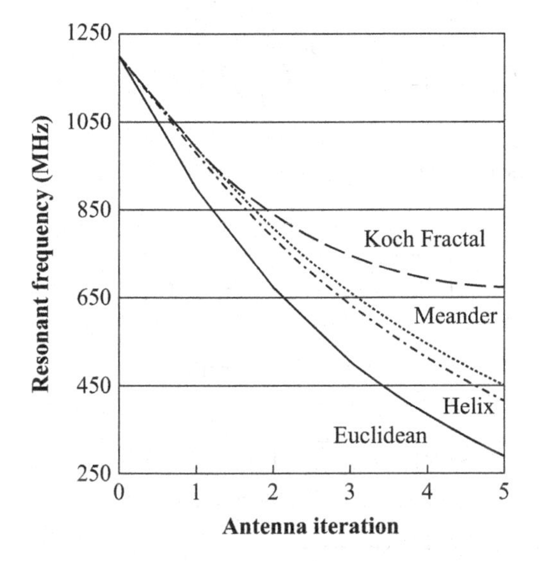


**Figure 7.90** Comparison of resonance frequency for three types of antenna in terms of iteration number ([39], copyright ©2003 IEEE).

structures differ depending on the curve structure and direction of the incident wave. Subsequently, resonance frequency differs depending on the polarization of the incident wave and relates with the electrical length of wire (stretched) residing in the square. For instance, for the horizontally polarized incidence wave the resonance frequency of the Peano curve structure is about one third of that for the vertical polarization incidence, because the electrical wire length is longer in the horizontally polarized wave than that of the vertically polarized wave. The antenna performance of the Hilbert curve structure is similar to that of the Peano curve structure. A Peano curve can fill a space much more than a Hilbert curve for the same number of iterations. This can be understood from Table 7.10.

The Peano curve structure is applied to an antenna of a dipole type as shown in Figure 7.91, in which the feed point is shown as an example [41].

In the same way, the Hilbert curve structure is applied to a dipole type as shown in Figure 7.92 [42]. Input impedance of the 2nd-order Hilbert structure is illustrated in

**Table 7.10** Total length (stretched) $S$ for Peano and Hilbert curves with respect to the iteration number ([40], copyright ©2006 John Wiley and Sons Inc.)

| | | |
|---|---|---|
| Peano | $S = (3^{2N} - 1)d$ | $d = L/(3^N - 1)$ |
| Hilbert | $S = (2^{2N} - 1)d$ | $d = L/(2^N - 1)$ |

Note: $L$ is the linear side dimension of the curve.

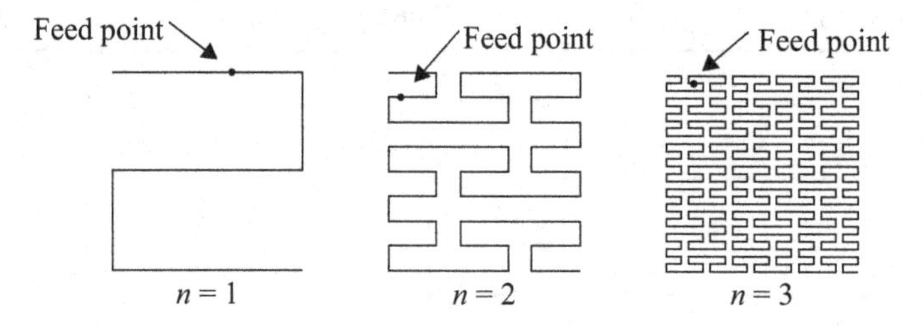


**Figure 7.91** Peano dipoles, orders $n = 1$ through 3 ([41], copyright ©2004 IEEE).

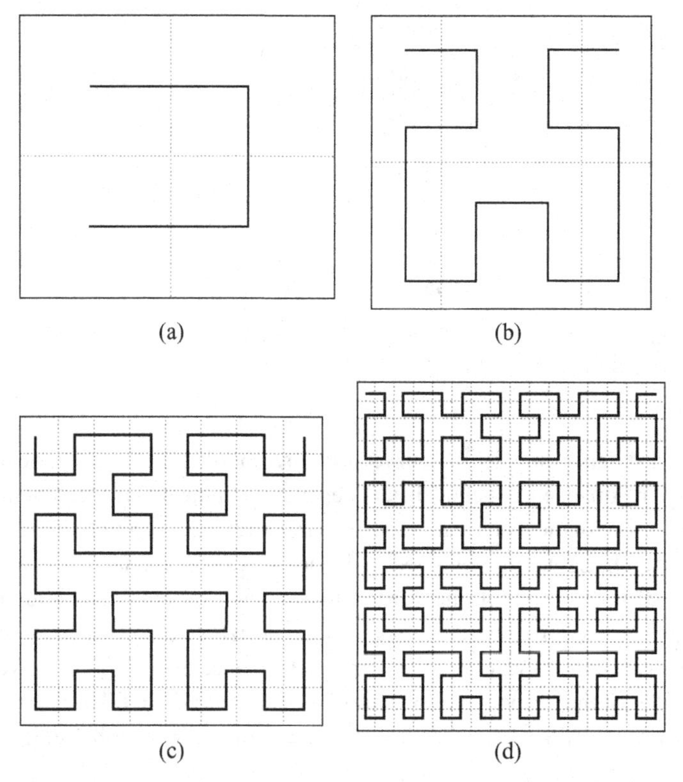


**Figure 7.92** Hilbert dipoles; (a) 1st order, (b) 2nd order, (c) 3rd order, and (d) fourth order ([42], copyright ©2003 IEEE).

**Table 7.11** Resonance frequency and input impedance of Hilbert antenna of 2nd and 3rd stages ([42], copyright ©2003 IEEE)

| Outer dimension (mm) | Fractal iteration order | Width of strip (mm) | Resonant frequency by RWG MoM code with substrate (MHz) | Impedance when resonating |
|---|---|---|---|---|
| 50 | 2 | 4 | 582.087 | 3.79829 |
| 50 | 3 | 4 | 415.516 | 0.95505 |
| 80 | 2 | 4 | 371.716 | 3.93187 |
| 80 | 3 | 4 | 263 | 0.98539 |
| 80 | 2 | 8 | 385.686 | 4.26708 |
| 80 | 3 | 8 | 277.2 | 1.04867 |

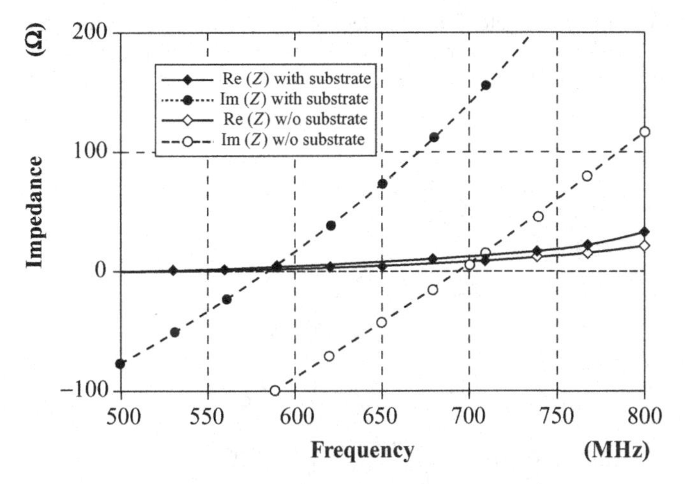


**Figure 7.93** Input impedance of 2nd-order Hilbert antenna fed at the center, with and without substrate; width = 4 mm and other dimensions = 50 mm ([42], copyright ©2003 IEEE).

Figure 7.93. The input impedance can be adjusted to about 50 Ω by selecting the feed position. Table 7.11 provides variation of resonance frequency and input impedance with respect to the iteration number, outer dimensions and width of strip. Current distributions along a 3rd-order Hilbert dipole printed on the substrate of $\varepsilon_r = 4.3$ and fed at the center at resonance frequency 582.087 MHz having the outer dimensions of 4.5 cm is shown in Figure 7.94. The current distribution is almost the same as that for the center-fed and very similar to that of a half-wave dipole resembling a sinusoidal wave, and does not vary significantly with change in the iteration number. Radiation patterns of the 3rd order Hilbert Dipole antenna is shown in Figure 7.95, which is nearly the same as that of a dipole antenna.

Figure 7.96 compares the size between a printed dipole antenna and the 2nd- and the 3rd-order Hilbert dipoles. The figure shows that the 2nd- and the 3rd-order Hilbert dipoles have dimensions of about 1/4 and about 2/11, respectively, of that of the dipole.

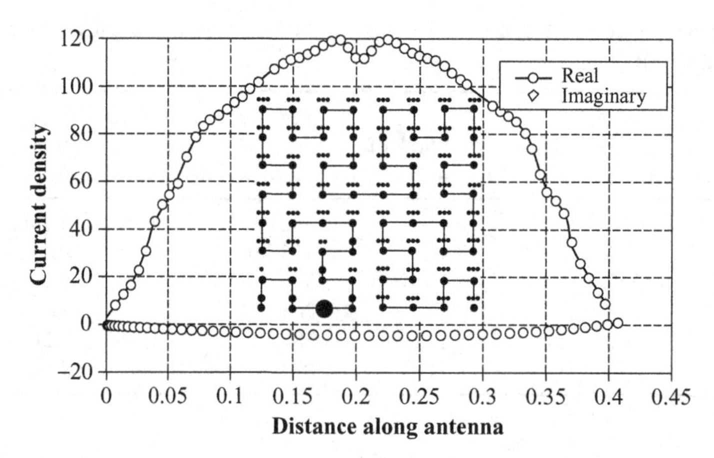


**Figure 7.94** Current distribution along a 3rd-order Hilbert antenna with substrate and fed at node 32 (a dot on the element) at resonance frequency (outer dimension is 4.5 cm and width is 2 mm) ([42], copyright ©2003 IEEE).

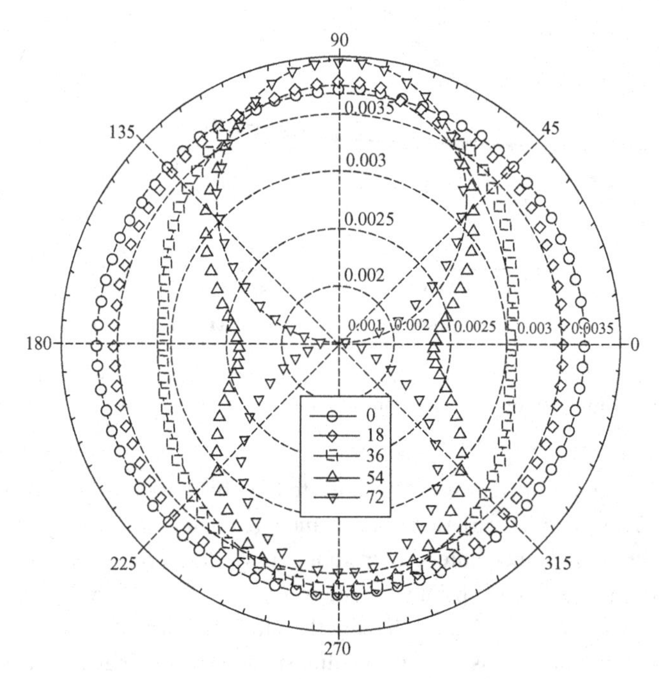


**Figure 7.95** Radiation pattern of the 3rd-order Hilbert antenna with substrate fed at node 24 at the resonance frequency ([42], copyright ©2003 IEEE).

### *7.2.2.1.2.2 PIFA using Peano curve elements*

A Peano curve structure is used for miniaturizing a PIFA element, replacing the horizontal element with planar Peano curve elements. The antenna has two planar Peano elements to cover dual bands, which are designed for applications to GSM in 900 MHz band and PCS in 1900 MHz band [43]. The geometry is shown in Figure 7.97, in which

**Table 7.12** Dimensions of optimized antenna shown in Fig. 7.97 ([43], copyright ©2006 IEICE)

| | $P_1$ | $L_1$ | $s$ | $x_1$ | $x_2$ | $x_3$ |
|---|---|---|---|---|---|---|
| A | 34.4 | 14.3 | 6 | 4.6 | – | 5.4 |
| B | 32.9 | 13 | 6 | 5.1 | 1.2 | – |
| C | 40 | 12.5 | 7.7 | 5.6 | 4.5 | 5.1 |

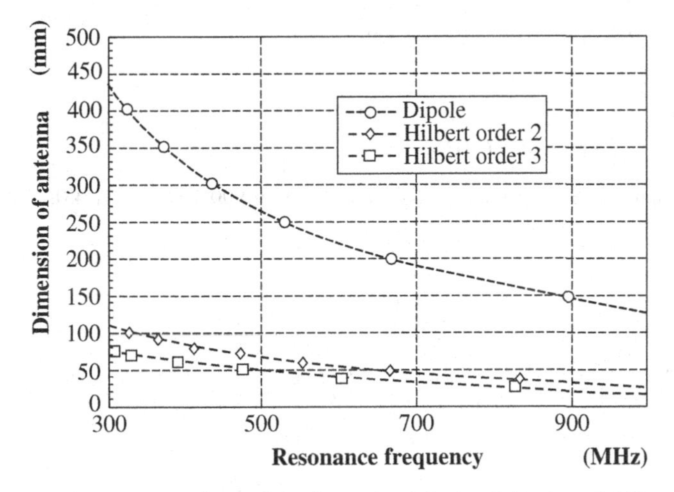


**Figure 7.96** Comparison of the dimension of printed dipole and Hilbert antenna at the same resonance frequency ([42], copyright ©2003 IEEE).

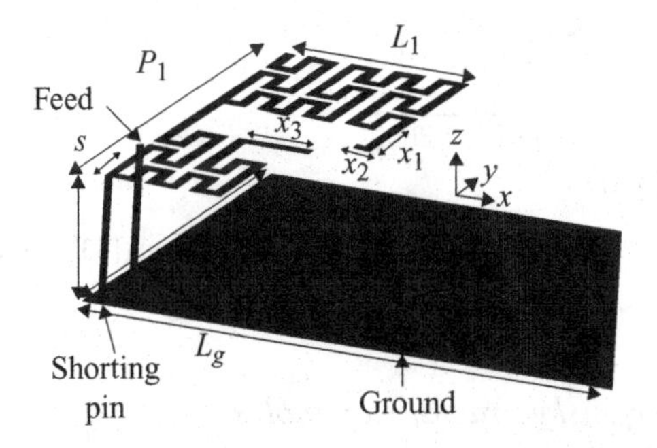


**Figure 7.97** Dual-band 2nd-order Peano PIFA ([43], copyright ©2006 IEEE).

dimensional parameters are given. Optimized antenna dimensions are listed in Table 7.12, in which A and B give parameters for an antenna with height = 8.25 mm and C gives for that $h = 8$ mm and the one side element is replaced with a rectangular patch. Computed VSWR of the antenna in the 900 MHz band and that of the antenna in the 1900 MHz band are illustrated in Figures 7.98 and 7.99, respectively. By using a Peano element, the PIFA can be miniaturized about 50% in volume compared with a conventional PIFA.

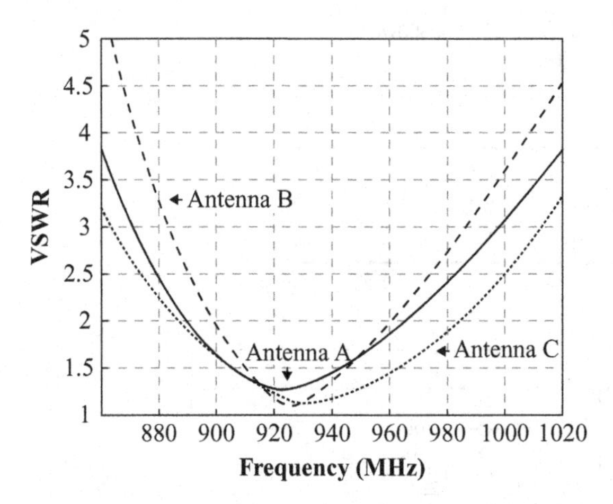


**Figure 7.98** Computed VSWR for antennas A, B, and C near 900 MHz band ([43], copyright ©2006 IEEE).

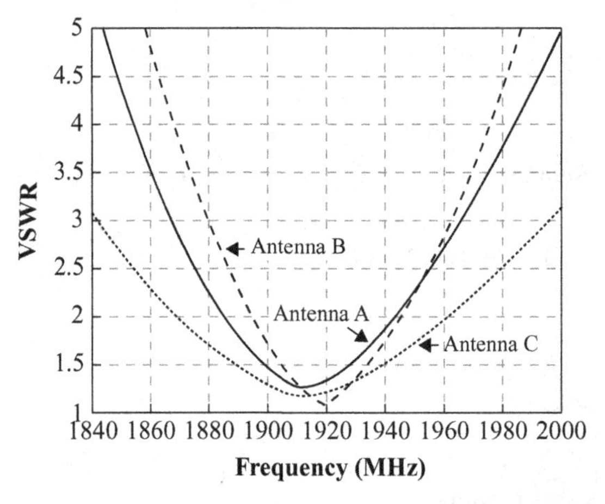


**Figure 7.99** Computed VSWR for antenna A, B, and C near 1920 MHz band ([43], copyright ©2006 IEEE).

### *7.2.2.1.3 Minkowsky and Sierpinsky Fractal monopoles*

#### *7.2.2.1.3.1 Planar Minkowsky Fractal monopole*

A Minkowsky Fractal monopole is generated starting from a Euclidean square by replacing each of the four segments of the initial square with the generator shown, and repeating it for an infinite number of times as illustrated in Figure 7.100 [44], which shows only from the initiator to the third iteration structure. To utilize space-filling ability, a first-order planar Minkowsky loop is fabricated (Figure 7.101) and its performances are reported [44]. The perimeter length $l_p$ is given by

$$l_{pn} = (1 + 2w/3)^n l_{p(n-1)} \tag{7.59}$$

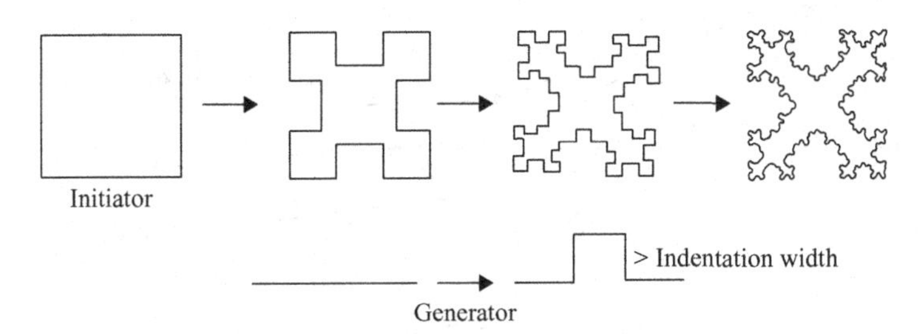


**Figure 7.100** Iterative generation procedure for Minkowsky island fractal; transition from a generator is shown below ([44], copyright ©2002 IEEE).

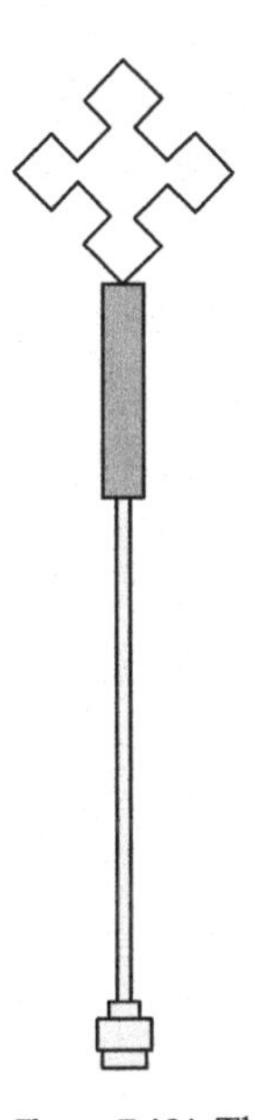

**Figure 7.101** The first-order Minkowsky fractal loop antenna ([44], copyright ©2002 IEEE).

where $n$ is the iteration number and $w$ is the indentation width. Scaling factor for resonance for a $\lambda/4$ square loop for various numbers of iteration and for varying indentation widths is shown in Figure 7.102, which provides design curves for resonance frequency with respect to number of iteration and indentation width. The resonance frequency is reduced as the iteration number increases and the indentation number as well.

#### *7.2.2.1.3.2 Comparison of planar Minkowsky fractal monopoles with other space-filling-curve monopoles [45]*

Planar monopoles using various space-filling curves such as (a) Minkowsky fractal, (b) Hilbert curve, (c) Koch fractal, (d) tee-fractal, (e) box-meander line, and (f) meander line, having planar area of 15.7 cm × 15.7 cm (246.5 cm$^2$), respectively are shown in Figure 7.103(a), (b), (c), (d), (e), and (f). The physical and resonance properties are

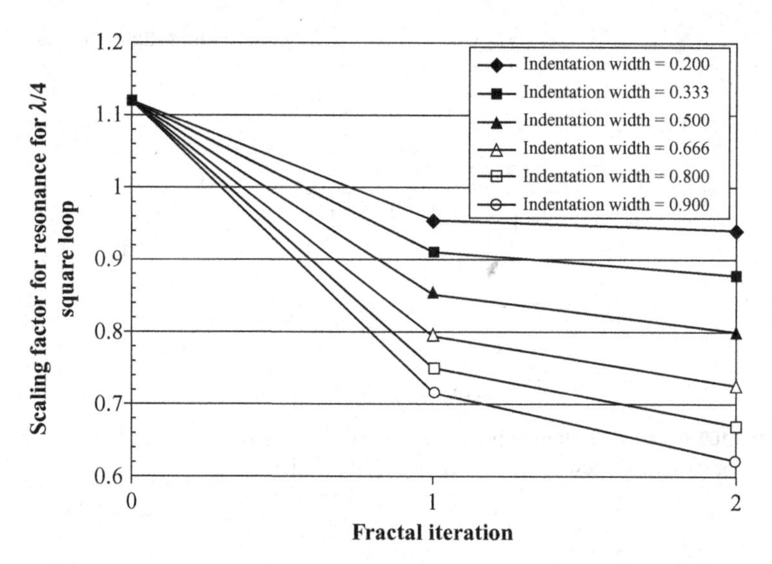


**Figure 7.102** Scaling factor for design of Minkowsky loop antenna for various indentation width in terms of iteration number ([44], copyright ©2002 IEEE).

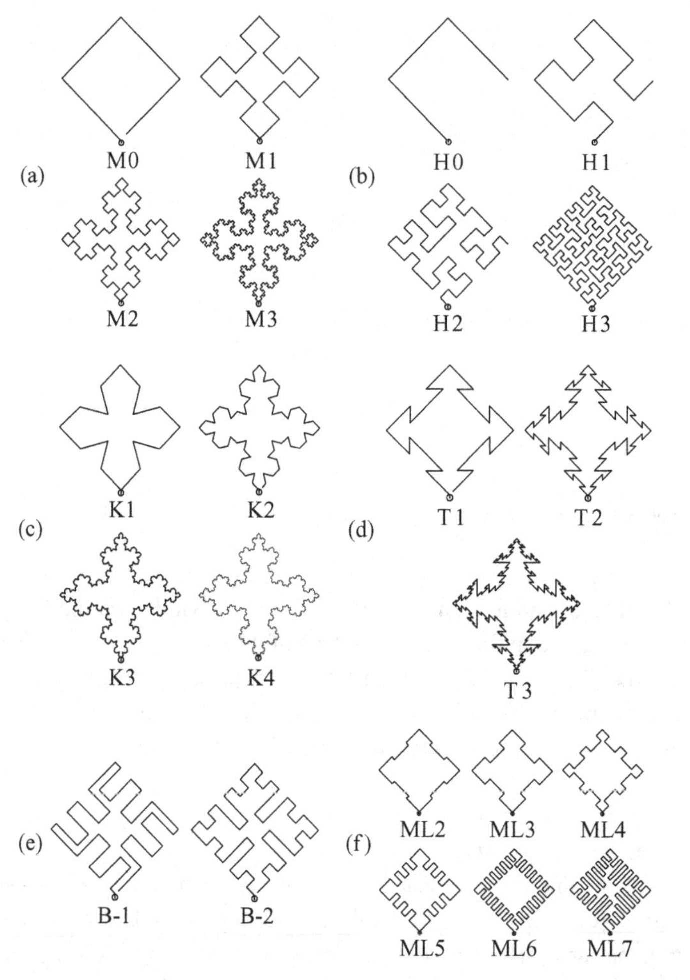


**Figure 7.103** Various types of planar fractal monopole: (a) mod-Minkowsky fractal, (b) Hilbert curve fractal, (c) Koch fractal, (d) tee-fractal, (e) box-meander line, and (f) meander line ([45], copyright ©2003 IEEE).

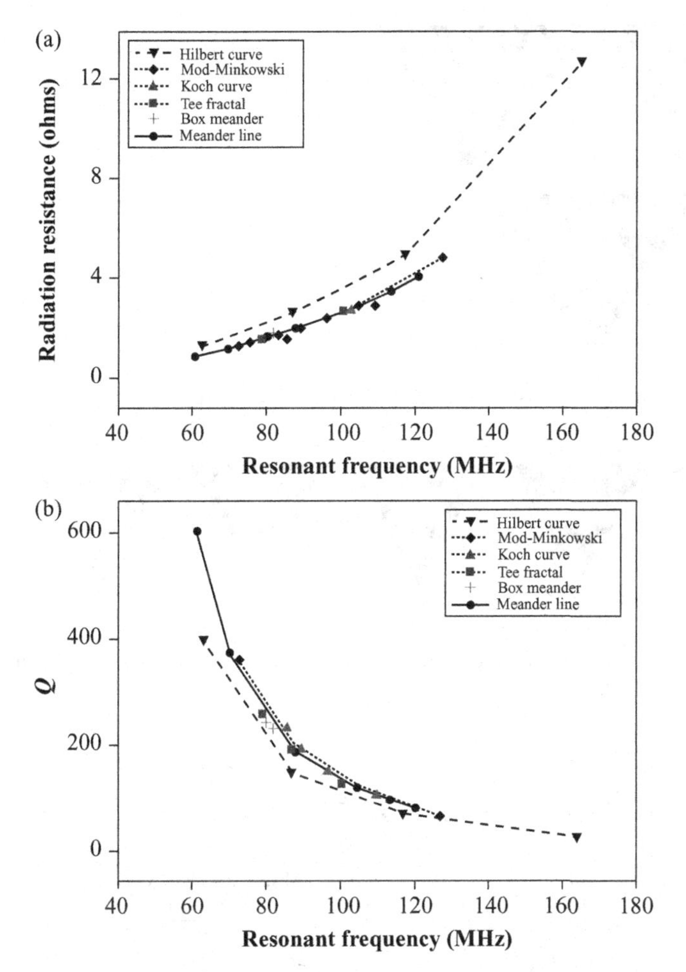


**Figure 7.104** Radiation resistance and $Q$ of six types of planar monopoles shown in Figure 7.103: (a) radiation resistance and (b) $Q$ ([46], copyright ©2009 IEEE).

compared in terms of resonance frequency, radiation resistance, efficiency, and $Q$ for various iteration numbers. Radiation resistance and $Q$ of these antennas are illustrated in Figure 7.104(a) and (b), respectively. The analysis disclosed that there are no significant differences in the resonance properties of these antennas, although antenna geometry and total wire length differ, indicating that their resonance behavior is primarily established by their occupied area [45]. This analysis provides useful suggestions for designing planar monopoles using space-filling curves.

*7.2.2.1.3.3 Sierpinsky Fractal monopole*

Sierpinsky Fractal structures (Figure 7.105 [46, 47]) were discovered by W. Sierpinsky in 1916. The Sierpinsky Fractal is generated starting from an equilateral triangle by removing the central triangle with vertices located at the midpoints of the sides of the original triangle, and repeating it for the three remaining triangles and then nine

**Figure 7.105** Sierpinsky fractal structure ([47], copyright ©2003 IEEE).

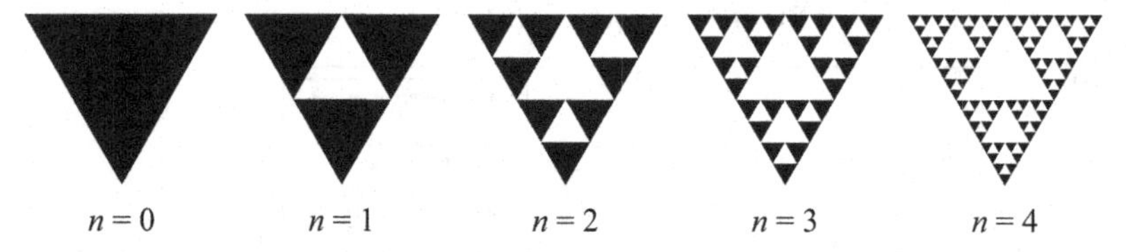


**Figure 7.106** The first four stages in construction of a conventional Sierpinsky gasket fractal ([47], copyright ©2003 IEEE).

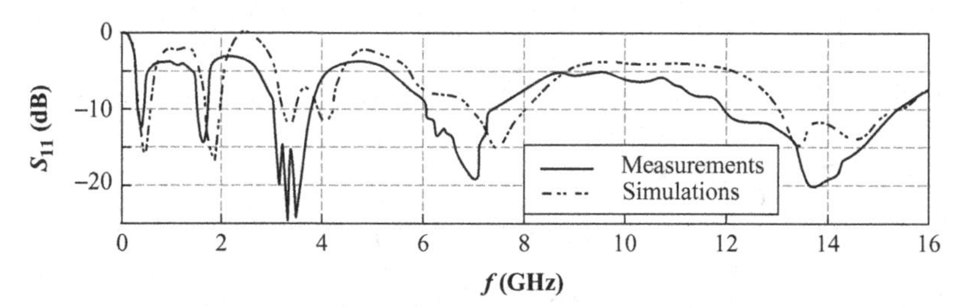


**Figure 7.107** Return-loss characteristics relative to 50 Ω ([47], copyright ©2003 IEEE).

remaining triangles, and so on to form a sieve of triangles as shown in Figure 7.105 [46]. The Sierpinsky sieve (or gasket) of triangles is, when applied to an antenna structure, characterized by a certain multiband behavior owing to its self-similar shape. A monopole antenna based on the Sierpinsky gasket has been shown to be an excellent candidate for multiband applications. The simulated and measured input reflection coefficient ($S_{11}$) of a four-stage Sierpinsky structure (Figure 7.106) is depicted in Figure 7.107 [46]. The antenna dimensions are: height $h = 88.9$, $h_n/h_{n+1} = 2$, flare angle $\alpha = 60°$, thickness of the substrate $h_s = 1.588$, permittivity $\varepsilon_r = 2.5$, and the iteration number $n = 4$ (in mm for all dimensional parameters). The antenna parameters are provided in Table 7.13, in which frequency band, resonance frequency, bandwidth, reflection coefficient, ratio of iteration segment, and height of triangle for each segment in terms of wavelength, are provided. A five-order Sierpinsky monopole shown in Figure 7.108 [47] was fabricated and measured and the simulated reflection coefficient is shown in Figure 7.109

**Table 7.13** Sierpinsky gasket fractal monopole parameters ([46], copyright ©2009 IEEE, and [47], copyright ©2003 IEICE)

| Iteration | Band number | $S_{11}$ $f_m$ [GHz] | Bandwidth [%] | [dB] | $f_{rn+1}/f_{rn}$ | $h_n/\lambda_n$ |
|---|---|---|---|---|---|---|
| 0 | 1 | 0.52 | 7.15 | 10 | 3.50 | 0.153 |
| 1 | 2 | 1.74 | 9.04 | 14 | 2.02 | 0.258 |
| 2 | 3 | 3.51 | 20.5 | 24 | 1.98 | 0.261 |
| 3 | 4 | 6.95 | 22 | 19 | 2.00 | 0.257 |
| 4 | 5 | 13.89 | 25 | 20 | – | 0.255 |

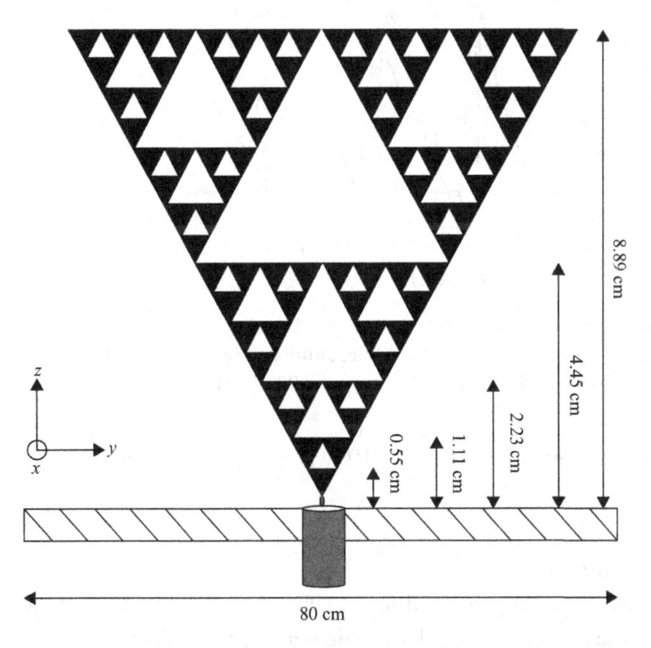


**Figure 7.108** A five-order Sierpinsky monopole ([47], copyright ©2003 IEEE).

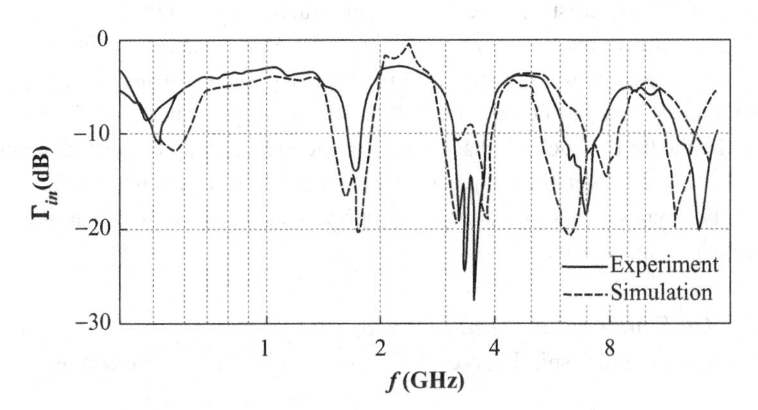


**Figure 7.109** Input reflection coefficient relative to 50 Ω ([47], copyright ©2003 IEEE).

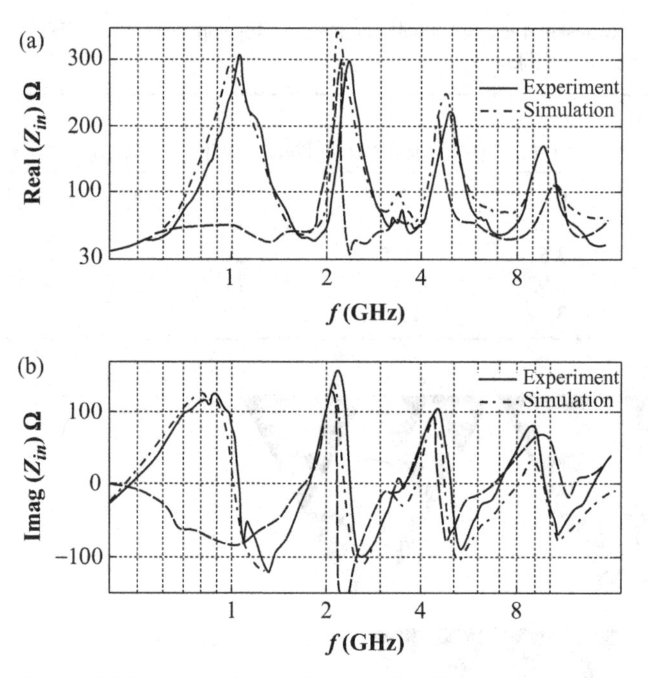


**Figure 7.110** Input impedance of a five-order Sierpinsky monopole: (a) real part and (b) imaginary part (solid line: measured, dashed line; simulated by using FDTD, and dashed-dotted line; simulated by using DOTIG4 ([47], copyright ©2003 IEEE).

Input impedances are given in Figure 7.110, where (a) shows resistance and (b) shows reactance.

### *7.2.2.1.4 Spiral antennas*

Antenna performances of dipoles and loops generally depend on the frequency; however, there are some antennas like self-complementary antennas which have frequency-independent characteristics. Other types, such as spiral antennas also have frequency-independent characteristics. The antenna structure is essentially determined by angle, not length, with which the antenna exhibits constant impedance characteristics. Ideally the spiral antenna of infinite structure is frequency independent; however, a practical antenna cannot have infinite structure, the frequency dependence on its impedance being limited by the finite structure; that is, the outer length of spiral curve and the impedance becomes constant over the frequency higher than that determined by the outer spiral length. Two types of spiral antenna are representative: equiangular spiral antenna and Archimedean spiral antenna.

#### *7.2.2.1.4.1 Equiangular spiral antenna*

The basic equiangular spiral curve shown in Figure 7.111 is expressed by

$$r = r_0 e^{a\varphi} \tag{7.60}$$

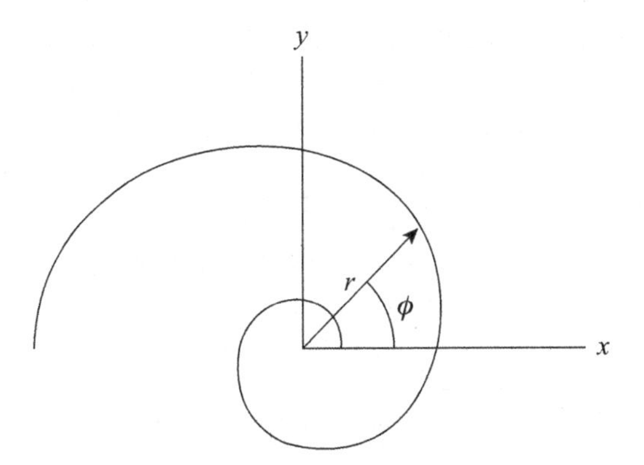


**Figure 7.111** Equiangular spiral curve ([48]).

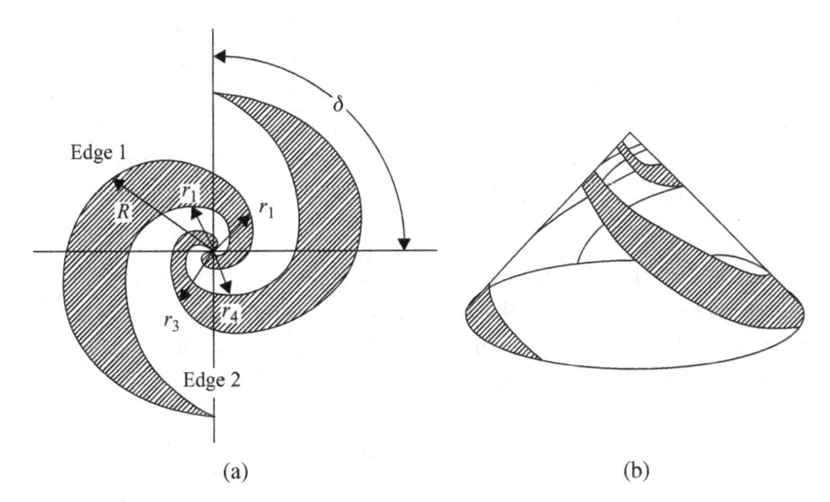


**Figure 7.112** (a) Planar equiangular spiral antenna (self-complementary case with $\delta = 90^\circ$ ([48]) and (b) a conical equiangular spiral structure.

where $r_0$ denotes radius for $\varphi = 0$, and $a$ is a constant controlling the flare rate of the spiral [48]. The spiral in the figure is right-handed, but left-handed spirals can be generated using negative values of $a$, or simply turning the spiral in the opposite direction. A planar equiangular spiral antenna is created by using the equiangular spiral curve shown in Figure 7.112 (a). Each edge of the spiral element has a curve basically expressed by (7.60). For instance, $r_1 = r_0 e^{a\varphi}$, $r_2$ is the curve rotated through angle $\delta$, so $r_2 = r_0\ e^{a(\varphi-\delta)}$, $r_3$ has the curve of $\delta = \pi$, then $r_3 = r_0 e^{a(\varphi-\pi)}$, and $r_4$ is the curve further turned from $r_3$, so $r_4 = r_0 e^{a(\varphi-\pi-\delta)}$.

Figure 7.112(b) depicts a conical equiangular spiral structure.

The flare rate $a$ is more conveniently expressed by using expansion factor $\varepsilon$, which is

$$\varepsilon = r(\varphi + 2\pi)/r(\varphi) = r_0 e^{a(\varphi+2\pi)}/r_0 e^{a\varphi} = e^{a2\pi}. \tag{7.61}$$

A typical value for $\varepsilon$ is 4, and then $a = 0.221$. The upper frequency $f_u$ is determined by the feed structure. The minimum radius $r_0$ is about $\lambda/4$ at $f_u$ for $\varepsilon = 4$. A nearly

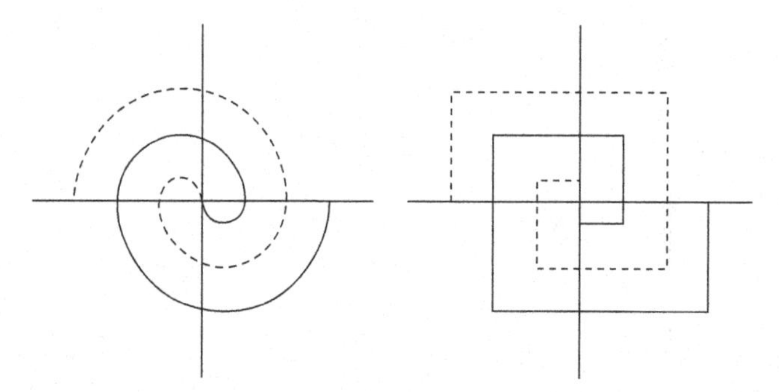

**Figure 7.113** (a) A circular type Archimedean spiral antenna and (b) a square type Archimedean spiral structure.

equivalent criterion is a circumference in the feed region of $2\pi r_0 = \lambda_u = c/f_u$. The lowest frequency $f_L$ is set by overall radius $R$, which is roughly $\lambda L/4$ ($\lambda L = c/f_L$), and the circumference $C$ of a circle, which encloses the entire spiral, is used to set $f_L$ by taking $C = 2\pi R = \lambda_L$. When $a = 0.221$, $R$ is $r(\varphi = 3\pi) = r_0\, e^{0.221(3\pi)} = 8.03\, r_0$ and equals $\lambda_L/4$. Meanwhile, $r_0$ equals $\lambda_u/4$, then, the bandwidth defined by $\lambda_L/\lambda_u = 8.03$, meaning about 8:1 bandwidth.

### *7.2.2.1.4.2 Archimedean spiral antenna*

Figure 7.113 illustrates the planar Archimedean spiral antenna [48], having two spiral curves, which are linearly proportional to the polar angle rather than exponential for the equiangular spiral, represented by

$$r = r_0(1 + \varphi/\pi) \text{ and } r_0(1 + \varphi/\pi - p). \tag{7.62}$$

The outside circumference in this case is one wavelength (a half wavelength for the outer half-circle) and an antenna with three turns is shown in Figure 7.113(a). Since the radiated fields produced by the two spirals are orthogonal, equal in magnitude, and 90 degrees phase difference, the wave is left-hand circular polarization (LHCP). The left-hand winding of the spiral determines the left-hand sense in the wave, which is viewed as radiation out of the page, and the opposite sense of wave is viewed as radiation toward the other side of page, that is RHCP.

Figure 7.113(b) illustrates a square Archimedean spiral modified from circular type.

The spiral produces a broad main beam perpendicular to the plane of the spiral; however, in many practical applications, a unidirectional beam is preferred. This suggests the use of a ground plane or cavity, over which the spiral is placed. The latter is called a cavity-backed spiral antenna. Since by using metallic material for the cavity, it is natural that the frequency performance is altered, leading to use of absorbing material loaded into the cavity to reduce the frequency variation.

The typical performance parameters for a cavity-backed Archimedean spiral antenna are HP (Half power beamwidth) = 75°, |AR| = 1 dB, G = 5 dB over 10:1 bandwidth or more [48]. The input impedance is approximately 120-Ω resistive.

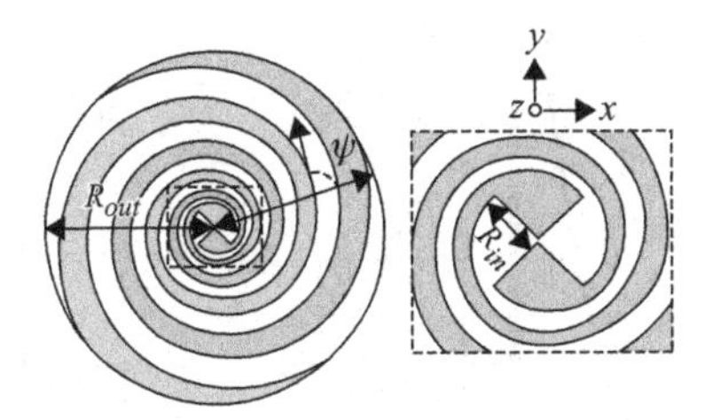


**Figure 7.114** Geometry of truncated two-arm equiangular spiral antenna ([49], copyright ©2007 IEEE).

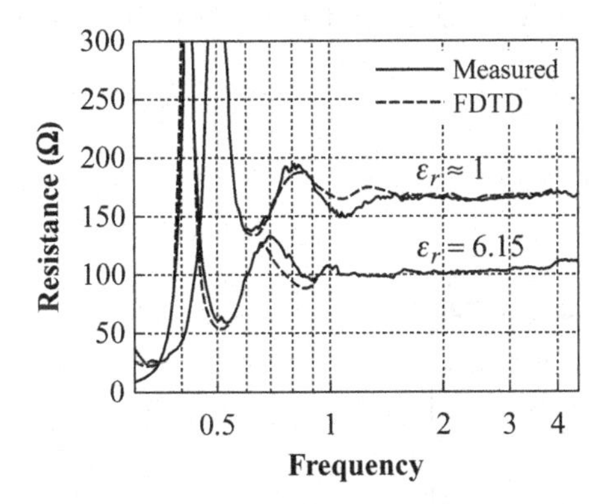


**Figure 7.115** Measured and FDTD simulated resistance for antenna type A and B ([49], copyright ©2007 IEEE).

### *7.2.2.1.4.3 Antenna performance and design of spiral antenna*

#### *7.2.2.1.4.3.1 Input impedance*

A model considered here is depicted in Figure 7.114 [49], which shows major design parameters $R_{in}$, $R_{out}$, and $\Psi$. Here two cases, A and B, where a spiral element is embedded on two different substrates, are treated. The dimensional parameters (see Fig. 7.114) are $R_{in} = 3$ mm, $R_{out} = 0.114$ m, $\Psi = 79°$. The antenna type A uses substrate modeled as a 0.1 mm polyester film layer $\varepsilon_r = 3.2$ on top of a foam layer $\varepsilon_r = 1$, while the antenna type B uses substrate modeled as a uniform dielectric with a thickness of 1.27 mm and $\varepsilon_r = 6.15$. Input impedances measured and calculated by using FDTD simulator are depicted in Figure 7.115, showing resistance, and Figure 7.116(a) and (b), showing reactance for the type A and the type B, respectively. Impedance characteristics tend to show peculiar behavior having three distinct regions. The outer truncation of the spiral curve dominates the performance in low-frequency regions, the inner truncation determines that in higher-frequency behavior, and the shape of the spiral curve itself relates to that of the intermediate-frequency regions. There is a region where a band of nearly constant impedance is seen between the erratic behavior at the upper and lower frequencies. This is the specific behavior of the spiral antenna as a frequency-independent antenna and this region is referred to as the "operating band." The input impedance $Z$ in this region is 188 Ω, when the spiral has self-complementary

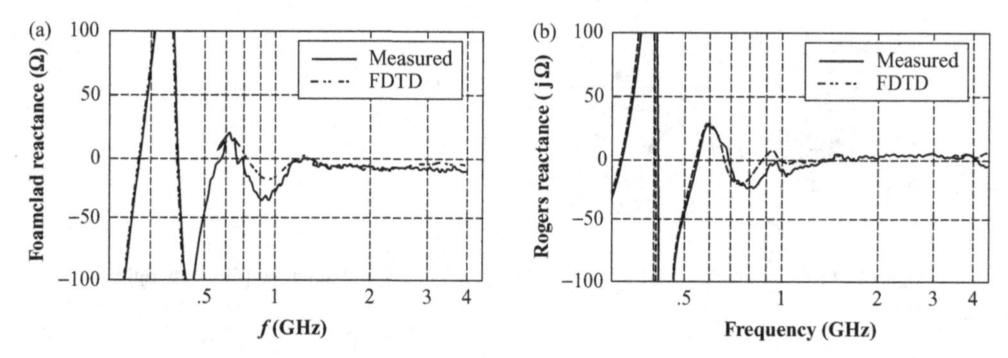


**Figure 7.116** Measured and FDTD simulated reactance: (a) for type A and (b) for type B ([49], copyright ©2007 IEEE).

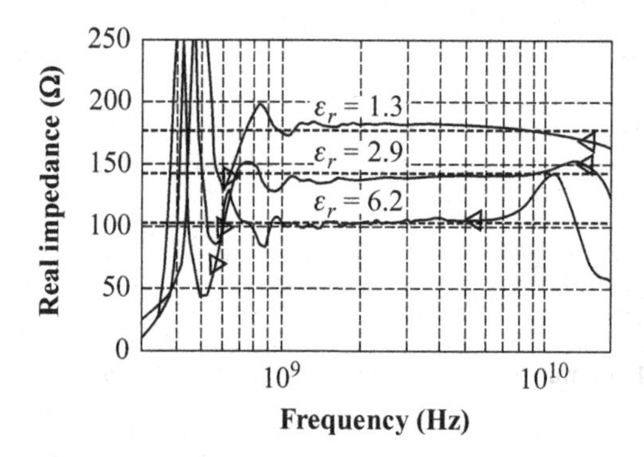


**Figure 7.117** Real impedance of spirals designed to be 188 Ω, 148 Ω, and 108 Ω (triangular markers denote the edges of the operating band) ([49], copyright ©2007 IEEE).

structure [50], which is given by

$$Z = (1/2)Z_0 = (1/2)\sqrt{\mu_0/\varepsilon_0} \tag{7.63}$$

where $Z_0$ is the free space impedance. When the antenna is embedded on the substrate of $\varepsilon_r$, the impedance $Z$ is given by using effective permittivity $\varepsilon_{eff}$ as

$$Z = (1/2)\sqrt{\mu_0/\varepsilon_{eff}} = (1/2)\sqrt{\mu_0/\varepsilon_r\varepsilon_0} = (1/2)Z_0/\sqrt{\varepsilon_r}. \tag{7.64}$$

The impedance as a function of frequency is illustrated in Figure 7.117, which is calculated for the spiral of type A on the substrate of thickness of 1.27 mm with different $\varepsilon_{eff} = 1.3$, 2.9, and 6.2. The lower-frequency region is dominated by a series of resonant peaks, the middle-frequency region is approximately constant, and higher-frequency regions contain either a resonant peak or a region where the real impedance decreases with frequency. In the figure, marker arrows denote the edge of the operating band and the dashed lines show the calculated characteristic impedance, which is the average in the operating band. The spiral impedance can be evaluated numerically over the range

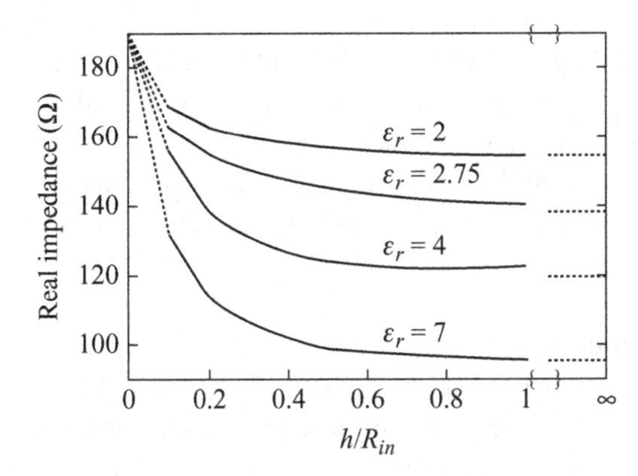


**Figure 7.118** Real impedance with respect to $h/R_{in}$ for $\varepsilon_r = 2, 2.75, 4$, and 7 ([49], copyright ©2007 IEEE).

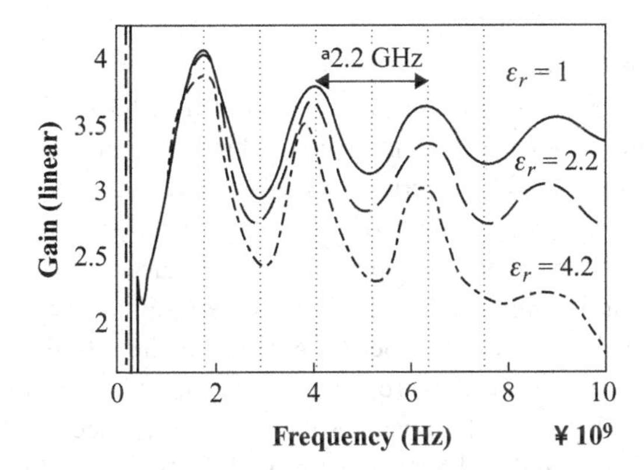


**Figure 7.119** Simulated gain for a spiral with $\varepsilon_r = 1, 2.2$, and 4.2 ([49], copyright ©2007 IEEE).

$0.1 < h/R_{in} < 1$ for $\varepsilon_r < 7$ in Figure 7.118. The impedance for smaller $h/R_{in}$ tends to $(1/2)\,Z_0$, corresponding to no substrate case (meaning in free space), and for larger $h/R_{in}$, to $(1/2)\,Z_0/\sqrt{\varepsilon_r}$.

### *7.2.2.1.4.3.2 Gain*

Measured and FDTD simulated gains of the spiral antenna of $\Psi = 79°$, $R_{in} = 2.5$ mm, $R_{out} = 0.12$ m are illustrated in Figure 7.119, where a sequence of gain variation for various substrates with $\varepsilon_r$ from lowest to highest 1, 2.2, and 4.2 is given. Gain $G$ can be evaluated as

$$G = \pi\sqrt{\varepsilon_{act}} \tag{7.65}$$

where $\varepsilon_{act}$ is effective permittivity of the active region ring in the spiral. The active region is a part that contributes to radiation, which occurs in a curved structure having circumference one wavelength or in a straight wire or edge with length of one-half wavelength. Equation (7.65) indicates that the spiral radiates with a constant gain of $\pi$ in free space. However, all three gains have a periodic ripple with period of approximately 2.2 GHz, which is the inverse of time delay $t_d = 0.45$ ns that is caused in the radiated field by the wave travelling on the spiral surface. This arises because of the truncation of the spiral, which is 12 cm from the feed. The time delay which causes ripple tends to be approximately $R/c$, where $R$ is the radius of the spiral and $c$ is velocity of light in free space.

*7.2.2.1.4.4 Radiation patterns*

Radiation patterns of the spiral antenna shown in Figure 7.114 ($\varepsilon_r = 4.2$) are illustrated in Figure 7.120 [49], where only upper hemisphere patterns are given, as the lower hemisphere patterns are the same. The patterns are seen to have alternative contraction and expansion along the $z$-axis that corresponds to the maxima of the gain ripple. In the dielectric case, it becomes further complicated as the frequency becomes higher, because of deflection of radiated power into side lobes.

*7.2.2.1.4.5 Unidirectional pattern*

In many practical spiral antenna applications, a unidirectional pattern is required. For this purpose, a conducting shallow cavity is placed behind a spiral antenna. However, since simple use of such a cavity deteriorates the inherent wideband property of the spiral antenna, an absorber may be applied to the outer spiral arms to restore the wideband property [50]. The absorber will suppress reflection of waves caused by the truncation of outer spiral arms. Figure 7.121 illustrates an example of an equiangular spiral antenna, to which a ring-shaped absorber (R-ABS) is applied. In the figure, (a) shows an exploded view, (b) the side view, (c) spiral arms, and (d) a magnified view of near the input terminals. Figure 7.122 illustrates the input impedance; (a) $R_{in}$ (resistance), (b) $X_{in}$ (reactance) with and without cavity, (c) with R-ABS. The antenna used here is a two-arm equiangular spiral antenna having parameters provided in Table 7.14, which is backed by a conducting cavity of height $H_{cav}$ (= antenna height), and diameter $D_{cav}$. Axial ratios with and without R-ABS are shown in Figure 7.123. As these figures show, fairly wideband characteristics can be seen with use of R-ABS.

The absorber size can be reduced only to cover near the edge of the outer spiral arms so that no or almost no reflection appears at the ends of the outer spiral arms. Figure 7.124 illustrates an example of the use of arc-shaped ABS (A-ABS) [50]. Input impedance $Z_{in}$ ($= R_{in} + jX_{in}$) and axial ratio for a case with A-ABS are shown in Figure 7.125(a) and (b), respectively, and radiation patterns are depicted in Figure 7.126, in which (a) and (b) show with A-ABS, but for the case where $\varphi_{arc} = 90°$ and $180°$, respectively, and (c) shows without A-ABS ($\varphi_{arc} = 0°$).

*7.2.2.1.4.6 Miniaturization of spiral antenna*

Spiral antennas can be miniaturized by lowering phase velocity of the waves travelling on the spiral arms from the feed point. However, since a spiral antenna has an inherently wide bandwidth property, the phase velocity at a single resonance frequency cannot be taken

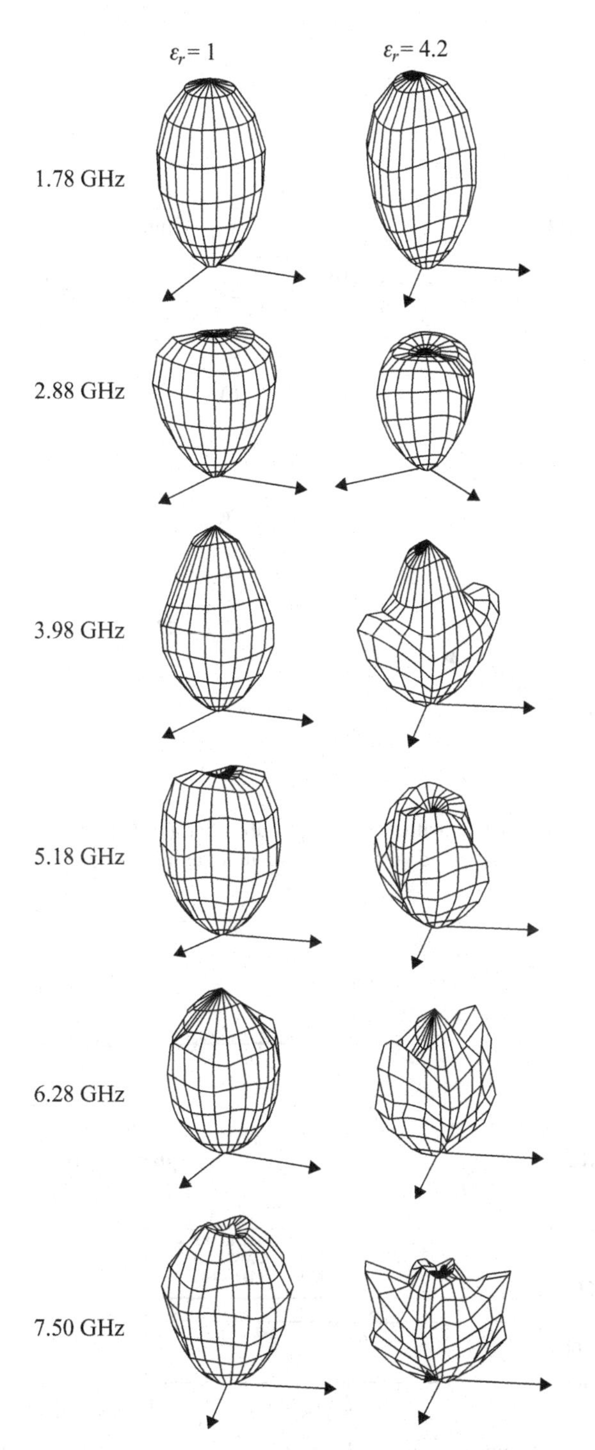


**Figure 7.120** Radiation patterns of antenna shown in Figure 7.114 on the upper hemisphere for cases $\varepsilon_r = 1$ and 4.2 ([49], copyright ©2007 IEEE).

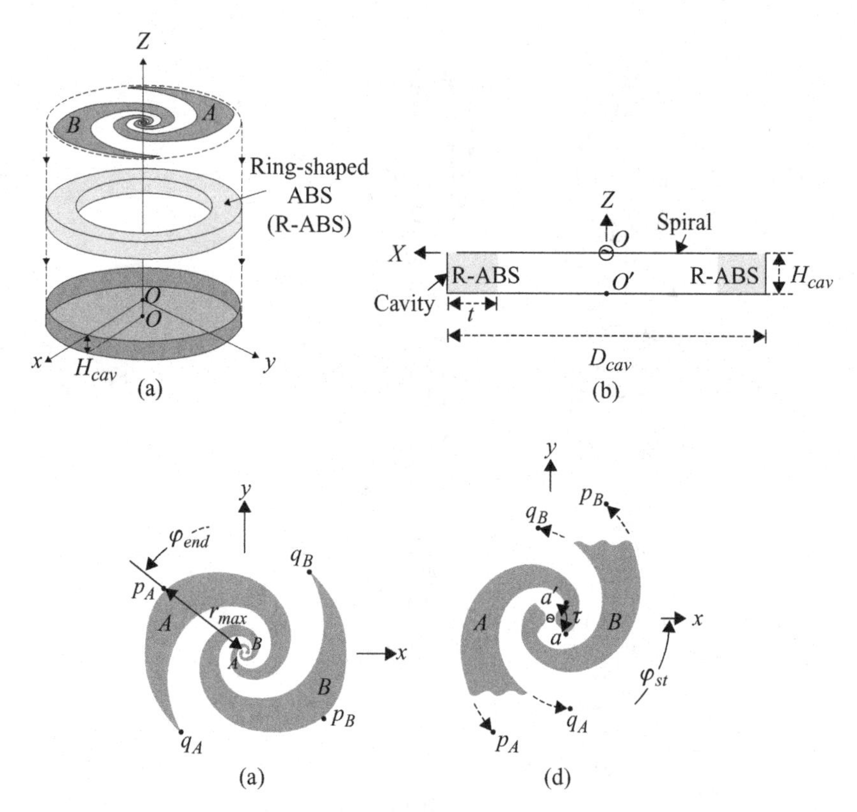


**Figure 7.121** Equiangular spiral antenna with a ring-shaped absorber: (a) exploded view, (b) side view, (c) spiral element, and (d) magnified view around the input terminals ([50], copyright ©2008 IEEE).

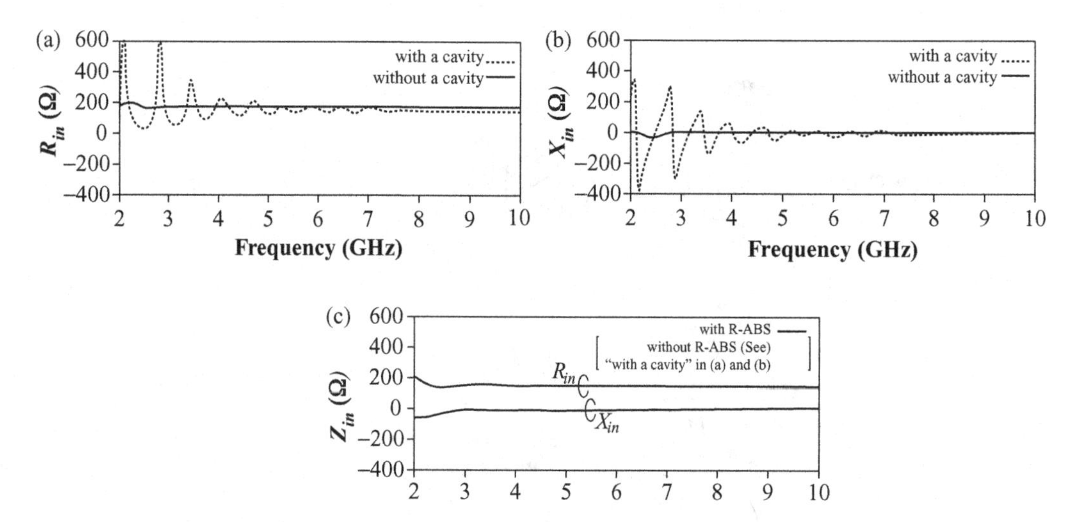


**Figure 7.122** Input impedance of equiangular spiral antenna; (a) $R_{in}$ resistance (b) reactance (dotted line, with cavity and solid line, without cavity), and (c) $R_{in}$ and $X_{in}$ with A-ABS ([50], copyright ©2008 IEEE).

**Table 7.14** Geometrical parameters ([50], copyright ©2008 IEEE)

| symbol | value | unit |
|---|---|---|
| $r_0$ | 1.5 | mm |
| $a_s$ | 0.35 | rad$^{-1}$ |
| $\varphi_{st}$ | $-0.25\pi$ | mm |
| $\varphi_{end}$ | $2.806\pi$ | rad |
| $\tau$ | $0.5\pi$ | rad |
| $D_{cav}$ | 120 | mm |
| $H_{cav}$ | 7 | mm |

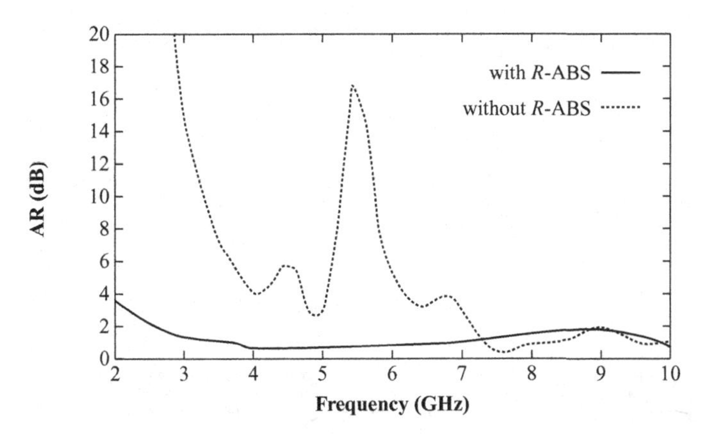


**Figure 7.123** Axial ratio with and without R-ABS ([50], copyright ©2008 IEEE).

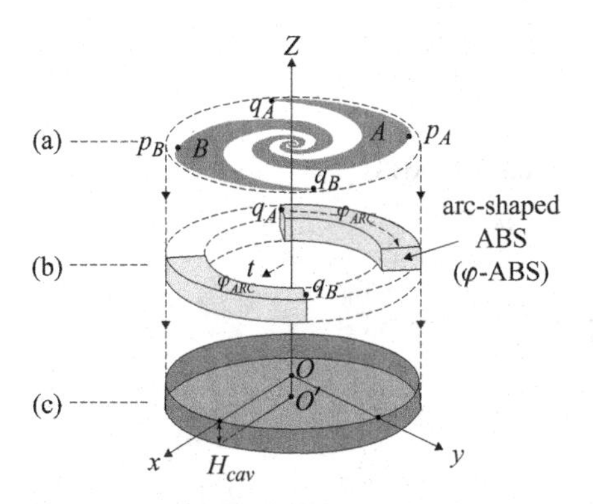


**Figure 7.124** Equiangular spiral antenna with arc-shaped absorber: (a) spiral element, (b) arc-shaped ABS, and (c) cavity ([50], copyright ©2008 IEEE).

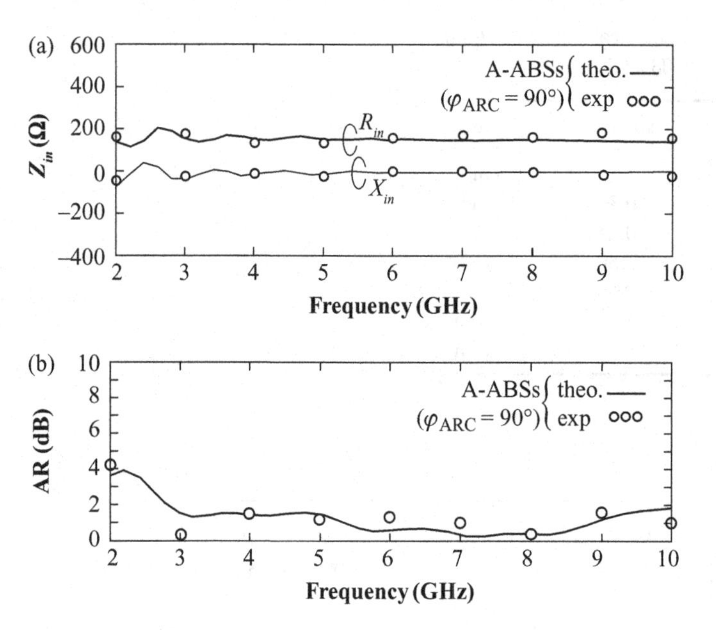


**Figure 7.125** (a) Input impedance $Z_{in} = (R_{in} + jX_{in})$ and (b) axial ratio with A-ABS ([50], copyright ©2008 IEEE).

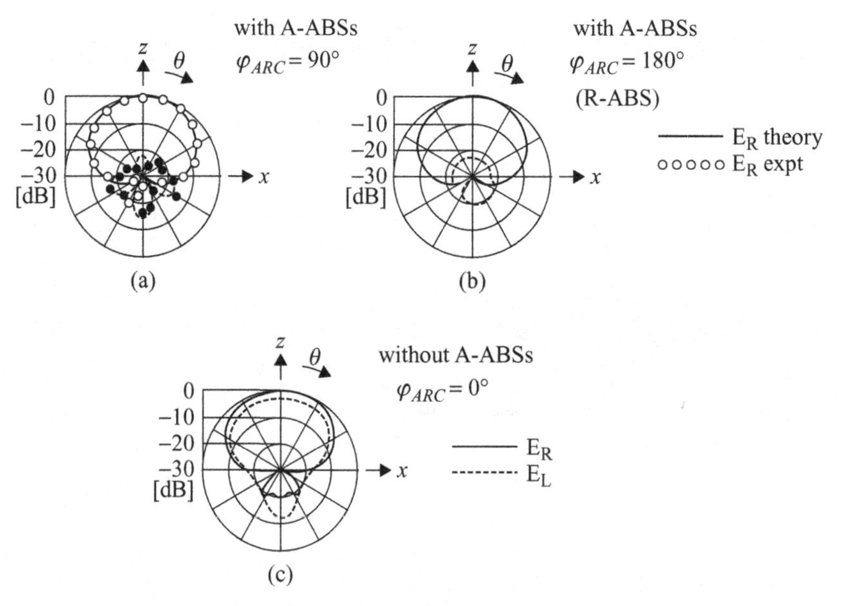


**Figure 7.126** Radiation patterns; (a) $\Phi_{arc} = 90°$, (b) $\Phi_{arc} = 180°$, and $\Phi_{arc} = 0°$ ([50], copyright ©2008 IEEE).

as a measure of miniaturization as in the case of narrow-band antennas. Miniaturization in cases of spiral antennas may be interpreted by reduction of the spiral arm length or by lowering operating frequency for the specified gain [51]. In practice, lowering the phase velocity of the wave travelling on the spiral arms can be realized by modifying the spiral arm structure, or by loading either resistance or reactance components on the spiral arms. There is still a need to optimize the antenna structure or way of loading to obtain minimum $Q$ or maximum gain for a wide frequency range with a given antenna size. In this case, techniques for proper matching to keep wideband characteristics are required as well.

### *7.2.2.1.4.6.1 Slot spiral antenna*

The slot spiral antenna has been studied extensively by J. H. Volakis and his colleagues [52–64]. Their work includes development of design algorithms to optimize the antenna geometry, feed, loading, and so forth, and demonstrates techniques of miniaturization and practical models. As spiral antennas have inherently wide bandwidth, miniaturization should be realized without losing the initial wideband characteristics. Loading of resistance, reactive components or materials is a common method for miniaturizing antennas. When applying the loading method to spiral antennas, specific care must be taken to keep wideband characteristics. Practical methods are, for example, modifying antenna structure by meandering or coiling, using dielectric materials in superstrate as well as substrate, and so forth. Wideband matching can be done, for instance, by tapering resistive loading on the spiral arms, by infinite coaxial balun along with spiral arms, by the broadband hybrid, and so forth. Since the slot radiates to both front and back sides normally to the spiral surface, the antenna is often placed on a thin cavity to produce unidirectional radiation, which is desired in many cases of practical operation. The spiral slot structure has a feature that is different from usual spiral of conducting arms; since the source of radiation is magnetic current, the spiral element can be placed close to the metallic ground plane that attributes the positive image effect, resulting in increase of the antenna gain and thinning the antenna structure as well. Some examples are introduced below.

An example of a slot spiral antenna is illustrated in Figure 7.127 [50], which shows the radiating slot spiral antenna; (a) cross section, (b) underside view, and (c) top view with microstrip balun. The antenna type is an Archimedean spiral, which is mathematically described as $r = a\theta + b$, where $r$ is the radian length from the origin, $a$ is the spiral growth rate, $\theta$ is the angular position, and $b$ is the initial radial offset from the origin. Geometries of the antenna are: $a = 0.28$ cm/radian, $\theta$ is in a range from 0.0 to 24.5 radians, $b = 0$, the maximum diameter $d_{max}$ of the radiating slot is 14.5 cm, and the slot width $w = 0.762$ mm. The substrate $\varepsilon_r = 3.38$ and the thickness of the substrate is 7.62 mm. The balun is constituted with microstrip line, which is placed on the metal part of the spiral toward the origin (center) of the spiral, using the radiating slot as its ground plane as Figure 7.127(a) and (c), respectively, show. At the periphery of the spiral, a coaxial connector is soldered to the microstrip line, and at the center of the spiral, the end of the microstrip line is shorted across the slot line to feed the slot. Termination of the slot

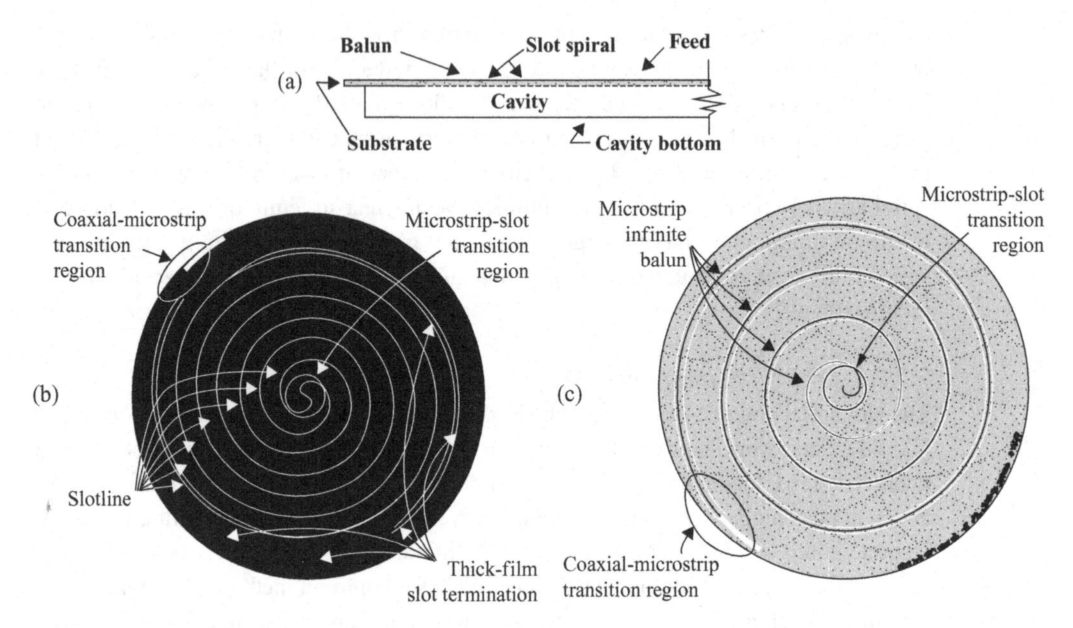


**Figure 7.127** Geometry of Archimedean spiral antenna backed with cavity: (a) cross section view, (b) underside view, and (c) top view ([51], copyright ©2001 IEEE).

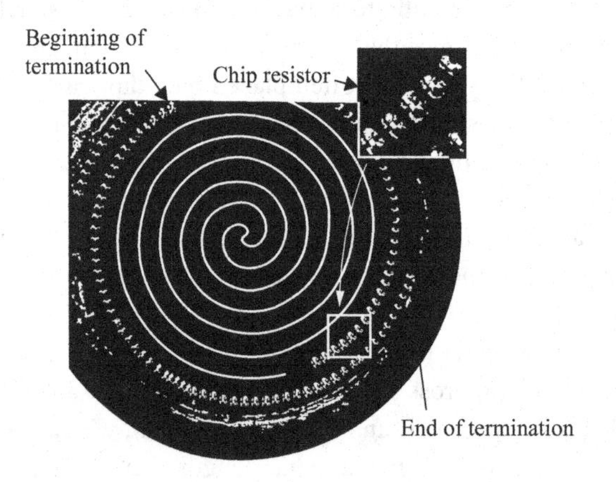


**Figure 7.128** Slot spiral termination with thick-film resistors ([51], copyright ©2001 IEEE).

is done by loading resistances on the slot line, forming an impedance transformer between the antenna impedance and the feed so as to ensure wideband characteristics and achieve desired axial ratio. The termination is implemented by using 60-Ω resistors, distributed equally along the overall transformer length. Figure 7.128 illustrates the implemented termination by using the thick-film resistor [54]. The diameter of the cavity is 14.9 cm, which can just accommodate the slots and the termination resistors. The depth $D$ of the cavity is set by the maximum desired frequency of the antenna, and at that point it

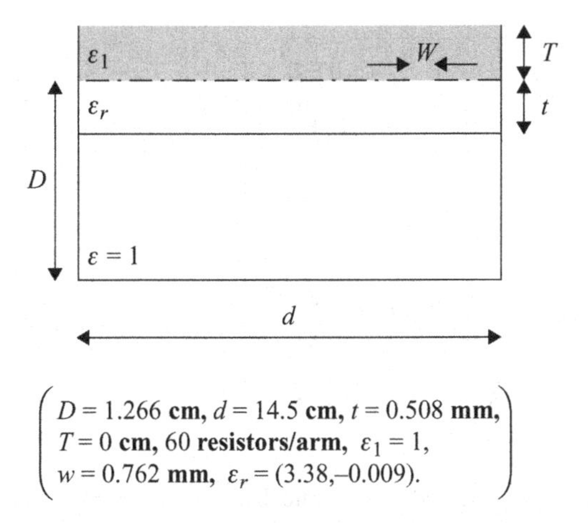


**Figure 7.129** Antenna geometry cross section, which shows loading with substrate of dielectric constant $\varepsilon_r$ and superstrate of dielectric constant $\varepsilon_1$ ([51], copyright ©2001 IEEE).

should be less than $\lambda_{highest}/4$ to prevent destructive interference. The minimum depth of the cavity is determined by the slot width $w$ and substrate dielectric constant $\varepsilon_r$. The actual depth of the cavity used is 6.35 mm ($\lambda_{lowest}/60$).

### *7.2.2.1.4.6.2 Spiral antenna loaded with capacitance*

This spiral antenna is miniaturized by either capacitive loading or inductive loading on the spiral arms. The capacitive loading is implemented by applying dielectric substrate and/or superstrate to the spiral surface. Figure 7.129 shows cross section of the antenna geometry in the cavity when a dielectric substrate with $\varepsilon_r$, having the thickness $T$, or a dielectric superstrate with $\varepsilon_1$, having thickness $t$, is used. $MF_{sm}$ (near-field miniaturization factor) is evaluated by taking the electrical lengths of spiral arms for various loadings $\theta_{loaded}$ and $\theta_{unloaded}$, given in [51] as

$$MF_{sm} = \sqrt{\theta_{loaded}/\theta_{unloaded}}. \tag{7.66}$$

Figure 7.130 shows $MF_{sm}$ as a function of substrate thickness with four values of $\varepsilon_r$. The phase progression on the slot line when dielectric superstrate (with $\varepsilon_1 = \varepsilon_r$) is added on the slot line is depicted in Figure 7.131, which shows variation of phase angle with respect to the slot line length for two different thickness $t$ (0 cm with white points, and 0.0508 cm with black points) and four different $\varepsilon_r$ (2.2, 3.38, 6.15, and 10.2) of the superstrate. This figure clearly gives an evidence of increase in the electrical length of the slot with increase in $\varepsilon_r$ of the superstrate. Figure 7.132 gives near-field miniaturization factor $MF_{sm}$ with respect to the superstrate thickness $t$ for four values of $\varepsilon_r$ (from 2.2 to 10.2). Gain of this antenna is about 0 dB around 1 GHz.

Miniaturization can also be done by loading either capacitive or inductive components on the slot line that increases electrical length of the slot line and thus lowers the operating frequency [58–63].

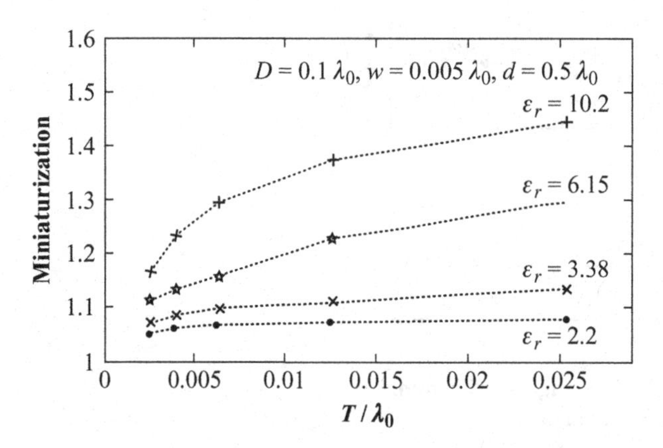


**Figure 7.130** Miniaturization factor with respect to substrate thickness for four values of $\varepsilon_r$ ([51], copyright ©2001 IEEE).

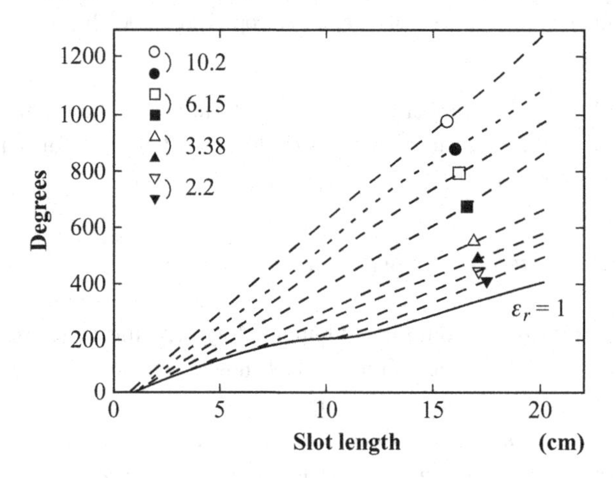


**Figure 7.131** Phase progression with respect to the slot line length for different superstrate thickness for four values of $\varepsilon_r$ from [51].

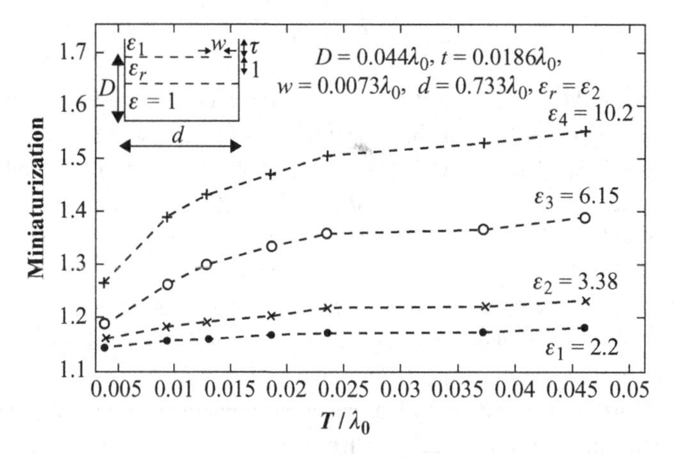


**Figure 7.132** Miniaturization factor with respect to superstrate thickness for four values of $\varepsilon_r$ ([51], copyright ©2001 IEEE).

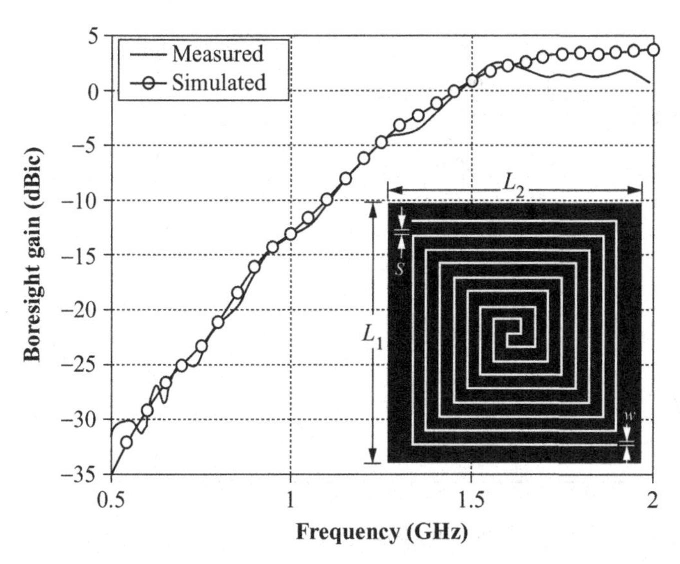


**Figure 7.133** Square spiral antenna geometry and gain with respect to frequency ([53], copyright ©2005 IEEE).

*7.2.2.1.4.6.3 Archimedean spiral antennas*

Another example of spiral antenna is the Archimedean spiral described in [53], in which feeding technique, single resistance loading, tapered dielectric superstrate, and so forth, are discussed and the performances are introduced. The antenna is shown in the inset of Figure 7.133. The geometrical parameters of the antenna are; slot width $w =$ 0.0762 cm, conductor width $S = 0.2286$ cm, and aperture dimensions are $5.715 \times 5.715$ cm, while the actual spiral has dimensions of $4.6482 \times 4.9530$ cm. To achieve unidirectional radiation, a square cavity having an inner dimension of 5.08 cm and a depth of 2.54 cm is used. The square spiral slot is printed on a 0.06096 cm thick substrate ($\varepsilon_r = 4.25 - j0.0595$). A 0–180° broadband hybrid is employed for feeding and 15 resistors using Klopfenstein taper [64] are used for the termination, instead of an infinite coaxial balun embedded on the surface of the spiral elements and feeding the spiral at its center. Figure 7.133 illustrates measured and simulated gain. In order to terminate with only a single resistor, the slot line is extended to the aperture edge, resulting in use of a 47-Ω resistor that can provide an effective termination in terms of both axial ratio and impedance matching. By this means gain is improved and is shown in Figure 7.134, which also shows the gain when resistive taper is used as a comparison.

Further improvement of gain is achieved by using high-contrast dielectric loading [53]. Photos of a tapered dielectric layer placed as the superstrate on the slot spiral are shown with cross sectional view in Figure 7.135(a), and the measured gains for various values of the superstrate $\varepsilon_r = 9$, 16, 30, and 90 are illustrated in Figure 7.135(b). Lesser thickness of the dielectric is used as $\varepsilon_r$ becomes larger. It was found that with higher $\varepsilon_r$, the gain tends to be higher. When −15 dB gain is taken as a reference, lowering frequency is observed from 884 MHz to 564 MHz as $\varepsilon_r$ increases from unity (no dielectric) to 90.

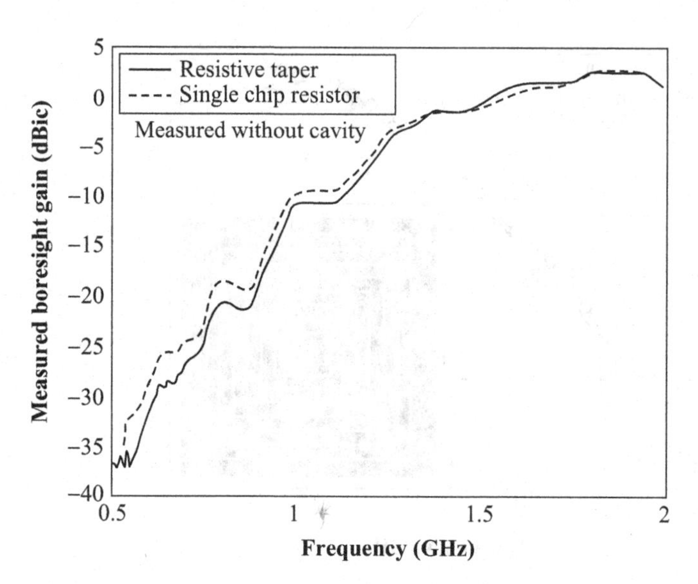


**Figure 7.134** Measured gain of the 2-inch square spiral antenna with a resistive taper and a single resistor termination ([53], copyright ©2005 IEEE).

This implies 36% size reduction or lowering of the operation frequency for the same size aperture without additional effort to improve impedance matching.

*7.2.2.1.4.6.4 Spiral antenna loaded with inductance*

Spiral arms are modified to increase their effective length by loading inductance on the arms. Inductive loading on the spiral wire arms can be formed with either planar meandering with rectangular or zigzag, or 3D coiling of the spiral line. Figure 7.136 illustrates a spiral element replaced with (a) planar meander line, and (b) loop-like winding (rectangular or conical) [58]. Gain of the 6-inch spiral antennas is evaluated to compare dependence of the spiral structures such as planar meandering spiral, volumetric spiral, and simple (straight wired) spiral as Figure 7.137 illustrates.

*7.2.2.2 Three-dimensional (3D) structure*

To improve antenna performance in a small antenna, efficient use of a volume, within which an antenna is contained, is considered. When an antenna takes a three-dimensional (3D) structure, the utmost 3D structure for this purpose is a sphere, and use of the periphery of the sphere is ideal. So far some attempts at placing helical elements on the periphery of a sphere have been made for efficient use of the periphery of a sphere so as to obtain the lower $Q$ compared with other types of antenna having an electrically small size [31].

Other types of structures than spheres are possible to realize small antennas having lower $Q$ in comparison with 2D structures of the same size, if the volume is within the reasonable dimensions for practical use. Typical antennas having such 3D structures are Koch tree, cylindrical helix, cube, and so forth.

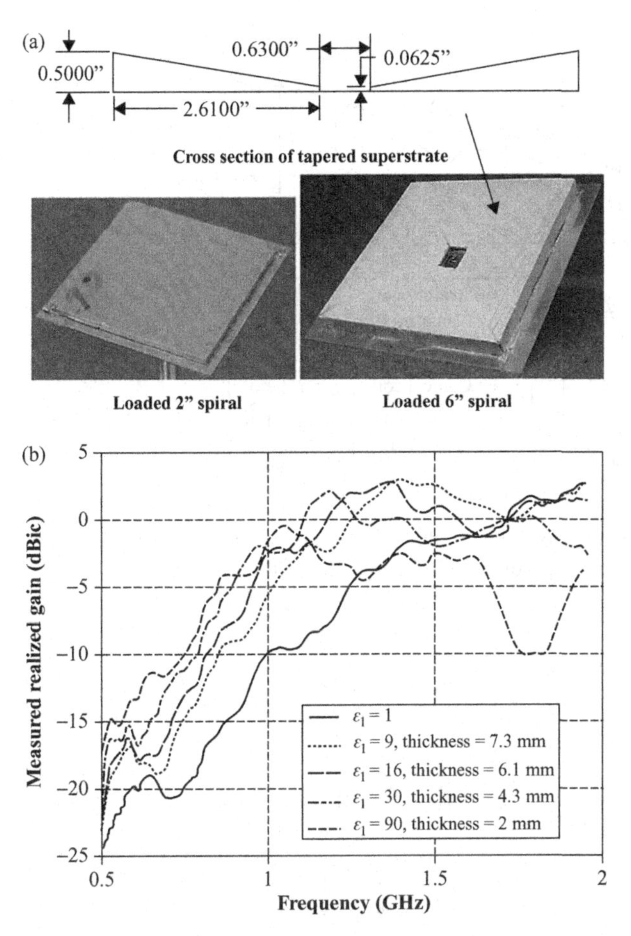


**Figure 7.135** (a) Cross section of tapered dielectric superstrate placed on the slot spiral and photos, and (b) variation of measured total circularly polarized gain for the 2-inch square spiral antenna depending on various dielectrics ([53], copyright ©2005 IEEE).

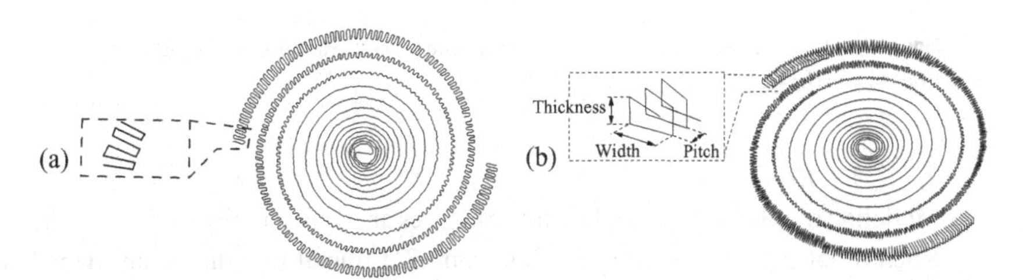


**Figure 7.136** Inductive loading of the 6-inch diameter spiral antennas: (a) with planar meandering and (b) with volumetric coiling ([58], copyright ©2006 IEEE).

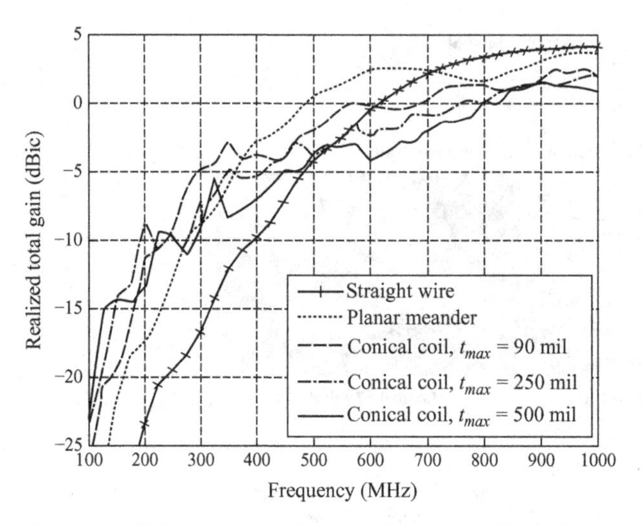


**Figure 7.137** Gain comparison between three types of spiral antennas; ordinary wire spiral, planar meandering, and volumetric coiling ([58], copyright ©2006 IEEE).

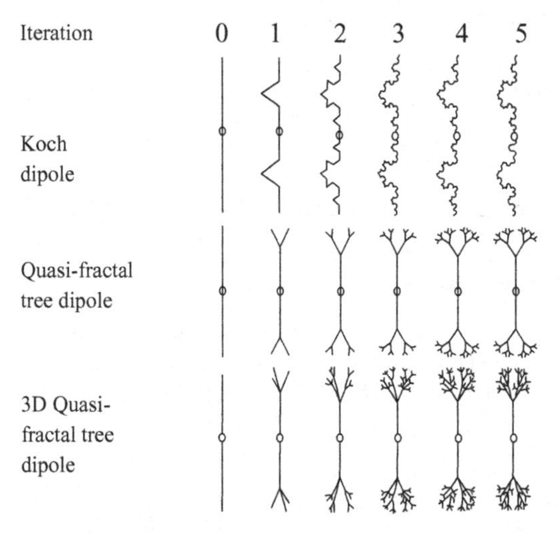


**Figure 7.138** Three types of fractals: Koch dipole, quasi-fractal tree dipole and 3D quasi-fractal tree dipole ([44], copyright ©2002 IEEE).

### *7.2.2.2.1 Koch trees*

Modeling dipoles with Koch Fractal structures are shown in Figure 7.138, where ordinary Koch dipole, quasi-fractal tree dipole, and 3D fractal tree dipole are depicted [44]. In the fractal tree, the top third of every branch is split into two sections and the 3D version has the top third of each branch split into four segments that are each one-third in length. The 3D fractal here implies only that the structure is not contained in a plane.

As one way for evaluating the performance, $Q$ of the antenna for each type is compared. Figure 7.139 provides $Q$ of (a) Koch dipole, (b) quasi-fractal tree dipole, and (c) 3D

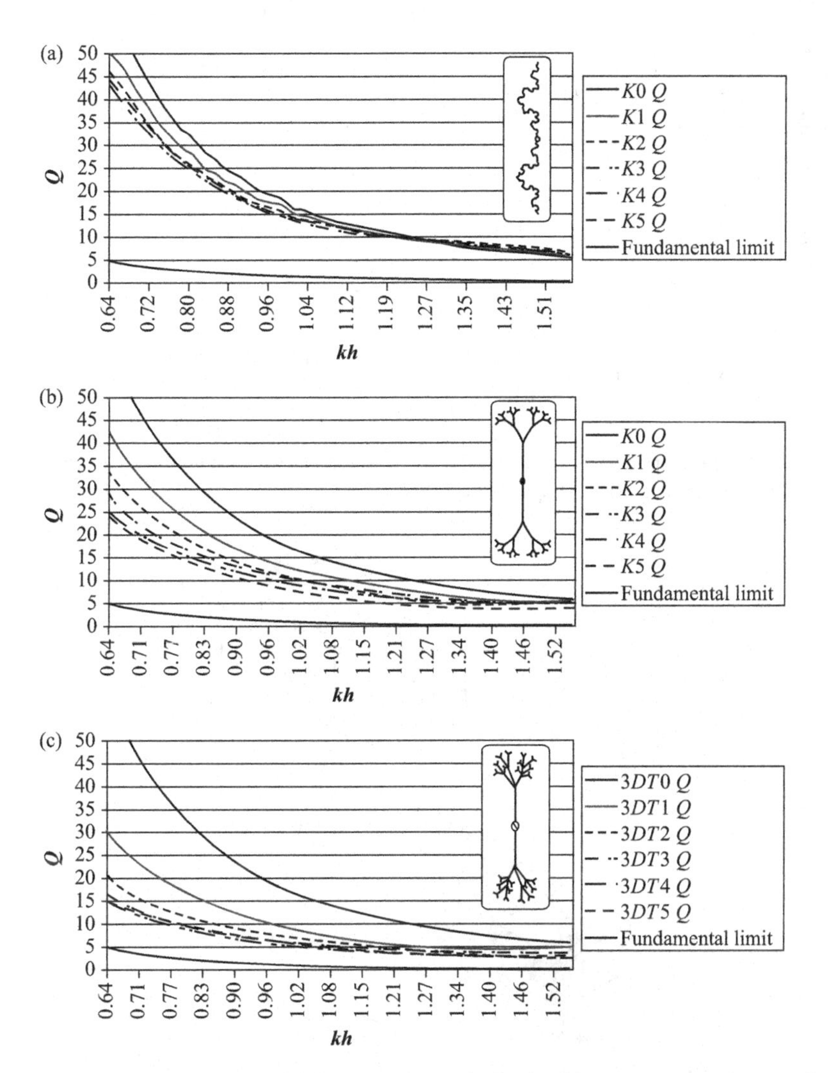


**Figure 7.139** Comparison of $Q$ between (a) Koch dipole, (b) fractal tree, and (c) 3D fractal tree ([44], copyright ©2002 IEEE).

quasi-fractal tree dipole, respectively. The 3D fractal tree shows the lowest $Q$, and the lowest resonance frequency as shown in Figure 7.140 [45], meaning that this type uses the given space most efficiently.

### *7.2.2.2.2 3D spiral antenna*

A very small spiral antenna to be used for bio-medical telemetry has been developed, and the design and realization of the antenna is introduced [65]. Antenna design started from a planar meander line structure, then its folded structure, and finally progressed to 3D spiral structure as shown in Figure 7.141. Geometry and dimensions of the optimized design of the antenna are shown in Figure 7.142. Reflection factor $|S_{11}|$

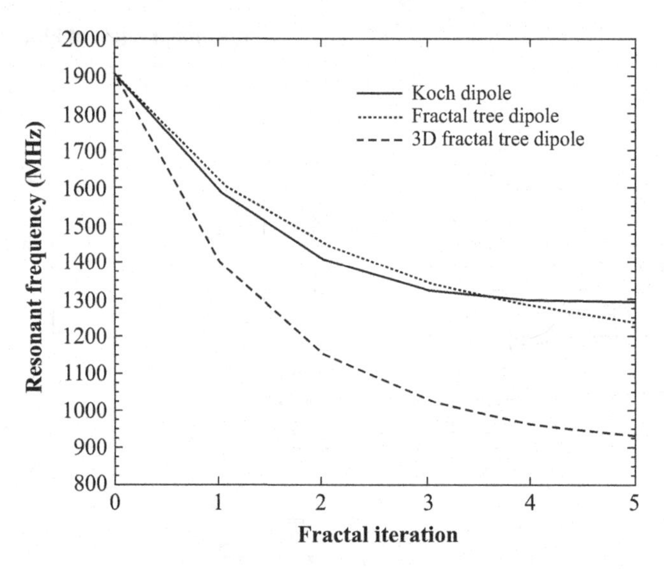


**Figure 7.140** Computed resonance frequency for three types of fractal dipole as function of the number of iterations ([45], copyright ©2003 IEEE).

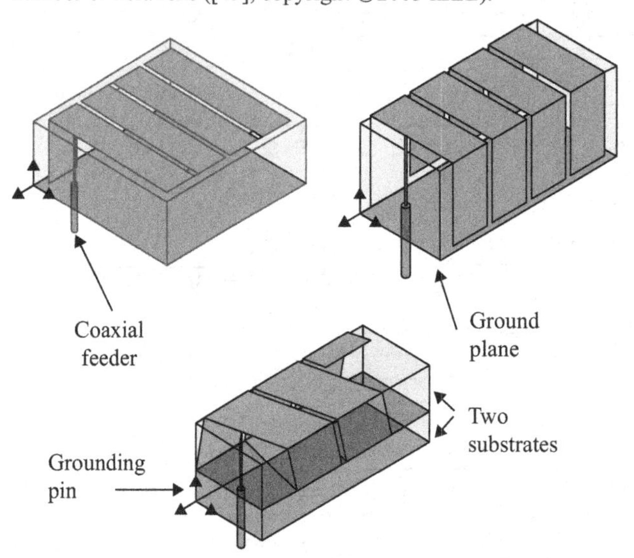


**Figure 7.141** Design step to develop 3D spiral structure ([65]).

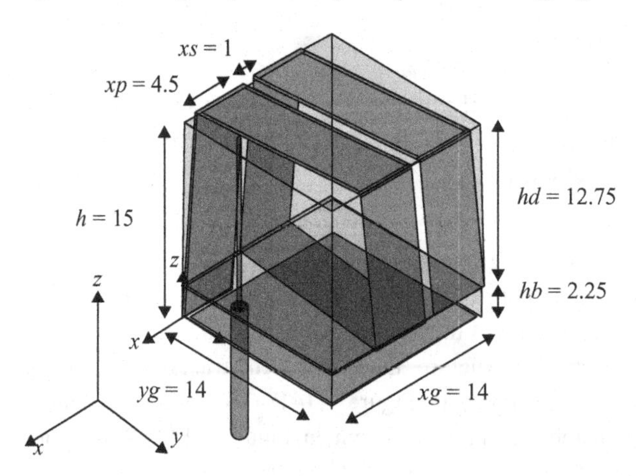


**Figure 7.142** Geometry and dimensions of the optimized design [65].

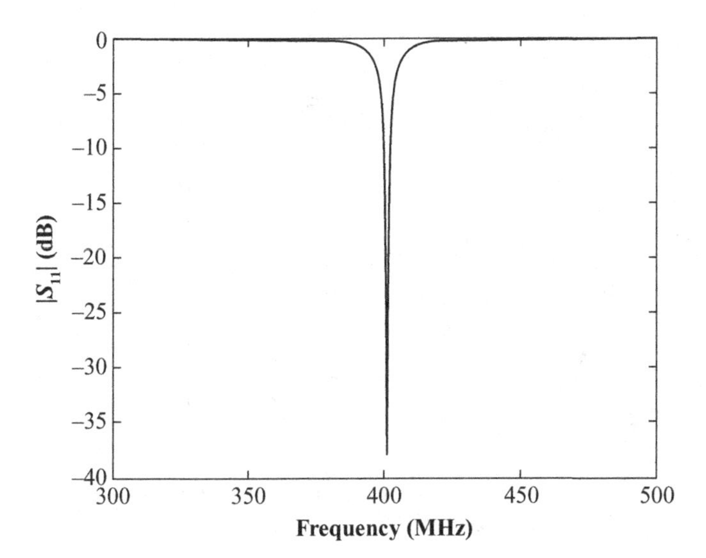


**Figure 7.143** $S_{11}$ performance [65].

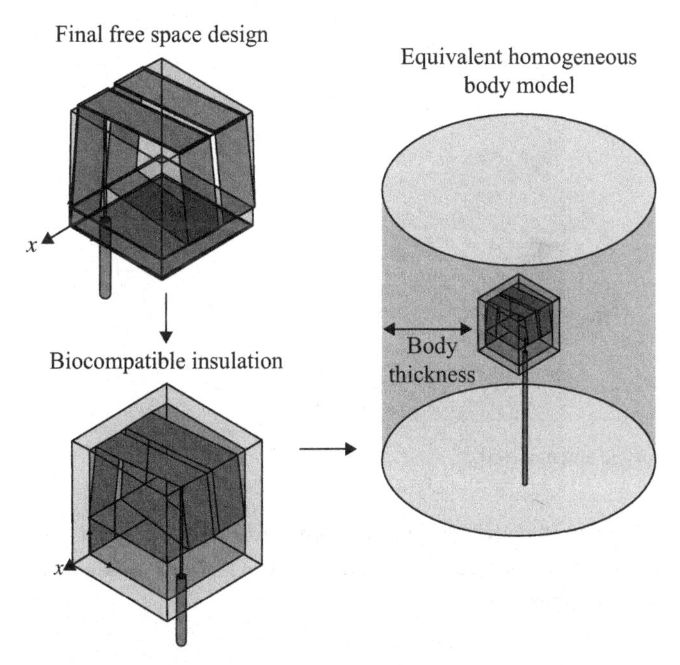


**Figure 7.144** Introduction of insulation layer to the antenna structure and equivalent model of the body [65].

is illustrated in Figure 7.143. Since the antenna is used in a human body, the antenna element should be insulated to avoid direct contact with the body tissues, and a biocompatible insulator with dielectric constant $\varepsilon_r = 3.2$ and $\tan\delta = 0.01$, is introduced. Figure 7.144 depicts the final free-space design and the biocompatible insulation model. Figure 7.145 illustrates variation of simulated $|S_{11}|$ depending on the thickness of the

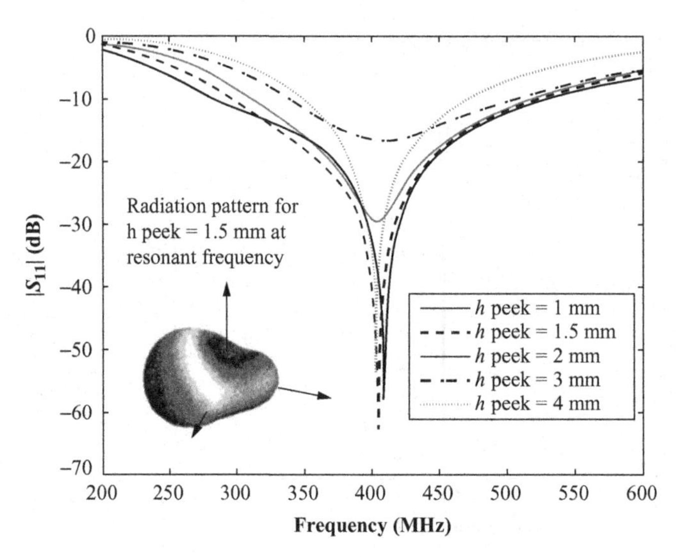


**Figure 7.145** Variation of $S_{11}$ depending on insulator thickness and 3D radiation pattern in free space [65].

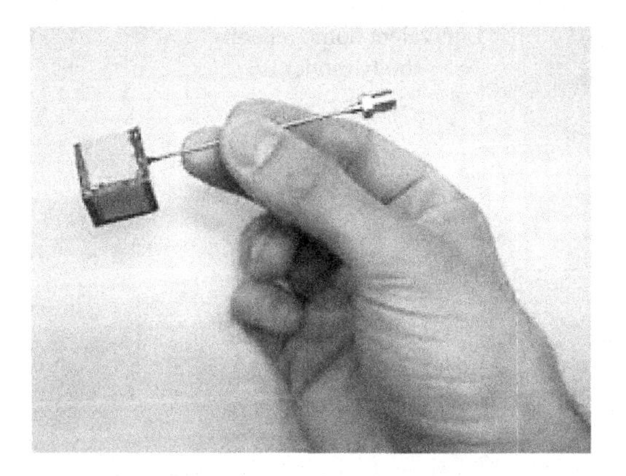

**Figure 7.146** Prototype antenna [65].

insulator and 3D radiation pattern for $h = 1.5$ mm (PEEK is the product name of the glue insulator [66]). The final structure is shown in Figure 7.146 and simulated and measured $|S_{11}|$ are depicted in Figure 7.147.

### *7.2.2.2.3 Spherical helix*

Wheeler stated [67] that the fundamental limitation on the bandwidth and the practical efficiency of a small antenna are related to the radiation power factor, which is defined as the ratio of the radiated power to the reactive power. It can be increased by utilizing as much as possible of the volume of a sphere whose diameter is equal to the maximum dimension of the antenna, when the antenna is restricted in its maximum dimension, but not in its volume. The spherical helix is considered as the most appropriate candidate

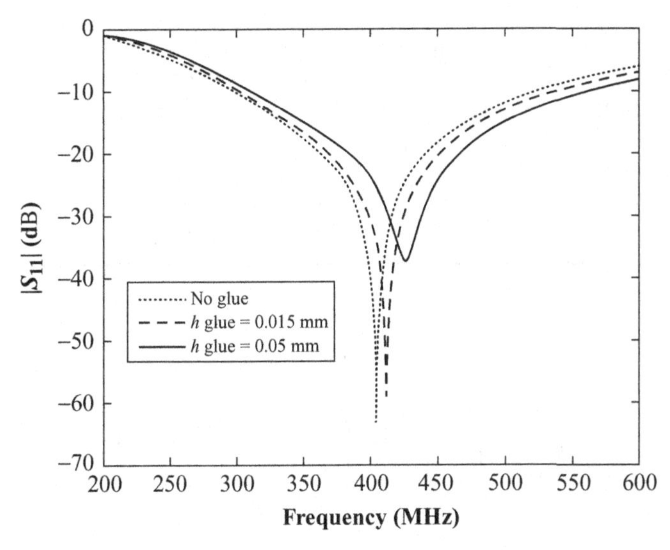


**Figure 7.147** Variation of $S_{11}$ depending on the thickness of glue layer [65].

for realizing such an antenna, as it can be constituted to occupy substantially nearly the whole volume of a sphere by wiring along the perimeter of the imaginary sphere having the radius $a$ corresponding to the maximum size of the antenna.

The radiation properties of several spherical helices have so far been studied [68–78]. The major antenna parameters are number of helical turns, number of helical arms, and the shape. Among the research work, Best did extensive study and demonstrated radiation properties of electrically small spherical helices [71, 72] and compared with that of a normal mode and cylindrical helix [73, 74]. In other papers, different types of antennas such as spherical wire [75] and spherical magnetic dipole antennas have been introduced [76–78].

### *7.2.2.2.3.1 Folded semi-spherical monopole antennas*

Radiation properties such as resonance frequency, radiation resistance, $Q$ and efficiency, of spherical helix monopole and dipole antennas are investigated with respect to the number of turns and number of arms [71]. The antenna model has a wire diameter of 2.6 mm and the maximum spherical radius (and height) is about 6 cm. Geometry of the antennas such as one-turn non-folded, one-turn two-arm, and one-turn four-arm folded spherical helices, respectively are illustrated in Figure 7.148(a), (b), and (c). With increase in number of arms, resonance frequency becomes higher, radiation resistance tends to be higher, $Q$ becomes lower, and radiation efficiency increases. They are observed in Table 7.15; (a) with two arms and (b) four arms. There is a difference in the height; in the four-arm antenna it is reduced to 5.77 cm from 5.89 cm in the other antennas. Figure 7.149 shows variation of $Q$ with respect to frequency and $ka$. Note the significance of the four-arm folded spherical helix, which shows the radiation resistance $R_a$ of about 43 Ω, suitable for matching to 50-Ω feed line, and the radiation efficiency of over 95%, and $Q$, which is the most notable result, being within 1.5 times

**Table 7.15** Antenna performances of folded spherical helix at resonance: (a) two-arm and (b) four-arm antenna. ([71], copyright ©2004 IEEE)

| No. of Turns | Arm Length (cm) | $f_R$ (MHz) | $R_A$ (ohms) | Efficiency (%) |
|---|---|---|---|---|
| | | (a) | | |
| ½ | 17 | 469.3 | 16.6 | 99.3 |
| 1 | 30.9 | 284.95 | 8.4 | 97.6 |
| 1½ | 45.07 | 203.8 | 4.7 | 94.5 |
| | | (b) | | |
| ½ | 17 | 515.8 | 87.6 | 99.6 |
| 1 | 30.9 | 300.3 | 43.1 | 98.6 |
| 1½ | 45.07 | 210 | 23.62 | 97.6 |

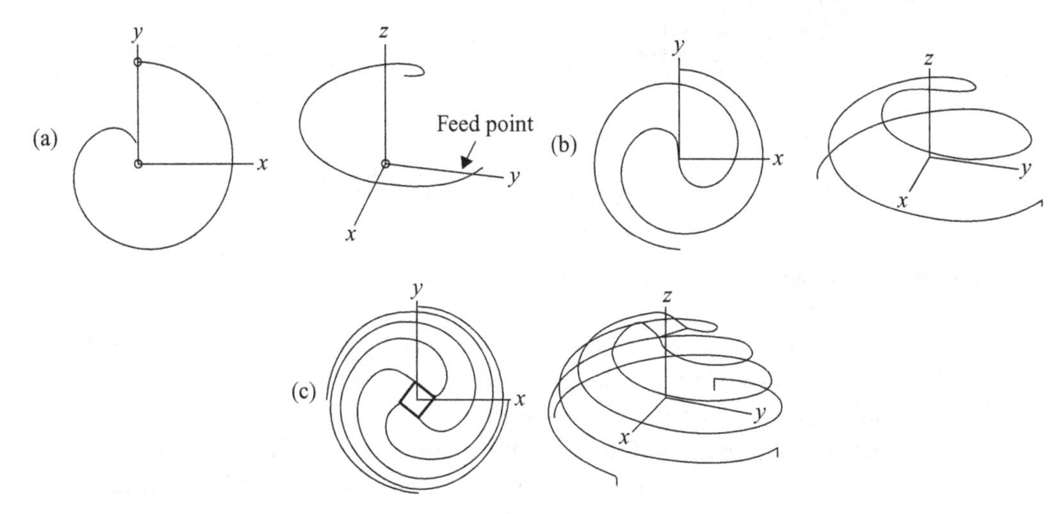


**Figure 7.148** One-turn spherical helix (a) non-folded, (b) two-arm folded, and (c) four-arm folded ([71], copyright ©2004 IEEE).

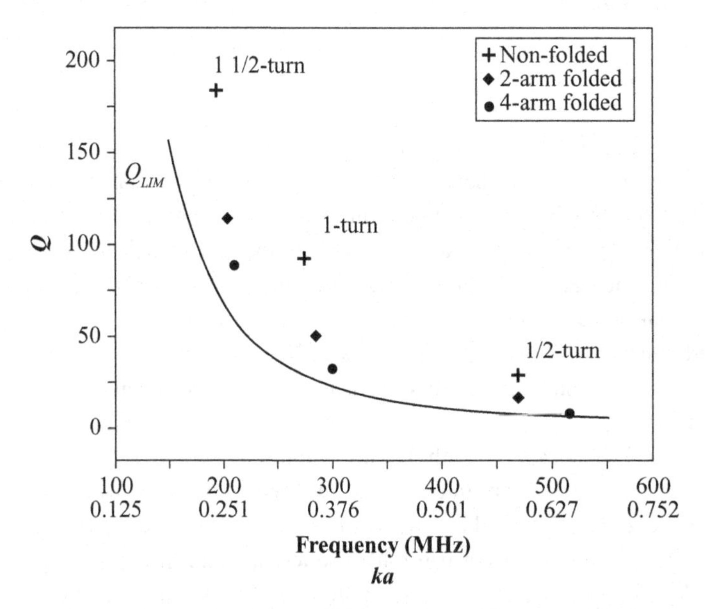


**Figure 7.149** Comparison of $Q$ of folded and non-folded spherical helix antennas ([71], copyright ©2004 IEEE).

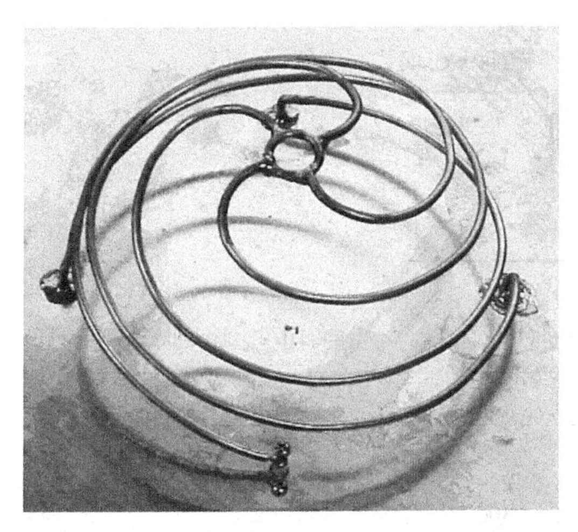

**Figure 7.150** Fabricated one-turn, four-arm folded spherical helix antenna ([71], copyright ©2004 IEEE).

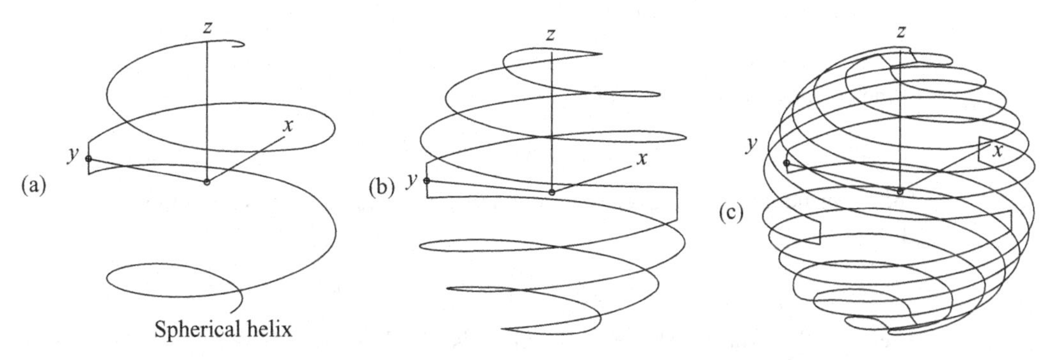


**Figure 7.151** (a) Spherical helix dipole, (b) two-arm folded spherical dipole, and (c) four-arm folded spherical dipole ([72], copyright ©2003 IEEE).

the fundamental limitation at a *ka* of approximately 0.38. A one-turn, four-arm spherical helix is shown in Figure 7.150.

Increasing the number of arms is not found to be good for obtaining self-resonance.

#### *7.2.2.2.3.2 Spherical dipole antenna*

Figure 7.151 depicts antenna models: (a) one-arm spherical helix, (b) two-arm folded helix dipole and (c) four-arm folded spherical dipole [72]. Geometries of antennas are provided in Table 7.16 and comparison of resonance properties for different type of antennas are given in Table 7.17, where $L$ is the overall length of antenna, $a$ is the radius of an imaginary sphere circumscribing the maximum dimensions of the antenna, $L_w$ is the total wire length in one dipole arm, $f_r$ is the resonance frequency, $R_r$ is the input resistance at the resonance, and $r$ is the radius of a sphere defining the effective volume

**Table 7.16** Geometry of antennas ([72], copyright ©2003 IEEE) (inserted in the text)

| | One-arm | Two-arm | Four-arm |
|---|---|---|---|
| $L$ (cm) | 8.22 | 8.25 | 8.36 |
| $a$ (cm) | 4.15 | 4.15 | 4.18 |
| $L_w$ (cm) | 58.23 | 62.8 | 65.53 |

**Table 7.17** Comparison of resonance properties of antennas ([72], copyright ©2003 IEEE)

| Antenna | $f_r$ (MHz) | $R_r$ (ohms) | $Q$ | $r/\lambda$ |
|---|---|---|---|---|
| Straight-Wire Dipole | 300 | 1.1 − j1015 | 950 | 0.0268 |
| Normal Mode Helix | 299.6 | 4.6 | 216.6 | 0.0438 |
| Spherical Helix | 300.2 | 2.2 | 143.9 | 0.0501 |
| 2-Arm Folded Spherical Helix | 299.6 | 10.3 | 101.1 | .564 |
| 4-Arm Folded Spherical Helix | 299.9 | 47.5 | 87.3 | 0.0592 |

* The small straight-wire dipole is not resonant near 300 MHz. Its performance at 300 MHz is presented for a relative comparison.

of the antenna. Here $r$ is given by [30]

$$r = (\lambda/2\pi)(9/2Q)^{1/3}.$$

The lower bound $Q_{lb}$ of an electric or magnetic dipole antenna is determined by using [32]

$$Q_{lb} = \eta[(ka)^{-3} + (ka)^{-1}] \tag{7.67}$$

where $\eta$ is the radiation efficiency. Then $Q_{lb}$ is calculated to be 57.3 at $ka = 0.263$. Since $Q$ of the four-arm spherical helix dipole is 87.3, it is about 1.52 times the lower bound on $Q$ for an electric dipole of $ka = 0.263$.

Radiation efficiency of the four-arm spherical folded dipole is 97.4%.

#### *7.2.2.2.3.3 Spherical wire antenna*

An antenna similar to a folded loop mounted on a sphere is introduced as a space-filling small antenna having the performance approaching the fundamental limitation. The antenna is composed of four arms which are arranged in a perpendicular geometry as shown in Figure 7.152, which illustrates (a) the antenna element along with the current flow and (b) a fabricated antenna [75]. With the perpendicular arrangement, radiators do not affect each other so that the cancellation between radiating arms is reduced, and also the scheme is effective to reduce the antenna size and at the same time lower the $Q$. Due to the orthogonal direction in the current flow as shown in Figure 7.152(a), coupling between antenna arms is minimized. The first vertical arm is on the diameter of a sphere, and has the feeding gap at the center. The second vertical arm, which is a 90° arc consists of a quarter of a circle on the sphere vertically. The third element is an arc which is

**Table 7.18** Measured and simulated results of SWA ([75], copyright ©2009 IEEE)

| Results | $f_0$(MHz) | $R_{rad}(\Omega)$ | $Q$ |
|---|---|---|---|
| Measurement | 374.6 | 56.62 | 9.48 |
| Simulation | 372.45 | 45.15 | 12.14 |

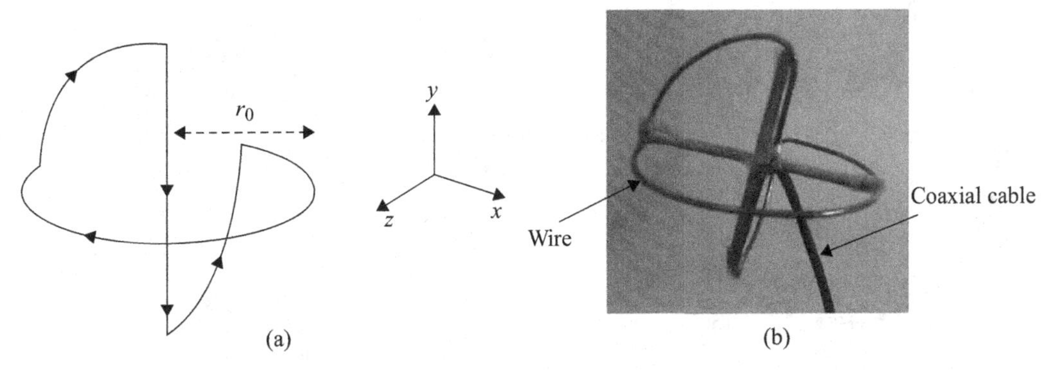


**Figure 7.152** Folded spherical wire antenna (SWA): (a) simulated antenna and (b) fabricated antenna ([75], copyright ©2009 IEEE).

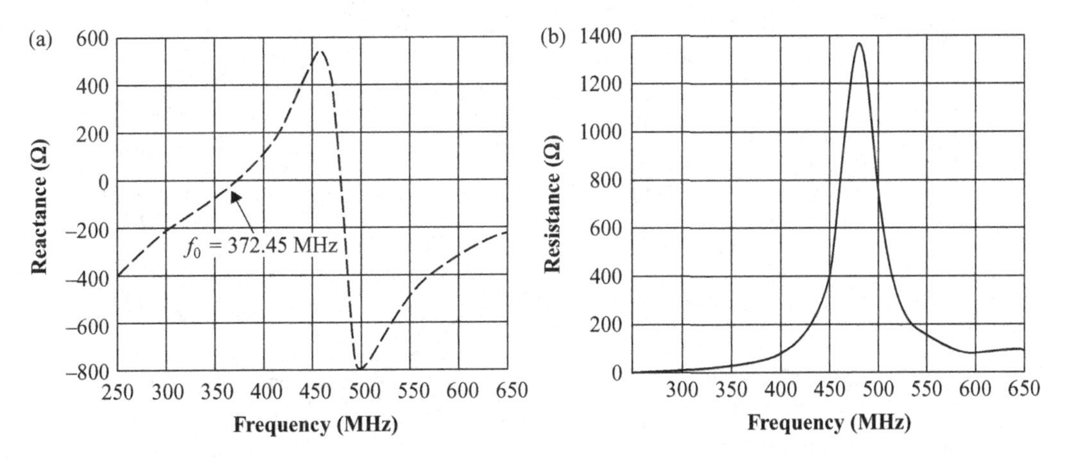


**Figure 7.153** Impedance of SWA: (a) resistance and (b) reactance([75], copyright ©2009 IEEE).

three quarters of a circle on the sphere horizontally. The fourth element is similar to the second arm and terminated at the end of the first arm. The sphere has diameter $2r_0 =$ 16 cm, the total wire length is 78.79 cm, and the wire radius is 1 mm. The simulated impedance is shown in Figure 7.153, where (a) gives reactance and (b) is resistance. From the figure, the resonance frequency is observed to be 372.45 MHz. Table 7.18 gives comparison between simulated and measured results. The $Q$ is low enough to compare to the fundamental limitation as the antenna size is as small as $kr_0 = 0.5$. Radiation

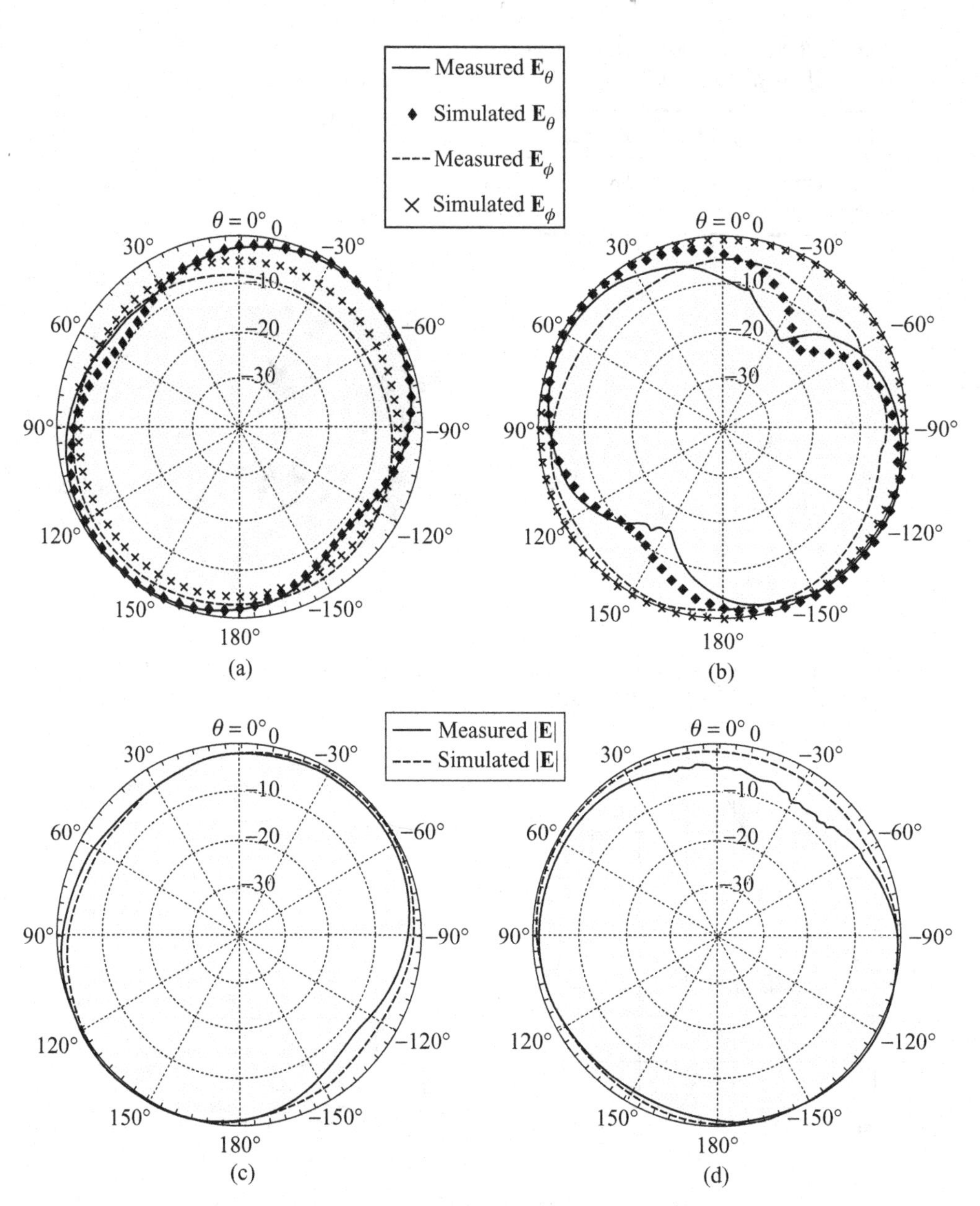


**Figure 7.154** Radiation patterns of SWA (a) *x–z* plane ($E_\theta$ and $E_\varphi$), (b) *y–z* plane ($E_\theta$ and $E_\varphi$), *x–z* plane ($|E|$) and (d) *y–z* plane ($|E|$) ([75], copyright ©2009 IEEE).

patterns are illustrated in Figure 7.154, in which (a) and (b) show $E_\theta$ and $E_\varphi$ patterns on the *x–z* plane and *y–z* plane, respectively, and (c) and (d) show $|E|$ patterns on *x–z* plane and *y–z* plane, respectively. They show almost isotropic patterns due to combined radiation from four arms in three directions which distribute energy in all directions and radiation of both horizontal and vertical polarizations for most directions.

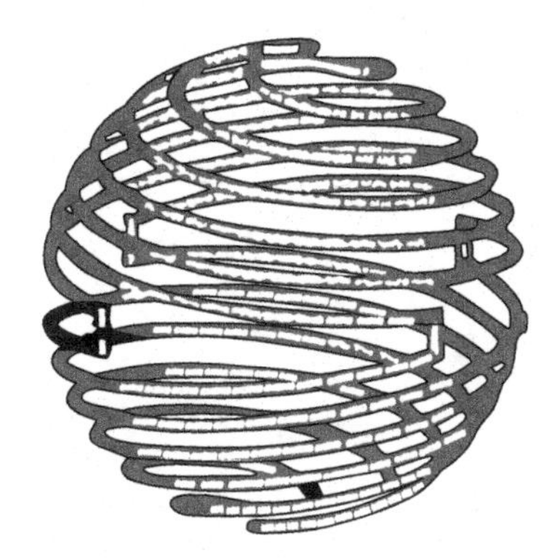

**Figure 7.155** Multi-turn folded spherical helix dipole antenna ([76], copyright ©2010 IEEE).

### *7.2.2.2.3.4 Spherical magnetic (TE mode) dipoles*

Magnetic (TE mode) dipoles can be designed to have complementary structure of electric (TM mode) dipoles. A similar folded-slot spherical helix that exhibits a magnetic dipole mode is developed by following the basic design concepts used with the folded spherical helix electric dipole [76]. The developed antenna has low VSWR, high radiation efficiency, and a low $Q$ at a small value of $ka$. Impedance, radiation patterns, and $Q$ properties are discussed.

Before describing the magnetic spherical helix, the electric spherical helix is introduced. As an example, a multi-turn folded spherical helix dipole is considered and the configuration is shown in Figure 7.155, which depicts a single, center-fed conductor, wound around the spherical structure. The outer radius of the sphere is about 4.22 cm and the conductor diameter is 2.6 mm. Self-resonance is achieved by adjusting the total length of the conductor, and impedance matching is done by increasing the number of conductors uniformly around the surface of the imaginary sphere and connecting them at the top and bottom of the sphere. With four folded arms shown in Figure 7.155, self-resonance is achieved at 300.7 MHz, where $ka = 0.266$, at which the resistance is equal to 49.7 Ω that can match to the load of 50 Ω, radiation efficiency is 97%, and a $Q$ is 84.2. This $Q$ value is about 1.53 times the lower bound of 55.2, consistent with the limit derived by Thal [77]. Radiation pattern is given in Figure 7.156, which indicates a dominant electric dipole or TM mode ($E_\theta$) pattern, having omnidirectional pattern in the horizontal plane.

Here a thin wall, hollow copper sphere having an outer radius of 4.3 cm is considered and a slot is inscribed within the sphere in a shape similar to the conductors of the folded spherical helix so as to form the complementary structure. Figure 7.157 illustrates a two-arm (double slot) version of the folded slotted spherical helix. The antenna is fed horizontally at the slot located on the $x$–$y$ (vertical) plane as shown in the figure. Similar to the electrical dipole version, the resonance frequency can be adjusted by increasing or decreasing the slot length. Radiation resistance can be increased by increasing the number of slots uniformly around the sphere. The two-arm version is self-resonant at a frequency of 294.86 MHz, but the radiation resistance is less than 1 Ω. $Q$ is 503. With increase

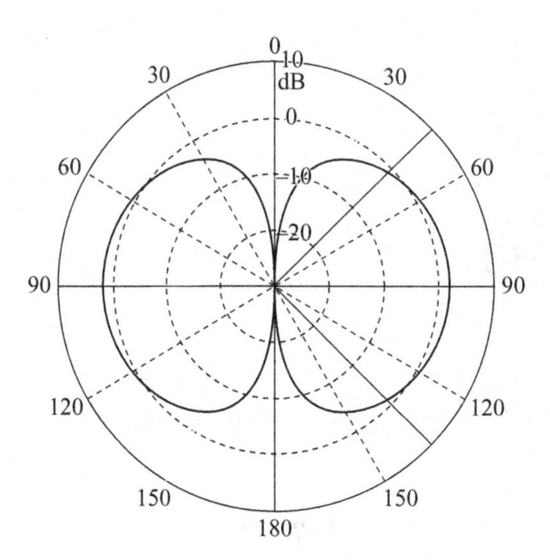


**Figure 7.156** $\theta$-sweep radiation pattern of the four-arm folded spherical helix (dominant polarization is $E_\theta$) ([76], copyright ©2010 IEEE).

**Figure 7.157** Two-arm folded spherical helix dipole ([76], copyright ©2010 IEEE).

in the number of arms, radiation resistance can be increased as shown in Table 7.19; however, with 16 slots self-resonance can be achieved. The slots are connected at the top and bottom of the sphere. The 4, 8, and 16-arm versions of the antenna are illustrated in Figure 7.158(a), (b), and (c), respectively. The performances are provided in Table 7.19, where the 16-arm version shows $Q$ of 148.5, that is about three times the lower bound. Radiation pattern of the 16-arm version is shown in Figure 7.159, which indicates a dominant magnetic dipole or TE mode ($E_\varphi$) pattern, having omnidirectional pattern in the vertical plane.

Three other modified spherical magnetic dipoles are developed [78]. Antenna configurations are illustrated in Figure 7.160, where (a) multi-arm spherical helix (MSH),

**Table 7.19** Comparison of performances for the 2, 4, 8, and 16-arm folded slot spherical antennas ([76], copyright ©2009 IEEE)

| No. Arms | Frequency (MHz) | *ka* | $R(\Omega)$ | $Q$ | VSWR |
|---|---|---|---|---|---|
| 2 | 294.9 | 0.266 | 0.62 | 502.7 | 81.2 |
| 4 | 274 | 0.245 | 2.17 | 327.1 | 23 |
| 8 | 309.1 | 0.279 | Not Resonant | 169.6 | 2.33 |
| 8 | 310.4 | 0.28 | 15.1 | 173.8 | 3.31 |
| 16 | 306.4 | 0.276 | Not Resonant | 148.5 | 1.22 |

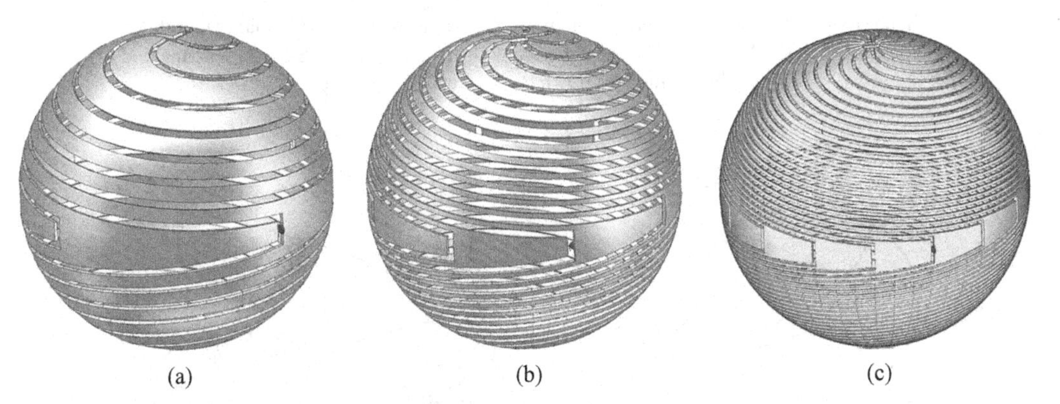


**Figure 7.158** (a) Four-arm folded spherical helix dipole, (b) eight-arm folded spherical helix dipole, and (c) 16-arm folded spherical helix dipole, ([76], copyright ©2010 IEEE).

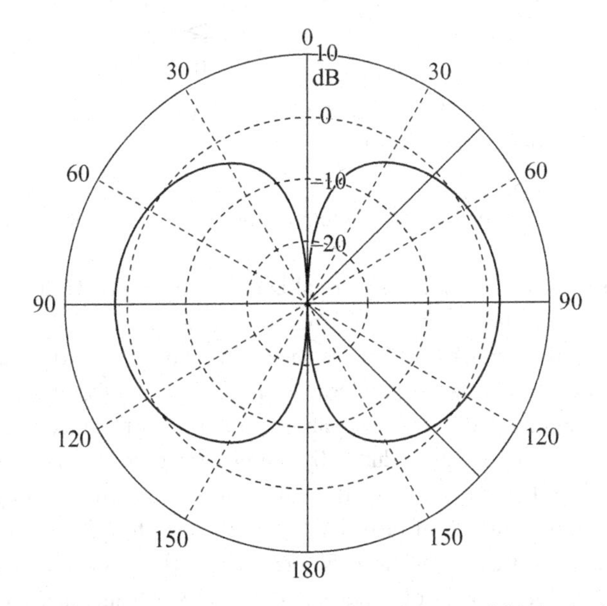


**Figure 7.159** $\theta$-sweep radiation pattern of 16-arm folded spherical helix dipole (dominant polarization is $E_\varphi$) ([76], copyright ©2010 IEEE).

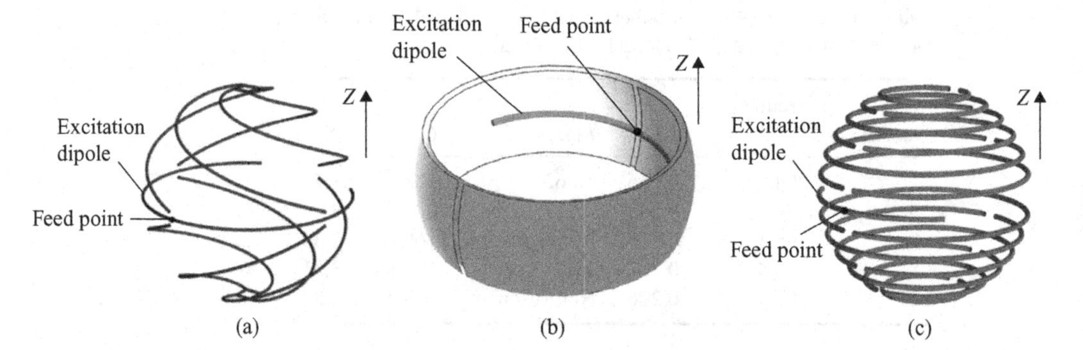


**Figure 7.160** Geometry of spherical magnetic dipole ($TE_{10}$ mode) antenna: (a) multi-arm spherical helix (MSH) antenna, (b) spherical split-ring resonator (S-SRR) antenna and (c) spherical split ring (SRR) antenna, ([78], copyright ©2010 IEEE).

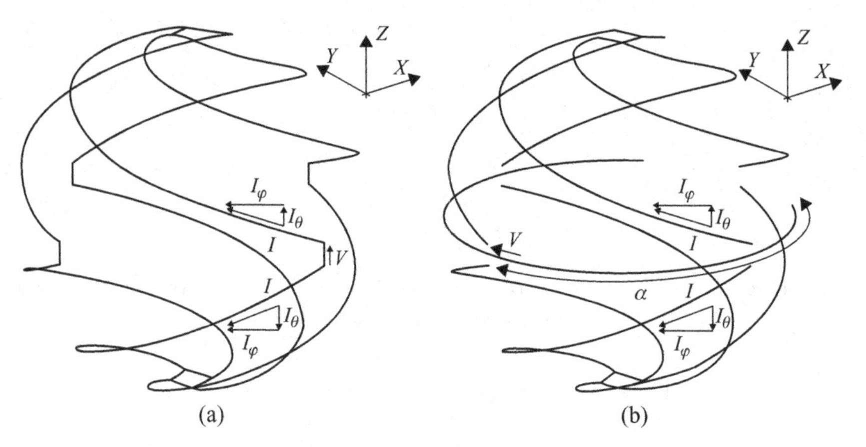


**Figure 7.161** Multi-arm spherical helix (MSH) antenna: (a) TM10 (electric mode) and (b) TE10 (magnetic mode) ([78], copyright ©2010 IEEE).

(b) spherical split ring resonator (S-SRR) antenna, and (c) spherical split ring (SSR) antenna are shown.

The $TE_{10}$ MSH is a modification of a spherical helix developed by Best [72]. In the $TM_{10}$ MSH antenna, the top and bottom of the arms are disconnected and a curved dipole is placed at the quarter of the antenna sphere as shown in Figure 7.161(a), where in (b) a $TE_{10}$ spherical mode is also shown as a reference. The antenna is fed at the midpoint of the driven dipole by applying driving voltage horizontally so as to cancel the far-field contribution from the $\theta$-component produced by the electric current ($I_\theta$), resulting in only fields of the desired $TE_{10}$ spherical mode remaining. The antenna is tuned to the resonance by changing the number of turns ($N_{turns}$) in the arm for given frequency $f_0$ and number of arms ($N_{arms}$) as is shown in Figure 7.162, where input impedance at $f_0 = 300$ MHz is plotted for antennas of $N_{arms} = 2, 4, 6$, and 8. All antennas have radius of 40 mm, and the wire radius is set to 0.5mm. The antenna occupies a spherical volume

**Table 7.20** Characteristics of the $TE_{10}$ MSH antenna ([78], copyright ©2010 IEEE)

| $N_{arms}$ | $N_{turns}$ | $\alpha$,° | $Q$ | $Q_{LB}$ | $Q/Q_{LB}$ | $TE_{10}$,dB | $TM_{11}$,dB | $TM_{20}$,dB | $Q_{LB}^{HO}$ | $Q/Q_{LB}^{HO}$ |
|---|---|---|---|---|---|---|---|---|---|---|
| 2 | 2.39 | 163 | 249.6 | 64.6 | 3.87 | −0.047 | −22.4 | −23.0 | 91.1 | 2.74 |
| 4 | 2.24 | 143 | 218.5 | 64.6 | 3.39 | −0.030 | −27.2 | −23.0 | 90.6 | 24.1 |
| 6 | 2.19 | 140 | 210.7 | 64.6 | 3.27 | −0.027 | −29.5 | −23.0 | 90.0 | 2.34 |
| 8 | 2.17 | 140 | 207.0 | 64.6 | 3.21 | −0.026 | −30.4 | −23.0 | 90.0 | 2.30 |

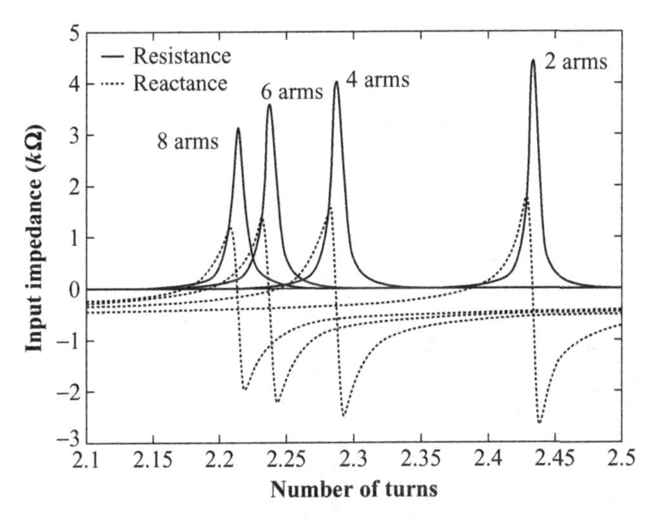


**Figure 7.162** Input impedance of the $TE_{10}$ MSH antenna as a function of number of turns in each arm ([78], copyright ©2010 IEEE).

of radius $a = 40.5$ mm, which corresponds to $ka = 0.254$ at 300 MHz. Table 7.20 summarizes the resonance state characteristics of the $TE_{10}$ MSH antenna for different numbers of arms. Here $Q_{LB}$ is the lower bound $Q$ and $Q_{LB}^{HO}$ is $Q_{LB}$ in the presence of higher modes.

The spherical split-ring resonator (S-SRR) antenna is composed of a spherical split-ring resonator and a curved driven dipole arranged so as to produce mostly $\varphi$-directed electric surface currents, thus the desired $TE_{10}$ mode. Since the antenna has symmetrical structure, it can be placed on the ground plane as Figure 7.163 illustrates. The S-SRR can be made electrically small, as the large area of the two spherical surfaces overlapped. The geometrical parameters are; $r_0 = a = 21$ mm, $t = 0.8$ mm, $g = 1$ mm, $r_{mnp} =$ 18 mm, and the radius of the monopole = 0.5 mm. To achieve lower $Q$, the S-SRR should cover the whole sphere ($\beta = \pi$); however, this results in blocking the magnetic flux over the antenna cross section and thus diminishes the electric current on the antenna surface. Therefore there is an optimum coverage to yield a minimum $Q$. It was found the angle $\beta = 73°$ is the optimum to obtain smallest $Q/Q_{LB}$. The resonance frequency depends also on the coverage and it becomes lowest when $\beta =$ 64°, which differs from the optimal $\beta$ for the smallest $Q/Q_{LB}$, although the difference

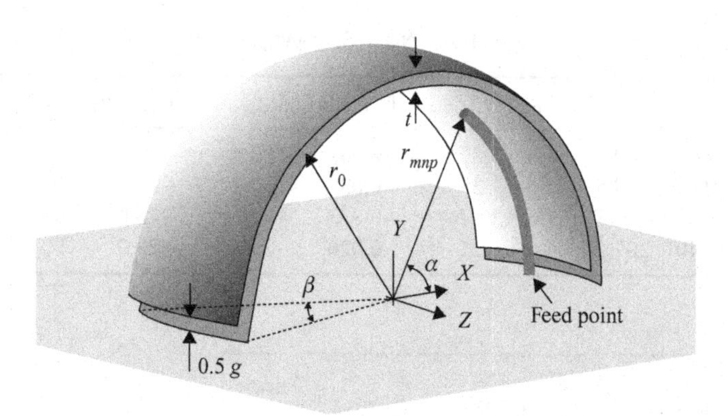


**Figure 7.163** Spherical split-ring resonator (S-SRR) antenna on a ground plane ([78], copyright ©2010 IEEE).

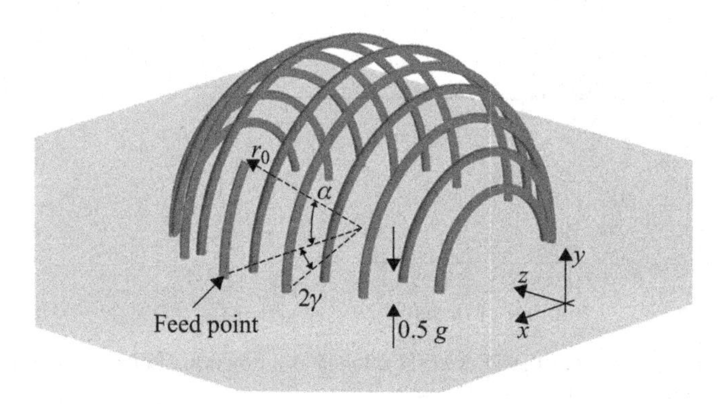


**Figure 7.164** Spherical split ring (SSR) antenna on a ground plane ([78], copyright ©2010 IEEE).

in the resonance frequency is minor, being only 1.5 MHz. For $\beta = 73°$, the resonance frequency $f_0$ is 297.0 MHz and $ka = 0.133$. The input impedance can be adjusted by the length of the driven dipole, and with $\alpha = 55°$, the input resistance at the resonance $R_0$ is 50 Ω for the optimal $\beta = 73°$. The lowest $Q/Q_{lb}$ obtained is around 3.4, which is nearly the same as that of MSH.

The spherical split ring (SSR) antenna consists of individual wire split rings distributed evenly in $\theta$ as shown in Figure 7.160(c). Every two neighbor rings are flipped with respect to each other and thus, operate as a conventional SRR. Multi-element SRR is constituted by combining rings with other rings so that uniform current distribution over the spherical surface is realized and the resonance frequency is lowered. The number of the rings is chosen to be odd, so that the antenna is driven at the central split ring, the length of which is adjusted to attain the input impedance to match the feed line. Since the SSR has symmetrical structure, it can be made in half and mounted on the ground plane as shown in Figure 7.164. The resonance frequency is changed by the number of split rings $N_{sr}$ and is determined by using the following expression

$$N_{sr} = 2\text{int}(90°/\gamma) - 1 \tag{7.68}$$

Table 7.21 Characteristics of the manufactured SSR antenna ([78], Copyright ©2010 IEEE)

| | $f_0$, MHz | $R_0$, Ω | efficiency | $Q$ | $Q_{LB}$ | $Q/Q_{LB}$ |
|---|---|---|---|---|---|---|
| simulated (SIE)[1] | 403.0 | 43 | 100 % | 564.0 | 165.6 | 3.41 |
| simulated (CST)[2] | 404.0 | 52 | 76 % | 462.3 | 124.9 | 3.70 |
| measured[3] | 403.0 | 51 | 73±2 % | 442.9 | 120.9 | 3.66 |

[1] PEC wires; infinite PEC ground plane.
[2] copper wires; infinite PEC ground plane.
[3] 1.5 m circular ground plane.

where $\gamma$ is an angular separation between two neighbor rings. A prototype SSR antenna is fabricated with $N_{sr} = 17$, wire diameter = 1.63 mm, driven dipole length $\alpha = 33°$. Simulated input impedance is 43 Ω and the resonance frequency is 403 MHz, which corresponds to the antenna electrical size $ka = 0.184$. Measured input impedance when the antenna is placed on the ground plane of the radius 1.5 m is 51 Ω. Simulated $Q/Q_{lb}$ is comparable with that of the MSH antenna. Performance parameters of the SSR are given in Table 7.21, where both simulated and measured results are provided.

*7.2.2.2.3.5 Hemispherical helical antenna*

In situations where antennas having compact structure, small size, and yet light weight, are urgently required in aerospace and mobile terminals, use of the hemispherical helix is desirable. New design of the hemispherical helix (HSH) antenna is explored and its wideband performance is introduced in [79]. The antenna is a 4.5 turn HSH with tapered strip radiating element. The width of the tapered element starts with 1 mm and ends with 4 mm, and the hemisphere radius is 20 mm. The antenna is fed at the side with non-linearly tapered matching section, and radiates a circularly polarized wave with wide beam width in the frequency range of 2.2–3.7 GHz. A prototype HSH antenna is constructed and measured for axial ratio, VSWR, and directivity. They are shown in Figure 7.165 through Figure 7. 167 along with simulated results, which have generally good agreement with measured results.

## 7.2.3 Uniform current distribution

### 7.2.3.1 Loading techniques

*7.2.3.1.1 Monopole with top loading*

The history of top-loaded antennas goes back to 1885, when Edison patented a communication system using a top-loaded antenna [80]. Hertz demonstrated electromagnetic waves in 1888 by using a dipole with large conducting plates attached to each end, which acted as a resonator with an inductance of the high-voltage-generator inductive coil. In the early days of communications, transmission of very low frequency (VLF) bands was accomplished by applying top-loading techniques to short monopoles that were subsequently adapted to LF, MF, and HF communications. In those days, most antennas

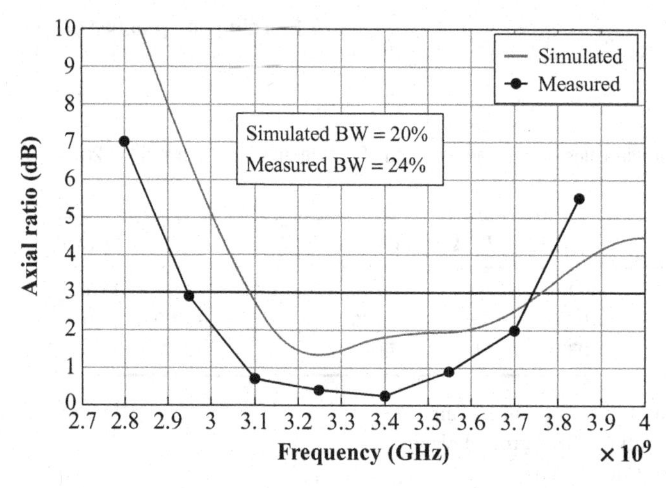


Figure 7.165 Measured and simulated axial ratio ([79], copyright ©2010 IEEE).

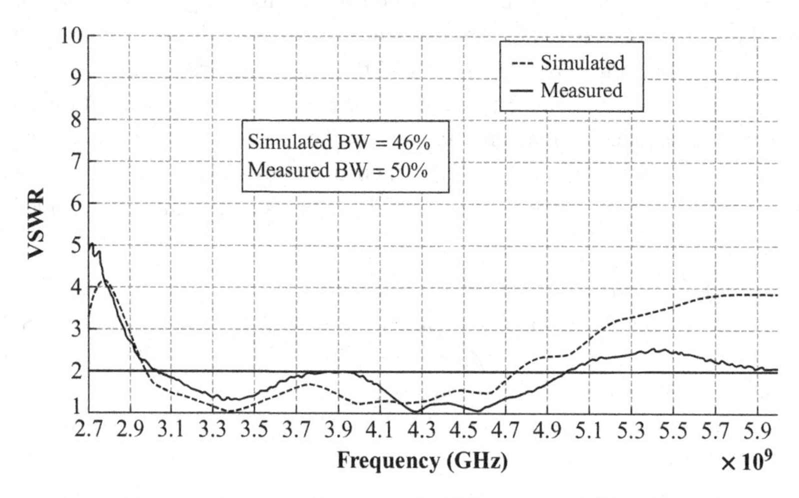


Figure 7.166 Measured and simulated VSWR ([79], copyright ©2010 IEEE).

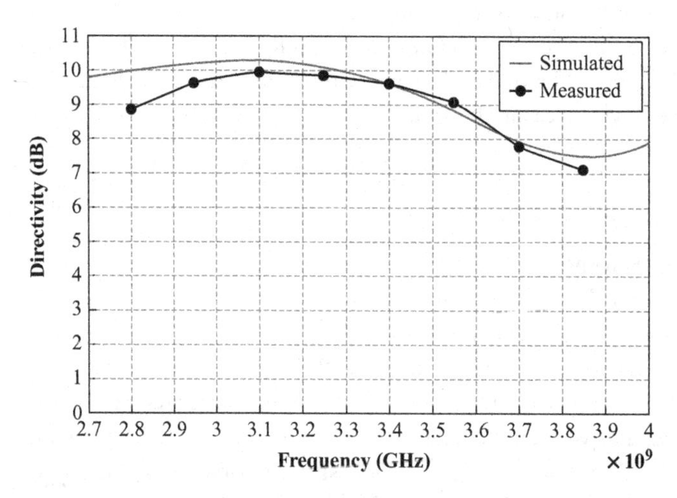


Figure 7.167 Measured and simulated directivity ([79], copyright ©2010 IEEE).

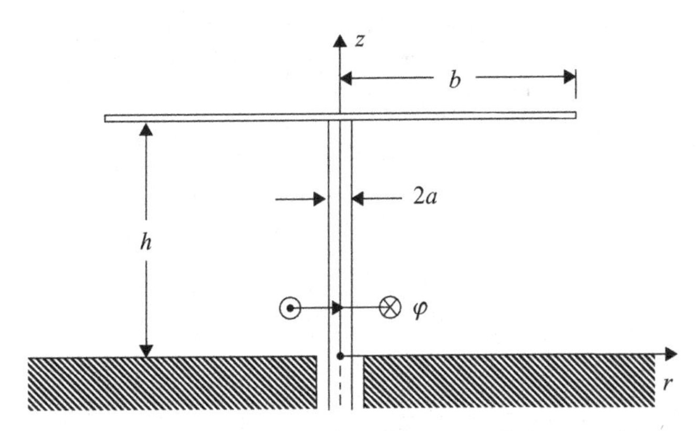


**Figure 7.168** Cross section of a disk-plate top-loaded monopole mounted on an infinite ground plane ([82], copyright ©2008 IEEE).

could be categorized as electrically small antennas and hence nowadays application of the top-loading technique to produce electrically small antenna is not necessarily new, but rather prevalent.

As an electrically small monopole has small radiation resistance and large capacitive reactance, addition of either a capacitive or inductive component to the monopole is implemented to make self-resonance and matching conditions feasible. Top loading was used as one of the most common means to realize capacitive loading on the monopole [81]. Practical inductive loading has been implemented by coil loading in the middle of a monopole element [82] and by modifying a linear monopole to either a meandered or helical wire structure.

Since top loading is effective to lower the resonance frequency, it is useful for reducing the antenna height and in turn for equivalently increasing the electrical length of a short monopole, assisting improvement of antenna performances. This increases radiation resistance and bandwidth even with the antenna of reduced size. The top loading can be ascribed to producing the uniform current distribution on the short monopole that yields significant improvements described above.

Top loading on a short monopole is implemented by a wire of L-shape (Inverted-L), T-shape, and crossed multi-elements, among others. A thin circular plate (disk) and its variations are also used as another common type of top loading. A circularly symmetric thin planar conductor top loaded on an electrically small monopole placed on the infinite ground plane shown in Figure 7.168 is representative [82]. In practice, the disk can take other forms, as Figure 7.169 illustrates: (a) wire-grid disk, (b) wired spiral, and (c) wire-grid spherical cap [83a, b]. There are of course dipole types as shown in Figure 7.169(d), where the top load is a wire grid, and also the monopole can be a meandered wire, and a helical wire as well. Figure 7.169(e) and (f) show examples of a wire-grid top-loaded helical dipole and a spherical-cap top-loaded helical dipole, respectively [83a, b].

In the planar disk model shown in Figure 7.168, currents on the monopole and the field they produce are treated as independent of the azimuthal angle $\varphi$. With dimensions $h = 0.3$ m, $b = 0.6$ m, and $a = 6.35$ mm, and the operating frequency $f_0 = 10$ MHz,

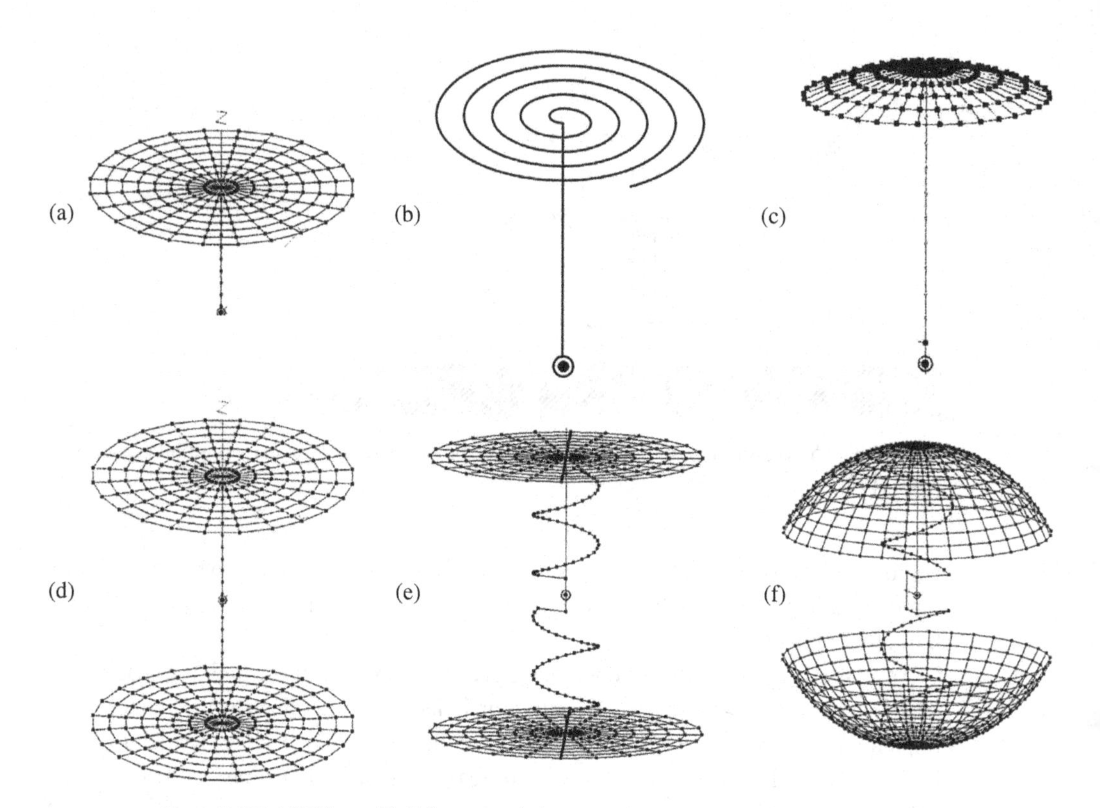


**Figure 7.169** (a) Wire-grid disk top-loaded monopole, (b) wired-spiral hat top-loaded monopole, (c) wire-grid spherical-cap top-loaded monopole, (d) wire-grid disk top-loaded dipole, (e) wire-grid top-loaded helical dipole, and (f) spherical-cap top-loaded helical dipole, ([83a, b], copyright ©2008 IEEE).

current and charge distributions on the monopole and the disk are shown in Figure 7.170 and Figure 7.171, respectively, in which the current on the monopole until $h = 0.01\lambda$ is observed nearly uniform. The input impedance ($= R + jX$) is depicted in Figure 7.172, where both calculated and measured results are shown. Radiation power factor $p_e$, radiation conductance $G_e$, and bandwidth $\Delta f$ along with applied voltage $V$ are illustrated as functions of the top-loaded ratio $b/h$ in Figure 7.173 and Figure 7.174, respectively. In these figures, $p_{e0}$ and $G_{e0}$, respectively, denote reference values for an antenna with $b = h$. The beneficial effect of top-loading is obvious from increases in bandwidth $\Delta f$ (proportional to $p_e$), and radiated power (proportional to $G_e$). Particularly the bandwidth with $b/h = 5$ is over 250 times as much when compared with that of $b/h = 0.02$, the unloaded monopole. Thesc figures suggest that top loading is vitally important to reduce the input voltage to a tolerable level and to achieve the required bandwidth.

Another type of top-loaded monopole is an electrically small monopole simply loaded with an open-circuited transmission line shown in Figure 7.175 [84].

The input impedance $Z$ is

$$Z = -jZ_0 \cot \beta b \tag{7.69}$$

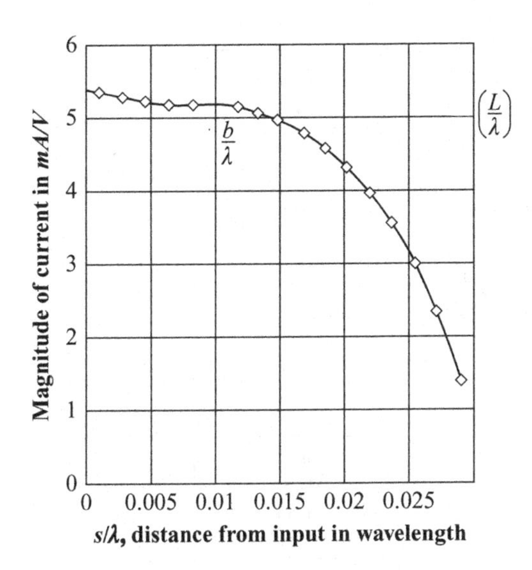


**Figure 7.170** Currents on a DLM with $h = 0.3$ m, $b = 0.6$ m, and $a = 6.35$ mm at 10 MHz ([82], copyright ©2008 IEEE).

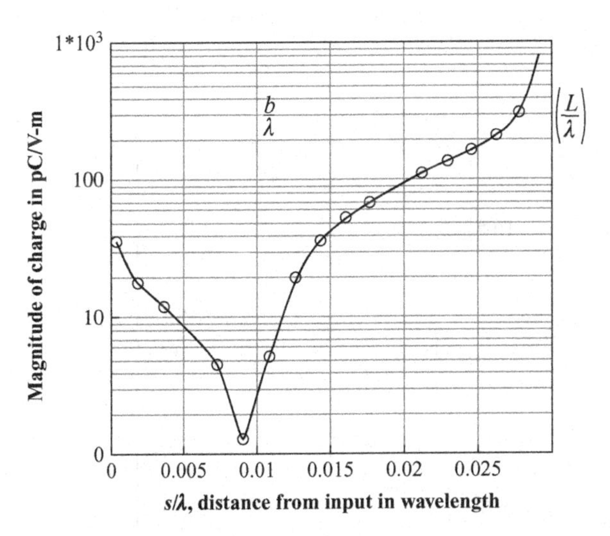


**Figure 7.171** Charge distributions on the same DLM as that in Figure 7.170 ([82], copyright ©2008 IEEE).

where $Z_0$ denotes the characteristic impedance of the metal shell and $b$ is its total length; that is, $b = h + r$. The metal shell acts as a series capacitance, by which $h$ can be extended to the effective height $h'$ as

$$h' = (1/\beta)\text{arc}\cot\{(Z_0/Z_{0e})\cot\beta b\} + H \tag{7.70}$$

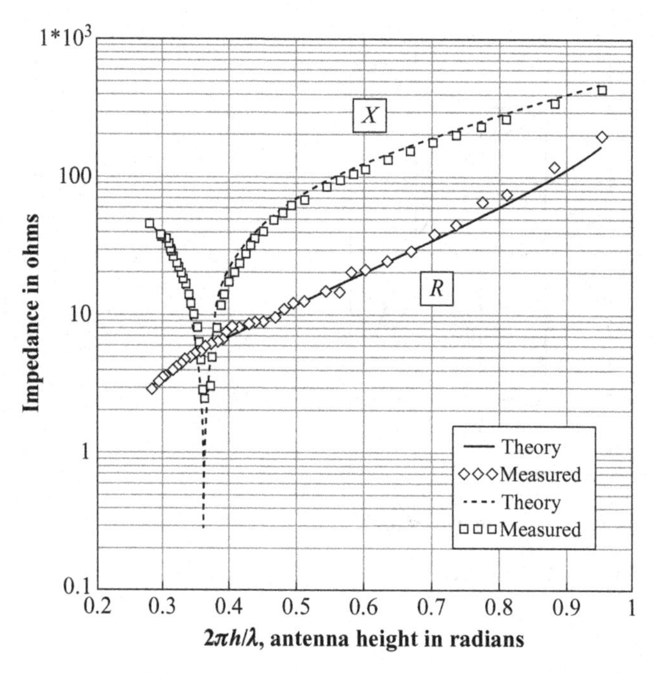


**Figure 7.172** Theoretical and measured impedance of a DLM with $h = b = 0.3$ m, and $a =$ 6.36 mm ([82], copyright ©2008 IEEE).

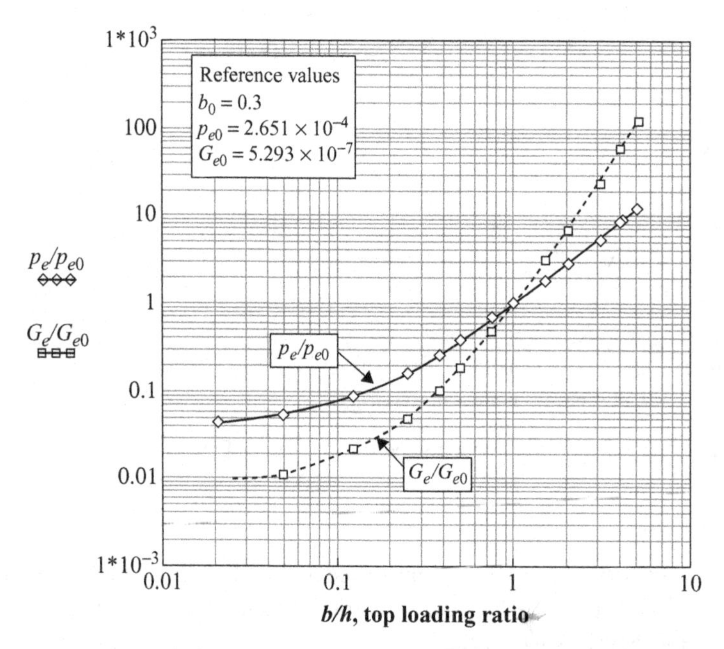


**Figure 7.173** Radiation power factor (normalized) $p_e$ and radiation conductance $G_e$ as function of $b/h$ for a reference antenna with $h = b = 0.3$ m, and $a = 6.35$ mm ([82], copyright ©2008 IEEE).

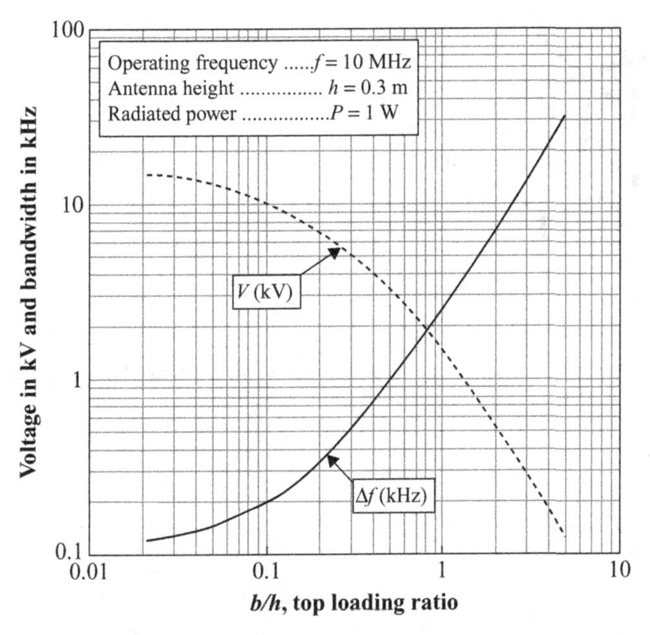


**Figure 7.174** Radiation bandwidth and applied voltage as function of $b/h$ for an antenna radiating 1 W at 10 MHz with $h = 0.3$ m, $a = 6.35$ mm ([82], copyright ©2008 IEEE).

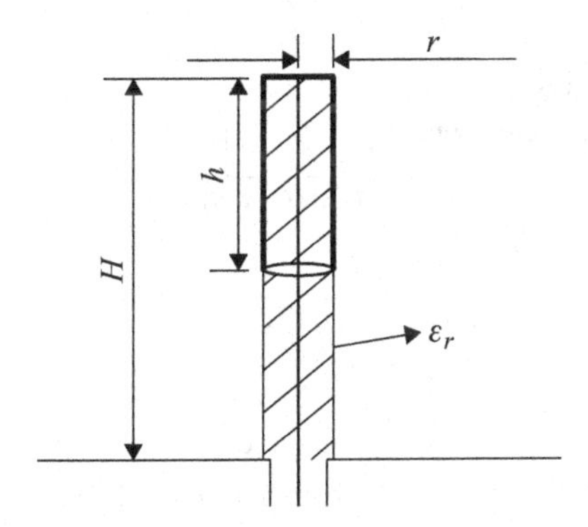


**Figure 7.175** Open-circuit transmission line loaded monopole antenna structure ([84], copyright ©2006 IEEE).

where $Z_{0e}$ is the effective characteristic impedance of the common monopole and $H$ is the height of the monopole. The metal shell also plays a role of matching circuit. The effective height $h'$ can be increased by lengthening $h$ and at the same time the resonance frequency $f$ is lowered.

Effectiveness of top-loading was shown by using a wire-grid disk top-loaded dipole that had as much as about 60% reduction in the antenna size [83]. Also about the same amount of size reduction was shown with a spherical-cap top-loaded dipole [83]. The dipole antenna sizes used in these examples were 18 cm and 19 cm in the wire-grid disk

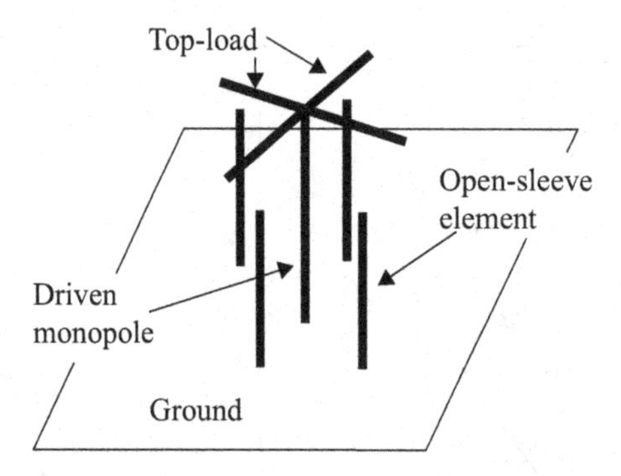


**Figure 7.176** Geometry of the cross-T line loaded antenna with sleeves ([85], copyright ©2006 IEEE).

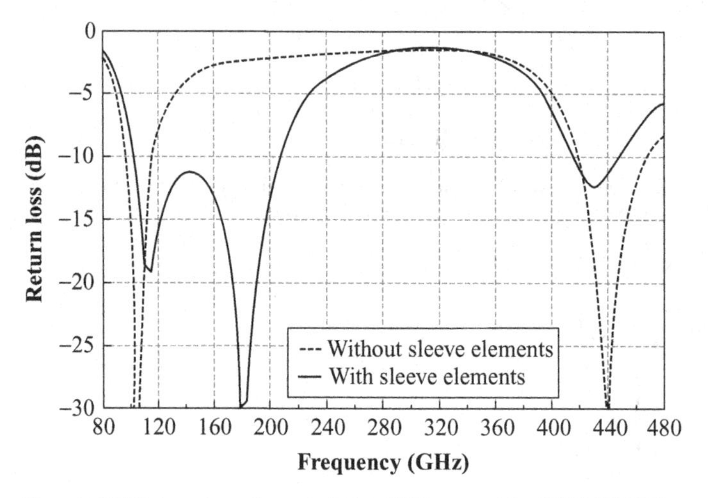


**Figure 7.177** Return-loss characteristics ([85], copyright ©2006 IEEE).

and spherical-cap dipole, respectively, for the resonance frequency of near 300 MHz. Antennas were matched to the load of 50 Ω by using a shunt stub at the feed point.

*7.2.3.1.2 Cross-T-wire top-loaded monopole with four open sleeves*

The antenna configuration is illustrated in Figure 7.176 [85]. The ground plane is assumed to be infinite. The geometrical dimensions of the antenna are: length of driven element $L_0 = 0.13\lambda_0$, length of the top-load element $L_1 = 0.035\lambda_0$, length of sleeve $L_2 = 0.1\lambda_0$, wire diameter $a = 0.015\lambda_0$, and distance between the driven element and the sleeve $R = 0.049\lambda_0$, where $\lambda_0$ is the wavelength at the resonance that is 100 MHz. Return loss is depicted in Figure 7.177 and its variation for different length $L_2$ of the sleeves is shown in Figure 7.178, where $L_2/\lambda_0$ is varied from 0.110 (case 1-1), 0.0105 (case 1-2), 0.095 (case 1-3), and 0.09 (case 1-4). The figure shows that decreasing the length $L_2$ of the sleeves increases the bandwidth.

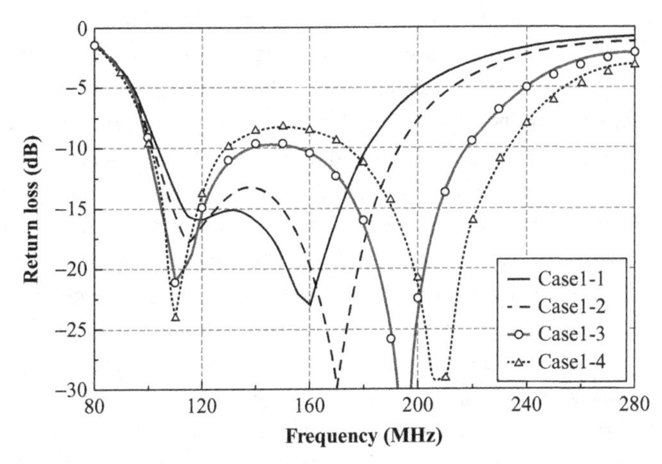


**Figure 7.178** Variation of return loss depending on the length of the sleeve ([85], copyright ©2006 IEEE).

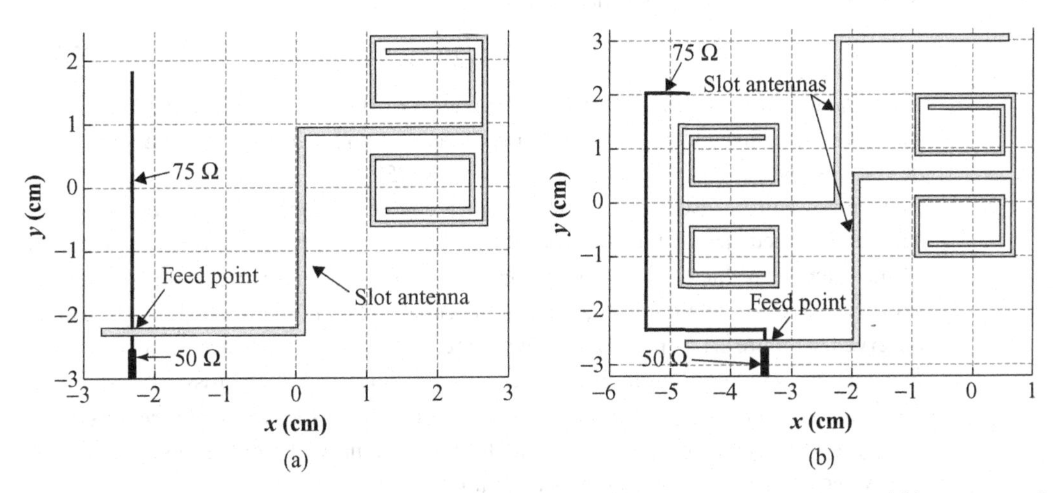


**Figure 7.179** (a) Geometry of single-element miniaturized slot antenna (SEA) and (b) geometry of double-element miniaturized slot antenna (DEA) ([87]).

### *7.2.3.1.3 Slot loaded with spiral*

As a different type of small-sized loaded antenna, a slot antenna terminated with spiral elements is introduced in [86] (Figure 7.179(a)). The radiating slot is designed to have length $\lambda_g/4$ and is terminated with two identical quarter-wavelength non-radiating spiral slots. $\lambda_g$ is the wavelength of the quasi-TEM mode supported by the slot line. In principle, as a resonant quarter-wavelength transmission line exhibits a short at one end reflected to an open at the other end, the non-radiating quarter-wavelength spiral slot shorted at one end behaves as an open at the other end. Therefore, a quarter-wavelength slot line shorted at one end and terminated by the non-radiating quarter-wavelength spiral should resonate and radiate the electromagnetic wave very efficiently. With this configuration, the antenna is reduced in size by half, and further reduction is accomplished by bending the radiating slot line. The antenna geometry along with dimensions is illustrated in Figure 7.179(a). The antenna occupies an area of about $0.15\lambda_0 \times 0.13\lambda_0$.

Since this antenna exhibits very narrow bandwidth, less than 1%, another parasitic antenna with the same configuration is placed in the remaining area in order to increase the bandwidth without significantly increasing the overall PCB (printed circuit board) size, as Figure 7.179(b) illustrates [87]. One of these two antennas is fed by a microstrip line, leaving the other one as a parasitic antenna. The parasitic antenna is coupled with the radiating slot at the elbow section, where the electric field is large. The magnetic currents on each antenna are in phase, so the radiation is enhanced. The two antennas are designed to resonate at the same frequency $f_{r1} = f_{r2} = f_0$, where $f_{r1}$ and $f_{r2}$ are the resonance frequencies of each antenna, and $f_0$ is the center frequency. In this antenna system, $S_{11}$, spectral response of the two coupled antennas, exhibits two nulls, as the coupling is adjusted strong enough to increase the bandwidth compared with that of a single slot antenna. The separation of these two frequencies is a function of the separation $s$ and distance $d$ of the overlapped elbow section of these two antennas. The coupling $k_t$ between these two antennas is defined as,

$$k_t = (f_u^2 - f_l^2)/(f_u^2 + f_l^2) \tag{7.71}$$

where $f_u$ and $f_l$, respectively, are the frequencies of the upper and lower nulls in $S_{11}$. The $k_t$ can be adjusted by varying $s$ and $d$, increasing with decrease in $s$ and increase in $d$. The designed resonance frequency $f_{r1} = f_{r2} = 850$ MHz; however, slightly different frequencies can be used to achieve a higher degree of control for tuning response. The input impedance of this antenna, for a given slot width, depends on the location of the microstrip line feed relative to one end of the slot and varies from zero at the short circuited end to a high resistance at the center. The optimum feed position can be observed in Figure 7.179(b), which shows the feed line, consisting of a 50-Ω transmission line connected to an open-circuited 75-Ω transmission line, which crosses the slot. The 75-Ω line is extended by $0.33\lambda_m$ beyond the strip-slot crossing to couple the maximum energy to the slot and also to compensate for the imaginary part of the input impedance ($\lambda_m$: wavelength of the wave in the strip line).

Antennas, two single-element antennas (SEA1 and SEA2) and a double-element antenna (DEA) (Figure 7.179(a) and (b)), are fabricated on a substrate of thickness 500 μm, having a dielectric constant of $\varepsilon_r = 3.5$ and a loss tangent of $\tan\delta = 0.003$, with a copper ground plane of 33.5 mm × 23 mm. The SEA1 is the constitutive element of the DEA and the SEA2 is an SEA having the same topology as the SEA1. Both calculated and measured $S_{11}$ of the DEA and the SEA2 are shown in Figure 7.180, which indicates bandwidth of 21.6 MHz (2.54%) in the DEA, being wider than 8 MHz (0.9%) of the SEA1 and 11.7 MHz (1.31%) of the SEA2. The measured gain of the DEA is 1.7 dB at 852 MHz, which is greater than that of the SEA1 of approximately 0.8 dB at 850 MHz. The antenna size of the SEA1 is $0.133\lambda_0 \times 0.154\lambda_0$, while that of the SEA2 and the DEA is $0.165\lambda_0 \times 0.157\lambda_0$.

By adding series inductive elements to a slot antenna, the size of the antenna can be further reduced. A dual-band small antenna is also developed by adjusting the coupling factor $k_t$ so as to create two separate frequencies in the $S_{11}$ response [87].

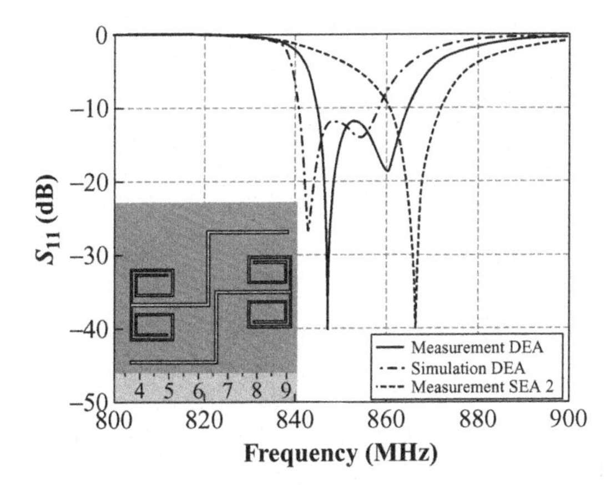


**Figure 7.180** Return loss of DEA and SEA2 [87].

## 7.2.4 Increase of excitation mode

By increasing the number of excitation modes (for instance, addition of a TM mode to a TE mode) enhancement or improvement of antenna performances such as gain, efficiency, bandwidth, and radiation pattern, can be expected. Composing an antenna with both TE and TM modes is one of the simplest ways. Examples are combination of a dipole with a loop, and a monopole with that of a dual slot that constitutes a self-complementary structure. Other types are a combination of an inductive element with a capacitive element in an antenna system that makes a conjugate structure. With the conjugate structure, the self-resonance condition can easily be achieved, even though the antenna has very small dimensions. A combination of an electric source with a magnetic source will bring out a conjugate structure as well as a complementary structure. A composite antenna system constituted with different types of antennas having different excitation modes is also used to create a conjugate antenna structure.

These means facilitate enhancement of the bandwidth, addition of functions such as multiband and multiple polarization, and so forth, in small-antenna design.

### 7.2.4.1 Self-complementary structure

The self-complementary structure can be implemented by combining two antennas, having complementary properties of each other. There are two types; one type has rotationally symmetric structure while another type has axially symmetric structure. In a planar structure, for example, a rotationally symmetric type is constituted from one arbitrary generating structure by rotating it 180 degrees with respect to the feed point. An example was shown previously in Figure 6.35. In contrast, an axially symmetric type is fabricated by combining an arbitrary structure on a half-infinite space with a structure complementary to it on the other half-infinite space, with axial symmetry to each other (Figure 6.35). In a practical fabrication, a half structure divided by the axis of symmetry is formed on a half-infinite PEC (perfect electric conductor) plate, while another half is formed with the same structure, but as spaces on another half-infinite PEC plate. Figure 6.34 illustrated this example, showing a monopole combined with a complementary slot on the ground plane.

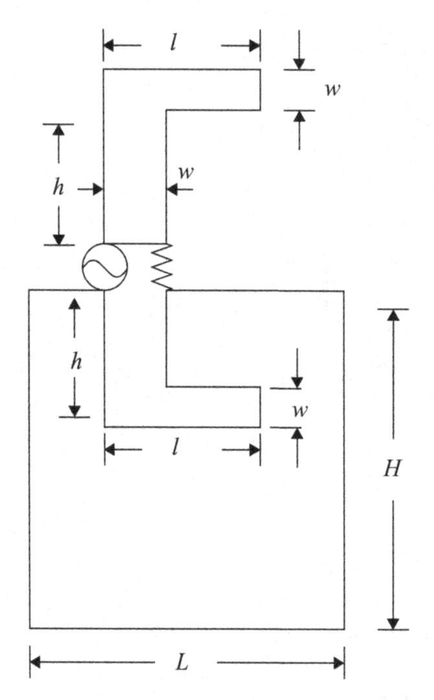


**Figure 7.181** L-shaped self-complementary antenna [from 87].

The self-complementary structure has an inherent frequency independent property, which is infinite bandwidth, when it is constituted with an infinite structure. Unfortunately, practical antennas can never be realized in an infinite structure, as a truncation can never be evaded, thus bandwidth must always be finite. Although the bandwidth would be limited, antennas with complementary structure may still have wide enough bandwidth for practical applications; that is, an antenna is practically useful when the bandwidth is reasonably wide to satisfy the requirement. Hence, the complementary concept, even with truncation in the antenna structure – always encountered in small antennas – is adapted as a useful means to attain an appreciable bandwidth for an antenna of very small size.

*7.2.4.1.1 L-shaped quasi-self-complementary antenna*

One of the most simple self-complementary antennas is a monopole combined with a dual slot (Figure 6.34) introduced in [88, 89], in which wideband performance was demonstrated, even with an antenna of small size. Since a practical antenna can never be composed with infinite structure, it should be referred to as quasi self-complementary. Instead of a monopole, an Inverted-L antenna is combined with a dual L-slot to compose a quasi self-complementary structure (Figure 7.181) [89, 90]. The input impedance of the antenna with dimensions of $l = 15$ mm, $h = 15$ mm, and $w = 4$ mm, shows broad frequency characteristics covering about 2 GHz to 10 GHz for VSWR less than two (Figure 7.182(a)). However, the ground plane size ($X \times Y$) was 46 mm $\times$ 60 mm, much smaller than one wavelength at the lowest frequency. Efficiency is sacrificed to a certain measure by using a load resistance 188 Ω at the side opposite of the feed

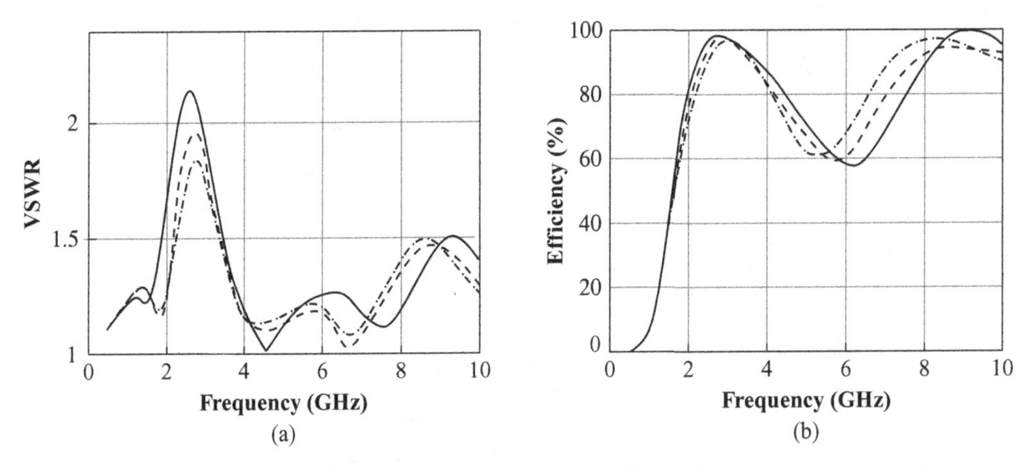


**Figure 7.182** (a) VSWR characteristics of L-shaped self-complementary antenna and (b) efficiency of L-shaped self-complementary antenna ([88], copyright ©2002 IEEE).

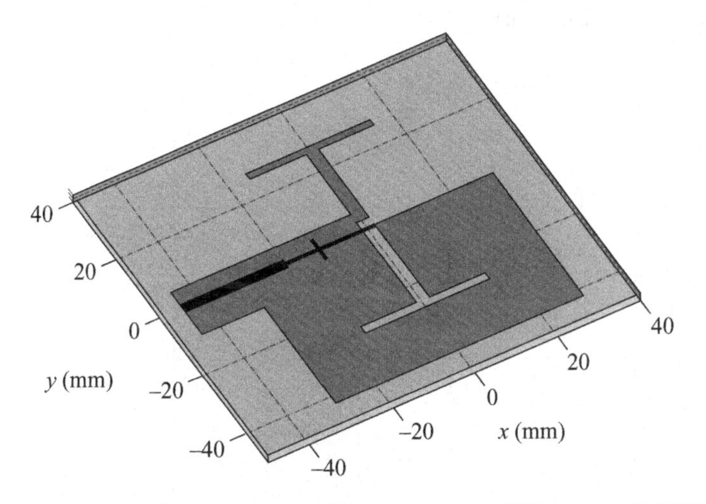


**Figure 7.183** Self-complementary H-shaped antenna ([91], copyright ©2007 IEEE).

as needed to maintain the complementary condition [89]. However, with this size of antenna, efficiency was observed as high as 70% almost over the frequency range of 2 GHz to 10 GHz (Figure 7.182(b)). The load resistance may be omitted to improve efficiency, if some amount of sacrifice in the bandwidth is allowed. If a smaller ground plane is used, as in practical applications to small mobile terminals, the bandwidth will become narrower, but still wide enough for practical applications.

### *7.2.4.1.2 H-shaped quasi-self-complementary antenna*

The antenna geometry is illustrated in Figure 7.183, in which a short-circuited microstrip line used for feeding the antenna and dimensions of antenna with the scale are shown [91]. A dielectric substrate used has $\varepsilon_r = 2.2$ and the thickness is 0.787 mm. Measured and simulated return loss and gain, respectively, are shown in Figure 7.184 and Figure 7.185. The figure demonstrates a wide bandwidth covering about 1.35 GHz to 3.2 GHz for the

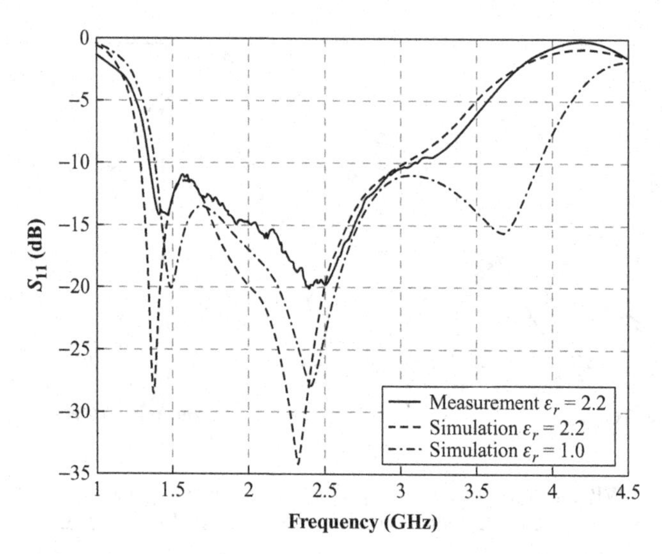


**Figure 7.184** Measured and simulated return loss of H-shaped antenna ([91], copyright ©2007 IEEE).

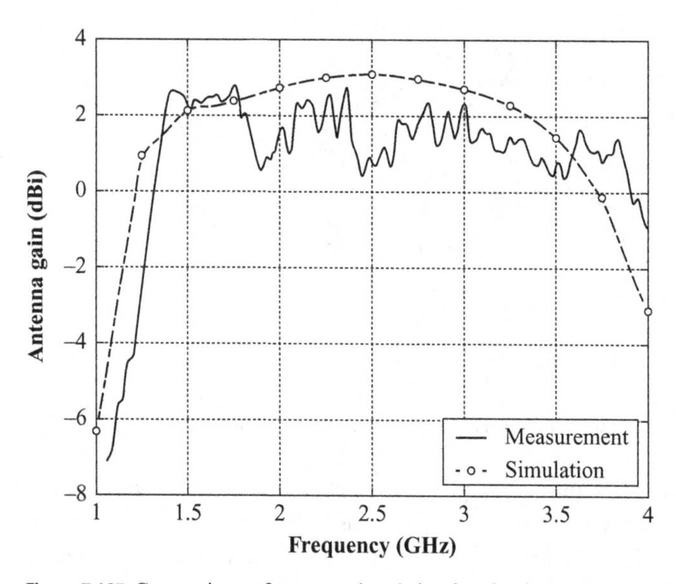


**Figure 7.185** Comparison of measured and simulated gain ([90], copyright ©2003 IEEE).

return loss below −10 dB with the substrate $\varepsilon_r = 2.2$, whereas when $\varepsilon_r = 1$, meaning no dielectric substrate, a considerably wider bandwidth (1.3–3.9 GHz) is obtained. The gain obtained is about 1 dBi over the frequency range of 1.3 GHz to 3.5 GHz.

### *7.2.4.1.3 A half-circular disk quasi-self-complementary antenna*

Antenna geometry with dimensional parameters is illustrated in Figure 7.186 [92], showing that a printed semi-circular disk on a dielectric substrate is combined with its

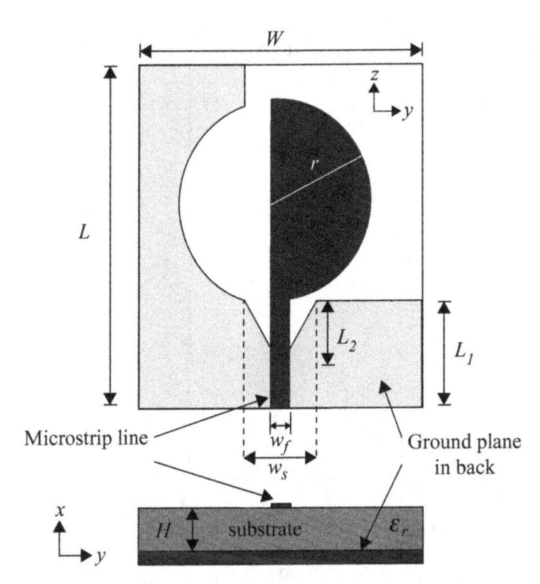


**Figure 7.186** Geometry of half-disk shaped quasi self-complementary antenna ([92], copyright ©2009 IEEE).

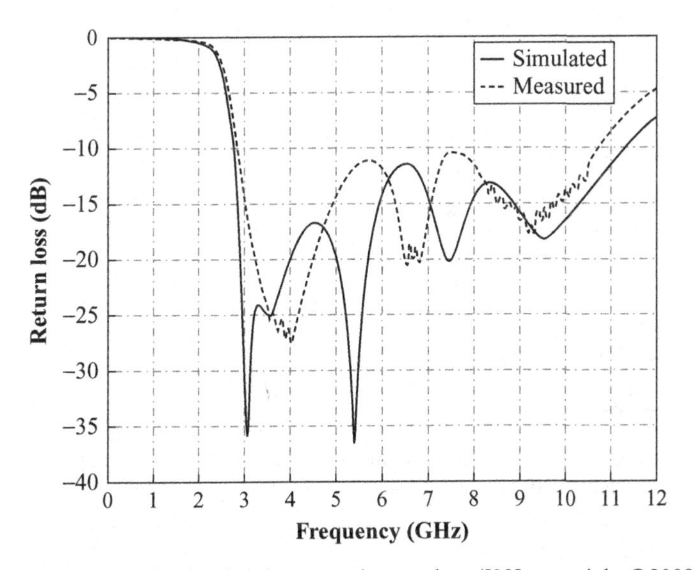


**Figure 7.187** Simulated and measured return loss ([92], copyright ©2009 IEEE).

dual slot to construct a quasi-self-complementary structure. A triangular notch is adopted at the feed point on the ground plane to improve the impedance matching. The substrate has thickness $H = 1.6$ mm and the relative permittivity $\varepsilon_r = 3.0$. Measured and simulated return loss are shown in Figure 7.187, which indicates fairly wide bandwidth, covering 3 GHz to 10.7 GHz for the return loss less than –10 dB. Simulated gain is depicted in Figure 7.188, in which a gain of 3 dB over the wide frequency range is observed.

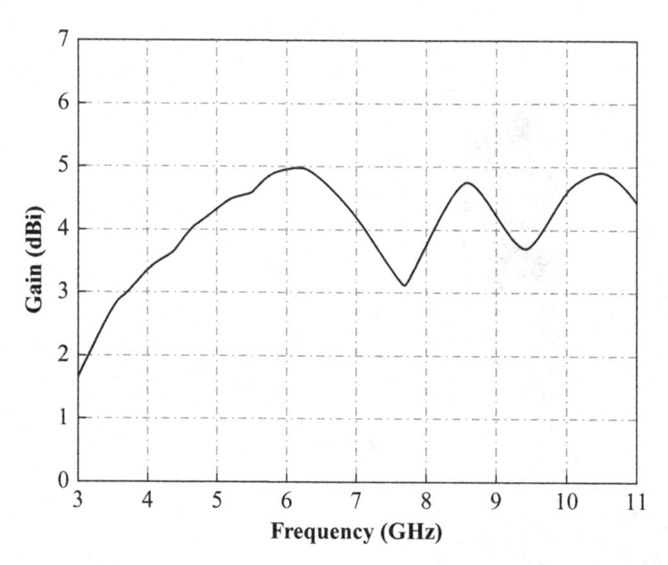


**Figure 7.188** Simulated peak gain ([92], copyright ©2009 IEEE).

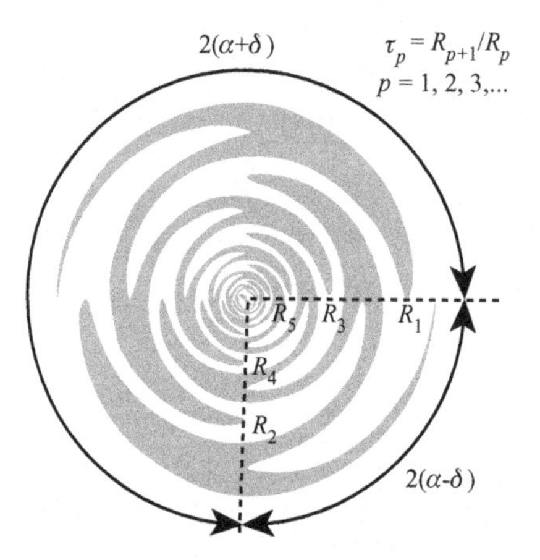


**Figure 7.189** A two-arm self-complementary sinuous antenna with $\tau = 0.725$, $\alpha = 30^\circ$ and $\delta = 45^\circ$ ([93], copyright ©2008 IEEE).

### *7.2.4.1.4 Sinuous spiral antenna*

A planar two-arm sinuous spiral antenna composed with self-complementary structure (Figure 7.189) was demonstrated to produce multiband and multi-polarized radiation over three or more octaves [93]. However, it was found that to achieve improved axial ratio, application of the self-complementary principle to a planar structure was not necessarily effective. Therefore, away from the planar structure, a 3D structure, a cone, on which a sinuous (or spiral) pattern is geometrically projected, was considered as a better choice. By adapting the concept of the self-complementary property on the

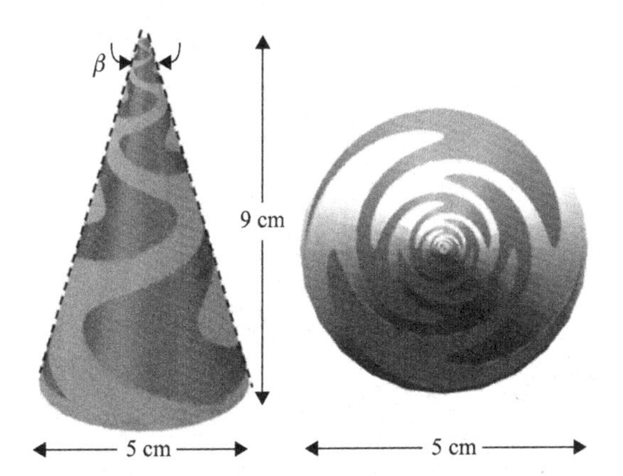


**Figure 7.190** An implemented self-complementary antenna on the cone ([93], copyright ©2008 IEEE).

cone, not only better axial ratio performance, but also impedance matching over a wide frequency range, was made possible. An implemented antenna is shown in Figure 7.190. The input impedance and axial ratio, respectively, are shown in Figure 7.191(a) and (b). Radiation of the antenna is uni directional as shown in Figure 7.192, where the performances at five bands are depicted along with the simulation software such as MoM, FEM, and (MoM + SEP) used for calculations.

### 7.2.4.2 Conjugate structure

#### *7.2.4.2.1 Electrically small complementary paired antenna*

An antenna system composed with two thick, short monopoles, which were fed through a complementary network, was studied theoretically and experimentally [94]. As Figure 7.193 shows, the complementary network, a 180-degree hybrid tee, was used, by which impedance of the antenna is modified to be complementary to that of another antenna. With the phase reversal, the capacitive impedance of a short monopole was changed to the inductive impedance, which is conjugate to that of the other short monopole so that the resonance condition could easily be obtained and matching was made feasible without extra matching network, even though the antenna size was very small. A developed model consisted of two identical monopoles, each of which had the height of one eighteenth of the wavelength at the lowest frequency, and 90-degrees conical section having one-half the total height for a length-to-diameter ratio of unity as Figure 7.194 shows. The antenna referred to as ESCP (Electrical Small Complementary Antenna) was designed for acceptable VSWR at the lowest frequency. Figure 7.195 illustrates VSWR of an ESCP in comparison with an individual monopole. In the paper [94] theoretical analyses were shown, in which the geometrical parameters to optimize the antenna performances were discussed. It was shown that an optimized configuration could yield substantial improvement in gain–bandwidth product at the expense of having a directional pattern over much of this bandwidth. The radiation

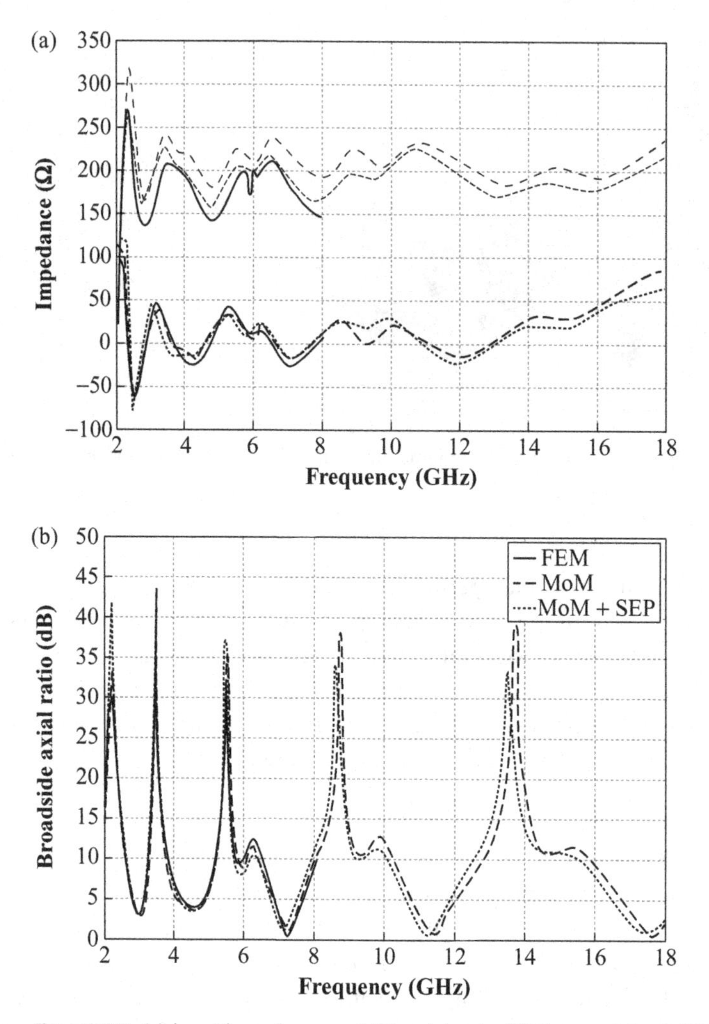


**Figure 7.191** (a) input impedance and (b) axial ratio ([93], copyright ©2008 IEEE).

patterns were also studied regarding the antenna system being in a phased array environment.

### *7.2.4.2.2 A combined electric-magnetic type antenna*

An antenna system composed with a dipole as an electric radiation (ER) source and a cavity-backed slot as a magnetic radiation (MR) source was designed for UWB systems. As a consequence of combining ER and MR sources, an extraordinarily wide bandwidth in terms of both the radiation pattern and the impedance was obtained [95]. The antenna geometry along with the coordinates and a prototype printed antenna along with the dimensions are illustrated in Figure 7.196(a) and (b) respectively. The ER source comprises the dipole arms 1 and 2, printed on the dielectric substrate, and is considered as a tapered slot antenna (TSA) at high frequencies, fed by the slot line 1. The MR source is a loop formed by cutting in the arm 1 of the ER source and fed by the slot line 2.

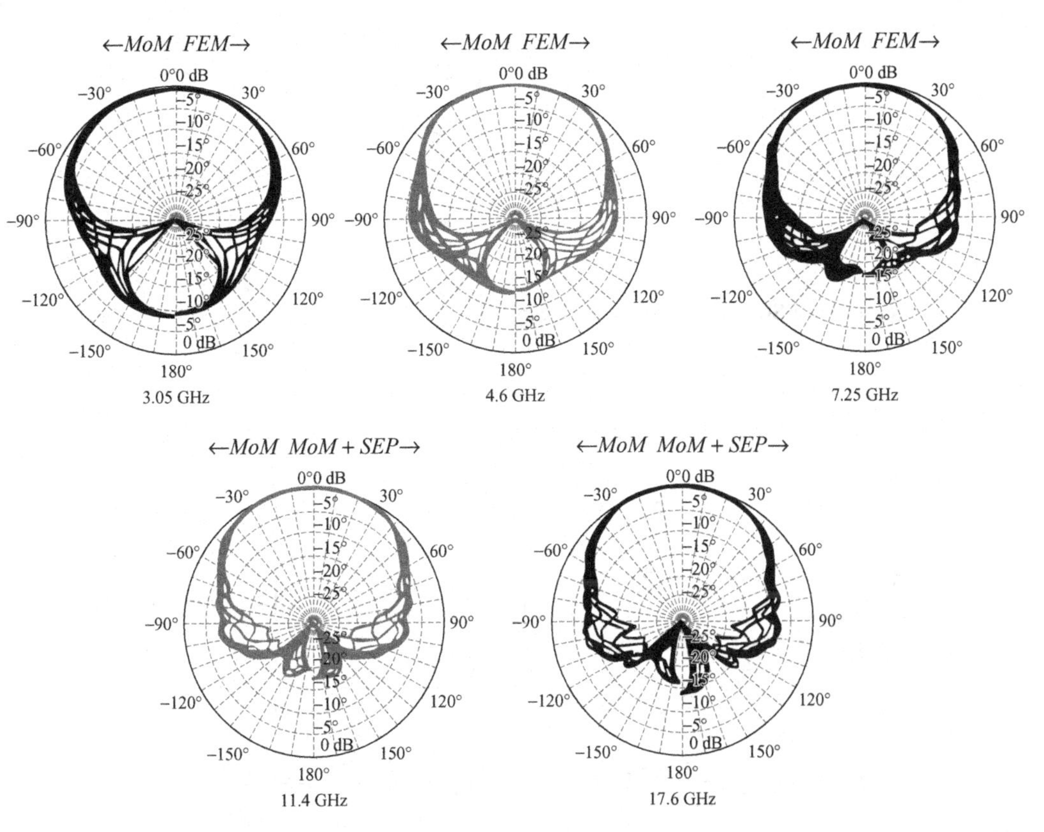


**Figure 7.192** Far-field patterns for five bands computed by MoM and FEM for lower bands and by MoM and MoM+SEP for higher bands ([93], copyright © 2008 IEEE).

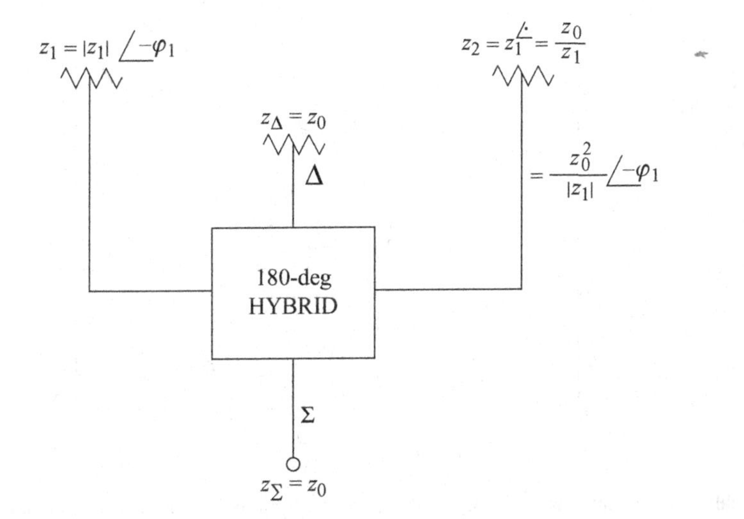


**Figure 7.193** Basic circuit for complementary feed ([94], copyright ©1976 IEEE).

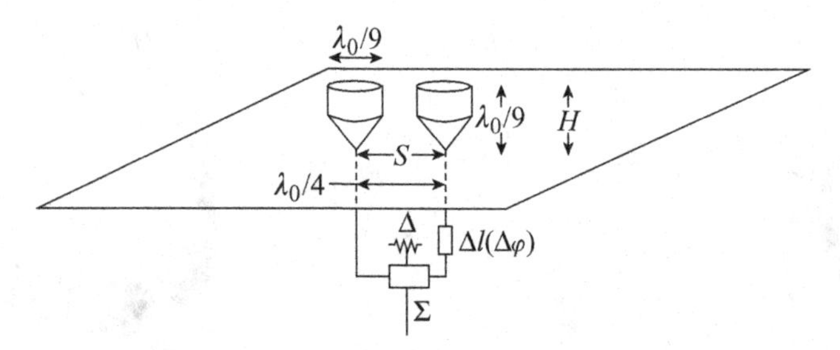


**Figure 7.194** ESCP configuration ([94], copyright ©1976 IEEE).

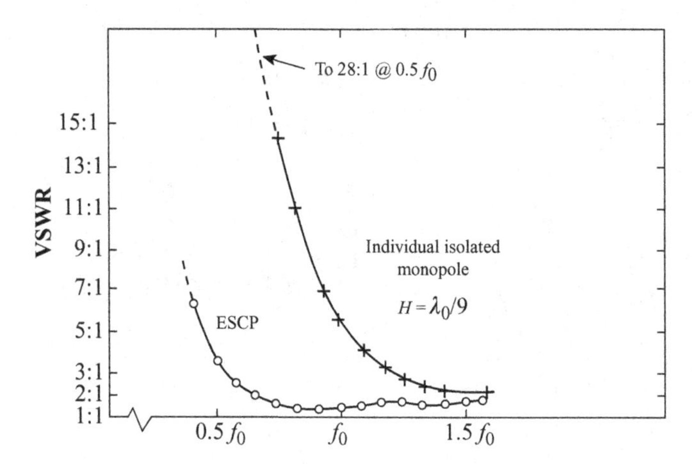


**Figure 7.195** ESCP impedance match characteristics ([94], copyright ©1976 IEEE).

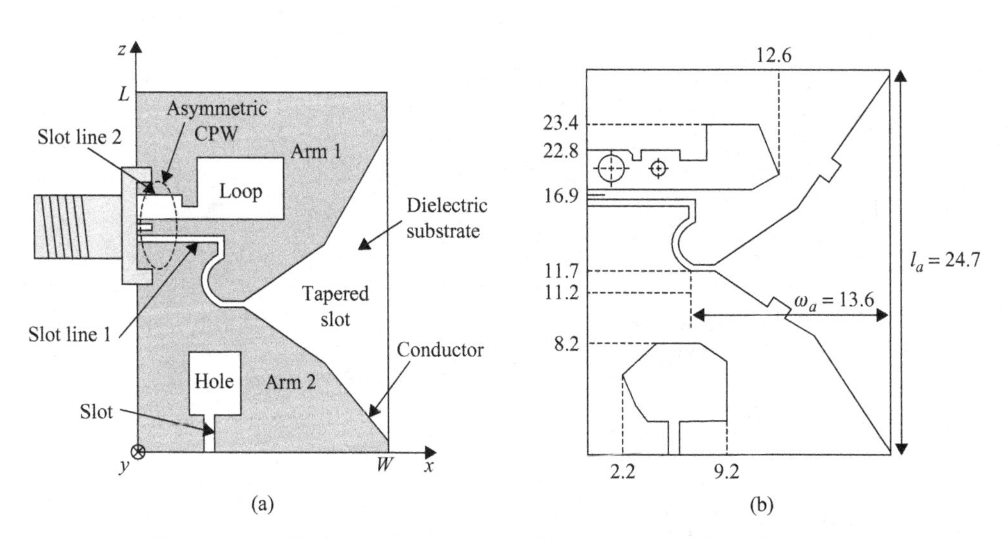


**Figure 7.196** (a) Basic antenna geometry of the printed combined electric and magnetic type antenna and (b) dimensions of a prototype antenna [95].

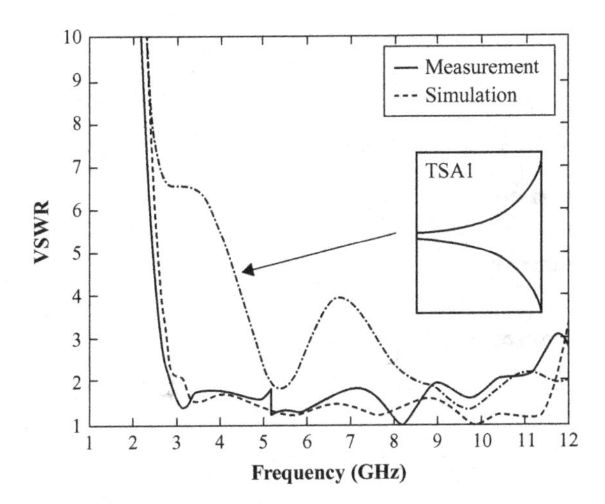


**Figure 7.197** Simulated and measured VSWR responses [95].

The slot lines 1 and 2 form an asymmetric coplanar waveguide (CPW). The loop acts as a matching stub for transmission from the CPW to the slot line 1 as was shown in the CPW-fed TSA design [96]. Figure 7.197 shows both simulated and measured VSWR characteristics, illustrating a wide frequency operation covering 2.82 to 10.6 GHz for VSWR of less than 2. In the figure, the simulated VSWR response of a TSA only is also shown along with geometry of the corresponding TSA as an inset.

It is noted that the VSWR response of the TSA takes on values higher than 2 over all frequencies except around 5 GHz, suggesting existence of uncompensated excess electric energy before combining the ER source with the MR source. As a consequence of combination of two sources, the excess energy of the ER source is compensated by the excess energy of the MR source, resulting in lowering the VSWR significantly. This implies, in other words, self-resonance in an antenna system, in which the time averaged electric and magnetic energies in the vicinity of the antenna system are balanced.

### 7.2.4.3 Composite structure

#### *7.2.4.3.1 Slot-monopole hybrid antenna*

A coplanar waveguide (CPW)-fed inductive slot antenna combined with a monopole antenna, featuring a dual-band operation is introduced in [97]. It is a common understanding that CPW antennas can provide relatively wide bandwidth, be easily integrated with surface-mount devices, and be designed to show dual-band operation [98, 99]. Figure 7.198 illustrates the antenna system, a CPW-fed inductive slot combined with a bifurcated L-shaped monopole, showing a dual-band operation as is depicted in Figure 7.199, which gives return losses of antennas, including those of a bifurcated L-shaped monopole (Figure 7.200) and a CPW-fed inductive slot (Figure 7.201) for a comparison. The antenna dimensions are given in each figure. The antenna is fabricated on the dielectric substrate having $\varepsilon_r = 4.4$ and thickness of 1.6 mm. The CPW line has a strip width of 4 mm and a gap width of 0.4 mm, corresponding to a 50 Ω

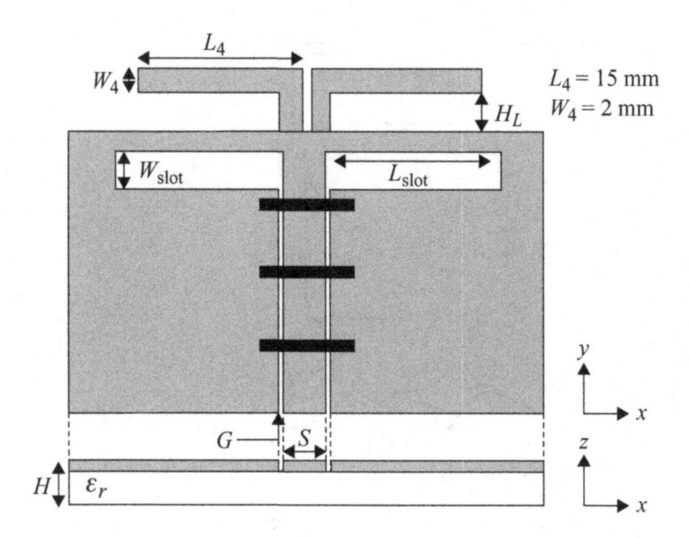


**Figure 7.198** Dual-band CPW-fed slot and L-shaped monopole antenna ([97], copyright ©2008 IEEE).

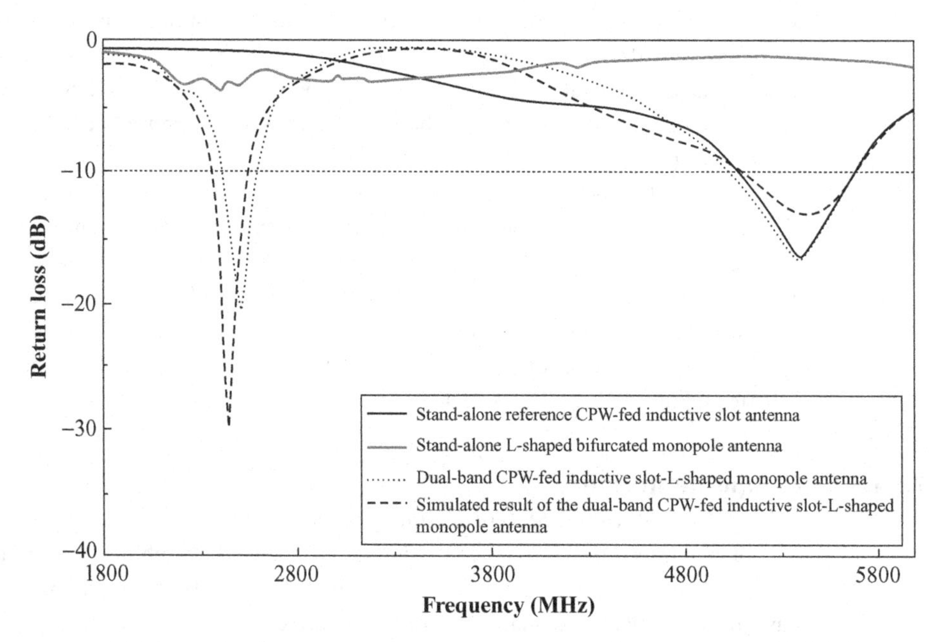


**Figure 7.199** Measured and simulated return-loss characteristics ([97], copyright ©2008 IEEE).

characteristic impedance. The stand-alone CPW-fed slot is designed for the higher band 5.4 GHz operation, while the bifurcated L-shaped monopole is for the lower band 2.4 GHz operation. Other types of monopole having bifurcated I-shapc or F-shape are designed and studied. Gain of the antenna is measured as 1.51 dB at 2.45 GHz and 4.91 at 5.2 GHz, and the relative bandwidth of the antenna is 11.4% at 5.3 GHz and 7.8% at 2.5 GHz, while the stand-alone CPW-fed slot antenna is 10.5%.

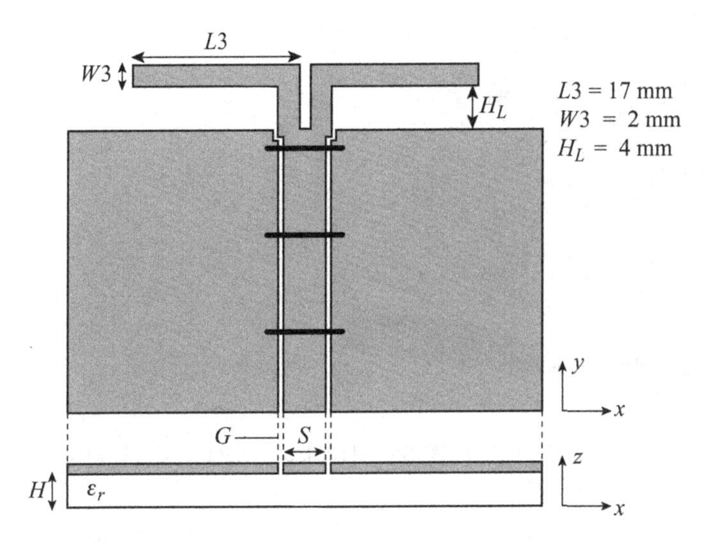


**Figure 7.200** L-shaped bifurcated monopole antenna ([97], copyright ©2008 IEEE).

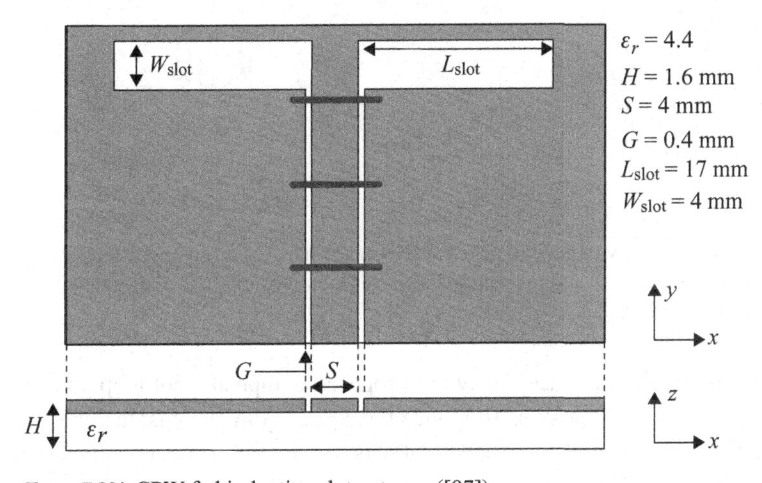


**Figure 7.201** CPW-fed inductive slot antenna ([97]).

### *7.2.4.3.2 Spiral-slots loaded with inductive element*

A miniaturized cavity-backed slot loop low-profile antenna was introduced in [100] (Figure 7.202). It is referred to as CBCSLA (Cavity-Backed Composite Slot Loop Antenna). The antenna is essentially a small magnetic loop, radiating similarly to that of a small electric dipole; that is, omnidirectional pattern in the horizontal plane with vertical polarization. The geometry is designed to be small and low height, having the diameter as small as $\lambda/10$ and the height less than $\lambda/100$. The design concept is to modify a simple slot loop; firstly by embedding it in a shallow cavity, secondly meandering it to reduce the size, and thirdly sectionalizing the structure into six $\lambda/2$ slots around the circle to accomplish resonance and achieve sufficient input impedance matching. The process of topology change is shown in Figure 7.203(a), (b), and (c). In the figure (c),

Figure 7.202 Photograph of fabricated CBCSLA ([100], copyright ©2008 IEEE).

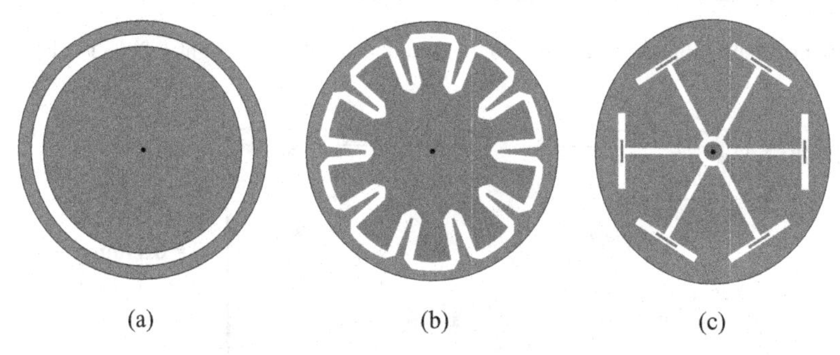


Figure 7.203 Topology of cavity-backed slot loop antenna; (a) original slot loop, (b) modified slot loop for size reduction, and (c) sectionalized slot loop for input impedance matching ([100], copyright ©2008 IEEE).

the geometry shows conceptually the proposed composite slot loop antenna, which is fed capacitively with a stub. To reduce the antenna dimensions, the edges of each slot are folded into a spiral-like shape that acts at the same time as inductive load on the edges. Figure 7.204 shows the antenna structure along with the antenna geometrical parameters. As can be seen in the figure, the coax feed is connected to the six Co-planar Waveguides (CPW), by which each of the slots is fed separately. At the edges of each CPW line, a corrugated stub is attached to control capacitance and thus improve impedance matching to the slot. Measured and simulated $S_{11}$ of the CBCSLA with the cavity height $h_c = 12.7$ mm is shown in Figure 7.205, and simulated radiation pattern is depicted in Figure 7.206.

### 7.2.5 Applications of metamaterials

Real metamaterials (MM) are not available in nature; however, equivalent media may be constructed of either resonant particles (RP) MM [101a, b, 102] or transmission lines (TL) MM [103a, b]. The RP MM may consist of periodical arrays of sub-wavelength thin wires (Figure 6.18) [101] or split rings (Figure 6.19) [102]. They provide either negative

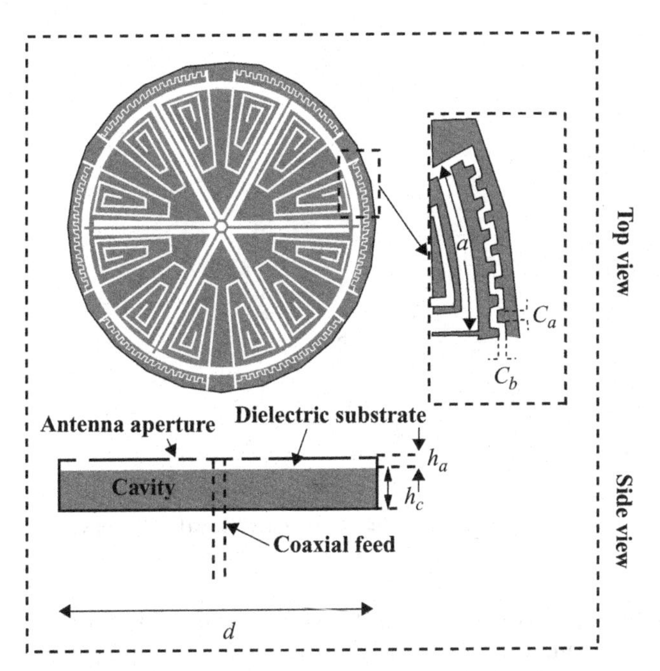


**Figure 7.204** Topology of the reduced-size CBCSLA and corrugated capacitive stub ([100], copyright ©2008 IEEE).

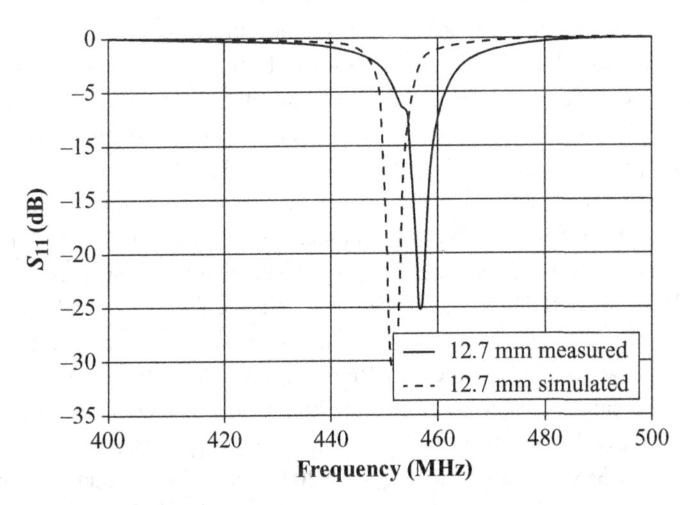


**Figure 7.205** Measured and simulated $S_{11}$ of the reduced-size CBCSLA with $h_c = 12.7$ mm. ([100], copyright ©2008 IEEE).

permittivity or permeability within their restricted frequency range, and a combination of them leads to a double negative medium (DNG). The TL MM may contain composite RH/LH transmission lines (CRLH TL) [103], which are obtained by cascading a sub-wavelength unit cell constituted of a series capacitance and a shunt inductance as the LH MM structure, and unavoidable parasitic shunt capacitance and series inductance existing in the practical circuit implementation as the RH MM structure. The RP

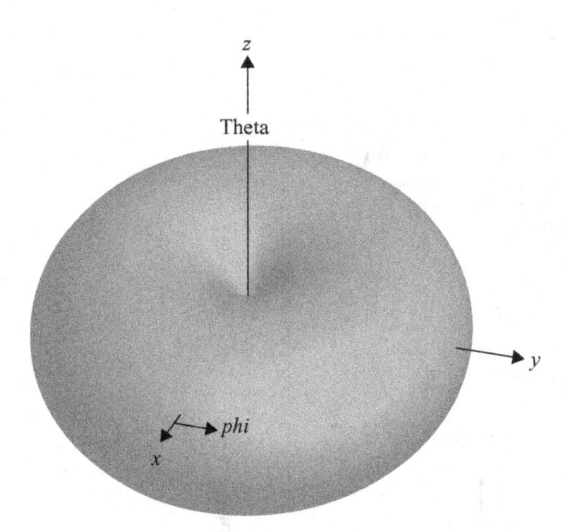


**Figure 7.206** Simulated radiation pattern of reduced-size CBCSLA ([100], copyright ©2008 IEEE).

MM has unfavorable properties; the typical ones are unsuitability for most microwave applications, narrow bandwidth, large loss, and necessity of volumetric formation. The TL MM, on the contrary, has advantages: applicability in microwave frequency regions, wide bandwidth, low loss, and suitability for implementing in not only planar, but also volumetric structures. They have been extensively studied, and various applications to electromagnetic devices and antennas have been introduced [104–106].

The MM has unique dispersive characteristics and phase delay of wave propagation in the media that are attributes of double negative constitutive parameters ($\varepsilon < 0$, $\mu < 0$). These MM properties can be effectively used to develop novel antennas. In the CRLH MM, as the frequency $\omega$ becomes higher, the wave number $\beta$ decreases, corresponding to increase in the wavelength $\lambda$, thus lowering the resonance frequency in the media. Hence, an appropriate design of the wave number $\beta$ in an antenna system constituted of an LH medium to obtain desired frequency $\omega$ renders the antenna size reduction.

In a CRLH TL MM of length $l$, constituted of a finite number $N$ multiple unit cells of length $p$, there would exist multiple resonance modes when length $l$ is a multiple of half a wavelength $\lambda_g$ (guided wavelength), that is, $l = n\lambda_g/2$, with $n = 0, \pm 1, \pm 2, \ldots \pm \infty$. A CRLH TL can be treated as effectively homogeneous media when the electrical length of the unit cell is smaller than $\pi/2$, that is, $p < \lambda_g/4$. The dispersion diagram of the CRLH TL MM is shown in Figure 7.207. The CRLH structure can support negative resonance ($n < 0$) in the LH region, because of transfer of the phase origin from frequency zero to the transition frequency $\omega_0$, and also zeroth-order resonance ($n = 0$) at $\omega_0$ in addition to the ordinary positive resonance ($n > 0$) in the RH region [107] (Figure 7.207 and Figure 6.49). Multiple resonances may occur at frequencies of $n\,\pi/N$. Therefore, resonance at two or more frequencies in a medium can be attained, indicating possibility of a multiband antenna design. When $n = 0$, that is the zero-order mode, $\beta = 0$, and $\lambda$ is infinite, but the group velocity $v_g$ is not zero, as the wavelength becomes infinite, and the amplitude and phase of the wave are the same anywhere in the media. This means that the resonance is not dependent on the dimensions of the resonator, but

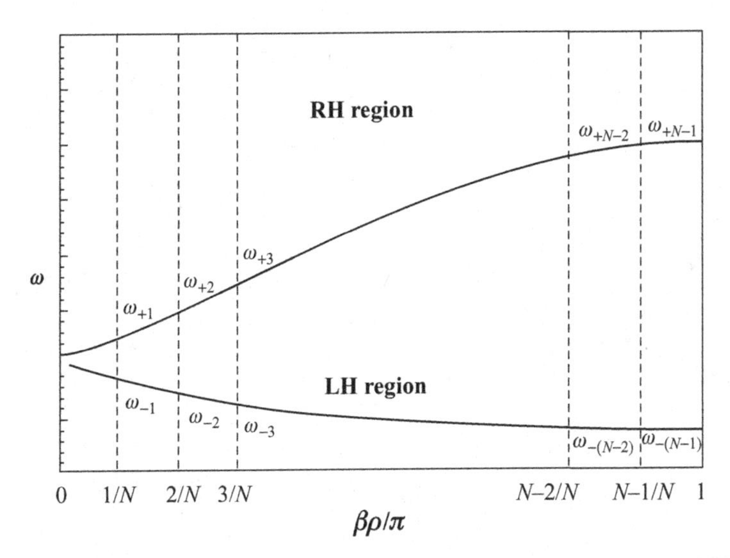


**Figure 7.207** Resonance spectrum of a CRLH structure ([116], copyright ©2008 IEEE).

solely on the values of lumped reactive components on the unit cells, and hence this property can be exploited to design an electrically small antenna.

In the DNG media, negative $\beta$ (the wave number) suggests that direction of the power propagation in the media is opposite to that of the phase propagation. By introducing such condition as negative $\beta$ in an antenna structure, a backward radiation can easily be produced, whereas an ordinary LW (Leaky Wave) antenna radiates only in the forward direction. A balanced-type CRLH TL [103a] can be used for a frequency scanned LW antenna, which is designed to produce backward radiation as well as forward by using the LH/RH property. By constituting a CRLH TL MM in 2D structure, a larger aperture is obtained and consequently a small antenna with enhanced gain can be designed. By dual feeding with phase delay circuits to the 2D CRLH TL MM, a circularly polarized small antenna can be realized.

Other than artificial materials, there are some materials which can exhibit negative permeability. For example, ferrite materials show negative-mu property near magnetic resonance, although the mu changes rapidly once reaching the positive peak value, turning to quickly descend to the negative lowest value through zero, and then gradually increasing to recover positive value. Then, over the frequency range in which the permeability takes negative values, the ferrite material can be used as a negative-mu (MNG) material. However, this frequency range is rather narrow, and that often checks the practical applications.

Meanwhile, as was described previously, real MNG materials have been realized with BaFe material and permalloy composite [108]. With the BaFe material, negative mu can be obtained in a lower-frequency region, 2 to 5 GHz, while the permalloy composite exhibits negative mu in higher-frequency regions, 9 to 18 GHz and beyond. Figure 7.208 provides variation of the permeability of the permalloy composite with respect to frequency. However, since these materials have some loss and a somewhat high permittivity, this problem has been targeted to be improved.

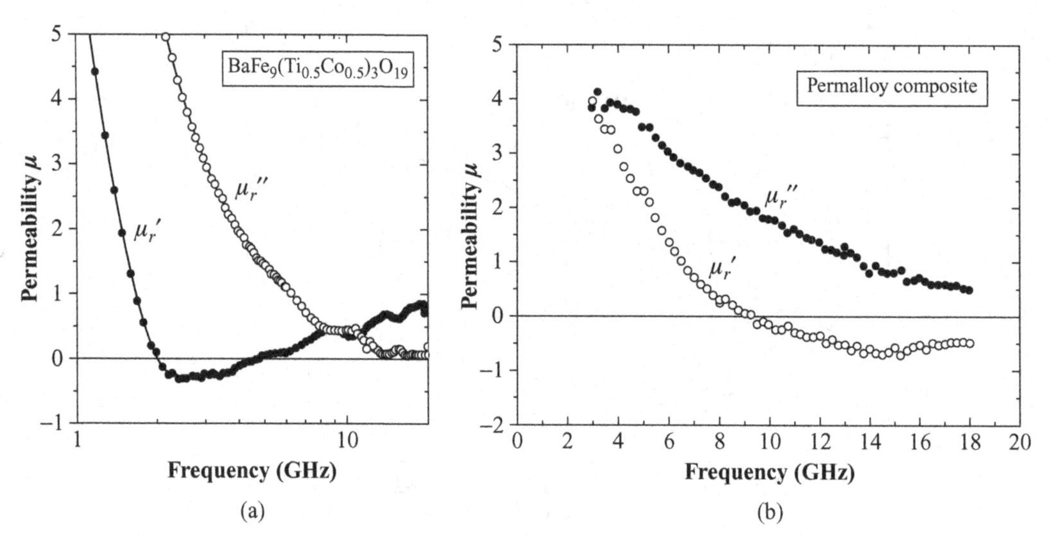


**Figure 7.208** Dispersion characteristics of magnetic composite materials; (a) BaFe material and (b) Permalloy composite material ([108], copyright ©2008 IEEE).

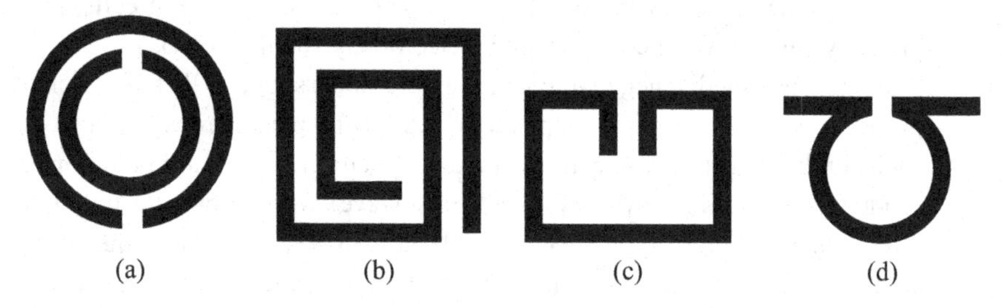


**Figure 7.209** Examples of MNG material unit: (a) SRR, (b) CLL, (c) SR, and (d) omega-shaped structure [101,109,110 and 111].

Artificial MNG materials can be implemented by other means; four examples are illustrated in Figure 7.209, (a) SRR (Split Ring Resonator) [101], (b) capacitive-loaded loop (CLL) [109], (c) spiral rings (SR) [110], and (d) omega shaped structure [111]. Meanwhile, artificial ENG (epsilon negative) material, as is mentioned above, can be constituted from an array of thin wires (Figure 6.18) [102]. Meander lines arranged in planar structure and helical windings are used as inductor materials, which can be a substitute for ENG materials for the purpose of space resonance or matching [109]. Applications of these artificial MMs to small antennas will be shown in the next section.

### 7.2.5.1 Applications of SNG (Single Negative) materials

#### *7.2.5.1.1 Mu-Negative (MNG) metamaterials (MM)*

##### *7.2.5.1.1.1 Circular patch antenna*

Implementation of a miniaturized circular patch antenna loaded with MNG MM has been introduced in [112 a, b], in which a thorough theoretical analysis of the magnetic field distribution underneath the patch, and the design of magnetic inclusions related

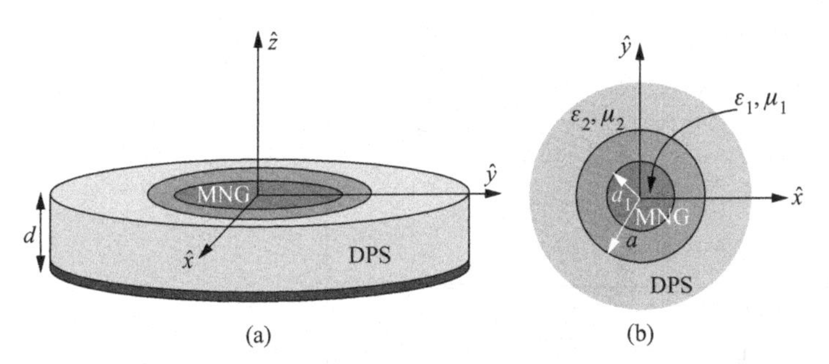


**Figure 7.210** Geometry of a circular patch antenna loaded with an MNG MM [112a and b].

with their location, arrangement, and alignment, are shown. To realize the MNG MM, multiple spiral ring resonators (MSRR) (Figure 7.209(c)) are employed as an appropriate selection, having the dimensions that fit with the limited thickness of the space between the patch and the ground plane, while their required resonance frequency is kept despite their small electrical size. The MSRR are aligned according to the expected preferred direction of the magnetic field at their location. Geometry of the patch antenna loaded with an MNG MM is illustrated in Figure 7.210, showing (a) the 3D view and (b) the top view of the antenna, which is partially loaded with an MNG MM underneath the patch. With this arrangement, resonant modes can be excited on the patch, even though the dimensions are significantly smaller than the operating wavelength. Even with the small dimensions, the MNG MM must provide an effective negative permeability in its complex near-field interaction with the feed and the patch. As the MNG MM has dispersion property, the permeability $\mu_1$ is given by assuming the Drude dispersion relation as

$$\mu_1 = \mu_0(1 - \omega_p^2/(\omega^2 - j\omega\delta)) \tag{7.72}$$

where $\mu_0$ is the free space permeability, $\omega_p$ is the magnetic plasma frequency, and $\delta$ is the damping factor. Here the MNG media required to design an antenna shown in Figure 7.210 for operating at 0.47 GHz is assumed to have the dispersion characteristic as shown in Figure 7.211, where the variation of the relative permeability $\mu_1/\mu_0$ with respect to the frequency is given [112a]. In this case, the geometrical parameters of the antenna used are: radius of the substrate $a = 20$ mm, radius of the MNG media $a_1 = 12$ mm, thickness of the substrate $d = 5$ mm, and $\delta = 0.01$ GHz. Then the required value of the relative permeability $\mathrm{Re}[\mu_1/\mu_0] = -2.17$ at 0.47 GHz. Geometrical sketch of the MNG integrated patch is depicted in Figure 7.212(a). In the figure, (b) illustrates arrangement of the MSR (Multiple Split Ring), viewing from the top of the patch. With this MSR arrangement, almost uniform field distributions are obtained, thus exciting the desired $TM_{11}$ mode effectively all over the patch. The presence of the MSR highly affects the current distributions underneath the patch, closing itself in an electrically small resonant loop, which produces the desired radiation patterns and gain [112a]. The return-loss characteristic is depicted in Figure 7.213, which shows matching features of the antenna at two frequencies. The antenna operates at higher frequencies (around 2.44 GHz, determined by the electrical size of the patch) as well as lower frequencies. At higher frequencies, the substrate behaves as a homogeneous material having the

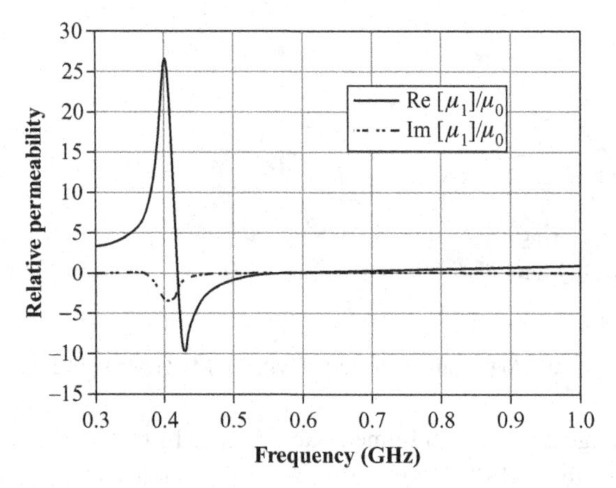


**Figure 7.211** Dispersion characteristic model of the MNG MM required to design the antenna of Figure 7.210 ([112a], copyright ©2008 IEEE).

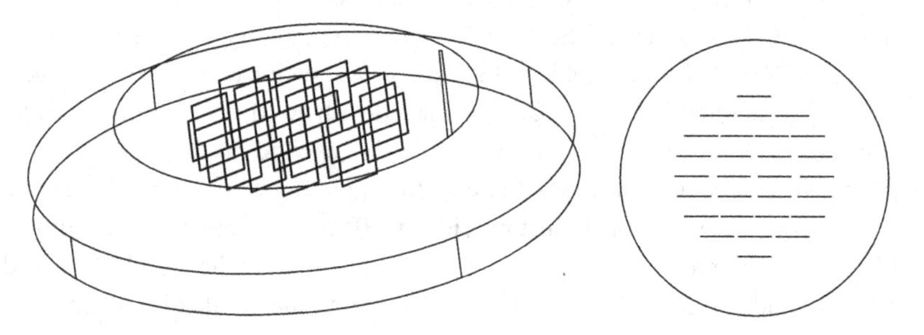

**Figure 7.212** Geometry of SR implementation inside the cylindrical core cavity underneath the patch: (a) 3D view and (b) top view of the patch, ([112a], copyright ©2008 IEEE).

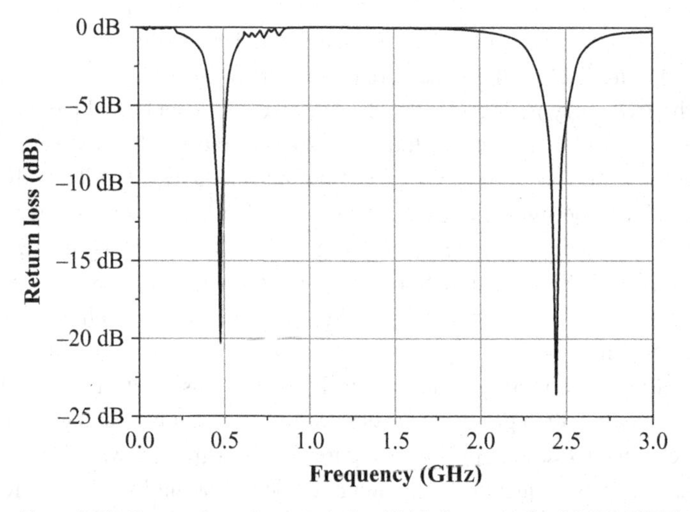


**Figure 7.213** Return-loss characteristics ([112a], copyright ©2008 IEEE).

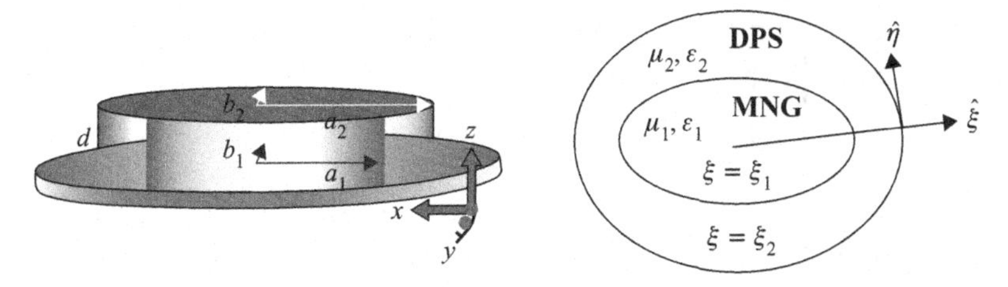


**Figure 7.214** Geometry of an elliptical patch antenna partially loaded with an MNG MM core surrounded by a DPS shell ([113], copyright ©2010 IEEE).

constitutive parameters $\varepsilon = \varepsilon_2 = 2.33\ \varepsilon_0$ and $\mu = \mu_2 = \mu_0$. The simulated resonance frequencies obtained by using the full-wave commercial code CST Microwave Studio are very close to the predictions using the theoretical cavity model, where the radius of the circular patch antenna is only 20 mm. Radiation patterns are nearly the same as that of an ordinary circular patch antenna. The ground plane in this model has a radius of 40 mm. Simulated gains and efficiencies at the two resonance frequencies, 0.47 GHz and 2.44 GH respectively, are 3.1 dBi, 6.3 dBi and 0.67, 0.92.

#### *7.2.5.1.1.2 Elliptical patch antenna*

An elliptical patch antenna loaded with MNG material was studied and showed the possibility of miniaturization of the antenna size and advantages of using the elliptical geometry compared to the use of circular geometry [113]. The antenna geometry is illustrated in Figure 7.214, which shows an elliptically shaped MNG material core of semi-axes $a_1$ and $b_1$ partially loaded underneath the elliptical patch and surrounded by an elliptically shaped DPS shell of semi-axes $a_2$ and $b_2$. The elliptical shape parameters are the semi-focal length $F = \sqrt{a_1^2 - b_1^2} = \sqrt{a_2^2 - b_2^2}$, and the eccentricity of the elliptical patch $e = \sqrt{1 - (b_2/a_2)^2} = 1/\cosh(\xi_2)(0 \leq e < 1)$. Here the Cartesian coordinates $x$ and $y$ are related with the elliptical system coordinates $\xi$ and $\eta$ by

$$\begin{aligned} x &= F \cos\xi \cos\eta \\ y &= F \sin\xi \sin\eta. \end{aligned} \tag{7.73}$$

Another parameter is the filling ratio $\Gamma$ = (volume of the material core)/(overall volume underneath the patch) $(0 < \Gamma < 1)$.

Based on the full-wave simulation, the antenna performances are analyzed. The parameters used are: surface area of the patch $A = 4\pi\ \text{cm}^2$, $\Gamma = 0.35$, $e = 0.7$, $\varepsilon_1 = \varepsilon_2 = \varepsilon_0$, and $\mu_2 = \mu_0$. Here, $\varepsilon_1$, $\mu_1$, and $\varepsilon_2$, $\mu_2$, respectively, are the constitutive parameters of the MNG and the DPS materials. Since the material has dispersive characteristic, different modes can be excited by the different set of magnetic plasma frequencies. By the selection of $\omega_{mp} = 1.286$ GHz, the odd mode is excited around the design frequency 0.5 GHz, while by the selection of $\omega_{mp} = 0.707$ GHz, the even mode at the same resonance frequency is excited. Simulated return loss $S_{11}$ is given in Figure 7.215, which indicates that near the design frequency, resonances are obtained in two cases with even and odd mode properties. In the figure, dual-band effect is observed in the odd-mode

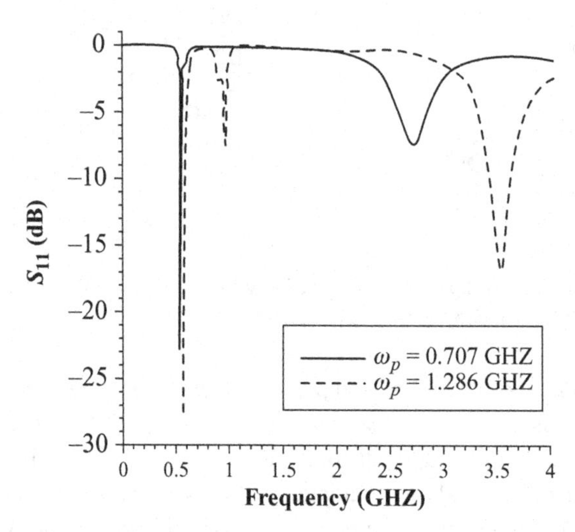


**Figure 7.215** Return-loss characteristics for even-mode (solid line) and odd-mode (dotted line) excitation of the elliptical patch antenna ([113], copyright ©2010 IEEE).

case. This is an attribute of the MNG material with $\mu_1 = -5.618$ at 0.5 GHz, and −1 at 0.91 GHz, respectively, obtained by using the Drude dispersion relation (7.72) with the magnetic plasma frequency $\omega_{mp} = 1.286$ GHz and 0.707 GHz, respectively. These correspond to the excitation of both the first odd and second even modes in the spectrum. In these cases, $\varepsilon_1 = 2\varepsilon_0$ is used. At higher frequencies, resonances with broader bandwidth are observed. These correspond to standard patch resonances for even mode (at 2.67 GHz) and odd mode (at 3.65 GHz) with $\mu_1 = \mu_0$, as the substrate permeability turns to become positive.

Realization of artificial MNG material may be approached with a technique introduced in the previous section employed for the circular patch [112]. Properly designed MSRR or MSR may perform similar magnetic resonance within a sub-wavelength volume, consistent with the miniaturized size of an antenna. Gain varies depending on the magnetic plasma frequency. The calculated gain normalized to the case of an ideal lossless MM at the resonance frequency (0.5 GHz) is shown in Figure 7.216. In this case, mismatch loss at the feed is excluded and only the effect of MM absorption, which concerns the damping frequency $\omega_\tau$ is included.

#### *7.2.5.1.1.3 Small loop loaded with CLL*

Artificial materials exhibiting negative mu implemented by Capacitive-Loaded Loops (CLL) are employed for miniaturization of a loop antenna [109]. The MNG interacts with the highly inductive near field produced by a small loop to attain resonance and also obtain matching in space, even though the antenna has an electrically small size, that is, $ka \leq 0.5$ ($a$: radius of a sphere circumscribing the maximum size of antenna). Some examples, using a semi-circular loop and a rectangular loop, are illustrated in Figure 7.217, which shows use of different types of capacitance in the CLL: (a) a spacing between a loop, (b) an inter-digital capacitor, and (c) a lumped component. Various types of antennas with different geometries and dimensions are designed, and

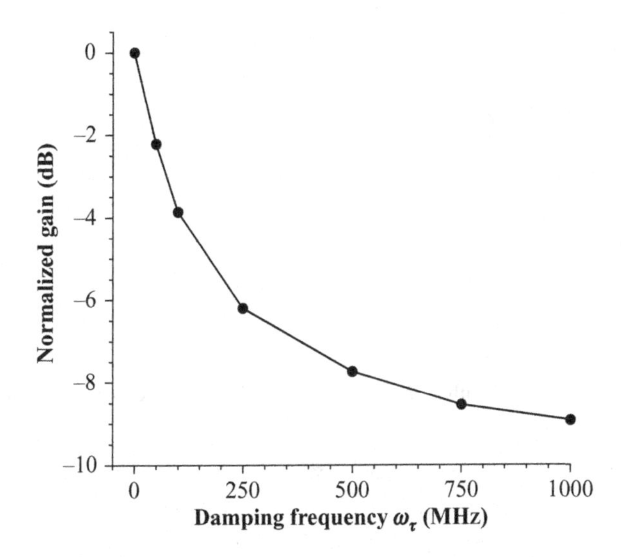


**Figure 7.216** Gain normalized to the case of an ideal lossless MM at the resonance frequency with respect to the damping frequency ([113], copyright ©2010 IEEE).

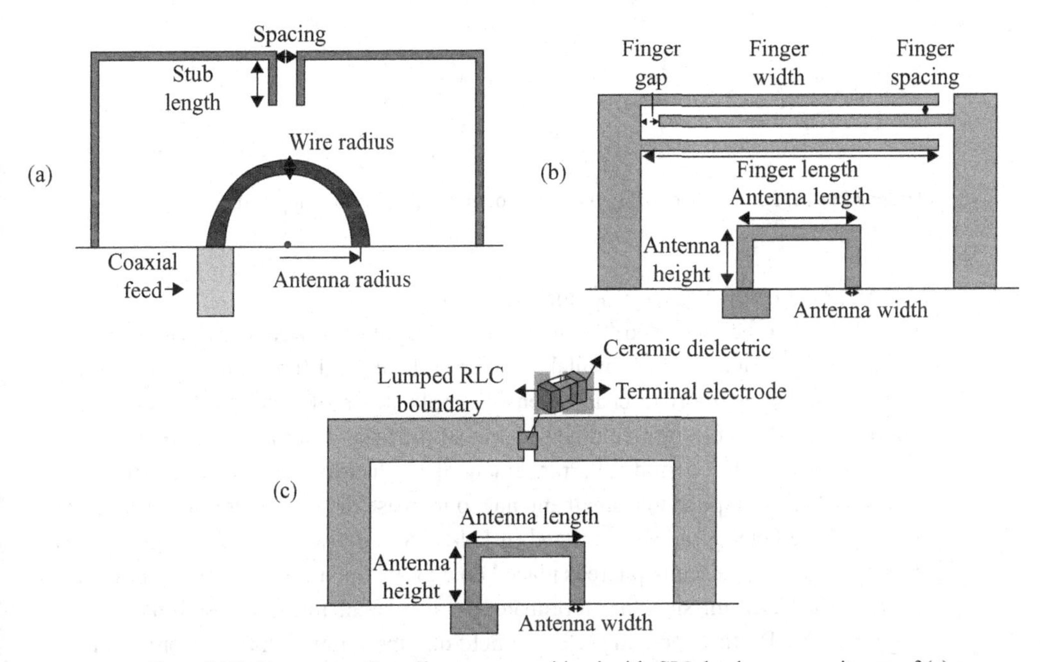


**Figure 7.217** Examples of small antenna combined with CLL having a capacitance of (a) a spacing between the loop, (b) an inter-digital capacitor, and (c) a lumped component ([109], copyright ©2008 IEEE).

their simulated performances (radiation efficiency, bandwidth, and $Q$, and so forth), in terms of their geometrical parameters and frequencies from 300 MHz to 6 GHz, are introduced. High radiation efficiency is obtained in most of the antennas even though the antennas have electrically small size, whereas the VSWR bandwidth is found to be generally very narrow, as low as 4% at maximum.

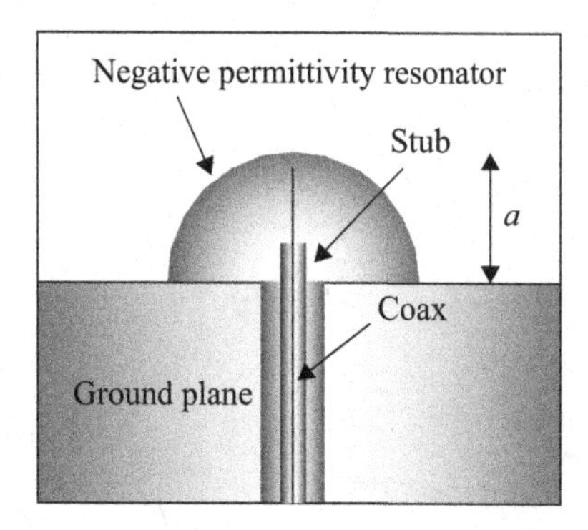


**Figure 7.218** A stub antenna loaded with a semi-sphere ENG MM ([114], copyright ©2006 IEEE).

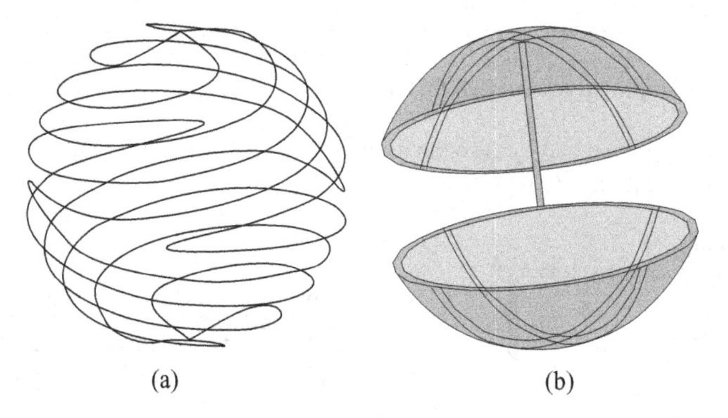


**Figure 7.219** (a) Folded spherical helix resonator and (b) spherical-capped dipole, ([114], copyright ©2006 IEEE).

### *7.2.5.1.2 Epsilon-Negative Metamaterials (ENG MM)*

ENG MMs can be employed for miniaturizing antennas as well as MNG MMs. However, because ENG MMs are available either in the optical frequency regions or using plasmas of the appropriate charge density, applications of ENG MM to antennas at microwave frequencies are confined to use of artificial materials. The representative artificial material is a periodical arrangement of conductors introduced in [101, 102].

An ENG MM applied to a small antenna to increase the bandwidth (lowering $Q$) and raising the efficiency is demonstrated in [114]. A negative permittivity metamaterial (i.e. ENG MM) of a semi-sphere is placed on a ground plane and fed through its center by a coaxial transmission line terminated with a small monopole stub as shown in Figure 7.218. By this geometry, electric field distribution within the sphere is uniform at the fundamental resonant mode and thus the small stub protruding into the center of the sphere provides a strong coupling between the coaxial transmission line mode and the resonant mode of the sphere. Through the analysis, it is shown that the spherical resonator composed of a semi-spherical ENG MM has a $Q$ that is only 1.5 times the Chu limit. This is comparable to the performance of the other types of small spherical antennas such as a folded spherical helix (Figure 7.219(a)) and a spherical-capped dipole (Figure 7.219(b)), which have $Q$ nearly 1.5 times the Chu limit, even with the size as

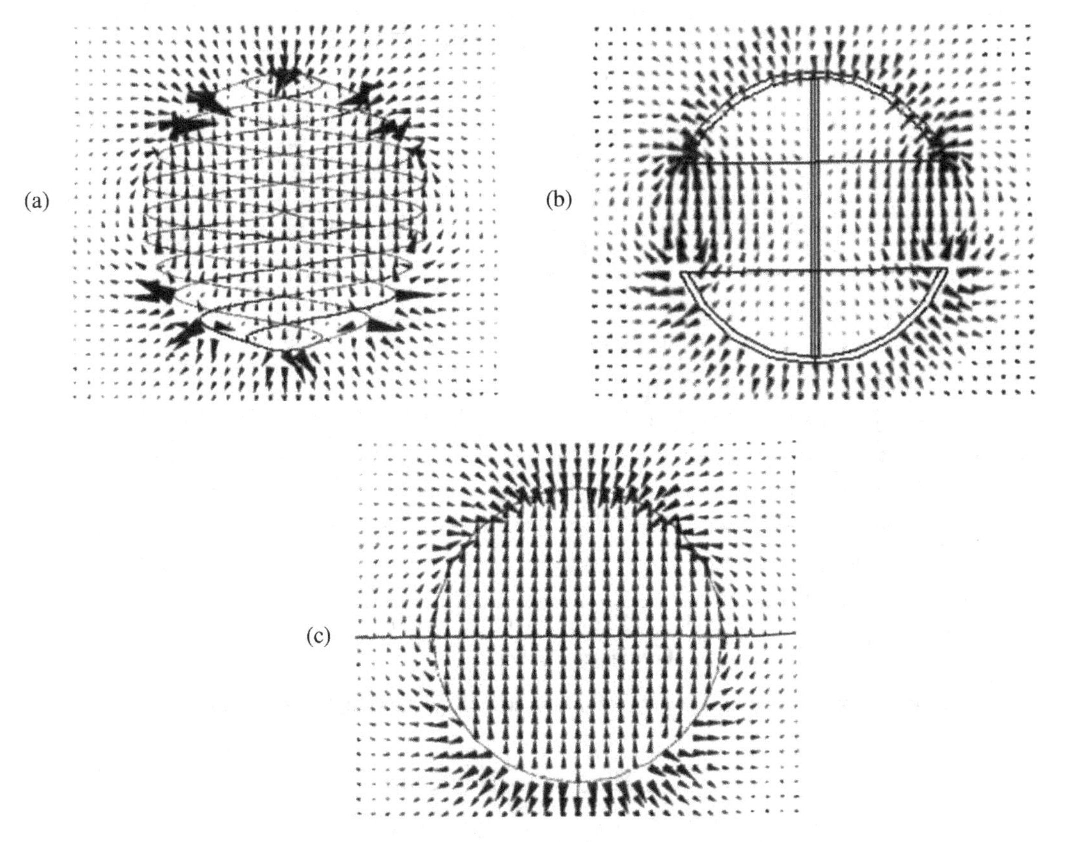


**FIGURE 7.220** Electric field distribution profiles (a) folded spherical helix, (b) spherical-capped dipole, and (c) ENG sphere ([114], copyright ©2006 IEEE).

small as around $ka = 0.3$. These verify the work of Wheeler, mentioning that optimal use of the sphere volume occupying an antenna may provide optimized bandwidth for the antenna, even though with electrically small size. The electric field distribution profiles of these three types are shown in Figure 7.220 [114], which shows for a case of (a) spherical helix, (b) spherical-capped dipole, and (c) ENG MM loaded semi-sphere. There is a similarity in the behavior of modes between that of the helix and that of the ENG MM sphere.

Simulation for the antenna model shown in Figure 7.218 is performed by using these parameters: 8 mm for the radius of the semi-sphere, 3 mm for the length of stub, which is optimized for good impedance matching, and the plasma frequency of 3.54 GHz for the ENG MM that obeys the Drude dispersion relation. The stub length 3 mm corresponds to $\lambda/50$ and the overall antenna size is characterized by $ka = 0.34$ (radius $a = \lambda/18.5$, where $\lambda = 148$ mm; the resonance frequency at 2025 MHz). Calculated return loss, and impedance are shown in Figure 7.221(a) and (b), respectively. Figure 7.222 depicts the simulated radiation pattern, which resembles that of an ordinary small monopole on the ground plane.

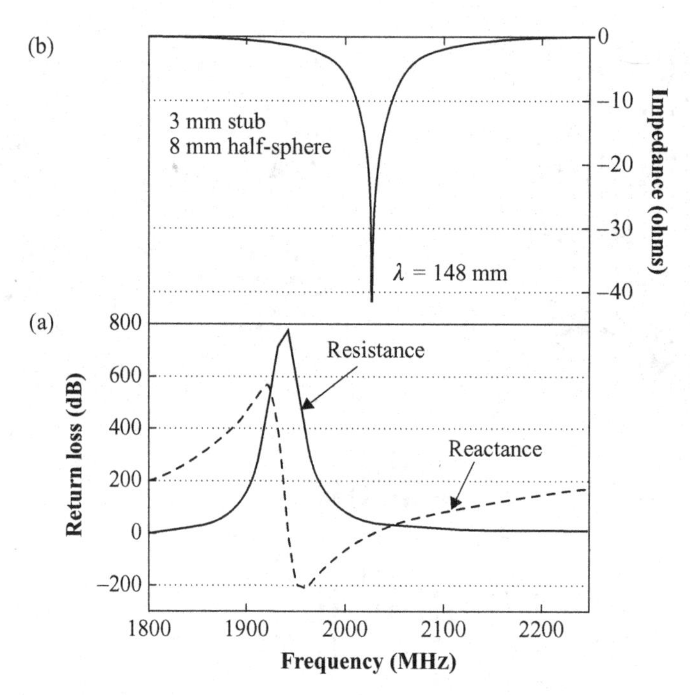


**Figure 7.221** (a) Return loss and (b) impedance ([114], copyright ©2006 IEEE).

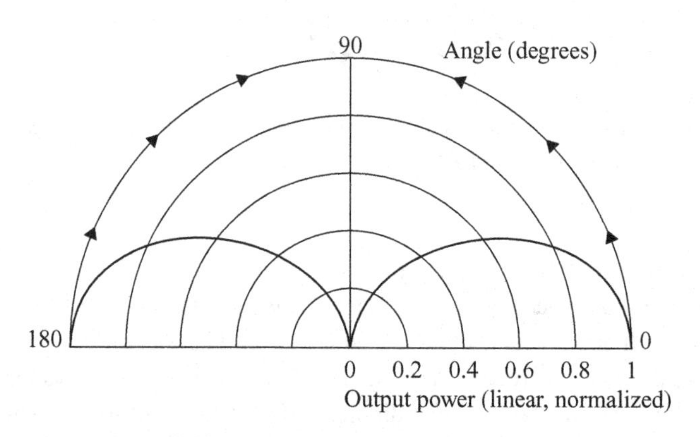


**Figure 7.222** Radiation pattern ([114], copyright ©2006 IEEE).

An ENG MM can be implemented by using an LH TL structure [115]. A compact TL MM antenna consisting of two unit cells, each of which comprises a microstrip TL loaded with five spiral inductors, is introduced in [115]. The antenna is illustrated in Figure 7.223, in which are shown (a) 3D schematic and (b) top view photograph of the fabricated antenna. Each of two arms works independently at its own frequency and corresponding to two frequencies designed to be merged into one passband so that wideband performance can be obtained. The TL MM unit cell employs a round spiral inductor (Figure 7.224(a)), which has dimensions of $d \times d = 4.5$ mm $\times$ 4.5 mm. The spiral element is connected to the ground through a via, to form a shunt inductance with a larger inductance, and is equivalently represented by a circuit shown in Figure 7.224(b),

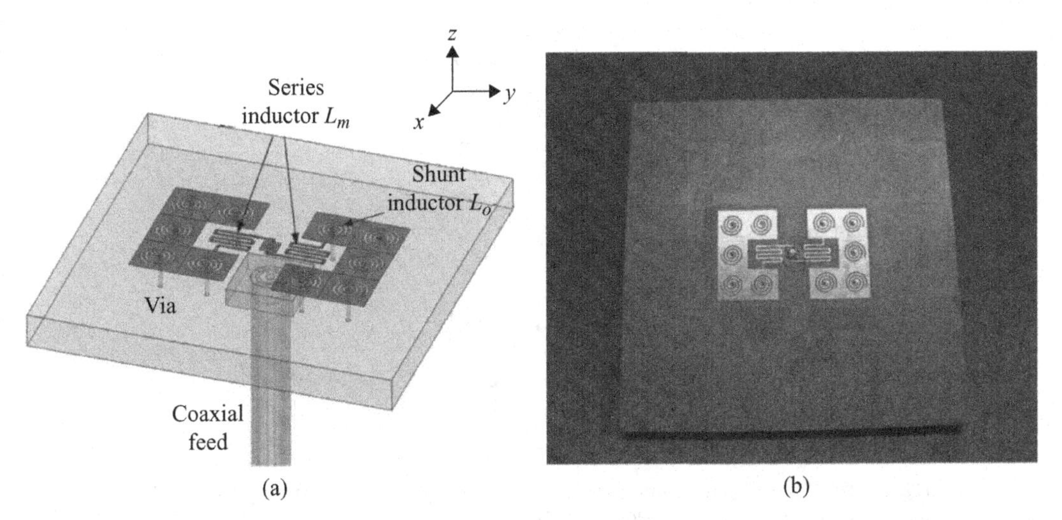


**Figure 7.223** The proposed two-arm TL-MM antenna with compact size and enhanced bandwidth: (a) 3D schematic and (b) top view photograph ([115], copyright ©2009 IEEE).

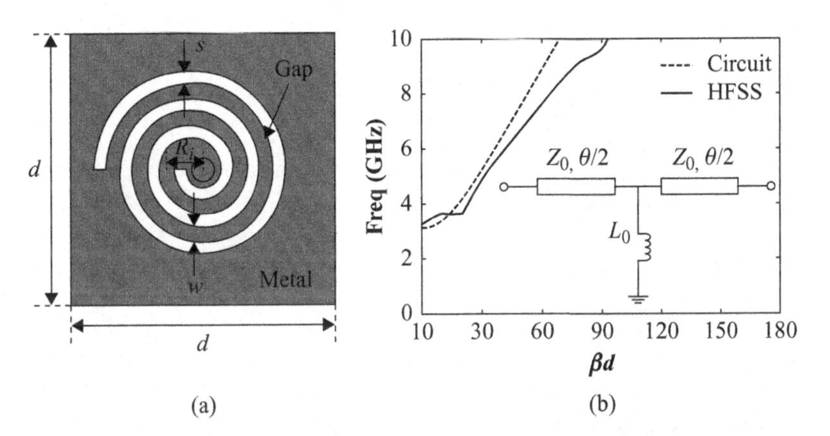


**Figure 7.224** Spiral-inductor-loaded TL-MM unit cell: (a) geometry of the cell and (b) transmission line representation ([115], copyright ©2009 IEEE).

in which the dispersion characteristics are shown, and $Z_0$ and $\theta$ denote the TL characteristic impedance and the electrical length per cell, respectively. The TL MM can be considered equivalently as an ENG MM with the corresponding plasma frequency $f_{sh}$ given by $1/(2\pi\sqrt{L_0 C})$, where $L_0$ is a combined inductance of the spiral and the via, and $C$ is the intrinsic capacitance of the TL. The metallic area of the antenna (Figure 7.223) is 22.5 mm × 13.5 mm and the size of the ground plane is 50 mm × 50 mm. The return-loss (–10 dB) bandwidth is about 100 MHz from 3.23 to 3.33 GHz, as a consequence of two resonances merged into the frequency range of 3.25 to 3.30 GHz. The radiation pattern is similar to that of a short monopole. Measured directivity and gain, respectively, are 2.61 dBi and 0.79 dBi, giving an efficiency of 65.8%. Effective extension of the bandwidth achieved by using a two-resonance TL MM with a small size is recognized by the comparison with a single patch antenna having the same bandwidth, which has a larger size than the TL MM antenna.

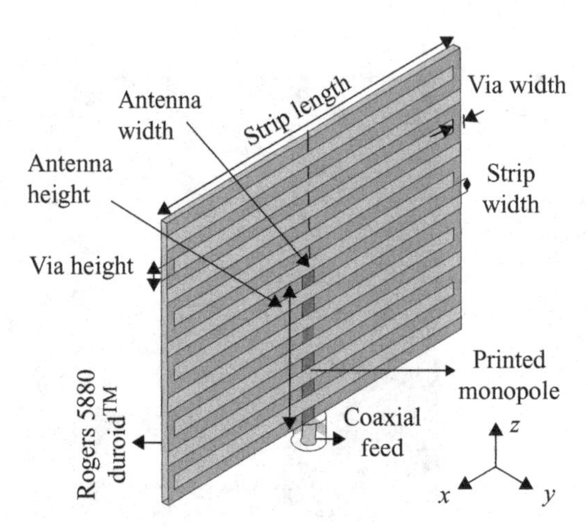


**Figure 7.225** Planar meander-line structure (an ENG equivalence) combined with a short monopole ([109], copyright ©2008 IEEE).

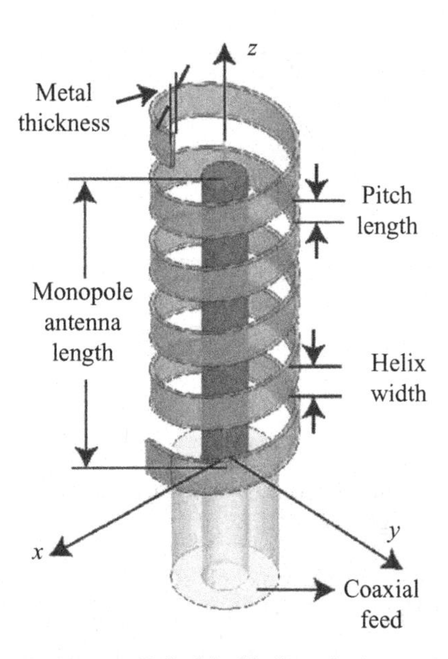


**Figure 7.226** Cylindrical helix strip (an ENG equivalence) combined with a short monopole ([109], copyright ©2008 IEEE).

Other examples of ENG MM applications to small antennas are use of a planar meander line structure and a cylindrical helix combined with a small monopole as shown in Figure 7.225 and Figure 7.226 [109], respectively. These planar meander line structure and cylindrical helix strip provide inductance effectively to interact with highly capacitive fields near the electrically small monopole to obtain matching in space. The behavior is similar to that of the ENG MM.

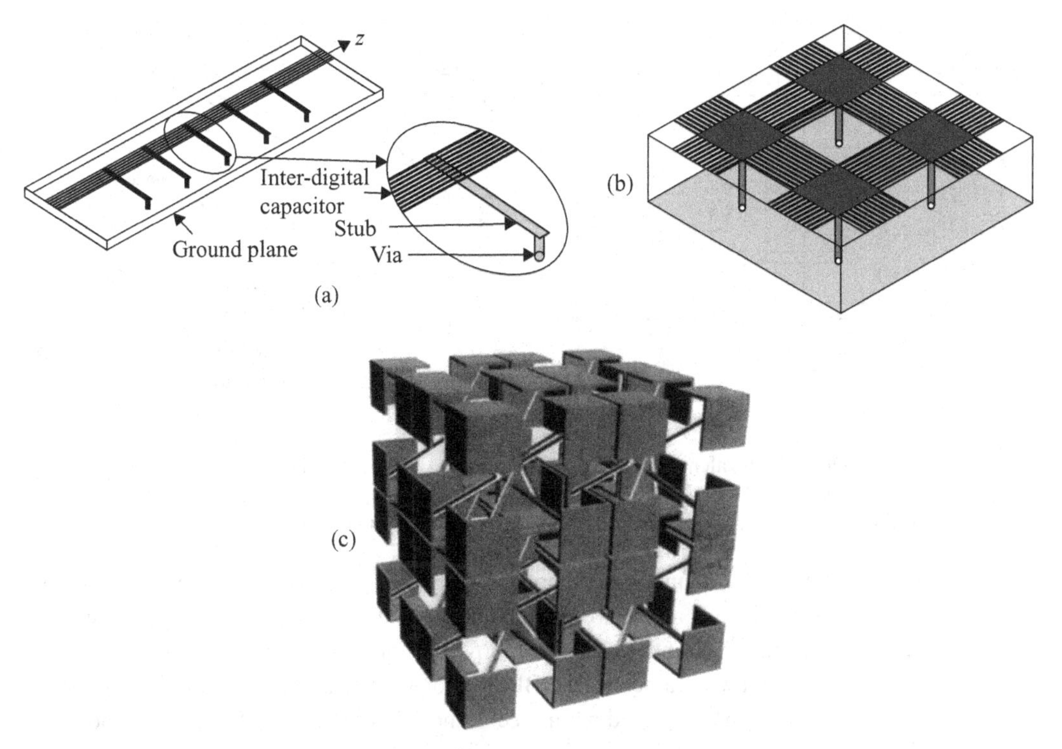


**Figure 7.227** CRLH (microstrip) materials of (a) 1D, (b) 2D, and (c) 3D structure ([116], copyright ©2008 IEEE).

### 7.2.5.2 Applications of DNG (Double Negative Materials)

The representative DNG MM is the CRLH TL MM (Composite Right/Left Handed Transmission-Line Metamaterial), which exhibits various significant functionalities and/or performances predicated on the notable dispersion properties and fundamental RH/LH duality, thus has drawn much attention of engineers for applying it to practical antennas [116]. The typical CRLH antennas are leaky wave (LW) antennas and resonant antennas. The LW antennas can provide full-space dynamic scanning capability, with various types of beams and actively shaped beams. The resonant antennas offer various performances such as multiband operation, and high efficiency and high directivity with zeroth-order mode constitutions. In addition, they have distinct features in implementation of CRLH antennas in planar, small, and compact dimensions that are an urgent requirement for various recently emerged small wireless systems.

Figure 7.227 illustrates three implemented structures: (a) 1D, (b) 2D, and (c) 3D, respectively, of periodic CRLH materials. The unit cell of the 1D structure, which is extendable to 2D or 3D structure, consists of a series resonant tank with a capacitor $C_L$ (inter-digital capacitor) and an inductor $L_R$ (parasitic), and a shunt anti-resonant tank with a capacitor $C_R$ (parasitic) and an inductor $L_L$ (a stub), which is depicted in the inset in Figure 7.227(a). The equivalent circuit of the unit cell is shown in Figure 7.228(a). The series tank components $C_L$ and $L_R$ are related to the metamaterial (MM) permeability

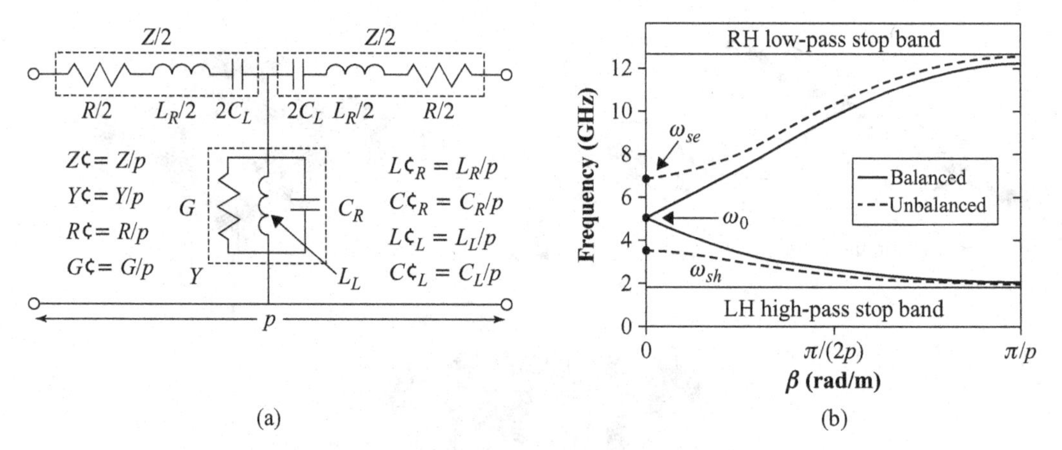


**Figure 7.228** (a) Equivalent circuit expression of CRLH TL unit-cell (primed variables represent per-unit-length and times-unit-length quantities) and (b) dispersion characteristics of the balanced and unbalanced CRLH TL ([116], copyright ©2008 IEEE).

$\mu = Z/(\mathrm{j}\omega p)$ and the shunt tank components $C_R$ and $L_L$ are related to the MM permittivity $\varepsilon = Y/(\mathrm{j}\omega p)$, where $Z$ and $Y$ are the impedance and the admittance, respectively, that characterize the MMs, and $p$ is the length (period) of the unit cell.

To describe these parameters, the Bloch–Floquet theorem is applied to the periodic structure constituted of cascading unit cells shown in Figure 7.228(a), and $Z$ and $Y$ are derived as

$$
\begin{aligned}
Z &= R + \mathrm{j}\{\omega L_R - 1/(\omega C_L)\} = R + \mathrm{j}\{(\omega/\omega_{se})^2 - 1\}/(\omega C_L) \to Z' p(p/\lambda_g \to 0) \\
Y &= G + \mathrm{j}\{\omega C_R - 1/(\omega L_L)\} = G + \mathrm{j}\{(\omega/\omega_{sh})^2 - 1\}/(\omega C_L) \to Y' p(p/\lambda_g \to 0)
\end{aligned}
\tag{7.74}
$$

where primed variables denote per unit length ($Z'$, $Y'$, $C'_R$ and $L'_R$) and times-unit length ($C'_L$, $L'_L$) quantities. $\omega_{se} = 1/\sqrt{L_R C_L} = 1/\sqrt{L'_R C'_L}$ and $\omega_{sh} = 1/\sqrt{L_L C_R} = 1/\sqrt{L'_L C'_R}$ are the frequencies of series and shunt resonances, respectively, and $(p/\lambda_g = \beta p/(2\pi) \to 0)$ indicates the infinitesimal limit of the unit cell to form the perfectly uniform TL or homogeneous MM structure ($\lambda_g$: guided wavelength). The Bloch impedance $Z_B$ obtained by the ratio of the periodic voltage and current at either port of the unit cell, and the specific propagation constant $\gamma_B = \alpha_B + \mathrm{j}\beta_B$ are expressed by

$$
\begin{aligned}
Z_B &\to Z(\omega) = Z_L = Z_R && (p/\lambda_g \to 0) \\
\gamma_B &\to \beta(\omega) = \omega/\omega_R - \omega_L/\omega && (p/\lambda_g \to 0)
\end{aligned}
\tag{7.75}
$$

where $Z_L = \sqrt{L_L/C_L}$, $Z_R = \sqrt{L_R/C_R}$, $\omega_R = \sqrt{L_R C_R}$, $\omega_L = \sqrt{L_L C_L}$, and the infinitesimal limit of the unit cell is also assumed.

Dispersion characteristics of the CRLH TL for the balanced ($\omega_{se} = \omega_{sh} = \omega_0$, $L_L = L_R$, $C_L = C_R$) and the unbalanced ($\omega_{se} \neq \omega_{sh}$) cases are shown in Figure 7.228(b).

The CRLH TL MM has equivalent constitutive parameters given by

$$
\begin{aligned}
\mu(\omega) &= \mathrm{j}Z'/\omega = L'_R(1 - \omega_{se}/\omega) \\
\varepsilon(\omega) &= -\mathrm{j}Y'/\omega = C'_R(1 - \omega_{sh}/\omega)
\end{aligned}
\tag{7.76}
$$

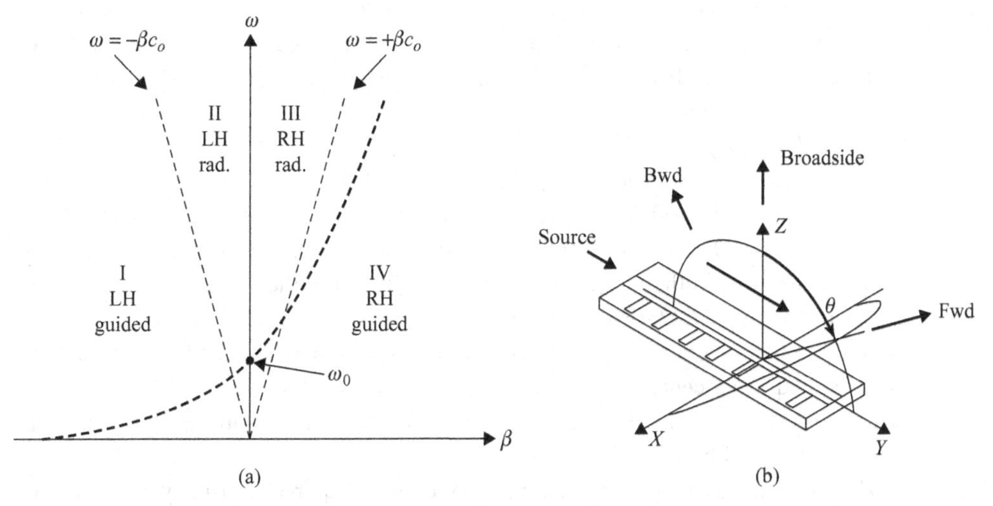


**Figure 7.229** Performance of CRLH LW antenna: (a) typical dispersion characteristics and (b) scanning performance ([103(b)], copyright ©2004 IEEE, [116], copyright ©2008 IEEE).

which are characterized by the pole at $\omega = 0$ and one zero at the plasma frequency ($\omega_{se}$, $\omega_{sh}$). These parameters may take values of positive, negative or smaller than unity. In case of both being zero/negative (balanced transmission frequency $\omega_0$), the refraction index is zero/negative.

#### *7.2.5.2.1 Leaky wave antenna [116]*

The CRLH MM exhibits unique dispersion characteristics that contain a radiation (or fast wave) region where $|\beta| < k_0 = \omega/c$ ($k_0$: free space propagation constant, and $c$: velocity of light), i.e, the phase velocity $v_p > c$, in addition to a guided (or slow wave) region ($\beta > k_0$) as shown in Figure 7.229(a). In the figure, these regions are categorized into four; I: LH guided, II: LH radiation, III: RH radiation, and IV: RH guided regions. An antenna constituted with an LW structure, on which a travelling wave travels faster than the speed of light ($v_p > c$), is referred to as a LW antenna. The radiation angle $\theta$ of the main beam of this antenna, defined from the normal direction, is given by

$$\theta = \text{arc}\sin\{\beta(\omega)/k_0\}. \tag{7.77}$$

In a conventional LW antenna, $\beta$ is positive and not dispersive, which implies the RH structure, where the radiation occurs in the direction $\theta > 0$, meaning only forward direction. In addition, broadside radiation ($\theta = 0$) is impossible, because it requires $\beta = 0$, that requires $v_g$ (group velocity; $\partial\beta(\omega)/\partial\omega) = 0$. This means that the guided structure is associated with a standing wave, whereas a travelling wave is necessary for the LW radiation. On the contrary, as a CRLH TL MM exhibits dispersion characteristics, where $\beta(\omega)$ varies from the region $\beta < -k_0$ to the region $\beta > +k_0$, the CRLH structure performs backfire-to-endfire frequency scanning. This can be understood by (7.77), where $\theta$ ($\beta = -k_0) = -90°$ (backfire radiation) and $\theta$ ($\beta = -k_0) = +90°$ (endfire radiation). In addition, if the CRLH structure has balanced resonances, transition of the transmission from the

LH regions to the RH regions with travelling wave ($v_g \neq 0$) is continuous at $\beta = 0$, providing unique broadside radiation that cannot be achieved by the conventional LW antenna. Figure 7.229(b) illustrates typical scanning operation of this type of antenna.

#### *7.2.5.2.1.1 Beam forming*

By varying circuit parameters $L_R$, $C_R$, $L_L$, and $C_L$, in the CRLH TL structure, electronic scanning can be achieved. The easiest method is to use varactors for capacitors. When an antenna is of 1D structure, scanning is only in one plane. In this case, the beam can be highly directive in the *y–z* plane, whereas it is broad in the perpendicular direction (*y–z* plane in Figure 7.229(b)). This is a fan beam.

A 1D CRLH structure can be extended to a 2D structure. The 2D CRLH structure, when excited in its center, can support a circular wave. When the wave exists in the CRLH dispersion regions II and III in Figure 7.229(a), it radiates in an LW manner and produces a conical beam.

When a maximum radiation in a unique direction is required, as usually necessary for point-to-point communications, a pencil-beam antenna is desired. The CRLH TL MM is useful for producing such pencil-beam scanning antennas [117, 118]. The antennas introduced in [117, 118] consist of arrays of LW elements using a combination of frequency tuning and phase-shift tuning to achieve pencil-beam scanning. By using 2D CRLH TL structures, pencil beams can be produced economically and flexibly compared to use of conventional phased arrays, which require complicated, lossy, burdensome matters in design, and 2D dispersive feeding networks.

#### *7.2.5.2.1.2 Active beam scanning*

Integration of active circuits along a CRLH TL structure is suitable to manipulate the magnitude of the signal along it as well as its phase, and active beam scanning is easily realized. The beam width of an LW antenna constituted of TL structure is controlled by its leakage factor, $\alpha_{lw}(\omega)$, which is the real part of the propagation constant $\gamma(\omega) = \alpha_{lw}(\omega) + j\beta(\omega)$. With passive structure, the leakage factor is fixed, and hence the effective aperture and the directivity cannot be increased, without otherwise extending the length of the LW structure. Meanwhile, since an active CRLH LW antenna, in which active circuitry is integrated, may have an unlimited effective aperture, it can provide an arbitrarily high directivity with single and simple TL excitation. Incorporation of amplifiers as repeaters into a CRLH LW antenna was reported in [119].

#### *7.2.5.2.2 Resonant antennas*

#### *7.2.5.2.2.1 Multiband antennas*

By reactively terminating a CRLH TL structure open to free space, by a short or an open circuit, a resonant CRLH antenna having effective wavelength and frequency response that are attributes of a CRLH MM property is obtained. The resonant modes of a CRLH structure of length $l$ are given by $l = n\lambda_g/2$, with $n = 0, \pm 1, \pm 2, \ldots \pm\infty$. The $n$ can be either positive (RH band) or negative (LH band) and even zero (at transition). Each positive ($n > 0$) resonance mode (at frequencies $\omega_{+n}$) has a twin negative ($n < 0$) resonance mode (at frequencies $\omega_{-n}$), and a zeroth-order ($n = 0$) mode exists at the transition frequency $\omega_0$ as was shown in Figure 7.207. A CRLH TL structure consisting

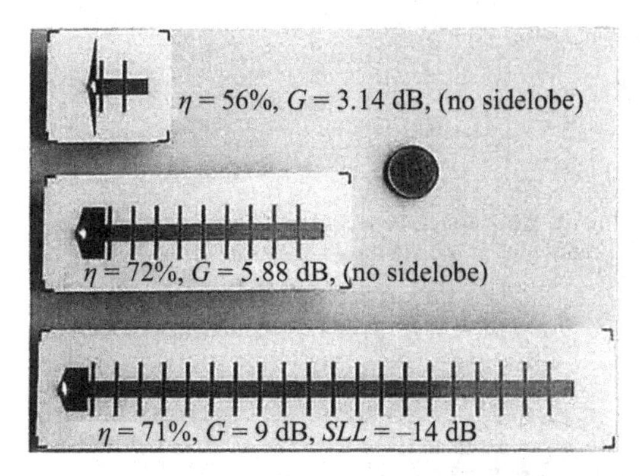


**Figure 7.230** Three CRLH zeroth-order microstrip resonant antennas of different sizes operating at 2.44 GHz with; (a) $\eta = 56\%$, $G = 3.14$ dB, (b) $\eta = 72\%$, $G = 5.88$ dB, and (c) $\eta = 71\%$, $G =$ 9 dB, and SLL (sidelobe level) $= -14$ dB. ([116], copyright ©2008 IEEE).

of $N$ unit cells has a finite number of $2N$ ($2N - 1$ in the balanced case) resonances, corresponding to $\beta_n p = \beta_n (l/N) = n\pi/N$. Figure 7.207 shows such a discrete spectrum of a CRLH resonator.

By utilizing the positive and negative resonance pairs of a CRLH structure, a dual-band resonant antenna is obtained. The antenna is back fed by a coaxial line at the off-center location for 50 Ω matching at one frequency. In principle, all of the ($2N -$ 1) resonances may be excited and matched to the source with proper excitation. Since the modes of each pair have the same guided wavelength and field distributions, input impedance of each mode has a similar value. By this means an efficient dual-mode operation can be achieved by a single resonator.

In principle, with higher-order CRLH TL structure, operation in multibands such as tri-band, quad-band, or even higher number of bands, is possible.

An example of this type of antenna is a square patch antenna, in which LH structures are partially filled [116].

#### *7.2.5.2.2.2 Zeroth-order antennas*

Unique application of a CRLH TL structure to antennas is of a zeroth-order mode, by which zeroth-order CRLH resonance ($l/\lambda_g = 0$, $n = 0$) is attained. Examples are shown in Figure 7.230, where three different-size microstrip antennas are illustrated [116]. The unit cell of the antenna uses an inter-digital capacitor and a stub inductor shorted at the end with a via as was shown in Figure 7.227(a). The size of a resonant CRLH antenna may be flexibly designed to attain required effective aperture and directivity, as the operating frequency is independent of the size, but is determined by LC unit-cell elements (Figure 7.228(a)). This feature can be used for designing either electrically small or larger antennas. In this type of resonant CRLH LW antenna, directivity can be enhanced by increasing the length of antenna at a given frequency. Gains and efficiencies, respectively, for each antenna of the three shown in Figure 7.230, are 3.14 dB and 56% for the smallest antenna, 5.88 dB and 72% for the middle-size antenna, and 9 dB and 71% for the longest antenna, respectively, all at 2.44 GHz.

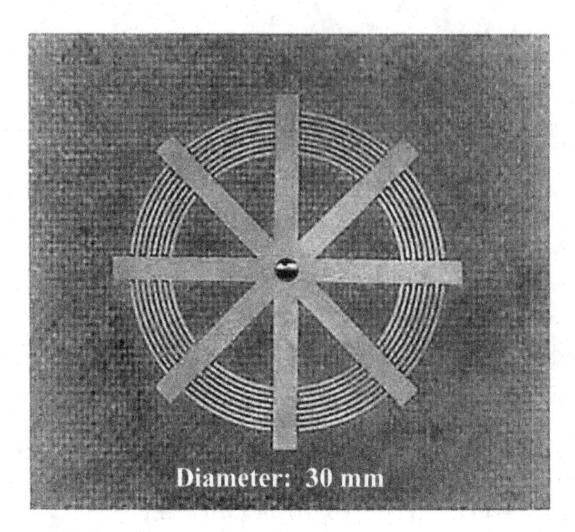


**Figure 7.231** A CRLH loop resonant microstrip antenna ([116], copyright ©2008 IEEE).

### *7.2.5.2.2.3 Electric/Magnetic plane monopoles*

A CRLH structure can be formed to have a closed circular loop configuration by folding a rectilinear CRLH structure as shown in Figure 7.231 [116]. This antenna in the zeroth-order mode can be operated either as an electric or magnetic monopole by feeding to excite either $\omega_{sh}$ or $\omega_{se}$ mode. In the $\omega_{sh}$ mode, $Y = 0$ (Eq. 7.74), and therefore the shunt paths are seen as open circuit, as the radial currents flowing on the stubs excite only $L_L$, while the overall shunt resonator current is zero. This indicates that this antenna structure acts as an electric monopole. In turn, in the $\omega_{se}$ mode, $Z = 0$ (Eq. 7.74), and therefore the series paths are seen as open circuit, and current flows exist only along the loop. This means that the antenna structure works as a magnetic loop. These two electric and magnetic monopoles are independent (uncoupled) from each other and may be excited simultaneously by using two different feeds (radial for $\omega_{se}$ and azimuthal for $\omega_{sh}$).

A monopole radiator may also be realized by a zeroth-order CRLH resonator in a patch configuration [116]. In Figure 7.232 a CRLH magnetic monopole patch operating in the CRLH zeroth-order resonance mode ($\omega_{sh}$) is illustrated. This is a complement to the magnetic dipole of a conventional patch. The monopole is formed as a result of the magnetic currents (uniform vertical electric field) on the periphery of the mushroom patch. It may be possible to realize an electric monopole patch operating in the CRLH zeroth-order mode $\omega_{se}$.

### *7.2.5.2.3 NRI (Negative Refractive Index) TL MM antennas*

When the top end of a quarter-wavelength monopole placed on a ground plane (GP) is folded back to the GPL (ground plane), the result is half a folded dipole, on which two mode currents, balanced and unbalanced, flow. As the balanced mode does not contribute to radiation and appears open at the input terminal, the unbalanced mode contributes to radiation, because the in-phase current flowing on both the feed terminal and its opposite side, becomes the meaningful part. The impedance seen at the feed terminal is four times that of the original monopole. Hence, the input impedance of a low-height monopole

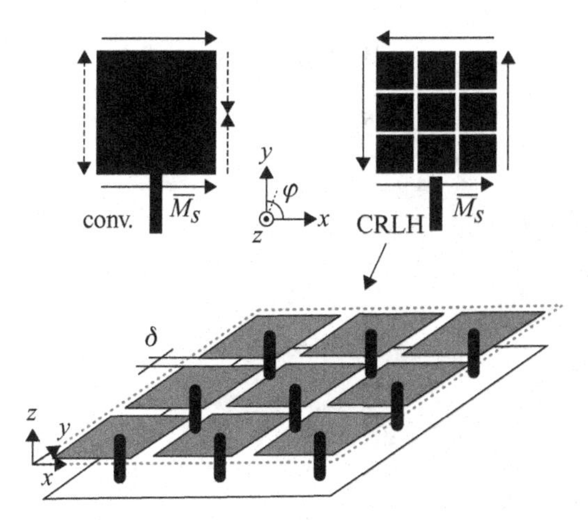


**Figure 7.232** A magnetic monopole microstrip patch antenna using the CRLH zeroth-order mode in a mushroom structure. ([116], copyright ©2008 IEEE).

can be increased by folding it and can be designed to match 50 Ω. When a low-height monopole is folded, in-phase current on the resultant folded dipole cannot be obtained. In this case, a phase shifter is inserted, replacing the short at the top of the folded dipole to obtain in-phase currents so that a high enough input impedance to match 50 Ω, even if the height is very low, and efficient radiation is achieved. A Negative Refractive Index (NRI) TL MM may be used as a substitute for such phase shifter and applied to realize a small, compact, yet efficient antenna.

*7.2.5.2.3.1 MM ring antenna*

An example of an NRI TL MM application to achieve a small antenna is a ring antenna operating at 1.77 GHz [120], which is constituted in a compact ($\lambda_0/11$ footprint; $\lambda_0$: operating wavelength) and low profile ($\lambda_0/28$ height). Figure 7.233 shows the antenna structure and its geometry. The antenna consists of two MM unit cells, which are synthesized by using conventional TL loaded with lumped series capacitors and shunt inductors in a dual TL topology. The MM unit cells are arranged in a ring structure implemented by microstrip technology, and designed to produce in-phase currents on the posts (vias) connected to the antenna through the inductances at each end of the TL structure. Lumped inductors and capacitors connected to the TL give rise to series and shunt resonance respectively by involving the parasitic capacitance and inductance of the TL. The short vertical posts are connected to the antenna surface through the lumped inductor at each end of the TLs, and act as the radiation element. Even though the length of the post (radiator) is very small ($\lambda_0/28$), the radiation resistance is adjusted to match 50 Ω, and $S_{11}$ obtained at the operating frequency of 1.77 GHz is nearly –30 dB. The lumped capacitance and inductance used, respectively, are 0.4 pF and 15 μH. The radiation pattern is similar to that of a short monopole, and gain and efficiency, respectively, are 1.245 dB and 54%. This is shown in Figure 7.234.

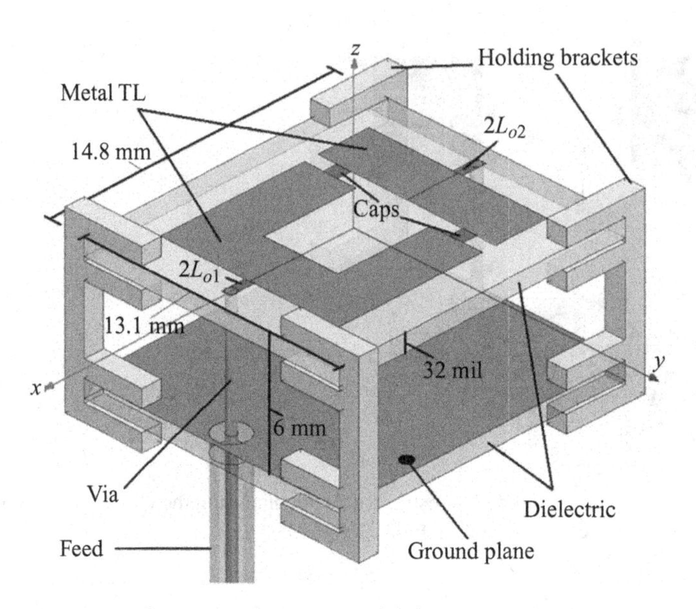


**Figure 7.233** Geometry of a metamaterial ring antenna ([120], copyright ©2005 IEEE).

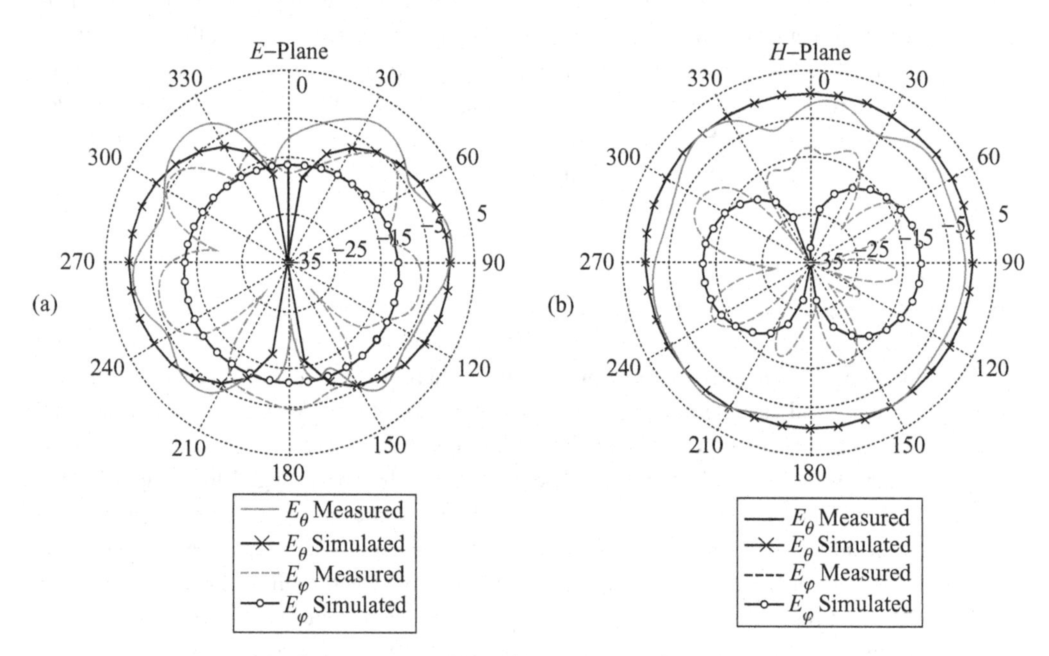


**Figure 7.234** Radiation patterns: (a) *E*-plane and (b) *H*-plane ([120], copyright ©2005 IEEE).

### *7.2.5.2.3.2 Multiple-folded monopole*

An electrically small NRI TL MM antenna operating at 3 GHz band is introduced in [121]. The antenna consists of four folded monopoles, to which NRI TL MM technology is applied to realize a small, compact, and broadband antenna with very small

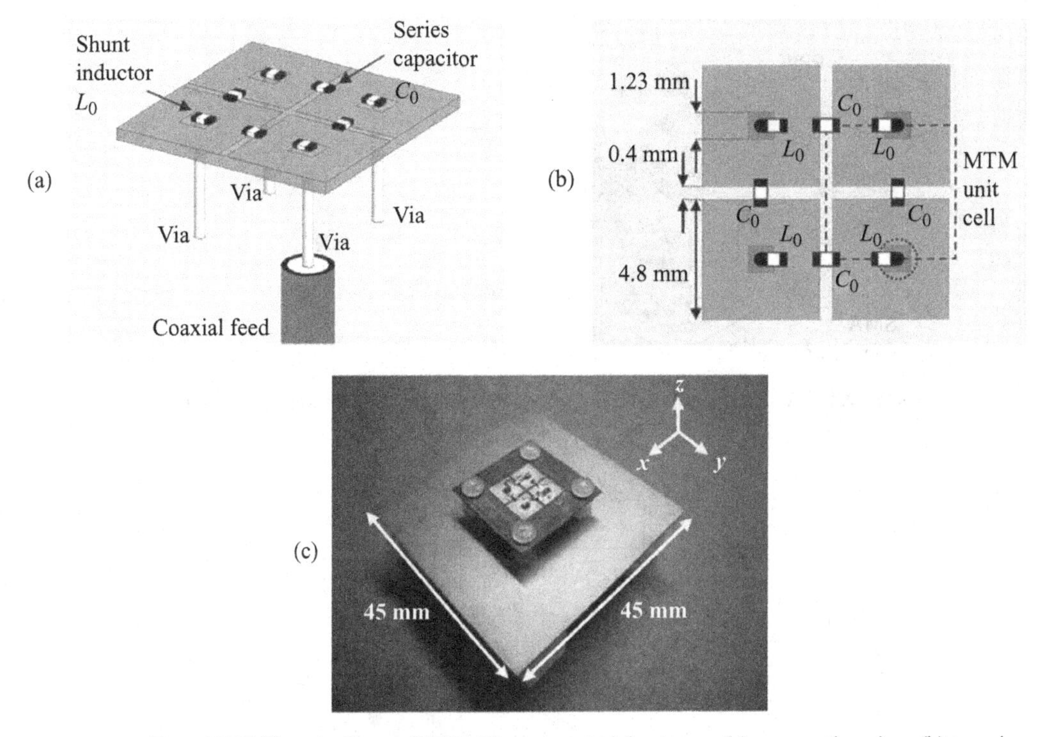


**Figure 7.235** Electrically small NRI-TL metamaterial antenna: (a) perspective view, (b) top view, and (c) 3D view ([121], copyright ©2008 IEEE).

dimensions ($\lambda_0/10 \times \lambda_0/10 \times \lambda_0/10$ over $0.45\lambda_0 \times 0.45\lambda_0$ ground plane; $\lambda_0$: the operating wavelength). Figure 7.235 illustrates the antenna; (a) perspective view, (b) top view, and (c) 3D view. The NRI TL MM unit cell is constituted of a conventional TL, to which lumped capacitors and inductors are loaded to ensure series and shunt resonances respectively with parasitic inductance and capacitance inherently existing in the microstrip TL structure. By adjusting the values of the loaded elements and the unit cell size, amplitude and phase of signals propagating along an NRI TL structure can be aligned to obtain in-phase excitation of monopoles so that efficient radiation can be achieved. The monopoles are top-loaded with a square plate, on which the NRI TL MM structure is constituted, and consequently, appropriate radiation impedance to match 50 Ω is attained even though the antenna height is very low. As can be seen in Figure 7.235(b), an NRI MM unit cell consists of a square plate, acting as a TL, to which a lumped inductance $L_0$ is connected, and a lumped capacitance $C_0$, bridging two TLs, thus resulting in an NRI TL MM.

Each of four square plates stands with a vertical post on a ground plane, one of which is the feeding post, and currents on each post are adjusted to be in-phase so that they are excited all in phase and efficient radiation can be obtained. It should be noted that use of folding monopole technology to increase the radiation resistance $R_r$, as was discussed in the previous section, is applied to this antenna structure, so obtained by placing an NRI MM between two posts.

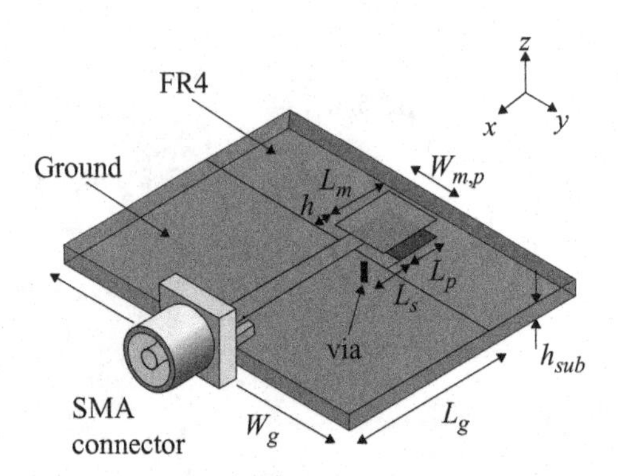


**Figure 7.236** NRI-TL metamaterial microstrip antenna ([122], copyright ©2002 IEEE).

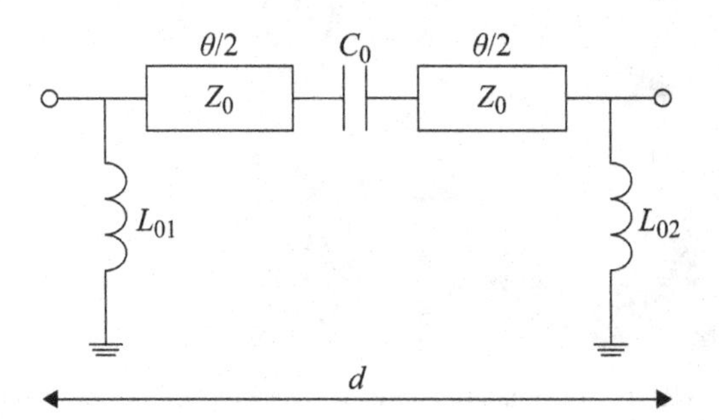


**Figure 7.237** NRI-TL metamaterial $\pi$-unit cell ([122], copyright ©2009 IEEE).

For an antenna of $N$ folded arms, the input impedance $R_{in}$ is given by

$$R_{in} = N^2 R_r. \tag{7.78}$$

This antenna has four posts and $R_r = 4\ \Omega$, if the post height is 5 mm, then $\lambda_0/20$ at 3 GHz, $R_{in}$ becomes 64 Ω. This is sufficiently close in value to 50 Ω for good matching. The antenna does not use a balun with the ground plane of 45 mm square. Simulated results by using HFSS show bandwidth of 42 MHz (−10dB), directivity of 1.31 dB and efficiency of 72.3%.

### *7.2.5.2.3.3 Dual-mode monopole antenna*

An NRI TL MM unit cell employed in a small monopole to achieve a folded monopole structure and combined with a ground plane that acts as a radiator so as to obtain dual-mode operation is introduced in [122]. The antenna geometry and the dimensional parameters are illustrated in Figure 7.236. The total size of antenna is 20 mm × 30 mm. The MM unit cell is implemented by placing a small rectangular patch behind the printed monopole patch, which is equivalently expressed by a Pi-network shown in Figure 7.237. The series capacitance $C_0$ is formed between these two patches and the shunt inductance $L_{01}$ is formed at the base of the monopole, while $L_{02}$ is formed by a thin inductive strip and a via that connects the patch behind the monopole patch and the ground plane (GP).

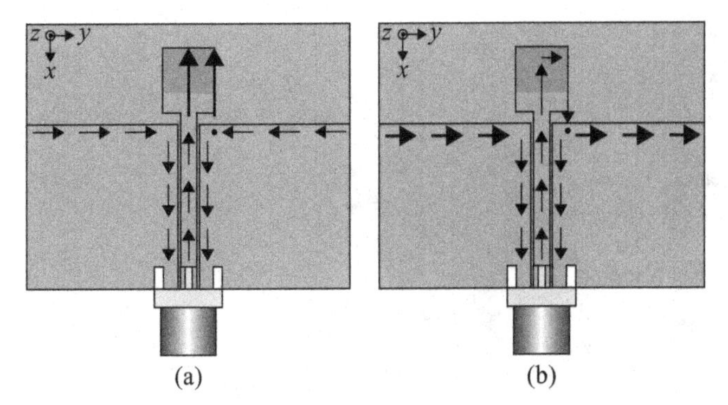


**Figure 7.238** Surface current distributions on the conductors of the MTM-loaded monopole antenna: (a) 5.5 GHz and (b) 3.55 GHz ([122], copyright ©2009 IEEE).

These elements are incorporated with TLs, which are comprised of the two monopole patches, to constitute a Pi MM unit cell. With this structure, currents flowing on these two patches can be arranged in-phase as Figure 7.238(a) shows, thus effectively a folded monopole structure is created, by which radiation resistance of the small monopole is increased. The original patch antenna (without MM loading) exhibits a single resonance at 6.3 GHz, whereas an MM-loaded monopole exhibits broadband dual resonance; these are a desired resonance at 5.5 GHz along with additional resonance at 3.55 GHz. The resonance around 3.55 GHz can be adjusted by changing the width of the GP.

At 3.55 GHz, the antenna acts as a dipole oriented along the $y$-axis, not as a folded monopole, because of the induced current on the top edge of the ground plane, which is shown in Figure 7.238(b). In the circuit shown in Figure 7.237, the TL section is negligible at the lower frequency and the circuit is simply transformed effectively into a series resonator, comprised of the capacitance $C_0$ and the inductances $L_{01} + L_{02}$. When the antenna is fed through $L_{01}$, the series resonator represents an MM loaded monopole and forms a short circuit, thus acting as balun for the currents on the GP. Since the currents on both sides of the feed line are in-phase, the top edges of the ground plane radiate just as a dipole with a different polarization (orthogonal) to that of the MM monopole, which operates at 5.5 GHz.

Measured –10 dB bandwidth is 4.06 GHz, from 3.14 to 7.20 GHz, and efficiency is on the order of 90% at both 3.5 GHz and 5.5 GHz. Bandwidth increase as a result of loading the NRI MM to a monopole antenna is verified by comparison with bandwidth of the unloaded monopole antenna, which is 2.48 GHz from 5.35 to 7.83 GHz in terms of the –10 dB bandwidth.

By means of NRI TL MM loading to a conventional microstrip patch antenna, a compact, broadband, dual-radiation-mode antenna, that is, a short folded monopole operating at 5.5 GHz and a small dipole antenna operating at 3.55 GHz, is realized.

*7.2.5.2.3.4 Tri-band monopole antenna*

A planar monopole, to which an NRI TL MM structure is loaded, and a defected ground structure is employed, operating in three modes, is described in [123]. (A Defected

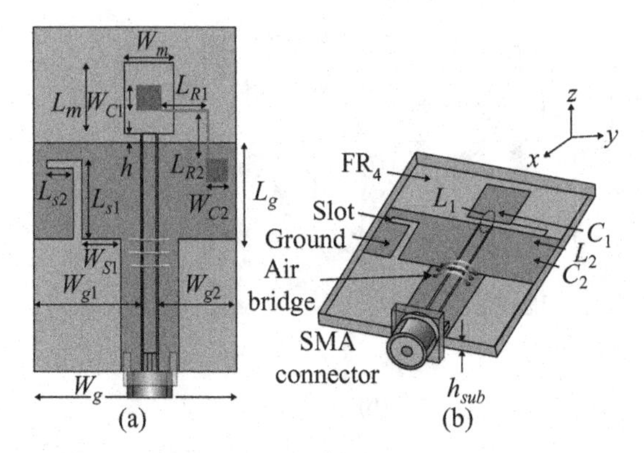


**Figure 7.239** Tri-band monopole antenna with single-cell MTM loading and a defected-ground plane: (a) top view and (b) 3D view. ([123], copyright ©2010 IEEE).

Ground Structure (DGS) is defined as a unit cell EBG or an EBG with limited number of cells and a period.) The antenna geometry and the dimensional parameters are illustrated in Figure 7.239, where (a) is top view and (b) is 3D view. A coplanar waveguide (CPW)-fed monopole antenna is loaded with a single NRI TL MM-based Pi-unit cell. The series capacitance $C_1$ is formed between the monopole on the top of the substrate and the rectangular patch placed opposite to the monopole. The TL MM cell is asymmetrically loaded with two shunt inductances, $L_1$ and $L_2$. As can be seen in Figure 7.239, $L_1$ is formed by the inductive strip to feed the monopole and $L_2$ is formed by the thin inductive strip that connects the rectangular patch beneath the monopole to the rectangular patch beneath the RH ground plane (GP). A capacitance $C_2$, formed between the rectangular patch and the RH GP, connects the shunt inductor $L_2$ to the ground. By appropriately adjusting these parameters in addition to the geometrical parameters of the antenna, current flows on the feed strip ($L_1$) to the monopole and the inductive strip ($L_2$) to the rectangular patch behind the monopole, can be made in-phase at the resonance frequency, thus the antenna performs effectively as a two-arm folded monopole. Figure 7.240(a) shows the current flows with arrows and indicates the antenna operates in the folded monopole mode. The resonance frequency in this case is around 5.0 GHz–6.0 GHz.

In this antenna structure loaded with an NRI TL MM, resonance occurs at lower frequency around 2.4 GHz–2.5 GHz in addition to the higher frequency with other modes. Major current flows on the GP are depicted in Figure 7.240(b), which shows that the currents along the CPW feed line are symmetric, while those on the upper edges of the GP are in-phase. This means that the currents on the edges of the GP contribute radiation and the antenna effectively performs as a dipole, which radiates in a direction orthogonal to that of the monopole mode. The in-phase currents on the top edges of the GP are produced by designing the series circuit comprising $C_1$, $C_2$, $L_1$ and $L_2$ to resonate at a frequency that overlaps with the dipole mode resonance, as the series circuit becomes short. The resonance frequency depends on the length of the current path, related with the size of the GP, $W_g + 2L_g$, (Figure 7.239). Based on the design consideration discussed above, the optimized loading patches have dimensions

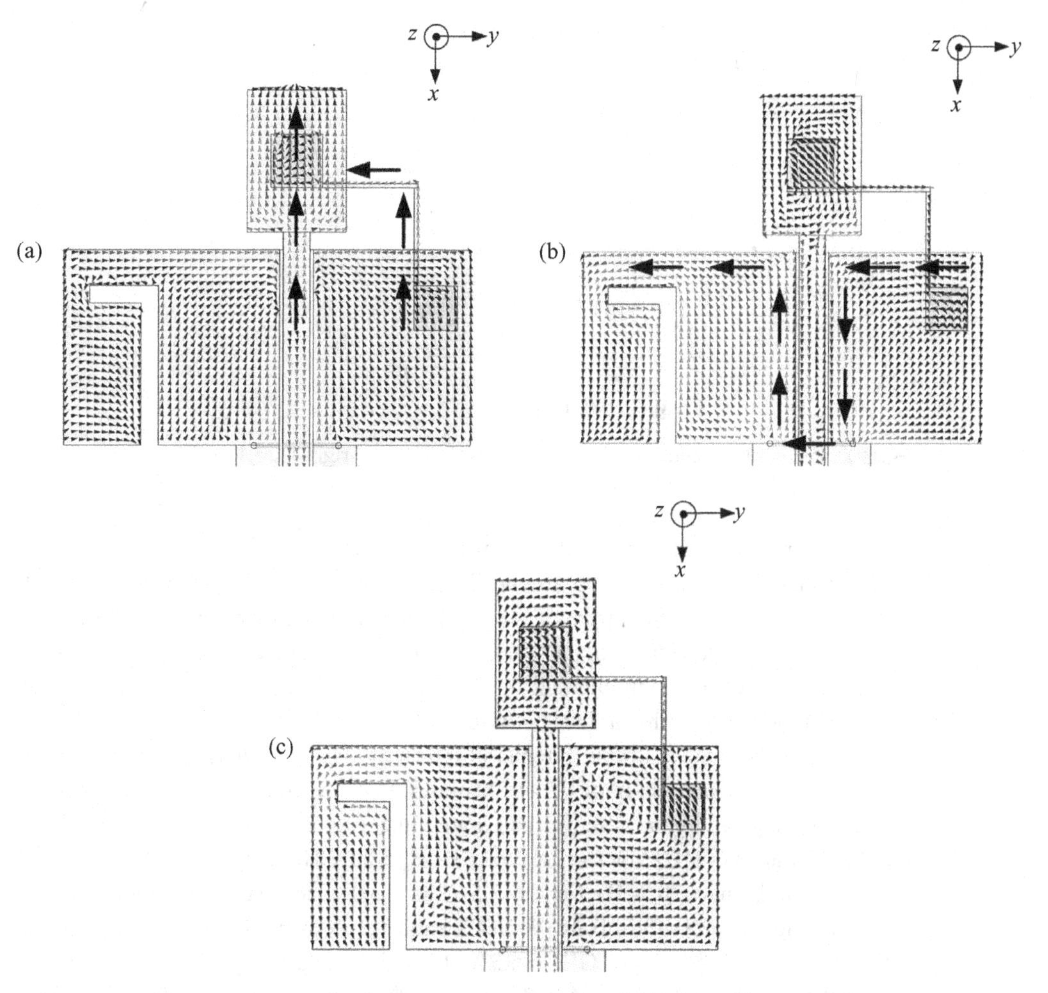


**Figure 7.240** Current distributions on the conductor of the tri-band monopole antenna shown in Figure 7.239: (a) folded monopole mode (5.80 GHz), (b) dipole mode (2.44 GHz), and (c) effect of defected ground plane (3.76 GHz) ([123], copyright ©2010 IEEE).

of 3.0 mm × 3.0 mm and 2.5 mm × 2.5 mm, respectively, and the two sections of thin strips have the same length of 5.5 mm and width of 0.25 mm. As the current flow takes a longer path than the length of the edge of the GP, a lower resonance frequency is achieved in comparison with the antenna introduced in the previous section [122], where the dipole-mode currents flow only on the top edges of the GP. This implies that a larger miniaturization factor is attained by this antenna structure.

In addition to the dipole mode, the antenna is designed to perform with a third mode, which is achieved by using a defected GP structure, formed by cutting an L-shaped slot out on the left-side of the CPW on the GP as shown in Figure 7.240. With this slot, resonance occurs around 3.5 GHz, while the dual-mode operation at around 2.5 GHz and 6.0 GHz is preserved. Current distributions on the GP at the resonance frequency

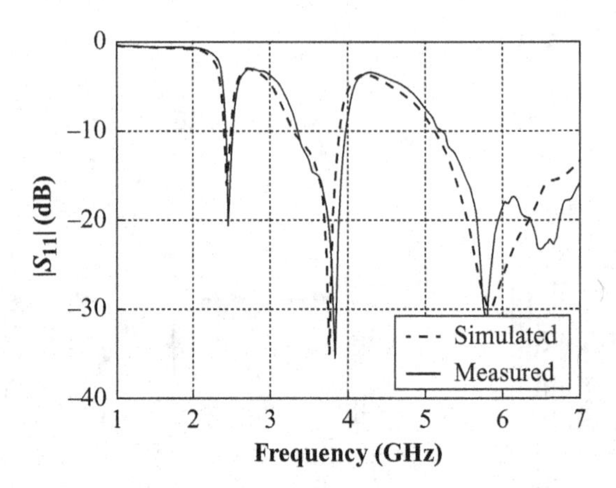


**Figure 7.241** Simulated and measured return loss ([123], copyright ©2010 IEEE).

3.76 GHz are shown in Figure 7.240(c), which illustrates a strong concentration of the currents that flow so as to wrap around the L-shaped slot on the left-side of the GP, thus taking a longer path of approximately $\lambda_g/2$ at the resonance of the third mode. The vertical and horizontal length of the slot $L_{s1}$ and $L_{s2}$, respectively, are adjusted to achieve a good impedance match throughout the 3.5 GHz band. The slot is located far enough away from the CPW so that it does not affect the balanced CPW mode.

Tri-band performance of this antenna is shown by the return-loss $S_{11}$ characteristics of a fabricated antenna as Figure 7.241 illustrates. In the figure, measured and simulated $S_{11}$ values are given by the dark line and dotted line, respectively. The measured −10dB bandwidths are 90 MHz for the 2.5 GHz band, 5.20 GHz to beyond 7 GHz for the 5.5 GHz band, and 620 MHz for the 3.5 GHz band. Each band corresponds to the lower and higher band of WiFi, and WiMAX band, respectively. The measured radiation efficiencies are 67.4% at 2.45 GHz, 86.3% at 3.5 GHz, and 85.3% at 5.5GHz. The full size of the antenna is 20.0 mm × 23.5 mm × 1.59 mm (or $\lambda_0/6.3 \times \lambda_0/5.3 \times \lambda_0/78.6$ with respect to the lowest resonant frequency of 2.45 GHz), including the GP size $L_g \times W_g = 11.0$ mm × 23.5 mm.

### *7.2.5.2.3.5 Multi-frequency and dual-mode patch antenna*

A microstrip square patch antenna, to which a CRLH structure is partially filled, enabling multi-frequency and dual-mode operation with miniaturized dimensions, is introduced in [124]. The antenna geometry and the structure are shown in Figure 7.242. As can be seen in the figure, the CRLH structure is implemented by using mushroom structures, the unit cell of which is composed of metal patches with vias to the GP (ground plane), forming an inductance $L_L$, and gaps between adjacent cells, forming a coupling capacitance $C_L$. The LH performance of these cells is provided by these equivalent circuit parameters, $L_L$ and $C_L$, while the current flux ($L_R$) and the parallel plate capacitor between the metal patch and the GP ($C_R$) render the RH nature. Rectangular patch rather than square patch will be used, because by using its longer side, larger coupling between

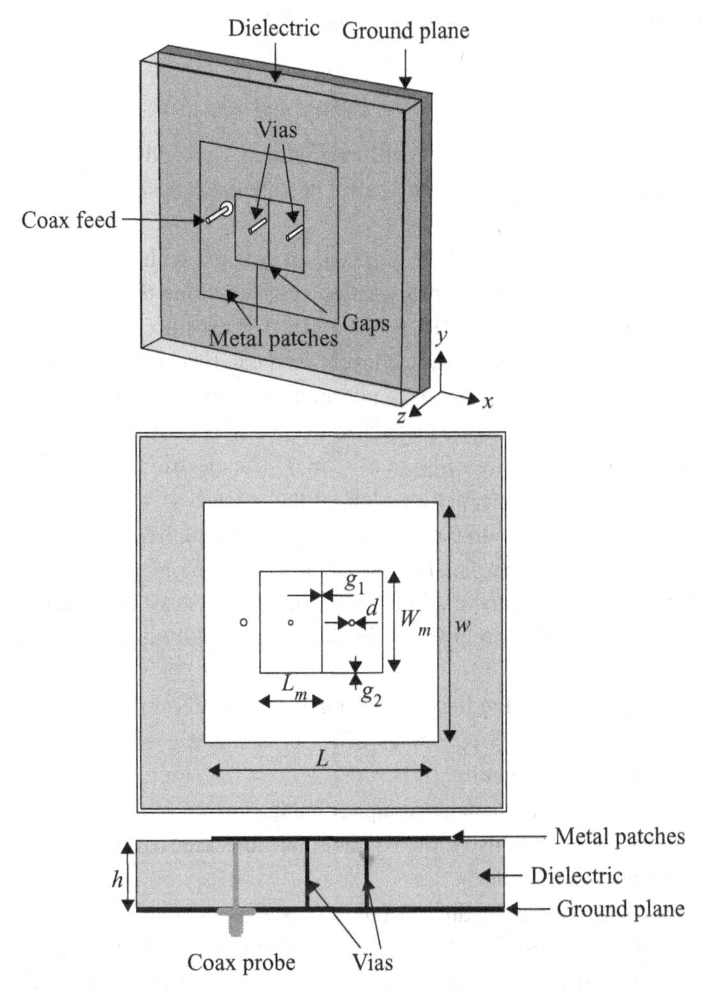


**Figure 7.242** Microstrip patch antenna partially filled with a 2 × 1 arrangement of CRLH cells ([124], copyright ©2008 IEEE).

adjacent cells is obtained, giving lower operating frequency and leading to miniaturization of the antenna.

Meanwhile, the square patch is a conventional microstrip antenna, which is modeled as an open-ended TL resonator, composed equivalently with a series inductance $L_{ms}$ and a shunt capacitance $C_{ms}$. The propagation constant $\beta_{RH}$ along this RH TL structure is expressed by

$$\beta_{RH} = \omega\sqrt{L_{ms}C_{ms}}. \tag{7.79}$$

The square patch partially loaded with a CRLH structure is treated as a combined TL of two RH and one LH TL structures. An open ended or short circuited RH + LH + RH TL works as a resonator that exhibits a behavior similar to a CRLH resonator at lower frequencies and a RH resonator at higher frequencies. The propagation constant along the LH section is given by

$$\beta_{LH} = -1/(p\omega\sqrt{L_L C_L}) \tag{7.80}$$

where $p$ denotes the length of the unit cell, smaller than $\lambda_g/4$ and $L_L$ and $C_L$, respectively, are inductance and capacitance per unit cell. The antenna is designed to operate at three modes, $n = \pm 1$ and 0.

At lower frequencies, where $|\beta_{LH}| > |\beta_{RH}|$, an antenna with a mode $n = -1$ can be obtained as $\beta_{-1} L = -\pi$ ($\beta_{-1}$: the propagation constant along the TL of the length $L$ at $n = -1$, and $L$: the length of the RH + LH + RH TL structure). The antenna operates here with a fundamental mode just as a conventional patch, which has two radiating slots having the same amplitude and opposite phase, thus the radiation pattern is dipolar.

As the frequency increases, $\beta_{LH}$ increases, while $\beta_{RH}$ decreases. Then, a mode $n = 0$ can be excited by making $\beta_{LH} + \beta_{RH} = \beta_0 L = 0$. The electric field distribution of this mode is constant, and so two radiation slots of the patch have the same amplitude and phase, leading to a monopole-like radiation at the resonance frequency $f_0$.

At frequencies higher than $f_0$, $|\beta_{LH}|$ is greater than $|\beta_{RH}|$ and thus the resonance condition for $n = +1$ mode can be achieved, where $\beta_{+1} L = +\pi$. This mode has the same field distribution and radiation property as the fundamental mode of the conventional patches.

Since the frequencies of the lower modes $n \leq 0$ depend on the CRLH (mushroom) structure, desired operating frequencies at lower bands are obtained by appropriate design of the mushroom structure.

For desired operating frequencies at higher bands, as the frequencies at higher mode $n = +1$ depend on the patch itself, the patch dimensions are so designed as the conventional patch without mushroom.

The proposed antenna having dimensions of $(L \times W) = 42$ mm $\times$ 42 mm, the substrate with $\varepsilon_r = 2.2$ and thickness $h = 10$ mm is used. The mushroom structure is comprised with a 2 $\times$ 1 cell array and has the dimensions ($L_m \times W_m$) 10.6 mm $\times$ 17.8 mm, the gap between the two mushrooms is 0.40 mm, and the separation gap between the LH structure and the microstrip patch is 0.2 mm. The diameter of the vias is 0.7 mm. The coaxial feed probe is placed 14 mm from the center and the size of the GP is 80 mm $\times$ 80 mm. These dimensional parameters are chosen to obtain the resonant frequencies at 1 GHz for the $n = -1$ mode, used in GSM (Global Systems for Mobile Communications), at 1.5 GHz for the $n = 0$ mode, used in navigation systems, and 2.2 GHz for $n = +1$ mode, used in UMTS (Universal Mobile Telephone Systems).

Experimental results are: for $n = -1$ mode ($f_0 = 1.06$ GHz), $D$ (directivity) = 4.5 dB, $G$ (Gain) = –3 dB, $E$ (efficiency) = 17.8%, and $BW$ (–6 dB bandwidth) = 3%; for the $n = 0$ mode ($f_0 = 1.45$ GHz), $D = 5.1$ dB, $G = 1$ dB, $E = 39\%$, $BW = 3\%$; and for the $n = +1$ mode ($f_0 = 2.16$ GHz), $D = 7.4$ dB, $G = 6.5$ dB, $E = 82\%$, and $BW = 13\%$. The lengths of the antenna in terms of the resonance frequency are $\lambda_0/6.74$, $\lambda_0/4.92$, and $\lambda_0/3.31$ for the mode $n = -1$, 0, and +1, respectively, indicating that at the lower modes, significant reduction of the antenna size is observed compared with the conventional $\lambda/2$ patch antenna. The antenna presents two dipole modes ($n = \pm 1$) and one monopole

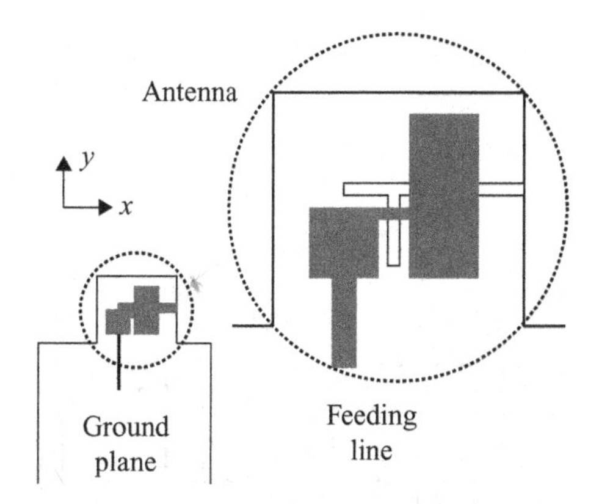


**Figure 7.243** Geometry of the RH/LH TLs loaded planar antenna ([125], copyright ©2008 IEEE).

mode ($n = 0$), and radiation patterns at the different operating modes are dipolar and monopolar.

A dual-frequency patch antenna, which aims at application to DCS at 1.8 GHz and UMTS at 2.2 GHz, was designed by using the same concept and an example is shown in [124]. In this case, a 2 × 2 CRLH structure is employed so that the two modes are excited, excluding $n = 0$ mode. The mushroom configuration was changed to be square and the number of cells was increased to four. The dimensions of this antenna are larger than the first one, with length of $\lambda_0/3.44$ at the lower frequency and $\lambda_0/2.83$ at the higher frequency. With increased size, efficiency of the antenna is made higher (60%) than the first one for $n = -1$.

### *7.2.5.2.3.6 Compact CRLH MM-loaded slot antenna*

A planar antenna utilizing cascaded RH/LH TLs to realize compact size is introduced in [125]. The antenna geometry is depicted in Figure 7.243, where the constitutive elements of the RH/LH TL are shown by black parts (printed capacitors) and slots (inductors). The RH TL consists of a series inductance and a shunt capacitance in a Pi-network structure, while the LH TL consists of a series capacitance and a shunt inductance in a T-network structure. They are connected with an open-circuit at the unconnected port of the LH TL, by which the TL works as a resonator. Along the power travelling direction, the RH and LH TLs have opposite phase property from each other, hence a ZOR (zeroth-order resonance) structure can be formed and the size of the antenna can be designed arbitrarily without regard to the wavelength since it is specified by the circuit parameters. The circuit elements are realized by using printed elements. The size of printed patches is designed to have proper capacitance against the ground, while the length of the metal traces or the size of slots on the ground plane is designed to create required inductance. The realized proposed antenna of straight type is shown in Figure 7.244. Layout of the antenna of L-shaped type is illustrated in Figure 7.245(a), in which dark lines show the top metal part and dotted lines give the bottom metal part. In the figure, (a) shows that three patches on the top metal part realize capacitances $C_1$, $C_2$, and $C_3$, respectively, and two narrow metal slots placed on the bottom metal, which

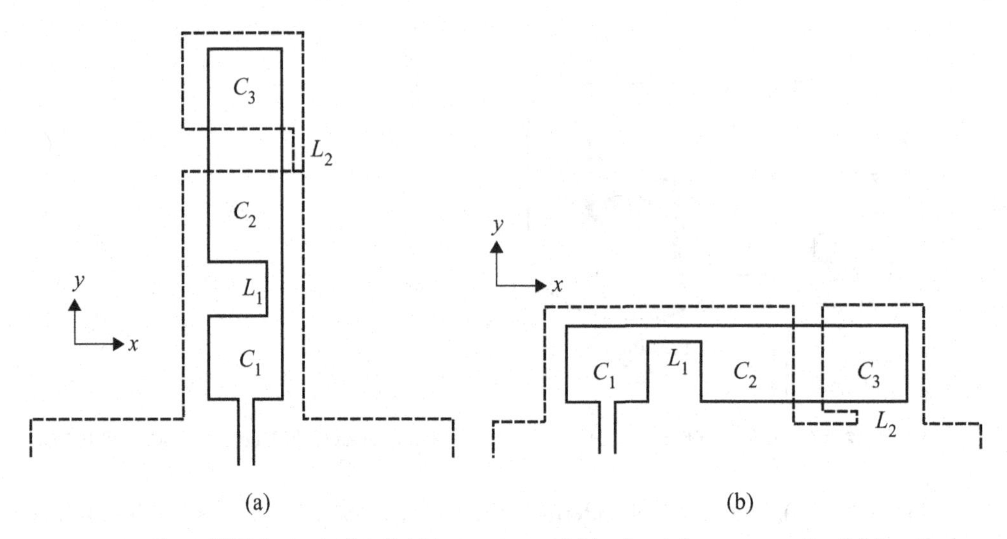


**Figure 7.244** Layout of straight type antenna, (a) horizontal arrangement and (b) vertical arrangement ([125], copyright ©2005 IEEE).

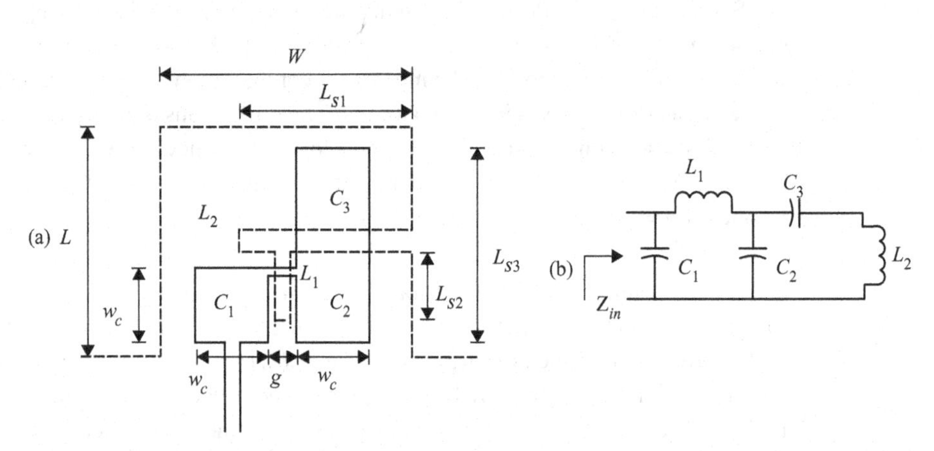


**Figure 7.245** Layout of L-shaped type (a) the composition of metamaterial components and (b) the equivalent circuit of a $\pi$-model for RH TL and a T-model for LH TL ([125], copyright ©2008 IEEE).

is considered as the ground plane (GP), produce inductances $L_1$ and $L_2$. A closed loop is formed by these elements, on which current flows in order of $C_1$, $L_1$, $C_2$, $C_3$, and $L_2$, and then back to $C_1$. The dimensional parameters (in mm) are $L = W = 11.5$, $w_c = 4$, $L_{s3} = 9.5$, $g = 1.3$, $L_{s1} = 7.2$, $L_{s2} = 3.2$, and the slot ($L_1$) width is 0.5. The size of the GP is 40 mm $\times$ 30 mm. All three capacitances are designed to have the same value 2.6 pF and both inductances have the same value of 1.62 nH. The equivalent circuit consisting of these circuit elements is given in Figure 7.245(b). Magnetic currents flow around $C_2$ and $C_3$, but currents on both sides have opposite phase and do not contribute to radiation, while the top sides of both elements operate as the radiating edges, and contribute to radiation. Since the field produced by $C_1$ is very weak, it is not taken into

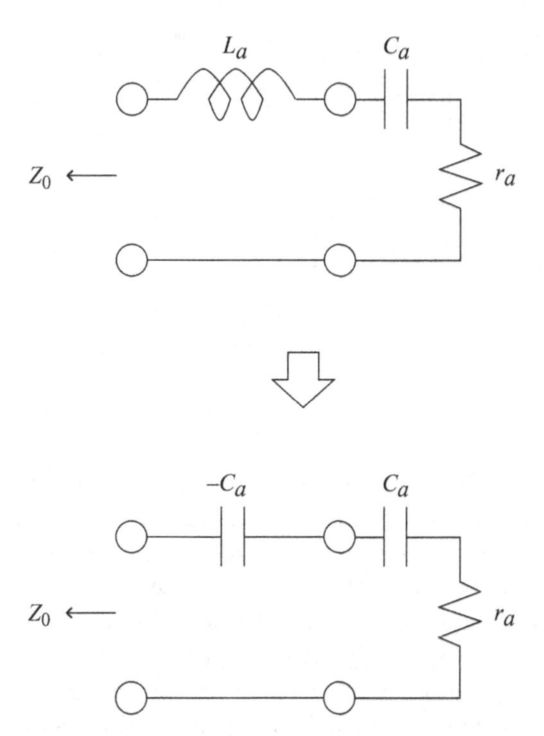


**Figure 7.246** Negative capacitance replacing a large inductance in the resonance circuit of a small antenna.

consideration. The operating frequency is 2.45 GHz. Measured results show a resonant frequency of 2.35 GHz with return loss of 23 dB and the –10 dB bandwidth of 4.5%. From the radiation patterns, the gain is evaluated to be 0.16 dBi.

The proposed antenna as fabricated by using small patches and slots could offer fairly good performance with a compact size of $\lambda_0/11$ square that is very small as compared with a conventional $\lambda_0/2$ antenna ($\lambda_0$: free space wavelength).

## 7.2.6 Active circuit applications to impedance matching

An antenna always needs impedance matching in practical use. As was mentioned frequently, matching at the feed terminals of an electrically small antenna is a crucial problem, because of the need to compensate its highly reactive impedance with low loss, and to transform its low resistive impedance to the load impedance (50 Ω). To overcome this difficulty, two methods have recently been disclosed; one is the use of active circuits and another is matching in the near field of a radiator, instead of matching at the feed terminals. The latter has been described in 6.2.5. Meanwhile, active circuits are applied, for instance, to cancel the large capacitance $C_a$ of a short dipole. In practice, as Figure 7.246 shows, a negative reactance $-C_a$ is used instead of a large inductance $L_a$ corresponding to the $C_a$ of a short dipole. In the figure, $r_a$ denotes the radiation resistance of the dipole. Since $L_a$ must be large enough to cancel $C_a$, it may have a large

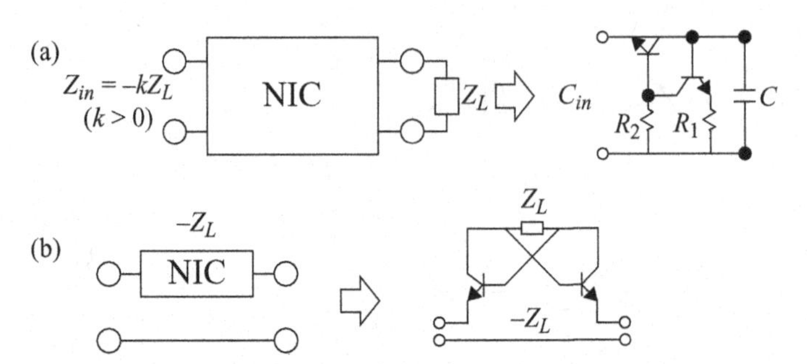


**Figure 7.247** Negative impedance converter (NIC) and example of practical implementation: (a) grounded type and (b) floating type.

loss that reduces the radiation efficiency, and hence, $-C_a$ substitutes for it. The negative capacitance $-C_a$ can be achieved by a negative reactance circuit, which is constituted of transistors. The negative reactance circuits can be realized by NIC as is illustrated in Figure 7.247, which shows two types of NIC; (a) grounded NIC and (b) floating NIC [126–134]. Examples of transistor circuits are shown on the right-hand side in the figure, where $C_{in}$ ($-C$) and $-Z_L$, respectively, are the capacitance converted from $C$ by the grounded NIC and the impedance converted from $Z_L$ by the floating NIC.

On the recent advent of NIC application to antennas, antenna engineers have acquired a pragmatic and effective design technique to develop a novel electrically small antenna having high efficiency and possibly wide bandwidth.

Active circuits, which do not follow Foster Reactance Theorem, are referred to as Non-Foster circuits. Another way to obtain proper matching in small antennas is the use of metamaterials (MM), which can represent negative reactance, instead of NIC. An MNG introduced in 7.2.5.1 is a typical example.

It must be noted, when negative reactance circuits are used, that the circuit should maintain stable condition, noise be kept as low as possible, and the linearity be kept as high as possible.

Regarding stability, an NIC needs inevitably to be open-circuit stable (OCS) at one port and short-circuit stable (SCS) at the other port. OCS means that the network is stable for any passive load on one side with the other port open, while SCS means that the network is stable for any passive load on one port with the other port short. The inherent conditional stability of an NIC constrains the magnitude of the impedances that can be connected to the OCS port and to the SCS port. This can be interpreted by the requirements $|Z_{L1}| > |Z_{in1}|$ and $|Z_{L2}| < |Z_{in2}|$. Figure 7.248(c) illustrates an NIC that has impedances $Z_{L1}$ and $Z_{L2}$, respectively, at each load terminal and impedances $Z_{in1}$ and $Z_{in2}$ at the input terminals of each side. In designing an NIC circuit, these requirements must always be met.

In addition, there is inconvenience in using the active circuit, as the transistor circuits are generally unidirectional, whereas an antenna is bi-directional, and hence the application is limited to either transmitter or receiver. In addition, transistor circuits need the bias supply, for which additional circuits are required within the antenna system.

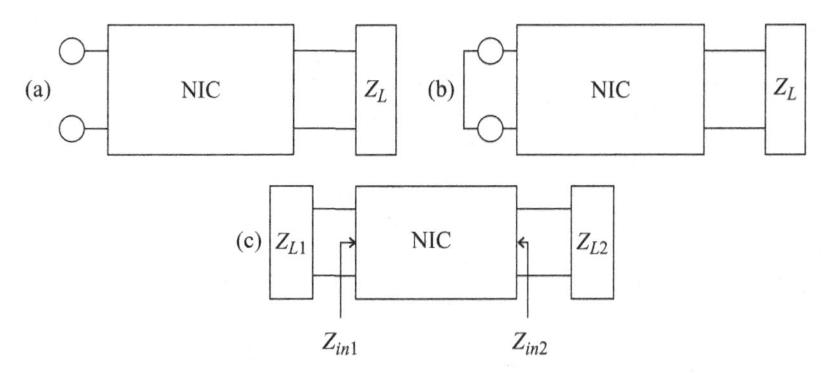


**Figure 7.248** NIC circuit and stability requirements: (a) shows OCS state, (b) gives SCS state, and (c) illustrates impedances that concern stability of an NIC; $Z_{L1}$ $Z_{L2}$ at the load and $Z_{in1}$, $Z_{in2}$ at the input terminals each side.

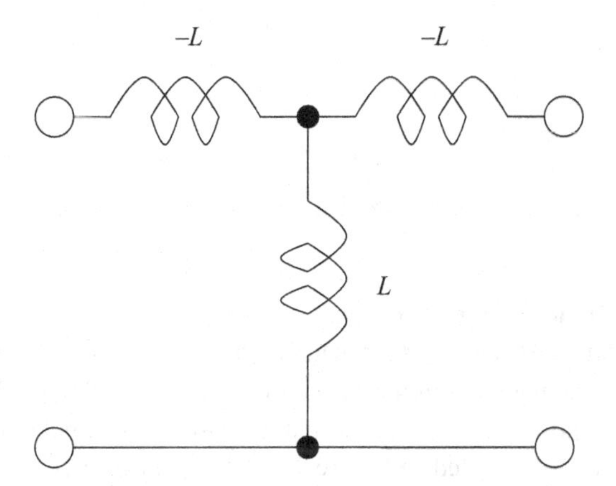


**Figure 7.249** Impedance transformer between a low resistance and the load (characteristic impedance $Z_o$).

Instead, an NMG material, if it could be available, is greatly advantageous, since the material enables bi-directional application feasibility that is favorable in antenna applications.

Nevertheless, there are distinct advantages of using negative circuits in antenna design, as they can overcome matching difficulty and thus assist in enhancing gain and efficiency in transmitter applications, even when the size of antenna is electrically small, while they may facilitate extension of bandwidth and improvement of signal-to-noise ratio in receiver applications.

NIC may also be used as a circuit that converts low radiation resistance of small antennas to the load (or source) resistance (50 Ω) after the reactance is compensated. For this purpose, a T-shaped inductance circuit, for example, may be used. Figure 7.249 depicts the circuit, which is formed of a shunt $L$ and two series $-L$s. Here, the $-L$ can be produced by an NIC. The matching circuit is then designed to employ NICs, which represent $-C_a$ and $-L_m$ as shown in Figure 7.250.

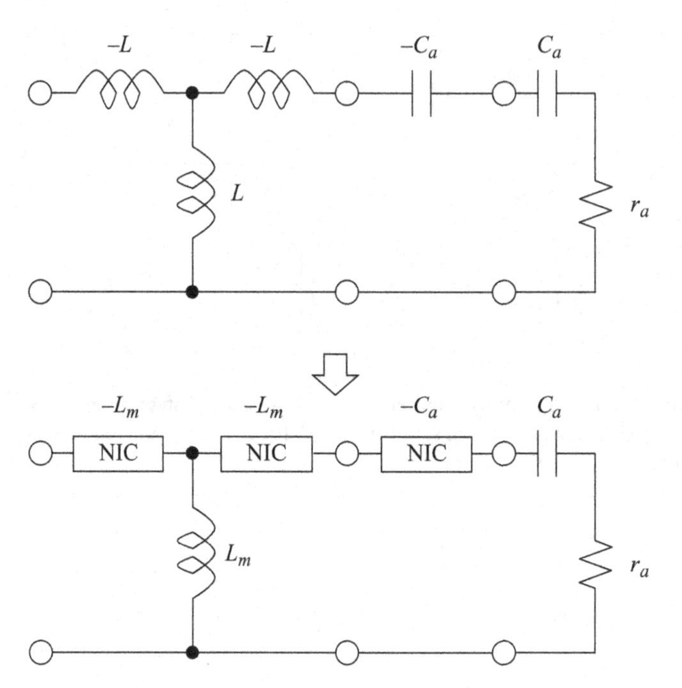


**Figure 7.250** Matching circuit and impedance transformer using negative reactance.

### 7.2.6.1 Antenna matching in transmitter/receiver

NICs are applied to matching circuits of antennas in the transmitter or the receiver [132, 133]. An example of antenna matching in the transmitter is introduced in [132], which describes a 15–30 MHz band transmitter, for which a two-foot monopole antenna is used. The matching circuit is illustrated in Figure 7.251, where two cases are shown: (a) passive matching and (b) active matching. In these circuits, an inductance $L = 1.8$ μH is used for tuning at 20 MHz with the antenna capacitance $C_a = 33$ pF and a serial L-C circuit, which derives 50 Ω of the load resistance $R_L$ from the low resistance 4 Ω at the output stage of the tuning circuit, is used. This 4Ω at the output of the tuning circuit is attributed to the loss of the inductance $L$, whereas the radiation resistance of the antenna at 20 MHz is 1Ω.

In order to transform 1 Ω to the source 50 Ω, a serial L-C network is used as shown in Figure 7.251, which is a different type of circuit to that shown in Figure 7.249.

Meanwhile, in the active matching, a floating NIC, which uses two transistors, each being a pair of npn-pnp, is used to cancel $L$ and $C_a$, to which the impedance transforming circuit is connected. Then the overall matching circuit will be that as shown in Figure 7.251(b).

The matching characteristics depend on the biasing of the NIC transistors. In [132], the matching characteristics are evaluated by $|S_{21}|^2$, which provides the ratio of the power delivered to the 1 Ω load to the power from the source (50 Ω), and is compared with that of the passive matching. In the active matching, two biasing cases, that is, conventional class A and class B biasing, are compared.

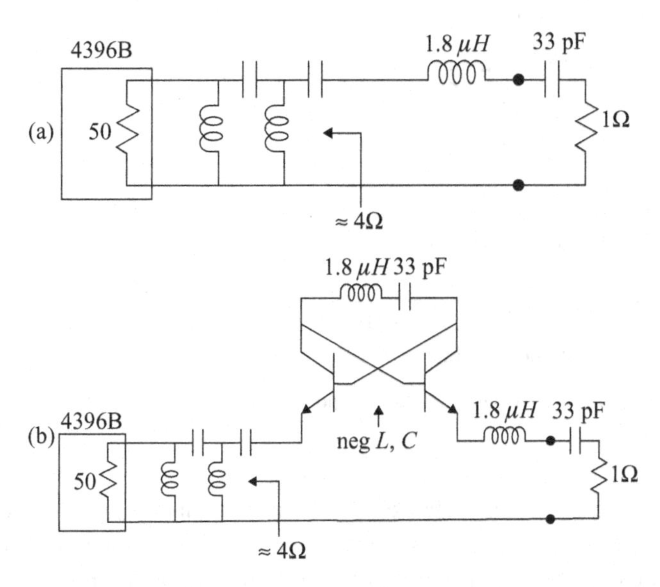


**Figure 7.251** Matching circuits for small antenna: (a) passive case and (b) active case [132].

The active matching showed power efficiency that exceeds 20 dB over that of the passive matching for the lower frequency band. The class B biasing demonstrates higher peak power delivery (about 24 dBm) than the best possible passive match near the center frequency (23 MHz) of the band, while the class A biasing shows nearly constant power level (15 dBm over the passive case) for the entire frequency band (15–30 MHz).

Application of NIC circuits to a receiver is discussed in [133]. A six-inch monopole antenna is used and the transmitter used for evaluation of the receiver performance is arranged to transmit the frequency of 20–110 MHz. The receiver has 4 dB noise figure achieved by using a low noise RF amplifier. In the matching circuit an NIC is used to represent $-C_a$ as shown in Figure 7.246. At 30 MHz, the antenna reactance –730 Ω was brought down to –14.6 Ω, not zero reactance, after the NIC. This results from the inevitable parasitic reactance. Measured signal-to-noise ratio (*S/N*) exhibits improvement Δ of 6 to a few dB over the entire frequency band. Figure 7.252 gives comparison of $S_1/N_1$ (active matching case) to $S_0/N_0$ (a case when the antenna and the receiver are directly connected).

### 7.2.6.2 Monopole antenna

Active circuit matching is described in [134], in which a case where an electrically small monopole placed on the infinite ground plane is considered. A two-port circuit model representing an antenna system is considered through simulation and the antenna performances, radiation efficiency, and bandwidth, of both passive and active matching cases are studied. The antenna considered here is a cylindrical monopole of 0.6 m in length and 0.010 m in diameter and the operating frequency is assumed to be 30 to 90 MHz. The matching circuit is basically the same as that shown in Figure 7.250, where an NIC is used to implement negative reactance corresponding to the series

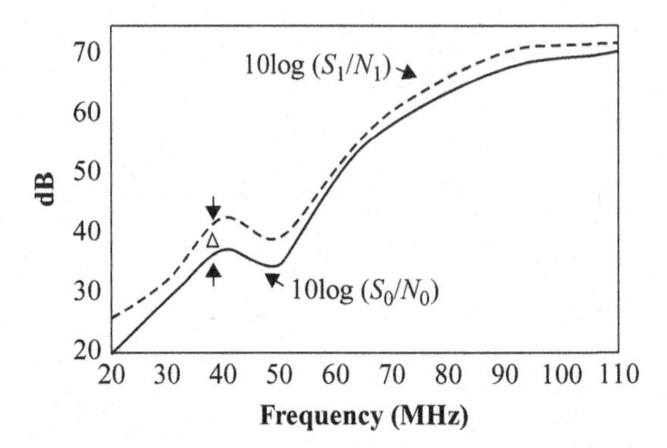


**Figure 7.252** Comparison of $S/N$ between passive and active matching cases [133].

antenna capacitance $C_a$ and inductance $L_a$ and an inductance $L_m$ of the transformer section that converts the small antenna resistance to 50 Ω (the load resistance). $L_m$ here is designed to equal $\sqrt{R_0 Z_0}/\omega_0$ where $Z_0$ is the desired impedance level, which is here 50 Ω. The active device used is a silicon bipolar NPN transistor. Simulated return loss and total efficiency of the antenna/matching network combination are illustrated in Figure 7.253, where (a) is the passive matching case and (b) is the active matching case. As can be seen in the figure, the bandwidth is extended from about 3 MHz (–3 dB efficiency) in the passive case to about 36 MHz to beyond 90 MHz (–10 dB return loss) in the active case. The total efficiency in the active case is better than 95% from about 36 MHz to above 90 MHz.

Another example of NIC application to a small monopole antenna is introduced in [135]. The antenna is a 3-inch wire monopole with 1.5 mm diameter placed on a finite ground plane (3 inch × 3 inch with 1 mm thickness) and a negative capacitor to make a 50 Ω load (generator) match to the antenna, for which an NIC is used. The frequency range considered is 1 MHz to 1 GHz. The total reactance of the antenna in the active matching case becomes lower than that of the antenna in the passive matching case. Consequently the transducer gain between the source and the antenna becomes 16.23 dB higher in the frequency range from 50 MHz to 644 MHz in the active matching case. This result is significant, since the antenna electrical length is very small; $\lambda/79$ at 50 MHz and $\lambda/6$ at 644 MHz.

### 7.2.6.3 Loop and planar antenna

To improve the bandwidth and reduce the size of the antenna, a non-Foster matching network is designed. A loop antenna is considered as an example. As a small loop antenna has different impedance behavior from that of a small dipole, the matching circuit must be optimized to meet the variation of the loop impedance. Then the circuit consisting of a shunt inductance and an optimized non-Foster network is taken into consideration [136]. Return loss of a 6-inch loop is depicted in Figure 7.254(a), where the inset shows the loop, and the return loss of the antenna with active matching network

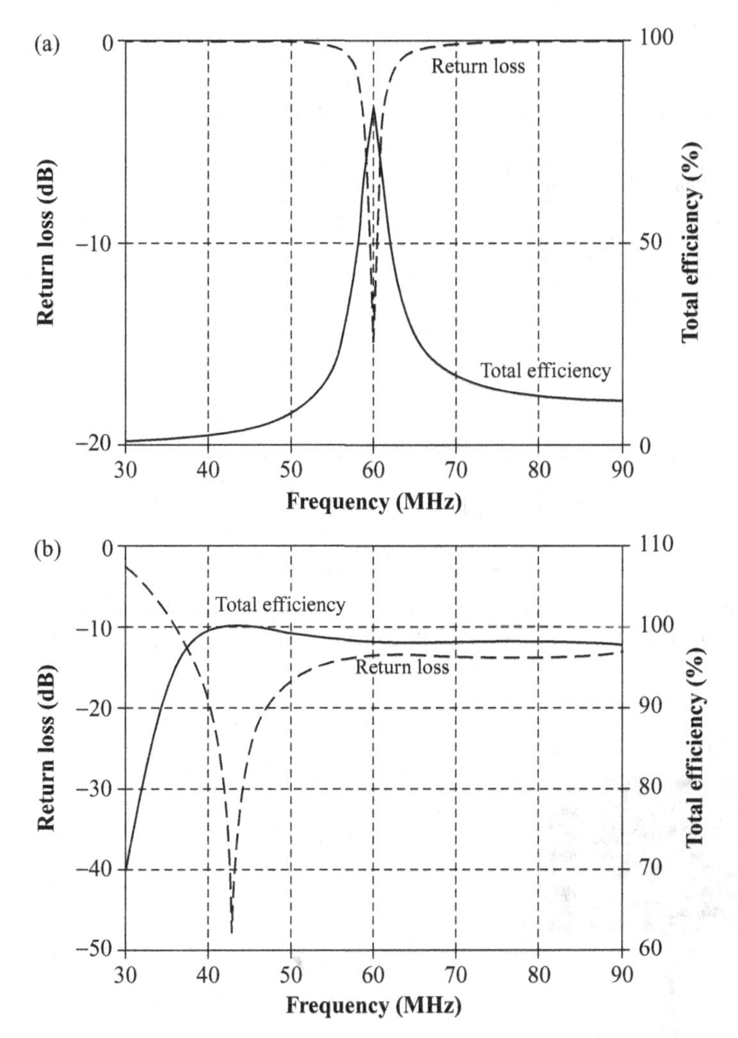


**Figure 7.253** Return losses: (a) passive matching case and (b) active matching case ([134], copyright ©2008 IEEE).

is given in Figure 7.254(b). The optimized non-Foster matching network is shown in Figure 7.254(c). From (a) and (b), increased bandwidth from 50 MHz to over 320 MHz can be observed after the non-Foster matching network is used.

In [136], a planar dipole antenna, to which a non-Foster matching network is applied, is treated. The antenna is printed on a very thin, flexible dielectric sheet, and the antenna size is $\lambda/4 \times \lambda/5$ at 250 MHz. It has a gain greater than 0 dB from 250 MHz to 1000 MHz, meaning 4 to 1 bandwidth. Since a simple matching circuit at the feed is not sufficient, two additional ports apart from the feed point within the antenna structure are defined, to which negative capacitances are added, producing a resistance around 50 Ω from 50 MHz to 300 MHz. With this resistance the optimized matching circuit is designed by using a non-Foster matching network, in which three negative inductances

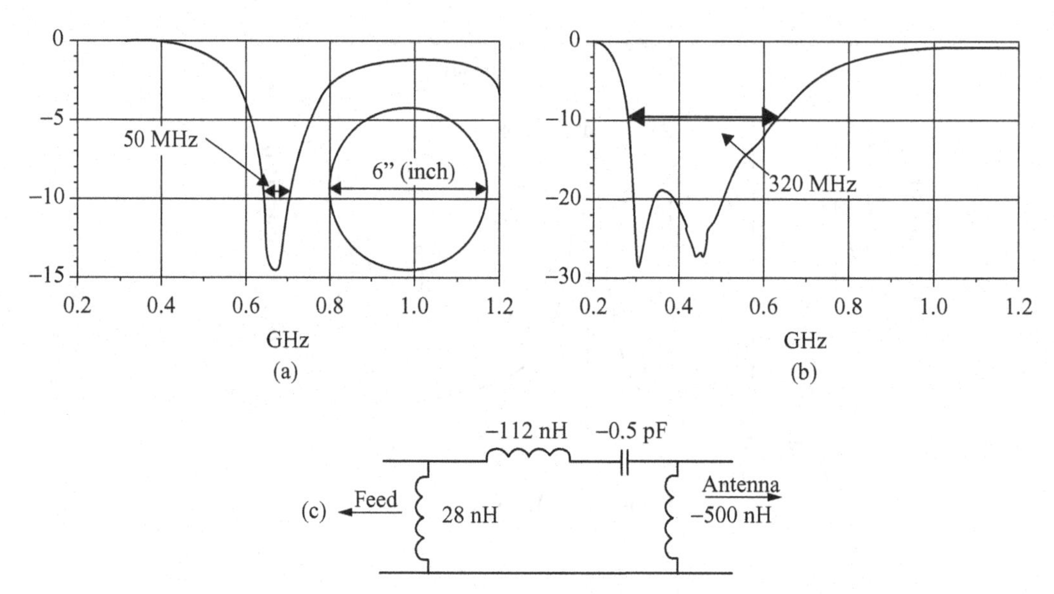


**Figure 7.254** Matching performance of a 6-inch loop: return loss in the case of (a) passive matching, (b) active matching, and (c) Non-Foster matching network ([136], copyright ©2009 IEEE).

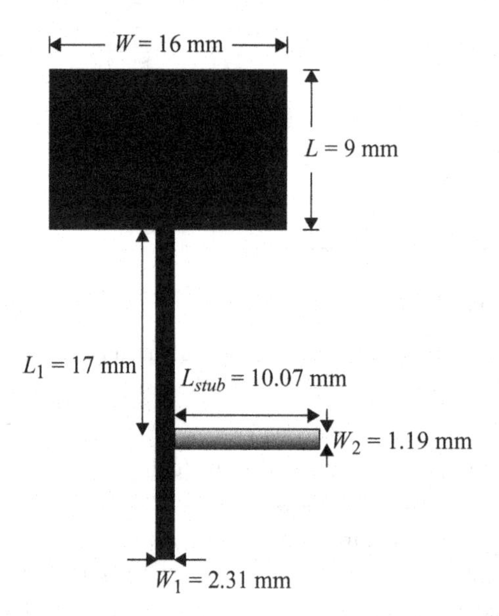


**Figure 7.255** The reference antenna ([137], copyright ©2007 IEEE).

and two negative capacitances are used. As a consequence, the imaginary part of the antenna impedance is cancelled from around 80 MHz to 275 MHz. This results in size reduction of the antenna having a $\lambda/12 \times \lambda/18$ planar dipole with a 3.5 to 1 bandwidth (80 MHz to 270 MHz) with gain greater than 0 dBi.

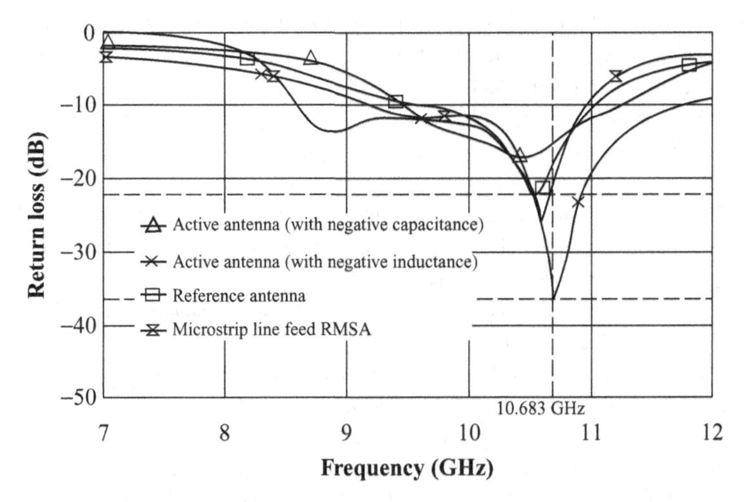


**Figure 7.256** Return loss in both passive and active matching ([137], copyright ©2007 IEEE).

### 7.2.6.4 Microstrip antenna

Either negative inductance or negative capacitance is applied to a rectangular microstrip antenna to enhance the bandwidth [137]. The patch dimensions are $W$ (width) = 16 mm, $L$ (length) = 9 mm and the ground plane size is a 50 mm square. The operating frequency is 10.55 GHz with $|S_{11}|= 21.5$ dB at the resonance frequency.

A floating negative inductance is realized by using two FETs inserted between the antenna and the source (transmitter). The bandwidth is increased from 12.2% in passive matching case to 24.5% in the active matching case.

In case of connecting a negative capacitance circuit to the output of the antenna, the bandwidth can be enhanced from 12.2% to 16.96%. The gain is increased from 6.6 dB to 9.2 dB. The reference antenna is illustrated in Figure 7.255, which shows that a matching stub is connected to the feed line. The return loss $|S_{11}|$of both active and passive matching cases are shown in Figure 7.256.

## References

[1] H. Nakano, H. Tagami, A. Yoshizawa, and J. Yamauchi, Shortening Ratios of Modified Dipole Antennas, *IEEE Transactions on Antennas and Propagation*, vol. 32, 1984, no. 4, pp. 385–386.

[2] L. C. Godara (ed.), *Handbook of Antennas In Wireless Communications*, CRC Press 2000, chapter 12.2.2.2, pp. 12–13–12–18.

[3] K. Noguchi, *et al.*, Impedance Characteristics of a Small Meander Line Antenna, *Transactions of IEICE* vol. JB, BII, no. 2, 1998, pp. 183–184.

[4] L. C. Godara (ed.), *Handbook of Antennas In Wireless Communications*, CRC Press 2000, Chapter 12.2.2, pp. 12–11–12–19.

[5] F. Kuroki and H. Ohta, Miniaturized Cross Meander-Line Antenna Etched on Both Sides of Dielectric Substrate, *International Symposium on Antennas and Propagation (ISAP)* 2006, Singapore a258 r266.

[6] H. Choo and H. Ling, Design of Planar, Electrically Small Antennas with Inductively Coupled Feed Using a Genetic Algorithm, *IEEE International Symposium on Antennas and Propagation* 2003, 22.1.

[7] C. W. P. Huang *et al.*, FDTD Characterization on Meander Line Antennas for RF and Wireless Communications, *Progress in Electromagnetics Research PIER*, 24, 1991, pp. 185–199.

[8] C-I. Lin, F-W. Chi, and K-L. Wang, Internal Meander Line Antenna for GSM/DCS/PCS Multiband Operation in Mobile Phones, *International Symposium on Antennas and Propagation (ISAP)*, 2006, Singapore a62 r93.

[9a] L. C. Godara (ed.), *Handbook of Antennas In Wireless Communications*, CRC Press 2000, chs. 12.2.3–12.2.5, pp. 12–27–12–34.

[9b] L. C. Godara (ed.), *Handbook of Antennas In Wireless Communications*, CRC Press 2000, chs. 12.2.6 and 12.2.7, pp. 12–34–12–39.

[10] K. Noguchi *et al.*, Increasing the Bandwidth of Meander Line Antennas Consisting of Two Strips, *Transactions of IEICE*, vol. JB2-B, 1999, no. 3, pp. 402–409.

[11] C. A. Balanis, *Antenna Theory, Analysis and Design*, 2nd edn., 1982, pp. 459–461.

[12] K. Noguchi, *et al.*, Impedance Characteristics of a Meander Line Antenna Mounted on a Conducting Plane, *IEICE National Convention*, B-1–106, 1999, p. 106.

[13] M. Takiguchi and Y. Yamada, Radiation and Ohmic Resistances in Very Small Meander Line Antennas of Less than 0.1 Wavelength, *Transactions of IEICE*, vol. J87-B, 2004, no. 9, pp. 1336–1346.

[14] Y. Yamada and N. Michishita, Efficiency Improvement of a Miniaturized Meander Line Antenna by Loading a High $\varepsilon r$ Material, *IEEE iWAT*, 2005.

[15] R. L. Bell, C. T. Elfving, and R. E. Franks, Near Field Measurement on a Logarithmically Periodic Antenna, *IRE Transactions on Antennas and Propagation*, vol. 8, 1960, pp. 559–567.

[16] P. E. Mayes, Balanced Backfire Zigzag Antennas, *1964 IEEE International Conference Record* pt. 1, pp. 158–165.

[17] S. H. Lee, Theory of Zigzag Antennas, Ph.D. Dissertation, Dept. of Electrical Engineering University of California, Berkeley, June, 1968, pp. 20, 31–33.

[18] S. H. Lee and K. K. Mei, Analysis of Zigzag Antennas, *IEEE Transactions on Antennas and Propagation*, vol. 18, 1970, no. 6, pp. 760–764.

[19] N. Inagaki, K. Tamura, and K. Fujimoto, Theoretical Investigation on the Resonance Length of Normal Mode Helical Antennas, *Technical Report of Nagoya Institute of Technology*, vol. 23, 1971, pp. 335–341.

[20] K. Fujimoto *et al.*, *Small Antennas*, Research Studies Press, 1987, pp. 59–75.

[21] N. Inagaki, T. Marui, and K. Fujii, Newly Devised MoM Analysis and Design Data for NMHA, *Technical Report of IEICE*, AP2007–194(2008–03) pp. 123–128.

[22] T. Endo, Y. Sunahara, and Y. Hoshihara, Resonance Frequency of Dielectric Loaded Normal Mode Helical Antenna, *IEICE Technical Report*, vol. 95, 1995, no. 535, pp. 1–6.

[23] J. S. Carreno and J. S. Solis, Broadband Log-periodic Normal Mode Helical Antennas, *IEEE APS International Symposium*, vol. 1, 2003, pp. 249–252.

[24] Y. Ogura, K. Asakawa, and T. Maeda, Folded Normal Mode Helical Antennas, *IEICE Technical Report*, vol. 104, 2004, no. 395.

[25] K. Noguchi *et al.*, Impedance Characteristics of Two-wire Helical Antenna in Normal Mode, *Electronics and Communications in Japan*, vol. 81, 1998, no. 12, pp. 37–44.

[26] S. R. Best, A Discussion on the Properties of Electrically Small Self-Resonant Wire Antennas, *IEEE Antennas and Propagation Magazine*, vol. 46, 2004, no. 6, pp. 9–22.

[27] H. A. Wheeler, Small Antennas, *IEEE Transactions on Antennas and Propagation*, AP-23, 1975, pp. 462–469.

[28] A. D. Yaghjian and S. R. Best, Impedance Bandwidth and Q of Antennas, *IEEE International Symposium on Antennas and Propagation, Digest*, vol. I, 2003, pp. 501–504.

[29] K-L. Wong, *Compact and Broadband Microstrip Antennas*, John Wiley and Sons, 2002, p. 5.

[30] H. A. Wheeler, Fundamental Limitation of Small Antennas, *Proceedings of IRE*, vol. 35, Dec 1947, pp. 1479–1484.

[31a] H. A. Wheeler, The Spherical Coil as an Inductor, Shield or Antenna, *Proceedings of IRE*, vol. 58, September 1958, pp. 1595–1602.

[31b] H. A. Wheeler, The Radian Sphere Around a Small Antenna, *Proceedings of IRE*, vol. 59, 1959, pp. 1325–1331.

[32] J. S. McLean, A Re-examination of the Fundamental Limits on the Radiation Q of Electrically Small Antennas, *IEEE Transactions on Antennas and Propagation*, AP-44, May 1996, pp. 672–675.

[33] N. Engheta and R. W. Ziolkowsky, *Metamaterials-Physics and Engineering Explorations*, John Wiley and Sons, 2006. p. 378.

[34a] C. P. Baliarda, J. Romeu, and A. Cardama, The Koch Monopole: A Small Fractal Antenna, *IEEE Transactions on Antennas and Propagation*, vol. 48, 2000, no. 11, pp. 1773–1781.

[34b] J. P. Gianvittorio and Y. Rahmat-Samii, Fractal Antenna: A Novel Antenna Miniaturization Technique and Applications, *IEEE Antennas and Propagation Magazine*, vol. 44, 2002, no. 1, pp. 20–36.

[34c] D. H. Werner and S. Ganguly, An Overview of Fractal Antenna Engineering Research, *IEEE Antennas and Propagation Magazine*, vol. 45, February 2003, no. 1, pp. 39–40.

[35] J. R-Mohassel, A. Mehdipour, and H. Aliakbarian, New Schemes of Size Reduction in Space Filling Resonant Dipole Antennas, *3rd European Conference on Antennas and Propagation*, vol. 23–27, 2009, pp. 2430–2432.

[36] H. K. Ryu and J. M. Woo, Miniaturization of Rectangular Loop Antenna using Meander Line for RFID Tags, *Electronics Letters*, vol. 43, March 2007, pp. 372–374.

[37] H. K. Ryu, S. Lim, and J. M. Woo, Design of Electrically Small, Folded Monopole Antenna using C-shaped Meander for Active 433.92 MHz RFID Tag in Metallic Container Application, *Electronics Letters*, vol. 44, 2008, pp. 1445–1447.

[38] C. Borja and J. Romeu, On the Behavior of Koch Fractal Boundary Microstrip Patch Antenna, *IEEE Transactions on Antennas and Propagation*, vol. 51, 2003, no. 6, pp. 281–291.

[39] S. R. Best, On the Performance of the Koch Fractal and Other Bent Wire Monopole, *IEEE Transactions on Antennas and Propagation*, vol. 51, 2003, no. 6, pp. 1292–1300.

[40] N. Engheta and R. W. Ziolkowsky, *Metamaterials Physics and Engineering Explorations*, John Wiley and Sons, 2006, pp. 378–381.

[41] J. Zhu, A. Hoorfar, and N. Engheta, Peano Antennas, *Antennas and Wireless Propagation Letters*, vol. 3, 2004, pp. 71–74.

[42] X. Chen, S. S-Naemi, and Y. Liu, A Down-Sized Hilbert Antenna for UHF Band, *IEEE International Symposium on Antennas and Propagation* 2003, pp. 581–584.

[43] H. Huang and A. Hoorfer, Miniaturization of Dual-Band Planar Inverted-F Antennas using Peano-Curve Elements, *International Symposium on Antennas and Propagation (ISAP)* 2006, a292 r206.

[44] J. P. Gianvittorio and Y. Rahmat-Samii, Fractal Antenna: A Novel Antenna Miniaturization Technique and Applications, *IEEE Antennas and Propagation Magazine*, vol. 44, 2002, no. 1, pp. 20–36.

[45] S. R. Best, A Comparison of the Resonant Properties of Small Space-Filling Fractal Antennas, *IEEE Antennas and Wireless Propagation Letters*, vol. 2, 2003, pp. 197–200.

[46] W. J. Krzysztofik, Modified Sierpinsky Fractal Monopole for ISM-Bands Handset Applications, *IEEE Transactions on Antennas and Propagation*, vol. 57, 2009, no. 3, pp. 606–615.

[47] D.H. Werner and S. Ganguly, An Overview of Fractal Antenna Engineering Research, *IEEE Antennas and Propagation Magazine*, vol. 45, 2003, no. 1, pp. 38–57.

[48] W. L Stutzman and G. A. Thiele, *Antenna Theory and Design*, 2nd edn., John Wiley and Sons, pp. 252–258.

[49] M. McFadden and W. R. Scott, Analysis of the Equiangular Spiral Antenna on a Dielectric Substrate, *IEEE Transactions on Antennas and Propagation*, vol. 55, 2007, no. 11, pp. 3163–3171.

[50] H. Nakano *et al.*, Equiangular Spiral Antenna Backed by a Shallow Cavity With Absorbing Strips, *IEEE Transactions on Antennas and Propagation*, vol. 56, 2008, no. 8, pp. 2742–2747.

[51] J. L. Volakis, N. W. Nurnberger, and D. S. Filipovic, A Broadband Cavity-Backed Slot Spiral Antenna, *IEEE Antennas and Propagation Magazine*, vol. 43, 2001, no. 6, pp. 15–26.

[52] M. W. Nurnberger and J. L. Volakis, Extremely Broadband Slot Spiral Antennas with Shallow Reflecting Cavities, *Electromagnetics*, vol. 20, 1996, no. 4, pp. 130–131.

[53] B. A. Kramer *et al.*, Design and Performance of an Ultra Wideband Ceramic-loaded Slot Spiral, *IEEE Transactions on Antennas and Propagation*, vol. 53, 2005, no. 7, pp. 2193–2199.

[54] D. S. Filipovic and J. L. Volakis, Broadband Meanderline Slot Spiral Antenna, *IEE Proceedings-Microwaves Antennas and Propagation*, vol. 149, 2002, no. 2, pp. 98–105.

[55] M. Nurnberger and J. L. Volakis, New Termination for Ultra Wide-band Slot Spirals, *IEEE Transactions on Antennas and Propagation*, vol. 50, 2002, no. 1, pp. 82–85.

[56] D. S. Filipovic and J. L. Volakis, Novel Slot Spiral Antenna Design for Dual- band/ Multi-band Operation, *IEEE Transactions on Antennas and Propagation*, vol. 51, 2003, no. 3, pp. 430–440.

[57] D. S. Filipovic and J. L. Volakis, A Flush Mounted Multifunctional Slot Aperture (Combo-antenna) for Automotive Applications, *IEEE Transactions on Antennas and Propagation* vol. 52, 2004, no. 2, pp. 563–571.

[58] B. A. Kramer *et al.*, Miniature Conformal Aperture with Volumetric Inductive Loading, *IEEE APS International Symposium*, 2006, Digest vol. 44, pp. 3693–3696.

[59] B. A. Kramer *et al.*, Miniature UWB Antenna with Enhanced Inductive Loading, *Small Antennas Novel Metamaterials, IEEE iWAT 2006 International Workshop*, pp. 289–292.

[60] C. C. Chen and J. L. Volakis, Spiral Antennas: Overview, Properties and Miniaturization Techniques, in R. Waterhouse, (ed.), *Printed Antennas for Wireless Communications*, John Wiley and Sons, 2007.

[61] M. Lee *et al.*, Distributed Lumped Loads and Lossy Transmission Line Model for Wideband Spiral Antenna Miniaturization and Characterization, *IEEE Transactions on Antennas and Propagation* vol. 55, 2007, no. 10, pp. 2671–2678.

[62] B. A. Kramer *et al.*, Size Reduction and a Low-profile Spiral Antenna using Inductive and Dielectric Loading, *IEEE Antennas and Wireless Propagation Letters*, vol. 7, 2008, pp. 22–25.

[63] J. L. Volakis, C. C. Chen, and K. Fujimoto, *Small Antennas-Miniaturization Techniques and Applications*, McGraw-Hill, 2010.

[64] R. W. Klopfenstein, A Transmission Line Taper of Improved Design, *Proceedings of IRE*, vol. 44, 1956, no. 1, pp. 31–35.

[65] J. Abadia *et al.*, 3D-Spiral Small Antenna Design and Realization for Biomedical Telemetry in the MICS band, Radioengineering, *Proceedings of Czech and Slovak Technical Universities and URSI Committees*, vol. 18, 2009, no. 4, pp. 359–366.

[66] Victrex® PEEK polymers, available at; www.victrex.com/en/products/victrex-peek-polymers/victrex-peek-polymers.php.

[67] J. H. Wheeler, Fundamental Limitations of Small Antennas, *Proceedings of IRE*, vol. 35, December 1947, pp. 1479–1484.

[68] J. C. Cardoso and A. Safaal-Jazi, Spherical Helical Antenna with Circular Polarization over a Broad Beam, *Electronics Letters*, vol. 29, 1993, no. 4, pp. 325–326.

[69] A. Safaal-Jazi and J. C. Cardoso, Radiation Characteristics of a Spherical Helix Antenna, *Proceedings of Institution of Electrical Engineers Microwave Antennas Propagation*, vol. 143, 1996, no. 1, pp. 7–12.

[70] H. T. Hui *et al.*, The Input Impedance and the Antenna Gain of the Spherical Helix Antenna, *IEEE Transactions on Antennas and Propagation*, vol. 49, 2001, pp. 1235–1237.

[71] S. R. Best, The Radiation Properties of Electrically Small Folded Spherical Helix Antennas, *IEEE Transactions on Antennas and Propagation* vol. 52, 2004, no. 4, pp. 953–960.

[72] S. R. Best, Low Q Electrically Small Linear and Elliptical Polarized Spherical Dipole Antennas, *IEEE Transactions on Antennas and Propagation*, vol. 53, 2003, no. 3, pp. 1047–1053.

[73] S. R. Best, A Comparison of the Cylindrical Folded Helix Q to the Gustafsson Limit, *EuCap* 2009, Berlin, pp. 2554–2557.

[74] S. R. Best, The Quality Factor of the Folded Cylindrical Helix, Radio Engineering, *Proceedings of Czech and Slovak Technical Universities and URSI Committee*, vol. 18, 2009, no. 4, pp. 343–347.

[75] A. Mehdipour, H. Aliakbarian, and J. Rashed-Mohassel, A Novel Electrically Small Spherical Wire Antenna With Almost Isotropic Radiation Pattern, *IEEE Antennas and Wireless Propagation Letters*, vol. 7, 2009, pp. 396–399.

[76] S. R. Best, A Low Q Electrically Small Magnetic (TE Mode) Dipole, *IEEE Antennas and Wireless Propagation Letters*, vol. 8, 2009, pp. 572–575.

[77] H. L. Thal, New Radiation Q Limits for Spherical Wire Antennas, *IEEE Transactions on Antennas and Propagation*, vol. 54, 2006, no. 10, pp. 2757–2763.

[78] O. S. Kim, Low-Q Electrically Small Spherical Magnetic Dipole Antennas, *IEEE Transactions on Antennas and Propagation*, vol. 58, 2010, no. 7, pp. 2210–2217.

[79] H. A. W. Alsawara and A. Safaai-Jazi, Ultrawideband Hemispherical Helical Antenna, *IEEE Transactions on Antennas and Propagation*, vol. 58, 2010, no. 10, pp. 3175–3181.

[80] J. Ramsay, Highlights of Antenna History, *IEEE Antennas and Propagation Society Newsletter*, December 1981, p. 8.

[81] C. W. Harrison, Monopole with Inductive Loading, *IEEE Transactions on Antennas and Propagation*, AP-11, 1963, pp. 394–400.

[82] T. L. Simpson, The Disk Loaded Monopole Antenna, *IEEE Transactions on Antennas and Propagation* vol. 52, 2008, no. 2, pp. 542–550.

[83a] S. R. Best and D. L. Hanna, A Performance Comparison of Fundamental Small-Antenna Designs, *IEEE Antennas and Propagation Magazine*, vol. 52, 2010, no. 1, pp. 47–70.

[83b] S. R. Best, Small and Fractal Antennas, in C. A. Balanis (ed.), *Modern Antenna Handbook*, John Wiley and Sons, 2008, 10.6.3.

[84] Z. Xing and X. Yadong, A Novel Electrically Small Monopole Load by Open-Circuited Transmission Line, *IEEE APS International Symposium* vol. 44, 2006, pp. 615–618.

[85] L. J-Ying and G. Y-Beng, Characteristics of Broadband Top-Loaded Open-Sleeve Monopole, *IEEE APS International Symposium* 2006, 157.7, pp. 635–638.

[86] K. Surabandi and R. Azadegan, Design of an Efficient Miniaturized UHF Planar Antenna, *IEEE Transactions on Antennas and Propagation*, vol. 51, 2003, no. 6, pp. 1270–1276.

[87] N. Begdad and K. Sarabandi, Bandwidth Enhancement and Further Size Reduction of a Class of Miniaturized Slot Antennas, *IEEE Transactions on Antennas and Propagation*, vol. 52, 2004, no. 8, pp. 1928–1934.

[88] P. Xu, K. Fujimoto, and L. Shiming, Performance of Quasi-self-complementary Antenna, *IEEE Antennas and Propagation Society International Symposium*, vol. 40, 2002, pp. 464–467.

[89] Y. Mushiake, *Self-Complementary Antennas*, Springer, 1996, pp. 40–42.

[90] Xu Pu and K. Fujimoto, L-shaped Self-complementary Antenna, *IEEE APS International Symposium*, vol. 3, 2003, pp. 95–98.

[91] R. Azadegan and K. Sarabandi, Bandwidth Enhancement of Miniaturized Slot Antennas Using Folded, Complementary, and Self-Complementary Realization, *IEEE Transactions on Antennas and Propagation*, vol. 55, 2007, no. 9, pp. 2435–2444.

[92] L. Guo, S. Wang, X. Chen, and C. Oarini, A Small Printed Quasi-Self-Complementary Antenna for Ultrawideband Systems, *IEEE Antennas and Wireless Propagation Letters*, vol. 8, 2009, pp. 554–557.

[93] M. C. Buck and D. S. Filipovic, Two-Arm Sinuous Antennas, *IEEE Transactions on Antennas and Propagation*, vol. 56, 2008, no. 5, pp. 1229–1235.

[94] K. G. Schroeder and K. M. S. Hoo, Electrically Small Complementary Pair (ESCP) with Interelement Coupling, *IEEE Transactions on Antennas and Propagation*, AP-24, 1976, no. 4, pp. 411–418.

[95] D-H. Kwon *et al.*, Small Printed Combined Electric-Magnetic Type Ultrawideband Antenna With Directive Radiation Characteristics, *IEEE Transactions on Antennas and Propagation*, vol. 56, 2008, no. 1, pp. 237–241.

[96] M. A. P. Lazzaro and R. Judaschke, A 150-GHz CPW-fed tapered-slot antenna, *IEEE Microwave Wireless Components Letters*, vol. 14, 2004, no. 2, pp. 62–64.

[97] X.-C. Lin and C.-C. Yu, A Dual-band Slot-Monopole Hybrid Antenna, *IEEE Transactions on Antenna and Propagation*, vol. 56, 2008, no. 1, pp. 282–285.

[98] W. S. C. Lin and W. R. Chen, CPW-fed Compact Meandered Patch Antenna for Dual-band Operation, *Electronics Letters*, vol. 40, 2004, no. 18, pp. 1094–1095.

[99] W. S. Chen and K. L. Wong, A Coplanar Waveguide-fed Printed Slot Antenna for Dual-frequency Operation, *IEEE Antennas and Propagation Society International Symposium*, vol. 2, 2001, pp. 140–143.

[100] W. Hong and K. Sarabandi, Low Profile Miniaturized Planar Antenna With Omnidirectional Vertically Polarized Radiation, *IEEE Transactions on Antennas and Propagation*, vol. 24, 1976, no. 4, pp. 411–418.

[101a] J. B. Pendry *et al.*, Low Frequency Plasmons in Thin-wire Structures, *Journal of Physics: Condensed Matter*, vol. 10, 1998, pp. 4785–4809.

[101b] S. Tretyakov, *Analytical Modeling in Applied Electromagnetics*, Artech House, 2003.

[101c] N. Engheta and R. W. Ziolkowsky, *Metamaterials: Physics and Engineering Explorations*, Wiley-IEEE Press, 2006, pp. 88–90.

[101d] W. Rotman, Plasma simulation by artificial and parallel-plate media, *IEEE Transactions on Antennas and Propagation*, vol. 10, 1962, no. 1, pp. 82–95.

[102a] J. B. Pendry *et al.*, Magnetism from Conductors and Enhanced Nonlinear Phenomena, *IEEE Transactions on Microwave Theory and Techniques*, vol. 47, 1999, no. 11, pp. 2075–2084.

[102b] D. R. Smith *et al.*, A Composite Medium with Simultaneously Negative Permeability and Permittivity, *Physical Review Letters*, vol. 84, 2000, no. 18, pp. 4184–4187.

[103a] C. Caloz and T. Itoh, Applications of the Transmission line Theory to Left-handed (LH) Materials, *USNC/URSI*, San Antonio, 2002.

[103b] A. Lai, C. Caloz, and T. Itoh, Composite Right/Left-Handed Transmission Line Metamaterials, *IEEE Microwave Magazine*, September 2004, pp. 34–50.

[103c] A. K. Iyer and G. V. Eleftheriades, Negative Refractive Index Metamaterials Supporting 2D Waves, *Proceedings of IEEE International Microwave Theory and Techniques* Symposium, Seattle, 2002.

[104] C. Caloz and T. Itoh, *Electromagnetic Metamaterials, Transmission Line Theory and Microwave Applications*, Wiley-IEEE Press, 2005.

[105] G. V. Eleftheriades and K. G. Balmain (eds.), *Negative Refraction Metamaterials: Fundamental Principles and Applications*, Wiley-IEEE Press, 2005.

[106] D. Guha and Y. M. M. Antar (eds.), *Microstrip and Printed Antennas*, John Wiley and Sons, 2011, chapter 11.

[107] D. Guha and Y. M. M. Antar, *Microstrip and Printed Antennas*, John Wiley and Sons, 2011, p. 365.

[108] T. Tsutaoka, *et al.*, Negative Permeability Spectra of Magnetic Materials, *IEEE iWAT* 2008, P202, pp. 279–281.

[109] A. Erentok and R. W. Ziolkowsky, Metamaterial-Inspired Efficient Electrically Small Antennas, *IEEE Transactions on Antennas and Propagation*, vol. 56, 2008, no. 3, pp. 691–707.

[110] J. D. Baena, R. Marques, and F. Medina, Artificial Magnetic Metamaterial Design Using Spiral Resonators, *Physical Review B*, vol. 69, 2004, 014402.

[111] C. R. Simovski and S. He, Frequency Range and Explicit Expressions for Negative Permittivity and Permeability for Isotropic Medium Formed by a Lattice of Perfectly Conducting Ω particles, *Physics Letters A*, vol. 311, 2003, p. 254.

[112a] F. Bilotti, A. Alu, and L. Vegni, Design of Miniaturized Metamaterial Patch Antennas with μ–Negative Loading, *IEEE Transactions on Antennas and Propagation*, vol. 56, 2008, no. 6, pp. 1640–1647.

[112b] A. Alu *et al.*, Subwavelength, Compact, Resonant Patch Antennas Loaded with Metamaterials, *IEEE Transactions on Antennas and Propagation*, vol. 55, 2007, no. 1, pp. 13–25.

[113] P. Y. Chen and A. Alu, Sub-Wavelength Elliptical Patch Antenna Loaded with μ–Negative Metamaterials, *IEEE Transactions on Antennas and Propagation*, vol. 58, 2010, no. 9, pp. 2909–2919.

[114] H. R. Stuart and A. Pidwerbetsky, Electrically Small Antenna Elements Using Negative Permittivity Resonators, *IEEE Transactions on Antennas and Propagation*, vol. 54, 2006, no. 6, pp. 1644–1653.

[115] J. Zhu and G. V. Eleftheriades, A Compact Transmission-Line Metamaterial Antenna With Extended Bandwidth, *IEEE Antennas and Wireless Propagation Letters*, vol. 8, 2009, pp. 295–298.

[116] C. Caloz, T. Itoh, and A. Rennings, CRLH Metamaterial Leaky-Wave and Resonant Antennas, *IEEE Antennas and Propagation Magazine*, vol. 50, 2008, no. 5, pp. 26–39.

[117] C. A. Allen, K. M. K. H. Leong, and T. Itoh, 2-D Frequency Controlled Beam-scanning by a Leaky Guided Wave Transmission Line Array, *IEEE International Symposium on Microwave Theory and Techniques* Digest, 2006, pp. 457–460.

[118] T. Kaneda, A. Sanada, and N. Kubo, 2-D Beam Scanning Plane Antenna Array Using Composite Right/Left-Handed Leaky Wave Antennas, *IEICE Transactions on Electronics*, vol. 89, 2006, no. 12, pp. 1904–1911.

[119] F. P. Casares-Miranda, C. Camacho-Ponalosa, and C. Caloz, High-Gain Active Composite Right/Left-Handed Leaky Wave Antennas, *IEEE Transactions on Antennas and Propagation* AP-54, 2006, no. 8, pp. 2292–2300.

[120] F. Qureshi, M. A. Antoniades, and G. V. Eleftheriades, A Compact and Low-Profile Metamaterial Ring Antenna with Vertical Polarization, *IEEE Antennas and Wireless Propagation Letters*, vol. 4, 2005, pp. 333–336.

[121] M. A. Antoniades and G. V. Eleftheriades, A Folded-Monopole Model for Electrically Small NRI-TL Metamaterial Antennas, *IEEE Antennas and Wireless Propagation Letters*, vol. 7, 2008, pp. 425–428.

[122] M. Antoniades and G. V. Eleftheriades, A Broadband Dual-Mode Monopole Antenna Using NRI-TL Metamaterial Loading, *IEEE Antennas and Wireless Propagation Letters*, vol. 8, 2009, pp. 258–261.

[123] J. Zhu, M. A. Antoniades, and G. V. Eleftheriades, A Compact Tri-band Monopole Antenna With Single-Cell Metamaterial Loading, *IEEE Transactions on Antennas and Propagation*, vol. 58, 2010, no. 4, pp. 1031–1038.

[124] P. J. Herritz-Martinez *et al.*, Multifrequency and Dual-Mode Patch Antennas Filled With Left-Handed Structures, *IEEE Transactions on Antennas and Propagation*, vol. 56, 2008, no. 8, pp. 2527–2539.

[125] Y.-S. Wang, M.-F. Hsu, and S.-J. Chung, A Compact Slot Antenna Utilizing a Right/Left-Handed Transmission Line Feed, *IEEE Transactions on Antennas and Propagation*, vol. 56, 2008, no. 3, pp. 675–682.

[126] T. K. Albee, *Broadband VLF Loop Antenna System*, US Patent no. 3953799, April 27, 1976.

[127] S. Koley and J. L. Gautier, Using a Negative Capacitance to Increase the Tuning Range of a Varactor Diode in MMIC Technology, *IEEE Transactions on Microwave Theory and Techniques*, vol. 49, 2001, pp. 2425–2430.

[128] A. Kaya, *et al.*, Bandwidth Enhancement of a Microstrip Antenna Using Negative Inductance as Impedance Matching Device, *Microwave and Optical Technology Letters*, vol. 421, 2004, pp. 476–478.

[129] J. G. Linvill, Transistor Negative Impedance Converter, Proceedings of IRE, vol. 41, 1953, pp. 725–729.

[130] T. Yanagisawa, RC Active Networks Using Current Inversion Type Negative Impedance Converters, *IRE Transactions on Circuit Theory*, vol. 4, 1957, pp. 140–144.

[131] H. Yogo and K. Kato, Cirucuit Realization of Negative Impedance Converter at VHF, *Electronics Letters*, vol. 10, 1974, no. 9, pp. 155–156.

[132] S. E. Sussman-Fort and R. M. Rudish, Non-Foster Impedance Matching for Transmit Applications, *IEEE iWAT*, 2006, pp. 53–56.

[133] S. E. Sussman-Fort, Matching Network Design Using Non-Foster Impedances, *International Journal of RF and Microwave Computer-Aided Engineering*, 2006, no. 16, pp. 135–142.

[134] J. T. Aberle, Two-Port Representation of an Antenna With Application to Non-Foster Matching Networks, *IEEE Transactions on Antennas and Propagation*, vol. 56, 2008, no. 5, pp. 1218–1222.

[135] K.-S. Song and R. G. Rojas, Electrically Small Wire Monopole Antenna with Non-Foster Impedance Element, *EuCAP* 2010, A02-1.

[136] S. Koulouridis and J. L. Volakis, Non-Foster Circuits for Small Broadband Antennas, *IEEE APS International Symposium* 2009, digest, pp. 1973–1976.

[137] A. Kaya and E.Y. Yukel, Investigation of Compensated Rectangular Microstrip Antenna With Negative Capacitor and Negative Inductor for Bandwidth Enhancement, *IEEE Transactions on Antennas and Propagation*, vol. 55, 2007, pp. 1275–1282.

# 8 Design and practice of small antennas II

## 8.1 FSA (Functionally Small Antennas)

### 8.1.1 Introduction

A Functionally Small Antenna (FSA) is an antenna system that has enhanced or improved performances without increasing the antenna dimensions. The FSA is constituted by (1) integrating or combining either radiating or non-radiating components into an antenna system so as to improve or enhance the antenna performance, and (2) adding some function to an antenna so that the antenna will perform with newly added function. The Functionally Small Antenna system is not necessarily dimensionally small; however, it can be referred to equivalently as a small antenna, because enhanced performances or added functions compare to a larger antenna that could not be accomplished otherwise without enlarging the antenna dimensions.

Components to be integrated or combined into an antenna structure are electronic devices (either passive or active). There are many cases where antenna elements, regardless of either linear or planar, are combined with other antenna elements to constitute an integrated antenna. An antenna composed with integrated structure is referred to as an Integrated Antenna System (IAS). An IAS containing active components is referred to as an active IAS.

### 8.1.2 Integration technique

There is a three-fold benefit of the integration techniques: (a) enhancement/improvement of antenna performances; (b) reduction of antenna dimensions; and (c) addition of some functions to an antenna system. The design target pertaining to (a) is enhancement/improvement of bandwidth, including multiband as well as wideband performance, and of gain or efficiency. Factors in (b) lead to miniaturization, and those in (c) include addition of functions such as variable tuning facility, reconfigurable operation as well as amplification, frequency conversion, oscillation, and so forth, all integrated into an antenna system.

#### 8.1.2.1 Enhancement/improvement of antenna performances

When designing a small antenna, to keep the bandwidth at least the same or increase it without degradation in gain and efficiency is the prime factor to consider. The integration technique may be employed as one of the most useful approaches for this purpose.

### *8.1.2.1.1 Bandwidth enhancement and multiband operation*

Fundamentally, there are a few methods to accomplish wideband or multiband performance: (a) increase the radiation modes or number of resonance frequencies; (b) modify the antenna geometry; and (c) change substrate parameters used in the antenna structure. It is also important to design the matching circuitry for wideband operation. For this purpose, feeding structure, ground plane parameters, and application of metamaterials as well as networks consisting of several tuning circuits are considered. Radiation modes can be increased by combining two or more radiation sources that can be implemented by combining, for example, an electric source and a magnetic source, and by increasing the number of currents flowing on the antenna element, which interact with each other to increase the number of resonance frequencies, resulting in wider bandwidth. Wideband performance is also achieved by way of coupling between two resonators; for instance, a patch, which is a resonator by itself, may be coupled with another patch to function over a wider frequency band as a consequence of interaction between two closely separated resonant frequencies. A typical example of combining an electrical source with a magnetic source is a complementary antenna that is implemented by a monopole and a slot. Another example is an antenna of conjugate structure. An increase in the number of currents with different lengths of paths on the antenna element results in increase in the number of resonance frequencies; that produces wideband or multiband operation. For this purpose, an area of the radiator surface is expanded so that the current will flow over a wider area, taking different lengths of paths. A typical example is a bow-tie antenna. The coupling concept is realized by cutting slots or notches (or slits) on the planar antenna surface, by which the current flow will take circuitous routes around the slots or slits, resulting in the wideband or multiband operation. The term wideband is usually defined as the relative bandwidth of greater than 10% in terms of the −10 dB reflection coefficient or VSWR $\leq 2$. Wideband antennas include antennas used in UWB systems, which operate in the frequency range from 3.1 GHz to 10.6 GHz.

There have been numerous papers introducing methods to accomplish wideband characteristics in small antennas. Wideband antennas are implemented practically by modification of antenna geometry, increasing area of planar antenna surface on which currents flow in different paths so that resonance occurs at different frequencies, creating wide bandwidth or producing multi-resonance. Various shaping of metallization such as circular, elliptical, rectangular, even irregular profiles have been explored. Arranging slots or slits (notches) on a planar antenna surface is another useful way to implement wideband or multiband antennas.

#### *8.1.2.1.1.1 Modification of antenna geometry*

The rectangular patch is an ordinal geometry of microstrip antennas. The shape can be modified to rectangle, triangle, circle, ellipse, and other forms such as L, E, and H shapes for obtaining wideband performance.

##### *8.1.2.1.1.1.1 E-shaped microstrip antenna*

For the purpose of bandwidth enhancement, two slots are incorporated in parallel into the antenna patch to form an E-shape, as shown in Figure 8.1, which illustrates (a) top view

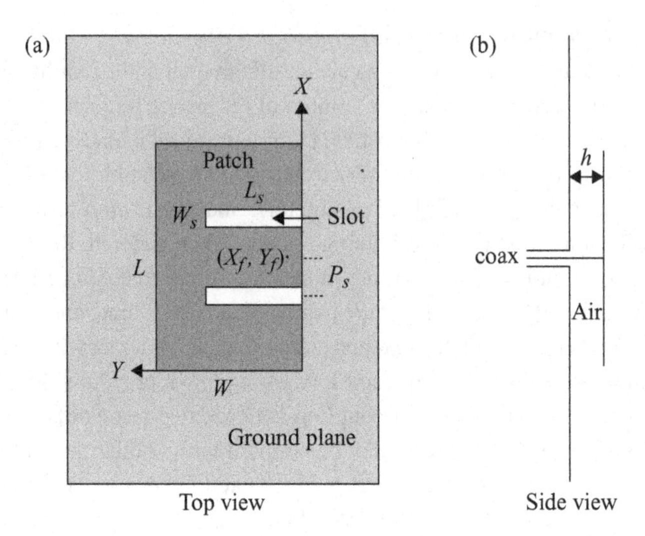


**Figure 8.1** Geometry of a wideband E-shaped patch antenna: (a) top view and (b) side view ([1], copyright © 2001 IEEE).

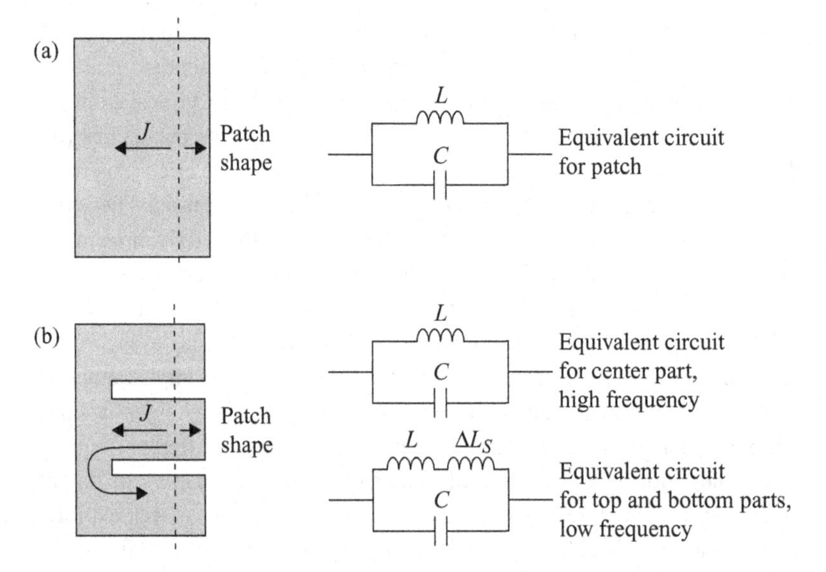


**Figure 8.2** Principle of wideband operation: (a) an ordinary rectangular patch antenna, (b) an E-shaped patch antenna and two equivalent circuits, one for the center part and another for two side parts of the patch [1].

and (b) side view [1]. In the figure, antenna geometrical parameters are also provided; for the patch, the length $L$, the width $W$, and the height $h$, and regarding the slot, the length $L_s$ and the width $W_s$, and the position $P_s$, distance from the center of the patch, and the feed position $(X_f, Y_f)$. The principle of the wideband operation is explained by using a model shown in Figure 8.2, in which (a) an ordinary rectangular patch and the equivalent LC circuit are shown for a comparison, along with (b) an E-shaped patch and the two equivalent LC circuits; one for the center part of the patch and another for the two side parts of the E-shaped patch. On the ordinary rectangular patch, current $J$ flows from the feed point to the top and bottom edges of the patch and these current path lengths determine the inductance $L$ and the capacitance $C$, as Figure 8.2(a) shows. On the other hand, a current on the E-shaped patch flows on the center part of the E-shape like an ordinary patch and other current flows into side parts of the E-shape taking the roundabout path, as Figure 8.2(b) shows. The center part of the E-shape patch can be modeled by a simple LC resonant circuit like the ordinary patch circuit, while the side part of the E-shape patch is modeled by a circuit with an additional series inductance $\Delta L_s$ to $L$ as shown in Figure 8.2(b). Hence, the equivalent circuit of the side part resonates at lower frequency than that of the center part. This means that a model of an E-shaped patch antenna changes from a single-resonance circuit to a dual-resonance circuit. Thus, coupling together of these circuits leads to wideband performance.

Essential parameters in the design of an E-shaped patch antenna are slot length $L_s$, slot position $P_s$, and slot width $W_s$. The slot length $L_s$ is the important factor to characterize the resonance frequency; when the $L_s$ is too short, the antenna has only one resonance frequency. As $L_s$ increases, another lower resonance frequency appears. The longer $L_s$ becomes, the lower the second resonance frequency will be. $P_s$ is a parameter that can adjust matching conditions properly. When $P_s$ becomes somewhat larger, the two resonance frequencies become distinct and a wideband match is obtained. However, as $P_s$ further increases, the $S_{11}$ between two resonance frequencies is larger than –10 dB. Then the antenna does not perform as a wideband antenna, but as a dual-band antenna. The third parameter $W_s$ is a useful parameter to adjust coupling and achieve good matching.

An E-shaped patch antenna was developed for use in wireless communications operating at 1.9–2.4 GHz [1]. The antenna geometrical parameters are $L = 70$, $W = 50$, $h = 15$, $X_f = 35$, $Y_f = 6$, $L_s = 40$, $W_s = 6$, and $P_s = 10$, where dimensions are all in mm. Calculated and measured $S_{11}$ are shown in Figure 8.3, in which $S_{11}$ of two other ordinary patch antennas (without slots) are provided for comparison. They have the same height and width as the E-shaped patch antenna, but the narrow patch antenna is one having the same length as the middle part of the E-shaped patch, and has a bandwidth about only 10%, while the wide patch is one with the same length as the E-shaped patch antenna and does not match well to 50 Ω. On the contrary, the E-shaped patch antenna has shown resonance at two frequencies, 1.9 GHz and 2.4 GHz, and the bandwidth is 30.3%.

It should be noted that the ground plane (GP) often becomes a critical factor when a platform to mount an antenna is small. In this example [1], the size of GP is chosen to be $14 \times 21$ cm, which is about eight times that of the patch size. By observing the current distributions on the GP obtained by calculation, most currents flow under the patch, but a larger area is required for lower frequencies than higher frequencies. It was concluded that a $1\lambda$ (1.9 GHz) square plate is sufficient for proper operation of the antenna.

Directivity of the antenna is calculated to be 8.5 dB at 2.4 GHz and 6.7 dB at 1.9 GHz. The measured radiation patterns agree well with the calculated results. In the E-plane, the 3-dB beamwidth is 42° at 2.4 GHz and 63° at 1.9 GHz. The peak cross polarization at 2.4 GHz is –25 dB, and is lower than –15 dB at 1.9 GHz. In the H-plane, the radiation patterns are similar to those of the E-plane, and the 3-dB beamwidth is 60° at both

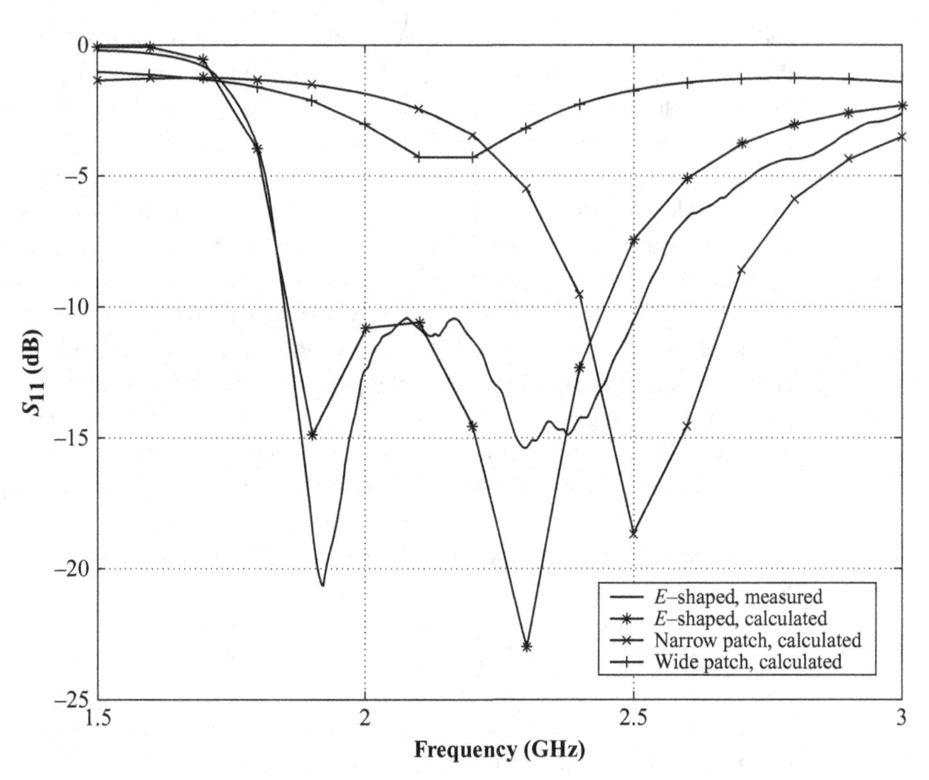


**Figure 8.3** $S_{11}$ of the E-shaped patch antenna, compared with simple patch without slots ([1], copyright © 2001 IEEE).

frequencies. The peak cross polarization is –7 dB at 50°, a rather high level that is due to the leaky radiation of the slots.

Another example shown in [1] is a wider band that was achieved by adjusting the parameters of the slots and the feeding point of an E-shaped patch antenna. The antenna has two distinct resonant frequencies at 2.12 GHz and 2.66 GHz and the frequency band ranges from 2.05–2.64 GHz, resulting in bandwidth as wide as 32.3%.

Another example of an E-shaped patch antenna introduced in [2] is designed to operate in the 800 MHz band and has wide bandwidth of 164 MHz. In order to obtain wider bandwidth, an LC circuit comprised of a small spiral, which produces an inductance, and a circular plate, which introduces a capacitance to compensate the inductance of the spiral and the feed line, thus producing a new resonance, is integrated into the feed circuit. This resonance frequency is very close to that of the E-shaped patch and thus wider bandwidth can be achieved.

An E-shaped patch antenna radiates a linearly polarized wave when currents on both side parts of the E-shaped surface are the same; however, when they are different as shown in Figure 8.4 [3], elliptically or circularly polarized (CP) waves are radiated. As can be observed in the figure, when the slot length varies, currents flowing on both sides of the E-shaped surface become unbalanced. When by this means, the $x$-directed current is made equal in magnitude to the $y$-directed current with a 90-degree phase difference,

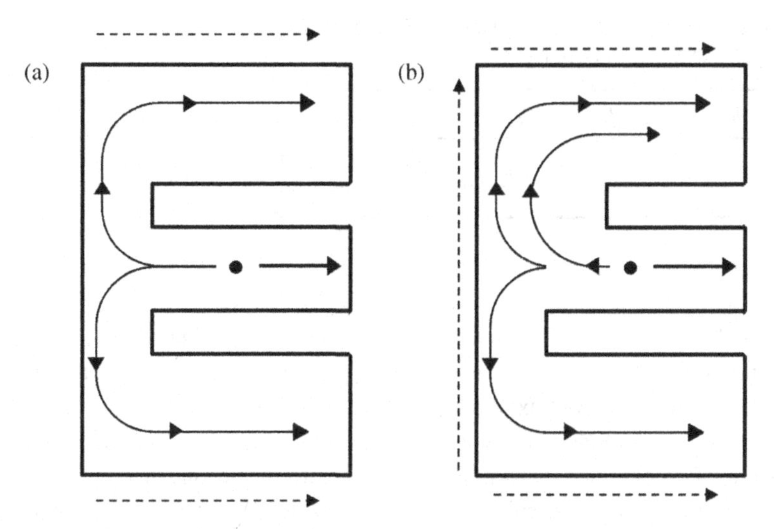


**Figure 8.4** Current flow across the E-shaped patch: (a) symmetrical flow and (b) asymmetrical flow ([2, 3], copyright © 2010 IEEE).

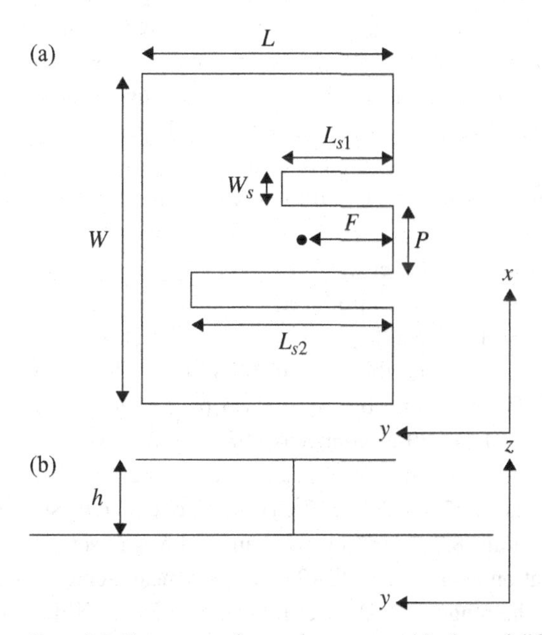


**Figure 8.5** Geometry of a patch antenna with slots of different length: (a) top view and (b) side view ([2, 3], copyright © 2010 IEEE).

an E-shaped antenna radiates a CP wave. The parameters to be considered in the design of a CP antenna are: each length of its two slots, $L_{s1}$ and $L_{s2}$; the slot width $W_s$; the separation $P$ between two slots and the location of the feed point $F$ with a given patch size; length $L$; and width $W$. The antenna geometry is shown in Figure 8.5, (a) top and (b) side views.

**Table 8.1** Dimensions of circularly polarized E-shaped patch [mm] ([3], copyright © 2010 IEEE)

| $\varepsilon_r$ | $h$ | $W$ | $L$ | $W_s$ | $L_{s1}$ | $L_{s2}$ | $P$ | $F$ |
|---|---|---|---|---|---|---|---|---|
| 1 | 10 | 77 | 47.5 | 7 | 19 | 44.5 | 14 | 17 |
| *2.2* | *6.7* | *63* | *33.5* | *4* | *27* | *6* | *20* | *10* |

The design procedure starts with an E-shape having the same length $L_{s1}$ and $L_{s2}$, then shortens either $L_{s1}$ or $L_{s2}$ to obtain circular polarization. Next change the probe position $F$ to have improved AR (axial ratio) and $S_{11}$, and align $L_{s2}$ or $L_{s1}$ for fine enhancement of AR as well as alignment with the $S_{11}$ band. Finally, the design procedure ends when desired CP is obtained, otherwise, iterate back to the second step and repeat the steps again to satisfy the requirements. In the second step, make $L_{s1}$ shorter with $L_{s2}$ fixed, if left-handed (LH) CP is required, whereas if right-handed (RH) CP is desired, follow the opposite procedure. When making $L_{s1}$ shorter, the resonance frequency becomes higher and at the same time AR decreases from larger values showing linear polarization to smaller values to show CP, as the $L_{s1}$ changes electrical path length of the $x$-directed current. In this process, there is a state where $L_{s1}$ takes optimum values for the minimum AR as the phase difference between orthogonal currents approaches 90 degrees. Beyond this critical value, the phase difference between the orthogonal currents departs from 90 degrees. The design goal is to achieve desired AR bandwidth, which is defined as the frequency range across the 3-dB AR level, and to obtain required impedance and AR bandwidth together.

A CP E-shaped patch antenna to cover the IEEE 802.11b/g band (2.4–2.5 GHz) is proposed and antenna performances are introduced [3]. Dimensions of the antenna are provided in Table 8.1 (in mm). The metallic patch is fabricated via milling a thin copper-clad substrate, which has $\varepsilon_r = 2.2$ and thickness of 0.787 mm. The patch is mounted above a ground plane of 200 × 95 mm by using a via of 0.75 mm radius, which is connected to the center pin of the 50-Ω connector. Simulated and measured $S_{11}$ are shown in Figure 8.6. The measured impedance bandwidth is 9.27% (2.34–2.57 GHz) while simulated bandwidth is 10.27% (2.31–2.57 GHz). AR obtained by simulation and measurement is given in Figure 8.7, by which AR bandwidth is observed to be 8.1% (2.28–2.45 GHz) in simulation, while 16% (2.3–2.7 GHz) in measurement. Bandwidths for $S_{11}$ and AR overlap in the range 2.34–2.57 GHz (9.27%). The maximum measured gain is 8.3 dBi with 15.5% 3-dB bandwidth. This can be observed in Figure 8.8, which depicts gain with respect to frequency. The bandwidth common to all of AR, gain, and $S_{11}$ is 9.27%.

The E-shaped patch antenna may have great potential to be applied in various purposes such as reconfigurable antenna, diversity antenna, etc. For example, implementation of a reconfigurable antenna becomes feasible by inserting RF switches in appropriate locations across each slot and turning them on or off to make the two slot lengths unequal, thereby producing CP radiation as a consequence.

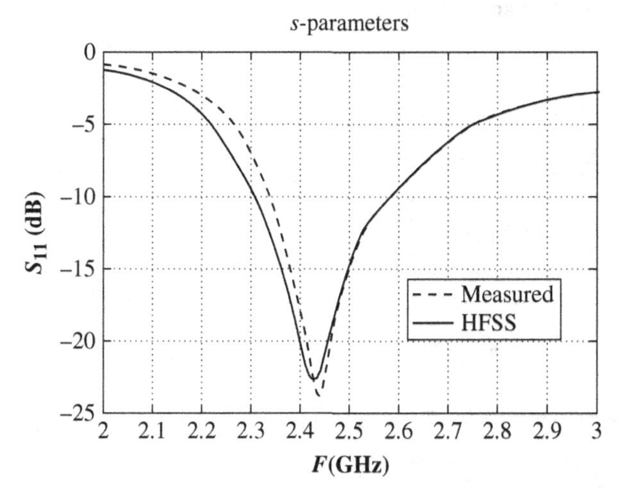


**Figure 8.6** Simulated and measured $S_{11}$ of the CP E-shaped patch antenna ([3], copyright © 2010 IEEE).

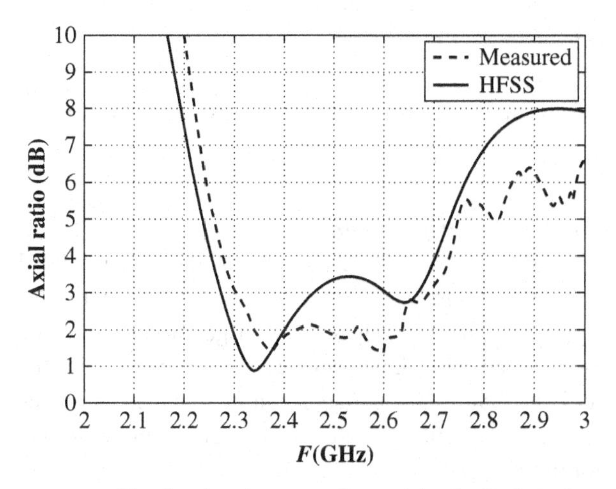


**Figure 8.7** Simulated and measured AR of the CP E-shaped patch antenna ([3], copyright © 2010 IEEE).

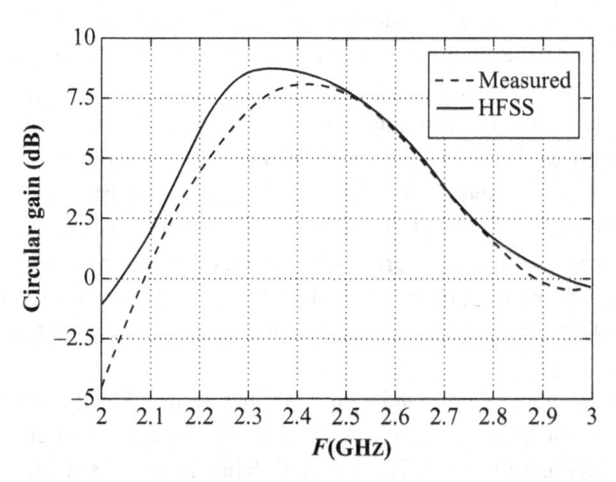


**Figure 8.8** Simulated and measured gain of the E-shaped patch antenna ([3], copyright © 2010 IEEE).

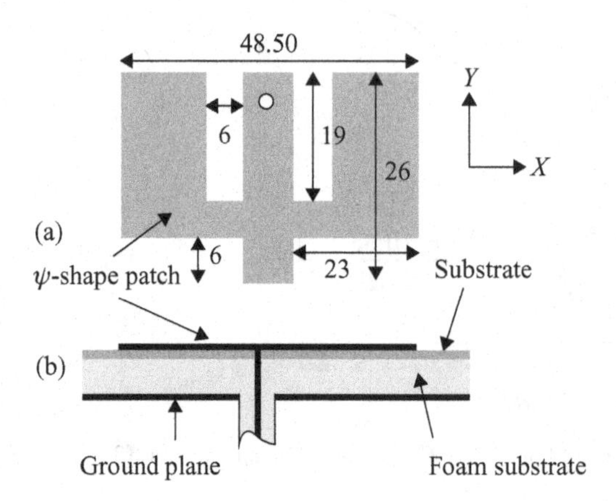


**Figure 8.9** A $\psi$-shape patch antenna modified from an E-shape patch antenna ([4a, 4b], copyright © 2009 and 2007 IEEE).

It can be also said that the E-shaped patch antenna is attractive for use in modern wireless systems.

### *8.1.2.1.1.1.2 $\psi$-shaped microstrip antenna*

An E-shape can be modified to a $\psi$-shape by rotating the E-shape 90 degrees and removing the bottom side out slightly, excepting the center part. A resultant $\psi$-shape patch is shown in Figure 8.9 [4a, 4b], where antenna geometry is illustrated; (a) top view and (b) side view, along with dimensions of slots in mm. With this structure, wider bandwidth over that of an E-shaped patch can be obtained as a consequence of improvement in the current distributions on the patch that attributes to removal of the bottom side conductor in the E-shaped patch.

The removed part at the bottom side of the patch acts in an important role for controlling the current distributions on the patch so that the bandwidth is increased. The $\psi$-shaped patch is etched on the substrate ($\varepsilon_r = 2.50$, $\tan\delta = 0.002$) of thickness $h = 0.33$ mm placed on the foam substrate ($\varepsilon_r = 1.06$, $\tan\delta = 0.0002$) of thickness $h = 6.0$ mm. Dimensions of the original rectangular patch are 48.5 mm in width and 26 mm in length, respectively, including two slots 6 mm wide and 19 mm long cut symmetrically around the patch center, and two conductor-removed parts at the bottom side of the patch, 6 mm long and 23 mm wide, resulting in a tail part. The ground plane size is 75 mm $\times$ 75 mm, which is determined by the parametric study.

The simulated and measured $S_{11}$ are shown in Figure 8.10, by which the simulated bandwidth is observed to be 55.02% (4.15 to 7.30 GHz) and measured bandwidth is 53.60% (4.10 GHz to 7.10 GHz), except that an $S_{11}$ of –9 dB is seen near 4.75 GHz.

These results show much wider bandwidth than that of an E-shaped patch antenna, introduced in the previous section as 32% [1]. A small discrepancy is observed between the simulated and measured results possibly due to fabrication error, mainly stemming

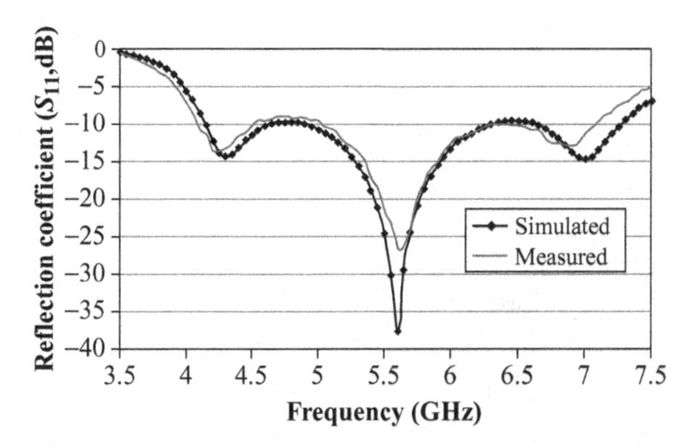


**Figure 8.10** Simulated and measured $S_{11}$ ([4a], copyright © 2009 IEEE).

from the difference in the foam thickness, increased from 6 mm to 6.4 mm, that is determined by the simulation, assuming somewhat different structures from the practical foam substrate. However, it can be said that they are in reasonable agreement, notwithstanding that discrepancy.

*8.1.2.1.1.1.3 H-shaped microstrip antenna*

For exciting a circular polarization (CP), another shape of patch antenna, an H-shaped one, is proposed [5]. The antenna geometry is shown in Figure 8.11 with top and side views. The antenna is a single H-shaped copper plate, 0.5 mm thick, suspended by using three nylon bolts over a very thin dielectric layer, having $\varepsilon_r = 3.38$, $\tan\delta = 0.0021$, and thickness 0.2 mm, which is identified as the feed layer, where a horizontal strip line is printed on this layer to feed the patch. The H-shape is formed from a square, on which two rectangular slots are cut away from the top and bottom sides of the square. The square has side $S = 37$ mm (0.30 $\lambda$ at 2.45 GHz) and the rectangular slot, which forms an H-shape, has size $a = 13$ and $b = 14$ mm, respectively. Generally, to excite a CP wave, a square patch is fed at a point on the diagonal line. With a small perturbation on the two sides of a square, CP can be excited, although with a somewhat narrow band of operation, whereas with a large rectangular perturbation a wideband CP operation can be produced.

To excite a left-hand-sense CP wave, the H-shape patch is fed through a probe placed at a point on the diagonal line of the square where a printed monopole is placed at 45 degrees with respect to the axis. The printed monopole is arranged toward the patch corner instead of the center, and used to compensate the feed inductance due to the long probe, thereby making better impedance matching easy. The patch is placed at height $H = 32$ mm (0.26$\lambda$) from the ground plane and the feed layer is placed at height $F_v =$ 24 mm, which equals the length of the vertical probe. The width and length of the horizontal monopole are $F_w = 2$ and $F_H = 5$ (in mm), respectively.

Measured return loss and boresight axial ratio (AR) are shown in Figure 8.12, where simulated results are also given for a comparison. It shows that bandwidth in terms of

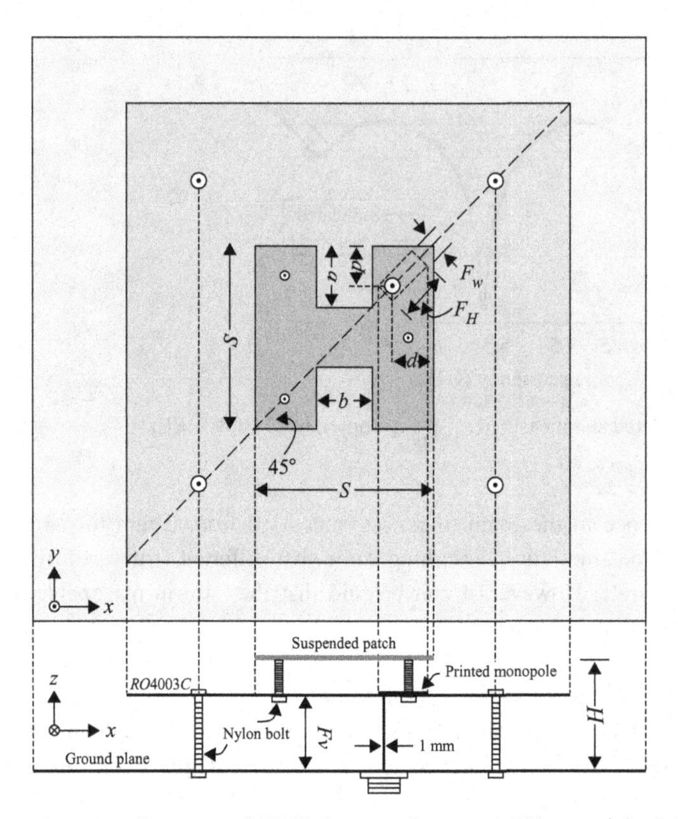


**Figure 8.11** Geometry of CP H-shape patch antenna ([5], copyright © 2010 IEEE).

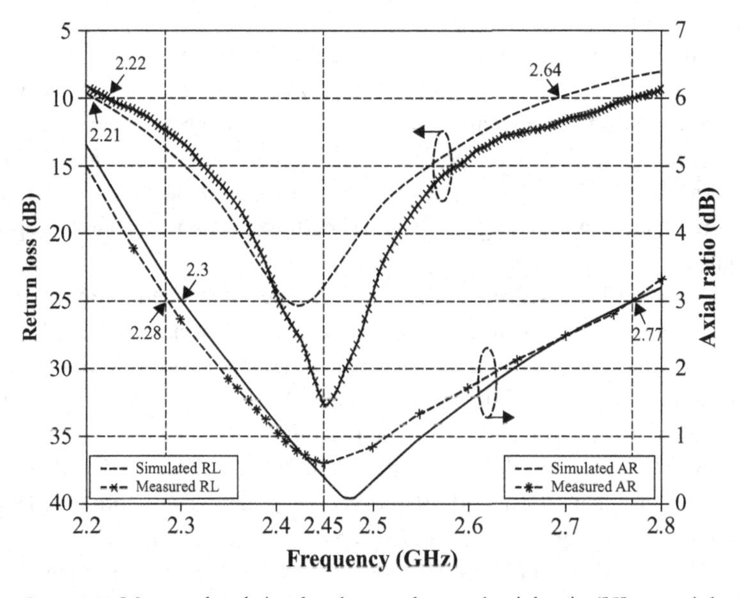


**Figure 8.12** Measured and simulated return loss and axial ratio ([5], copyright © 2010 IEEE).

10-dB return-loss is 2.22–2.64 GHz (17.2%) in the measurement and is 2.21–2.77 GHz (22.5%) in simulation. The 3 dB axial ratio bandwidth is 2.3–2.77 GHz (18.5%) in simulation and 2.28–2.77 GHz (19.8%) in the measurement. These wideband characteristics are attributed to the inclusion of a thick air substrate ($0.26\lambda$) in the design. This discrepancy between measured and simulated results is possibly attributed to the practical structures, which are not included in the simulation, and to fabrication tolerances. By using thick air substrate and low-loss mechanisms, the antenna can show the maximum gain of 5.7 dBic and high efficiency of 95% at 2.45 GHz.

#### *8.1.2.1.1.4 S-shaped-slot patch antenna*

A dual-band single-feed circularly polarized S-shape-slotted patch antenna is proposed in [6]. The antenna geometry is depicted in Figure 8.13, which shows (a) side view, (b) S-shape-slotted patch radiator, and (c) aperture-coupled feeding structure. An S-shaped slot is embedded at the center of the patch surface and fed by a microstrip line located underneath the center of the coupling aperture ground plane. The frequency ratio of the two frequencies can be controlled by adjusting the length of the S-shaped slot and attained finally to be 1.28. The measured 10-dB return loss bandwidth for the lower and higher bands are 16% (1.103–1.297 GHz) and 12.5% (1.568–1.577 GHz), respectively. Measured and simulated return loss are illustrated in Figure 8.14. The measured axial ratio bandwidth is 6.9% for the lower band (1.195–1.128 GHz) and 0.6% for the higher band (1.568–1.577 GHz). The measured gain is observed more than 5 dBc both lower and higher bands. Measured and simulated gain and axial ratio are shown in Figure 8.15(a) and (b), respectively. The overall antenna size is $0.46\lambda_0 \times 0.46\lambda_0 \times 0.086\lambda_0$ at $f_0 =$ 1.2 GHz.

The antenna exhibits good impedance matching, high gain, small dual-band frequency ratio, and wide CP (Circular Polarization) beamwidth. The antenna is suitable for GPS applications and array design.

#### *8.1.2.1.1.2 Use of slot/notch (slit) on planar antenna surface*

Embedding a slot/notch (slit) on a planar antenna is quite beneficial to implement a small, compact, low-profile, wideband antenna, yet with enhanced performances such as higher gain, circular polarization, and so forth. The shape, number, and location of slots or notches, are essential parameters in designing a wideband antenna, and in turn, the type, size, shape, and dimensions are the essential antenna parameters. The most popular types of antenna are Planar Inverted-L antenna (PILA), Planar Inverted-F antenna (PIFA), and various shapes of microstrip antenna (MSA) such as rectangle, triangle, circle, and so forth. Slots/notches may be embedded also on the ground plane as well as on the radiating surface of the patch to reduce the antenna size and generate multiband operation.

In a practical example, small mobile terminals employ slots/notches (slits) on a planar antenna (a PIFA is the most representative type of antenna) to enhance the bandwidth or increase the number of operating frequency bands without increasing the antenna dimensions, keeping the antenna structure compact. Modified PIFAs have been observed, on which slots or slits are embedded to the point where the structure is transformed to

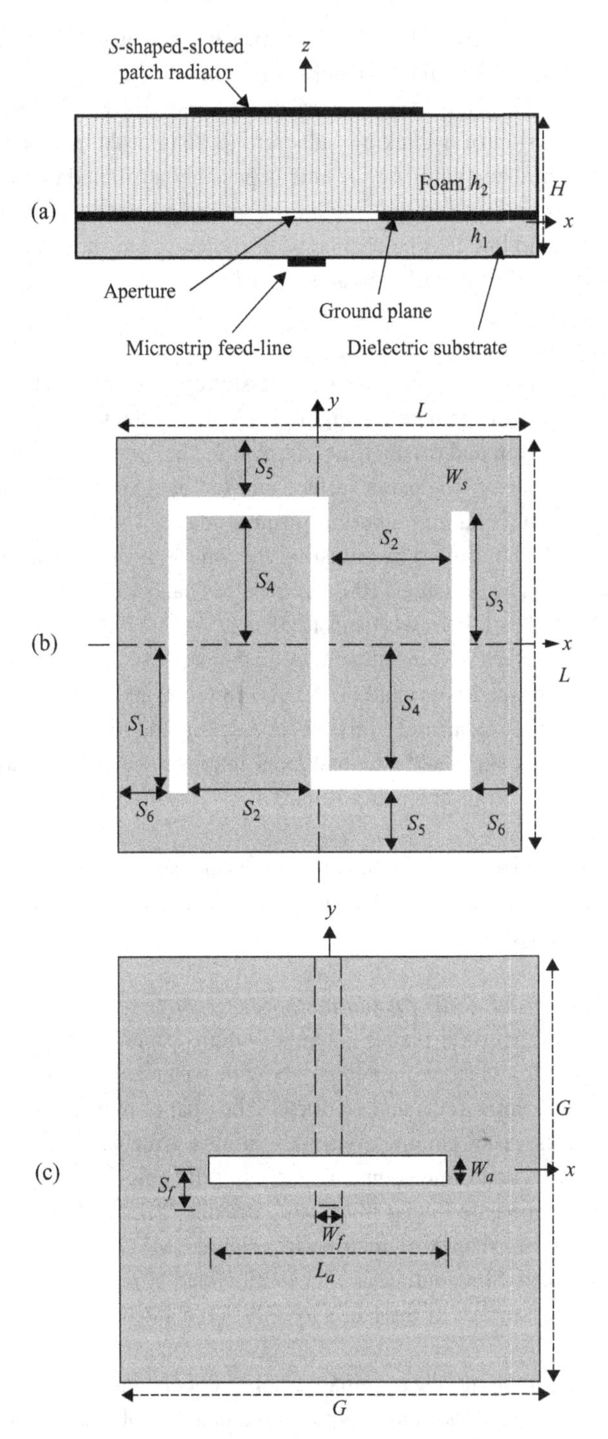


**Figure 8.13** Geometry of S-shaped-slot antenna: (a) cross sectional view, (b) S-shaped-slotted patch radiator, and (c) aperture-coupled feeding structure ([6], copyright © 2010 IEEE).

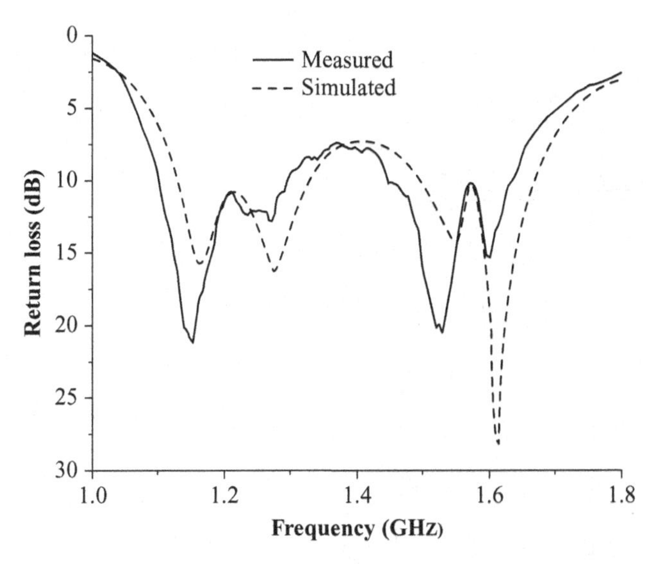


**Figure 8.14** Measured and simulated return loss ([6], copyright © 2010 IEEE).

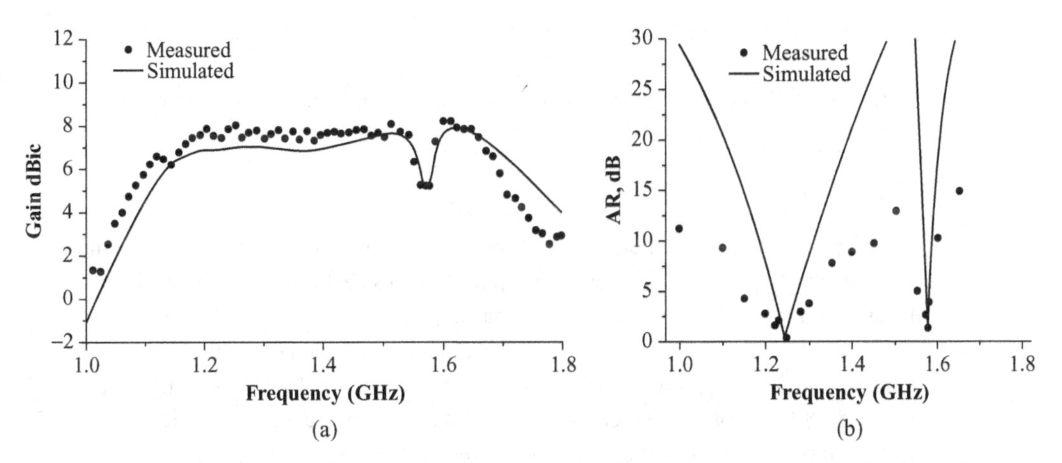


**Figure 8.15** (a) Simulated and measured gain at the boresight and (b) measured and simulated axial ratio at the boresight ([6], copyright © 2010 IEEE).

be so complicated that it may hardly be recognized as a PIFA. Use of slots or slits is, however, considered as the best means to realize a multiband antenna that accommodates the need of the recently deployed wireless communication systems, even with small size that can be fit within the extremely limited space in the small equipment.

*8.1.2.1.1.2.1 Microstrip slot antennas*

Microstrip monopole slot antennas with different shapes such as straight, L, and T, have been investigated and demonstrated that a wide bandwidth could be obtained by proper

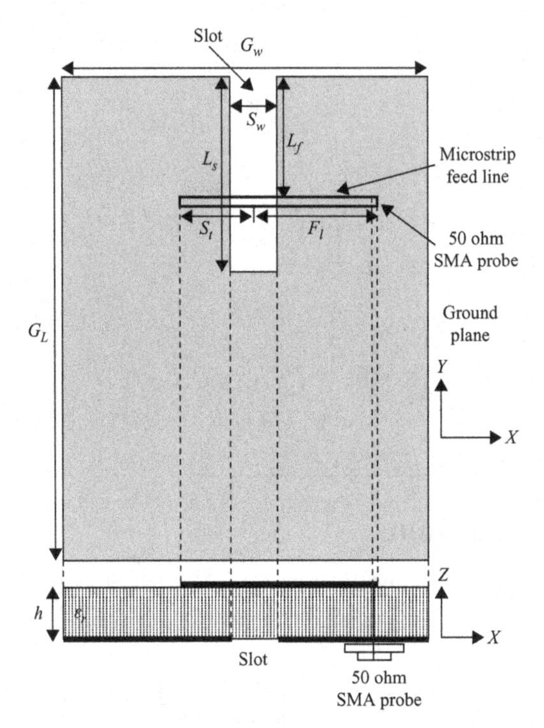


**Figure 8.16** Geometry of microstrip monopole slot antenna on a small ground plane fed by a microstrip line ([7a], copyright © 2005 IEEE).

selection of antenna parameters, length of feed line, position of the feed over the slot, and stub length [7a]. These antennas are placed on a small ground plane, which is about the size of a typical PC wireless adapter card.

Geometries of microstrip monopole slot antennas on a small ground plane fed with a microstrip line are illustrated in Figure 8.16 (straight), Figure 8.17 (L-shape) and Figure 8.18 (T-shape), in which dimensional parameters are also given. In Figure 8.17, three different feeding shapes are shown. The size of the ground plane is width $G_W = 50$ and length $G_L = 80$ (in mm). The slot in each antenna is open at its end, positioned at the center of the narrow ground plane edge, and fed with a 50-Ω microstrip transmission line.

The straight slot has length $L_s = 30$, width $S_w = 30$, and length of feed lines $F_1 = 24$, and $S_1 = 11$. In the optimized design, the substrate with $\varepsilon_r = 2.5$ and $\tan \delta = 0.001$, having thickness $h = 1.57$ is used (measurements in mm). The simulated return loss (–10 dB) bandwidth of this antenna is 58.8% (2.46 GHz to 4.51 GHz) as can be observed in Figure 8.19.

Regarding the L-slot antenna, simulated return loss bandwidth is shown in Figure 8.20, where comparisons depending on the feed shape, straight, inclined, and bent, are provided. The geometrical parameters of the antennas fed with different shapes of feed line, respectively, are $L_1 = 18.5$, $L_2 = 11.5$, $S_w = 7$, $G_w = 50$, $G_L = 80$ as

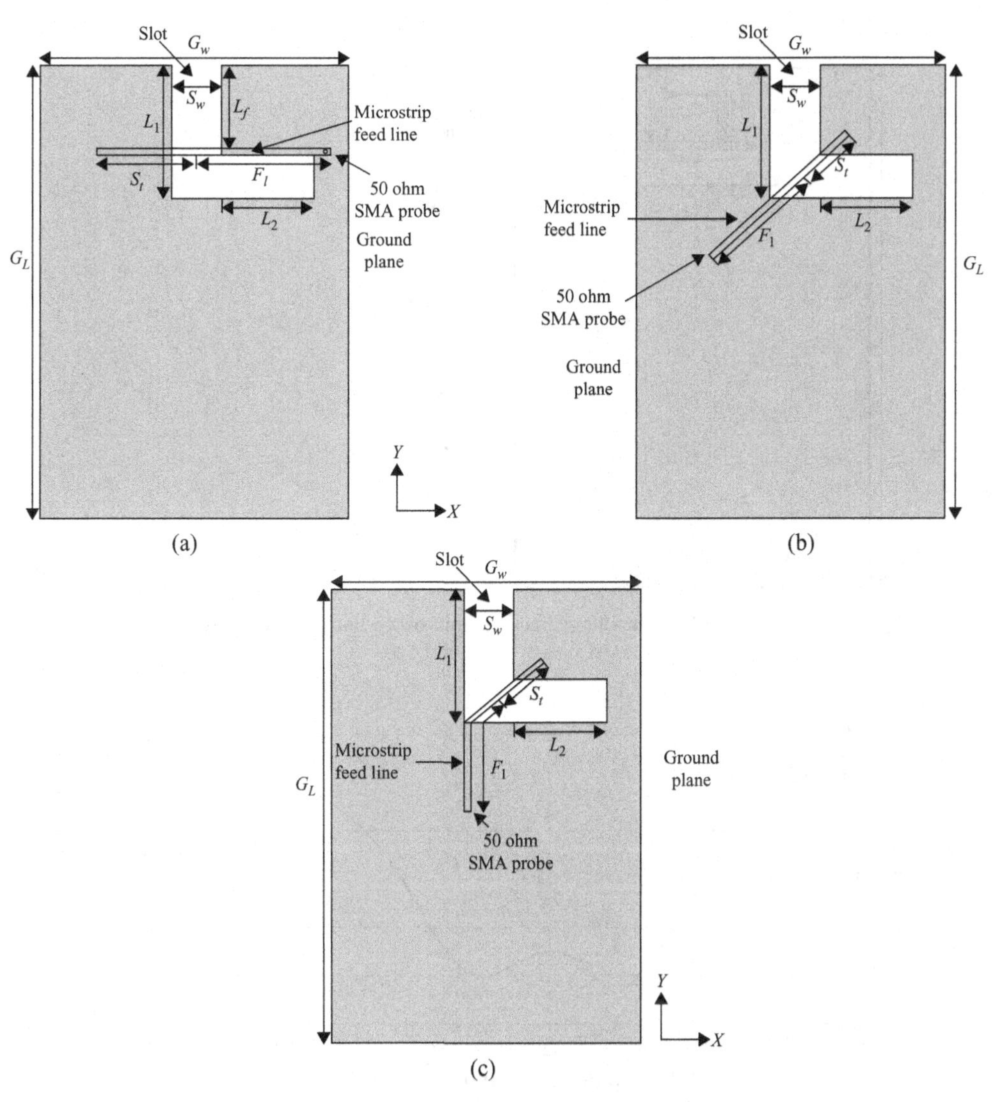


**Figure 8.17** Geometry of L-slot antenna fed by a line with different shape: (a) straight, (b) inclined, and (c) bent ([7a], copyright © 2005 IEEE).

common dimensions and $L_f = 11$, $F_t = 18.5$, $S_t = 9$ in the straight feed line, $F_t = 16.5$, $S_t = 7.5$ in the bent feed line, and $F_t = 17$, $S_t = 7.5$ in the inclined feed line. The substrate used is common to all antennas, with $\varepsilon_r = 4.5$ and $\tan\,\delta = 0.002$, having thickness $h = 0.81$ (in mm). The impedance bandwidths of antennas with straight feed line, bent feed line, and inclined feed line, respectively, are 82% (2.24 GHz to 5.36 GHz), 75.7% (2.48 GHz to 5.5 GHz) and 75.7% (2.47 GHz to 4.97 GHz). By replacing the substrate with a low loss one, with $\varepsilon_r = 2.5$ and $\tan\,\delta = 0.001$, having thickness $h = 0.79$, an improved bandwidth of 82% (2.42 GHz to 5.78 GHz) can be obtained.

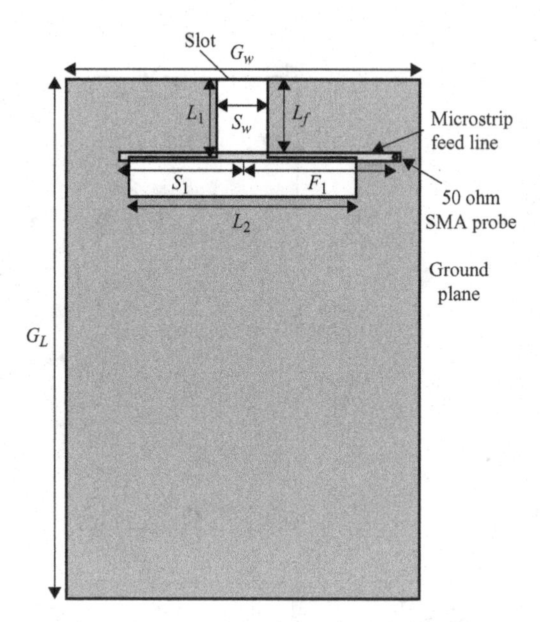


**Figure 8.18** Geometry of T-slot antenna fed by a microstrip line ([7a], copyright © 2005 IEEE).

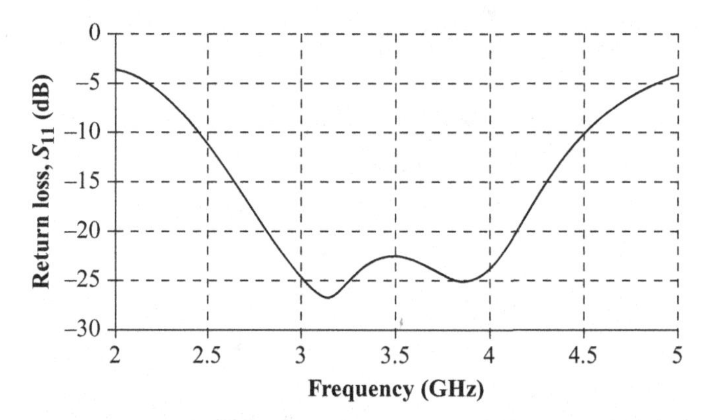


**Figure 8.19** Simulated return loss of the antenna shown in Figure 8.16 ([7a], copyright © 2005 IEEE).

With the T-slot, having $S_w = 7$, $L_f = 75$, $F_l = 23.5$, $S_t = 10$, $L_1 = 7$, $L_2 = 23$ mm, using the same substrate as the other slot-type antennas, simulated return loss is depicted in Figure 8.21. The impedance bandwidth of 78% (2.56 GHz to 5.8 GHz) is obtained, while 80% (2.74 GHz to 6.4 GHz) resulted with low-loss substrate.

A novel compact open-end slot antenna fed by a short-circuited microstrip line is introduced in [7b]. The antenna has reduced dimensions by virtue of placing two matching capacitors at both sides of an electrically small open-end slot. With the proposed configuration, the antenna resonance frequency is 2.45 GHz, having the slot of

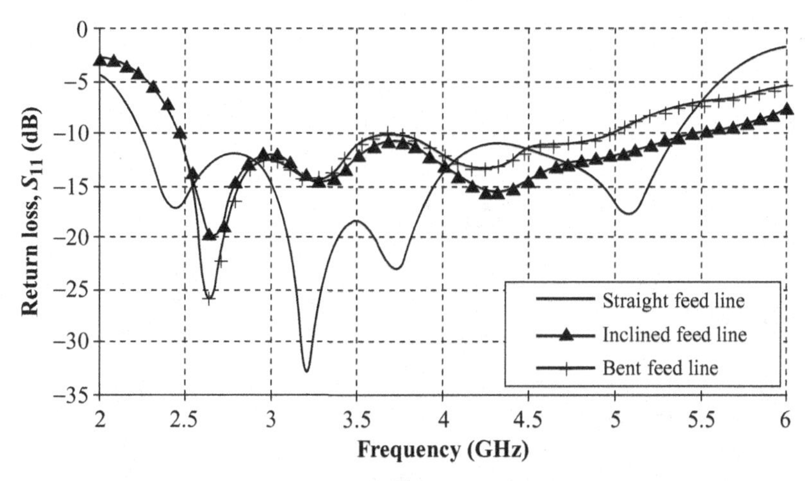


**Figure 8.20** Simulated return loss of the L-shape antenna shown in Figure 8.17 ([7a], copyright © 2005 IEEE).

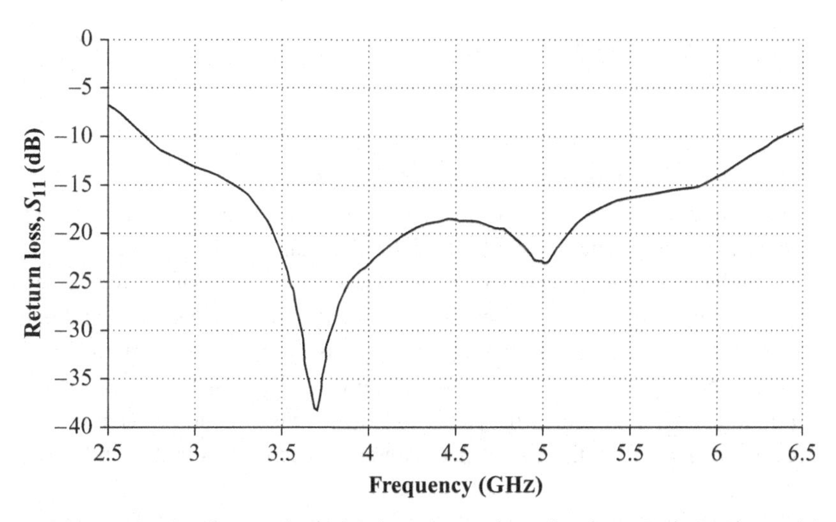


**Figure 8.21** Simulated return loss of T-slot antenna shown in Figure 8.18 ([7a], copyright © 2005 IEEE).

9 mm long and 1.5 mm wide, originally a 4.8 GHz antenna in a conventional design. The equivalent circuit model is developed, and input impedance of the antenna for variations in the feed position is calculated. The results of calculation agree well with full wave simulation. By adding two capacitances to the slot for cancelling the inductive reactance of the slot at lower frequencies, the resonance frequency can be lowered. The antenna length becomes shorter than one-eighth the wavelength at the operating frequency. The measured bandwidth of 109 MHz at 2.45 GHz and peak gain of 1.89 dBi demonstrate the performance of the proposed small-sized slot antenna design.

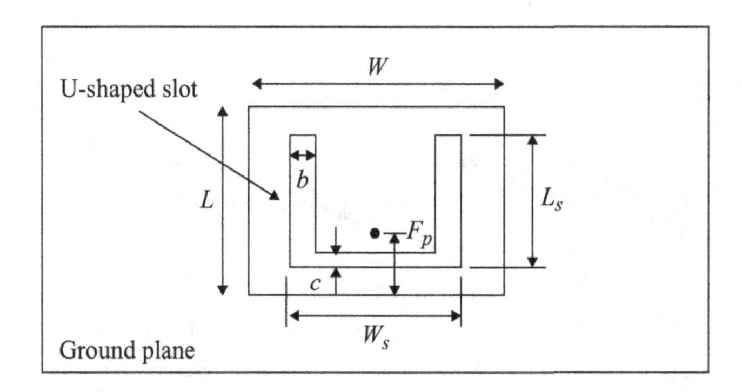


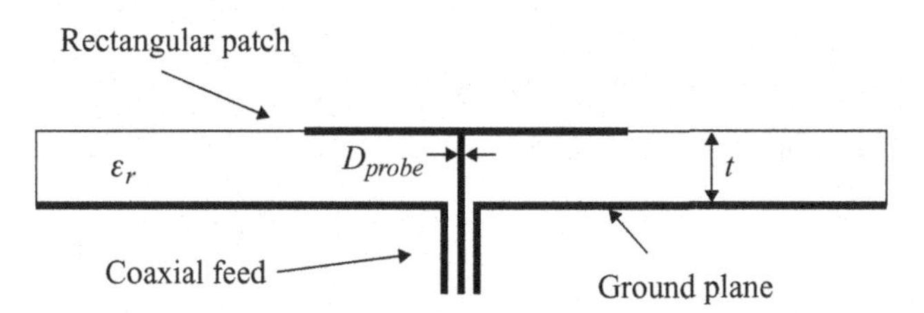


**Figure 8.22** The basic geometry of U-slot patch antenna ([8b], copyright © 2003 IEEE).

*8.1.2.1.1.2.2 Microstrip patch antennas*

*8.1.2.1.1.2.2.1 Patch antennas with U-slot*

A patch antenna with a U-slot was introduced in 1995 [8a] as an antenna capable of providing a wide impedance bandwidth in a range of 25–30%. Since then, a number of studies followed and recent interest has arisen to design a U-slot patch antenna for multiband operation and circular polarization as well as wideband operation. Size reduction is also a significant requirement, as the resonance length of a conventional patch antenna was about a half-wavelength, which was too large for small wireless portable devices. To meet the requirement, various techniques have been developed to decrease the size of the U-slot patch antenna, keeping its wide bandwidth unchanged. Typical examples are use of an L-shaped probe to feed [8b, 8c], a shorting wall [8b, 8d], a shorting pin [8b, 8e], or reduction of a U-shape to a half-U-shape [8b, 8f, 9], which corresponds to an L-shape.

The basic geometry of a U-slot antenna is illustrated in Figure 8.22, where dimensional parameters are given [8b]. Basically, the slot dimensions determine the resonance frequency as well as the patch dimensions. By appropriately designing slot and patch dimensions, wideband or multiband resonance can be attained. The U-slot cancels out the feed inductance and currents around the U-slot create additional resonance that makes the antenna broadband as a result of combination with the resonance of the patch.

A reduced-size U-slot patch antenna of the basic structure (Figure 8.22) by increasing the substrate $\varepsilon_r$ is introduced. Design parameters are provided in Table 8.2, which shows cases for three values of $\varepsilon_r$, including an air substrate case $\varepsilon_r = 1$. The operating

**Table 8.2** Design parameters for 900 MHz U-shaped patch antenna [cm] ([8b], copyright © 2000 IEEE)

| | $\varepsilon_r = 1.0$ | $\varepsilon_r = 2.33$ | $\varepsilon_r = 4.0$ |
|---|---|---|---|
| $W$ | 21.97 ($0.659\lambda_0$) | 12.40 ($0.372\lambda_0$) | 9.29 ($0.201\lambda_0$) |
| $L$ | 12.45 ($0.374\lambda_0$) | 8.96 ($0.269\lambda_0$) | 6.71 ($0.201\lambda_0$) |
| $W_s$ | 6.86 | 4.82 | 3.61 |
| $L_s$ | 8.22 | 6.20 | 4.65 |
| $b$ | 1.94 | 1.38 | 1.03 |
| $c$ | 0.89 | 0.69 | 0.52 |
| $F_p$ | 6.22 | 4.48 | 3.36 |
| $t$ | 2.69 ($0.081\lambda_0$) | 2.76 ($0.083\lambda_0$) | 2.40 ($0.072\lambda_0$) |
| $Dprobe$ | 0.3 | 0.34 | 0.17 |

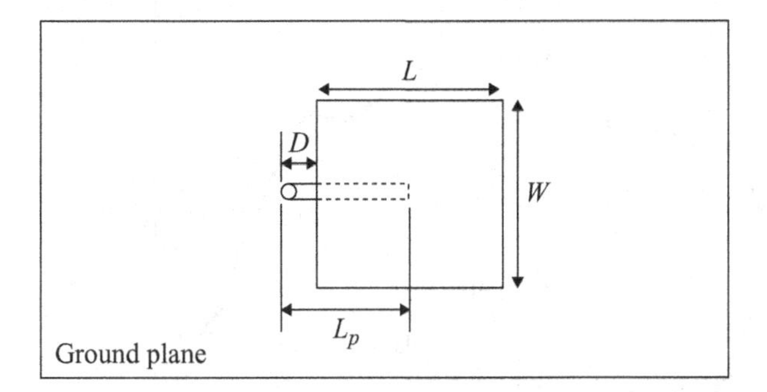


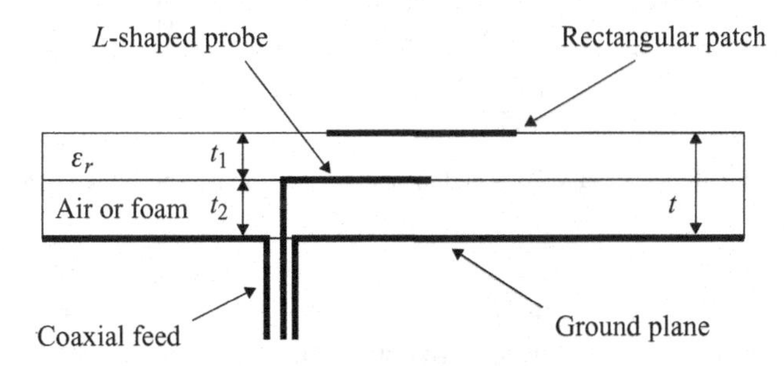


**Figure 8.23** Geometry of L-probe-fed patch antenna ([8b], copyright © 2003 IEEE).

frequency $f_0$ of the antenna is 900 MHz. The patch length $L$ is initially $0.374\lambda_0$ with $\varepsilon_r = 1.0$, but reduced to $0.201\lambda_0$ with increased $\varepsilon_r = 4.0$. The patch size, defined here as the normalized area in comparison with the U-slot patch area ($21.97 \times 12.45 = 273.53\ \text{mm}^2$) when $\varepsilon_r = 1$, is also decreased to 23% ($\varepsilon_r = 4.2$) of that with $\varepsilon_r = 1$. Size reduction is achieved with sacrifice of the bandwidth, being decreased to 22.1% ($\varepsilon_r = 4.2$) from 42% ($\varepsilon_r = 1$), although it is still substantially wideband.

By using an L-probe to feed the U-slot patch as shown in Figure 8.23 [8b, 8c], nearly the same amount of size reduction with the increase in $\varepsilon_r$ can be achieved,

**Table 8.3(a)** Design parameters of L-probe patches (mm) ([8b], copyright © 2000 IEEE)

| | $\varepsilon_r = 1.0$ | $\varepsilon_r = 2.32$ |
|---|---|---|
| $W$ | 15.0 (0.383$\lambda_0$) | 15.0 (0.315$\lambda_0$) |
| $L$ | 13.0 (0.332$\lambda_0$) | 13.0 (0.273$\lambda_0$) |
| $L_p$ | 7.0 | 7.6 |
| $D$ | 3.0 | 3.0 |
| $t_1$ | – | 5.0 |
| $t_2$ | – | 1.58 |
| $t = t_1 + t_2$ | 6.0 (0.15$\lambda_0$) | 6.58 (0.14$\lambda_0$) |
| $dprobe$ | 1.0 | 1.0 |

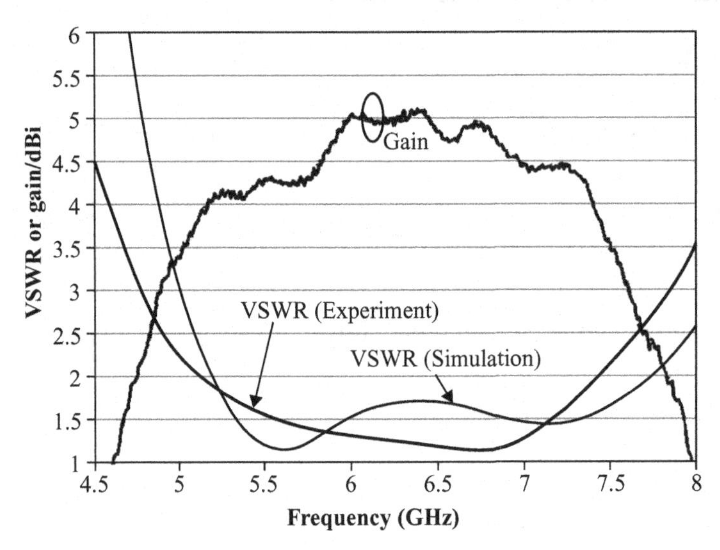


**Figure 8.24** Simulated and measured VSWR and measured gain of an L-probe fed patch antenna ([8b], copyright © 2003 IEEE).

while substantially wide bandwidth is kept. The L-probe-fed U-slot patch antenna is fabricated practically by embedding a horizontal arm on the foam substrate beneath the dielectric substrate, on which the patch is embedded. An L-probe-fed U-slot patch antenna with two-layer configuration to operate at the 6-GHz band is designed with the antenna parameters given in Table 8.3(a) for cases of two values of $\varepsilon_r$. The simulated and measured VSWR and measured gain for the case of $\varepsilon_r = 2.32$ are shown in Figure 8.24 and the results are provided in Table 8.3(b) with the patch dimensions.

As another way to reduce the size of the U-slot patch, a shorting wall is introduced on the substrate as shown in Figure 8.25 [8b, 8d], in which geometrical parameters are also given. A U-slot patch antenna with a shorting wall is fabricated on a substrate with $\varepsilon_r = 4.4$. The center frequency $f_0 = 2.5915$ GHz and the design parameters are; $W = 14$, $L = 14$, $W_s = 9$, $L_s = 12$, $b = d = 1$, $c = 2$, $F_p = 2$, $D_{probe} = 1$, and the thickness $t = 12.1$ (units are mm). The patch is a square with the sides measuring $0.121\lambda_0$ and

**Table 8.3(b)** Measured results of two-layer L-probe patch ([8b], copyright © 2000 IEEE)

| | Patch Dimensions (mm) | Thickness (mm) | $f_0$ (GHz) | Normalized $f_0$ | BW | Gain |
|---|---|---|---|---|---|---|
| 1.0 | 15.0 × 13.0 | 6.0 (0.15$\lambda_0$) | 7.66 | 1.0 | 36% | 6.5 dBi |
| 2.32 | 15.0 × 13.0 | 6.0 (0.15$\lambda_0$) | 6.3 | 0.82 | 36% | 4.5 dBi |

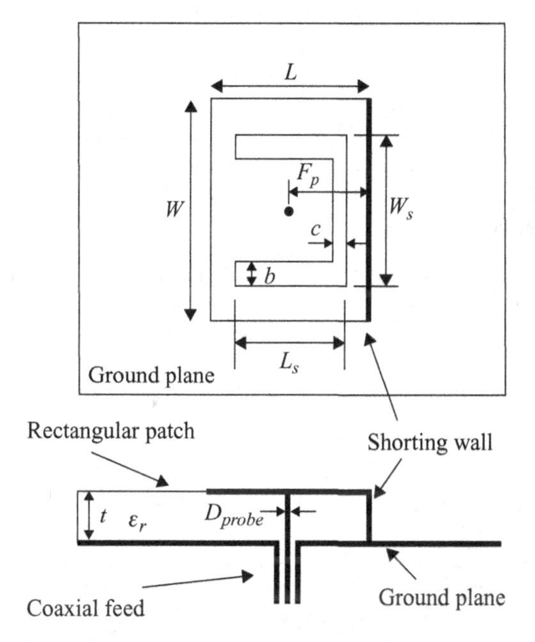


**Figure 8.25** Geometry of U-slot patch antenna with a shorting wall on a substrate ([8b], copyright © 2003 IEEE).

thickness of $0.105\lambda_0$. The area occupied by the patch is $0.01446\lambda^2$, a 94% area reduction compared to a square half-wave patch. Figure 8.26 illustrates the simulated and measured VSWR.

Another way to reduce the patch size is to use a shorting pin, which is added in close proximity to the probe feed as shown in Figure 8.27 [8b, 8e]. The shorting pin and the feed are placed at the opposite sides of the patch. The shorting pin couples equivalently capacitively with the resonance circuit of the patch, consequently effectively increasing the permittivity of the substrate. A U-slot patch antenna with a shorting pin is designed for operation at approximately $f_0 = 4$ GHz. The design parameters are; $W = 18$, $L = 15$, $W_s = 16$, $L_s = 10$, $b = 4$, $c = 4.5$, $F_s = 4$, $F_p = 14.5$, $D_{probe} = 1$, $D_{short} = 2$, and the thickness $t = 7$ (units are mm). Figure 8.28 illustrates simulated and measured VSWR and measured gain. The bandwidth (VSWR $\leq 2.0$) obtained through simulation is 30% and the average gain is 2 dBi. The patch is a $0.200\lambda_0$ by $0.240\lambda_0$ rectangle supported

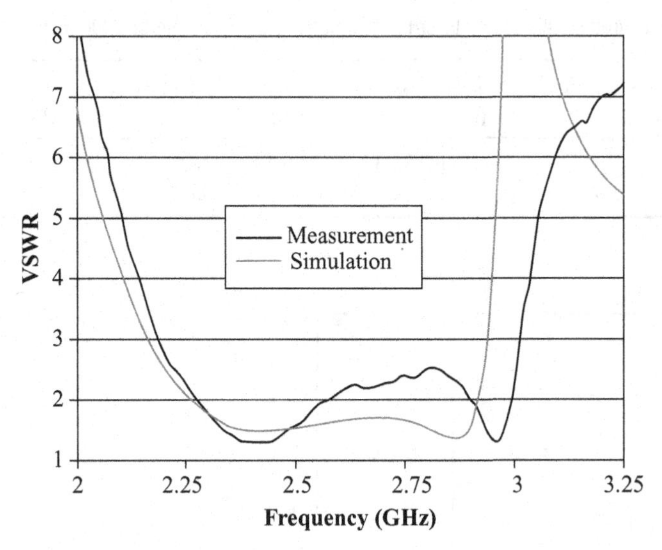


**Figure 8.26** Simulated and measured VSWR for the U-slot patch antenna with a shorting wall ([8b], copyright © 2003 IEEE).

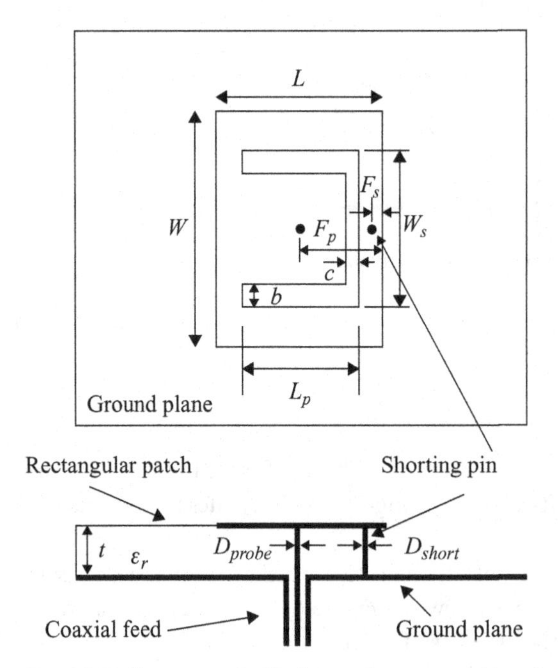


**Figure 8.27** Geometry of a U-slot patch antenna with a shorting pin ([8b], copyright © 2003 IEEE).

by a foam substrate with $0.093\lambda_0$ thickness. The area occupied by the patch is $0.048\lambda^2$, 80.8% smaller than that of a square half-wave patch.

A full-size U-slot is halved to form a half-U-slot or an L-slot by removing the patch area along the line of symmetry as is shown in Figure 8.29 [8b, 8f, 9]. In the figure,

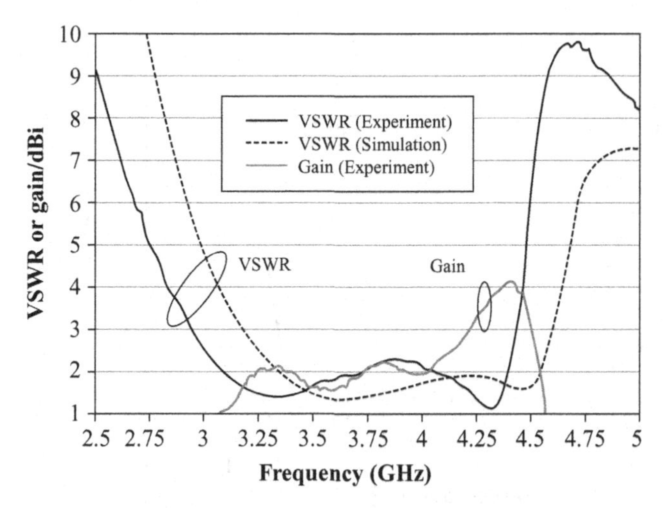


**Figure 8.28** Simulated and measured VSWR and measured gain of the U-slot antenna with a shorting pin ([8b], copyright © 2003 IEEE).

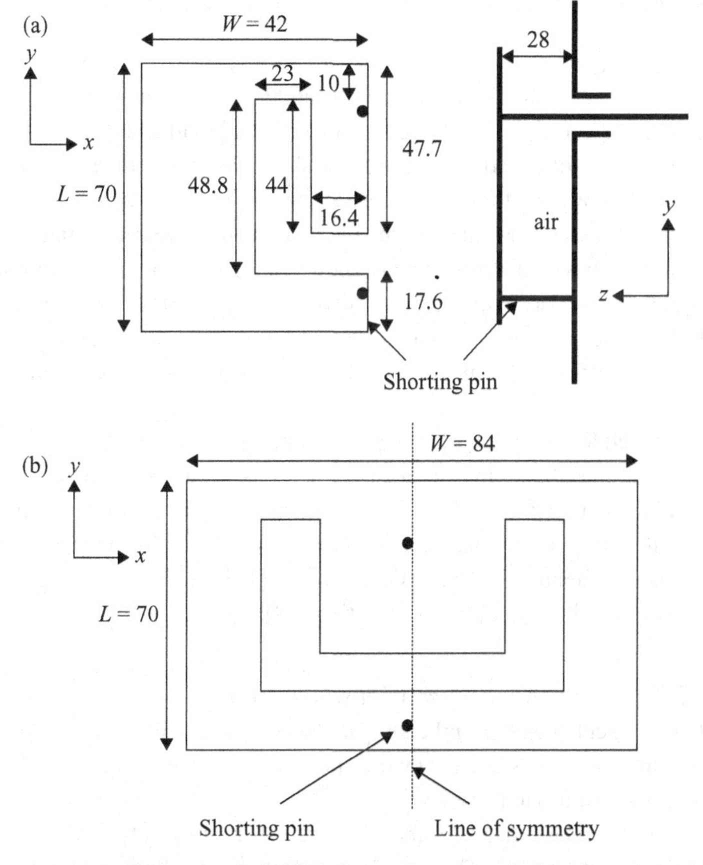


**Figure 8.29** Geometry of a half-U-slot patch antenna (in [mm], not to scale), (a) half-U-slot and (b) full-U-slot ([8f], copyright © 2005 IEEE).

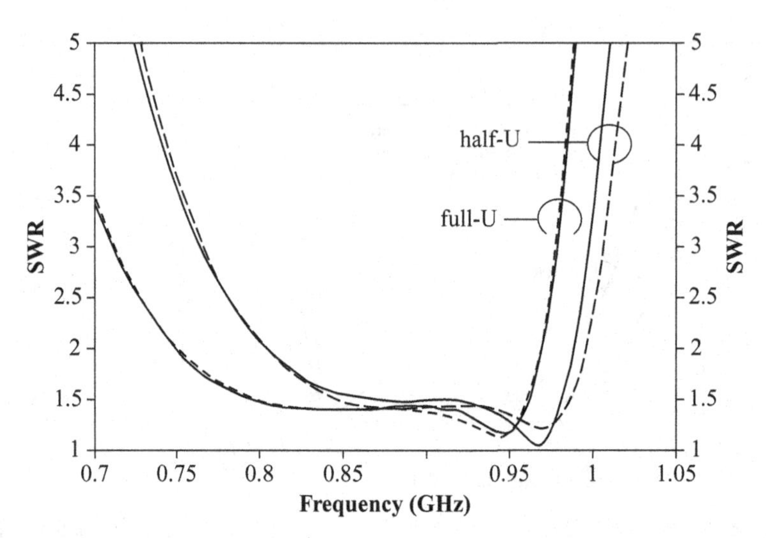


**Figure 8.30** Measured (dark line) and simulated (dotted line) SWR of the half-U-slot patch antenna with a shorting pin. Thin line and thin dotted line provide measured and simulated data, respectively, for the full-U-slot patch for a comparison ([8f], copyright © 2005 IEEE).

the dimensions of an example antenna with length $L = 70$ mm ($0.21\lambda_0$), and width $W = 42$ mm ($0.13\lambda_0$) ($f_0 = 0.9$ GHz) are given along with the size of the half-size slot. The feed probe line (radius of 2 mm) and the shorting pin (radius of 4.65mm) support the patch in air and they are located at the non-radiating edge of the half-U-slot patch. The ground plane is a square with side of $1\lambda_0$. Measured and simulated SWR of the half-U-slot patch with shorting pin are shown in Figure 8.30, where for comparison, those of a full U-slot patch antenna are also provided. In the figure, thick line and thick dotted line, respectively, indicate measured and simulated VSWR of the half-U slot patch and thin line and thin dotted line. respectively, give those of the full U-slot patch.

A U-slot embedded on a rectangular patch is modified to achieve various functions such as wideband, multiband, and circular polarizations. Figure 8.31 shows representative ones: (a) double U-slots [9, 10], (b) a U-slot on a square patch with truncation [9], and (c) an unequal arm U-slot patch [9, 11]. Parametric analysis of design for the U-slot rectangular patch antennas has been described in [12]. A U-slot can be applied to a triangular patch to achieve wideband operation [13].

### *8.1.2.1.1.2.2.2 Rectangular patch with square slot*

Bandwidth can be enhanced by embedding slots/slits with various shapes on the surface of the patch antenna as has been shown in previous sections. Similar methods can be applied to achieve multiband operation.

A square patch with a square slot fed by a microstrip line is a typical design example for wideband operation [14]. The antenna geometry with dimensional parameters is illustrated in Figure 8.32, where two types of feeding are shown: (a) with a fork-like

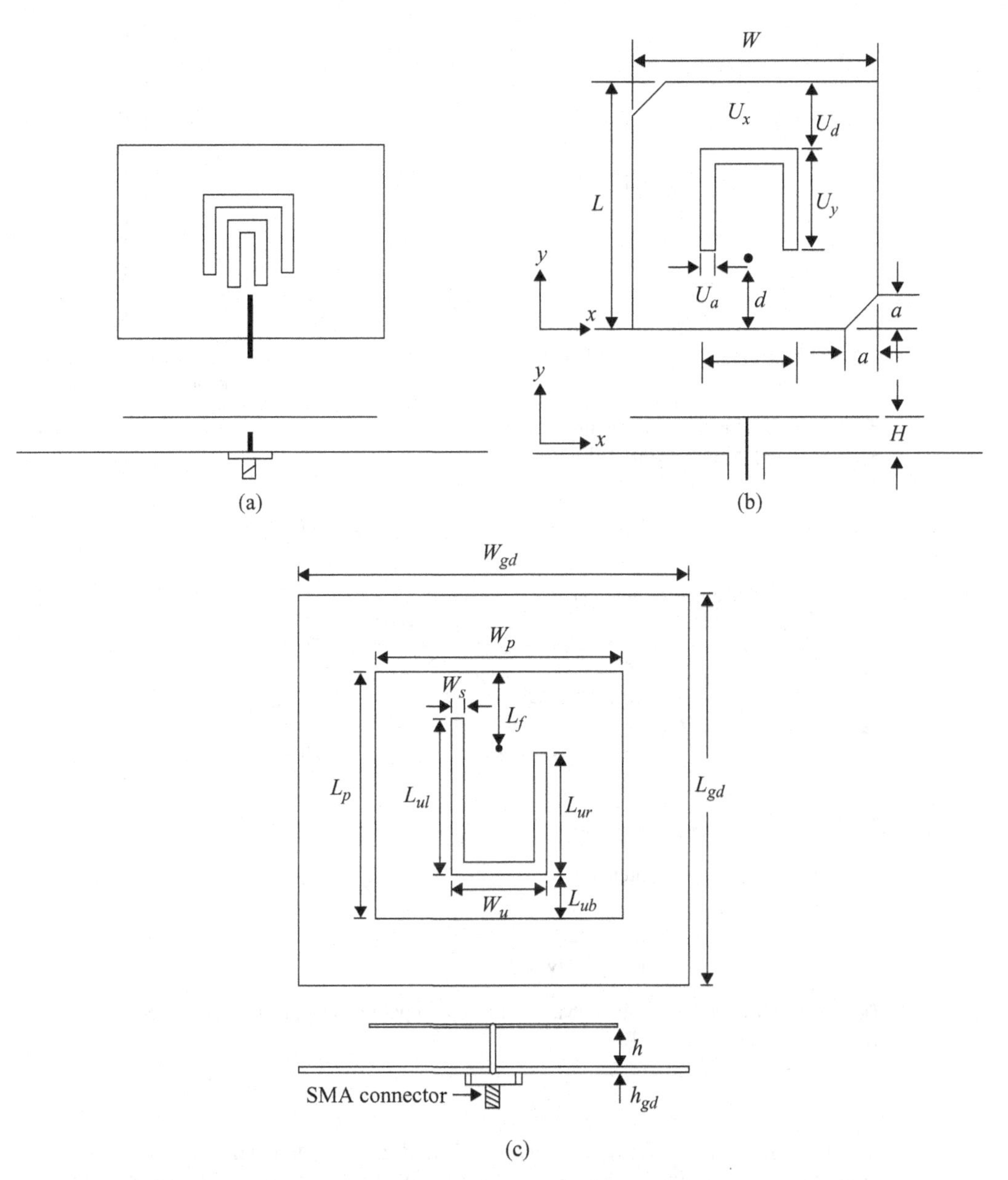


**Figure 8.31** Geometries of (a) the triple-band antenna with two U-slots, (b) perturbed patch antenna with U-slot, and (c) patch antenna with unequal length U-slot ([9], copyright © 2010 IEEE).

tuning stub, and (b) with a simple tuning stub, which is a conventional method of feeding. By properly selecting stub lengths $\ell_1$, $\ell_2$, and $\ell_3$, a good impedance matching across a widely enhanced bandwidth can be achieved. Return loss for three antennas with different stub lengths are shown in Figure 8.33, where comparison with that of the reference antenna is provided. Antenna parameters are: substrate $\varepsilon_r = 4.4$, thickness

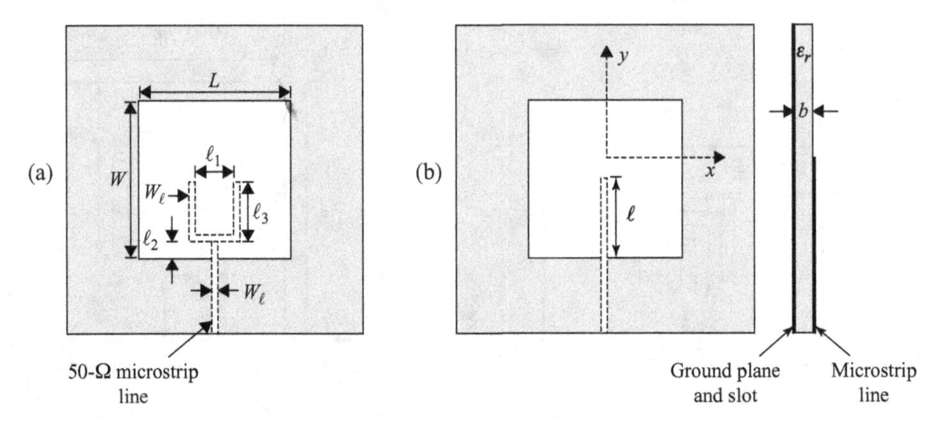


**Figure 8.32** A printed antenna with a square slot fed by (a) a fork-like stub and (b) a simple tuning stub ([14], copyright © 2001 IEEE).

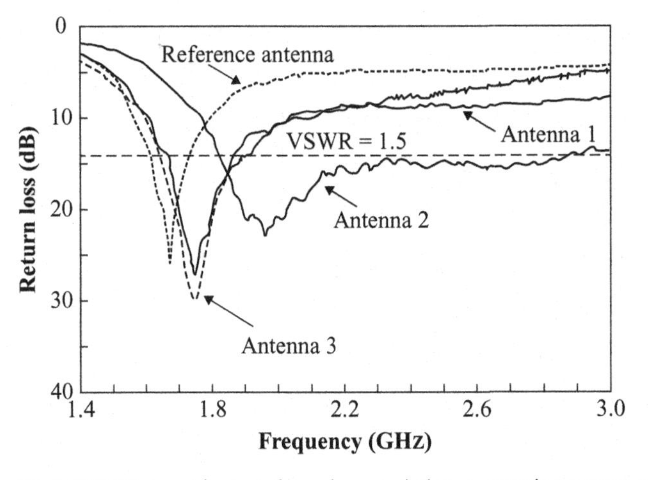


**Figure 8.33** Measured return-loss characteristics: comparing antennas 1–3 with reference antenna ([14], copyright © 2001 IEEE).

$h = 0.8$, antenna $L = W = 53.7$, $w_f = 1.5$, and ground plane size $= 110 \times 110$. The stub sizes are: for the reference antenna $\ell = 28$; for Antenna 1 $\ell_1 = 10$, $\ell_2 = 2$, and $\ell_3 = 20.6$; for Antenna 2 $\ell_1 = 15$, $\ell_2 = 2$, and $\ell_3 = 15.9$; and for Antenna 3 $\ell_1 = 15$, $\ell_2 = 0$, and $\ell_3 = 24.9$ (all in mm). Antenna 2 has widest bandwidth 1091 MHz in terms of VSWR $\leq 1.5$, followed by Antenna 3 with 268 MHz and next Antenna 1 with 197 MHz. They are wider than that of the Reference Antenna, 115 MHz, exhibiting significant improvement in the bandwidth. Within this wide bandwidth, the operating bandwidth with usable broadside radiation pattern is observed to be still wide, being about 580 MHz, and the peak antenna gain is about 5 dBi with variation of less than 1.5 dBi within the operating bandwidth.

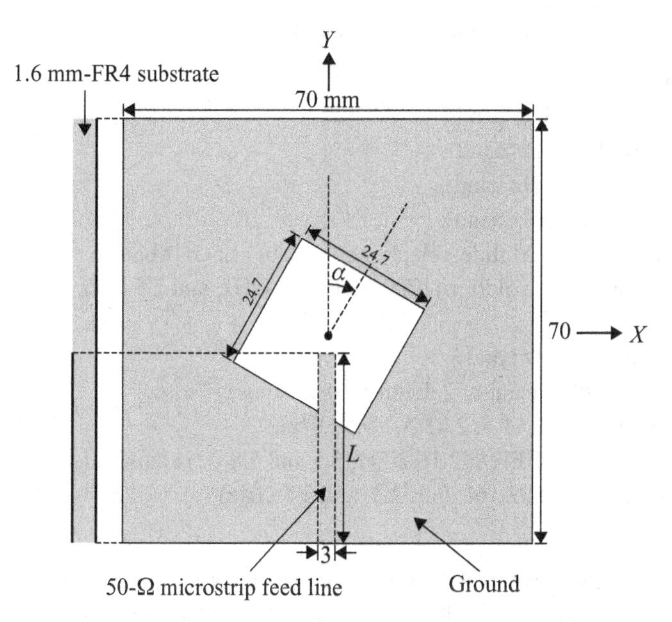


**Figure 8.34** Geometry of antenna with a rotated square slot fed by microstrip line ([15], copyright © 2005 IEEE).

When the square slot is rotated as shown in Figure 8.34, the bandwidth is further enhanced with proper selection of rotation angle $\alpha$ and length $L$ of the feed line [15]. With a square slot size of 24.7 mm designed for operation at 4 GHz as shown in the figure, nearly 2.2 GHz impedance bandwidth for –10 dB VSWR is obtained when $\alpha = 45°$ and $L = 31.5$ mm, This wide bandwidth is about four times that of the corresponding conventional microstrip line-fed wide-slot patch antenna.

*8.1.2.1.2 Multiband and wideband*

*8.1.2.1.2.1 Multiband antenna*

Recent small wireless equipment requires small antennas with not only compact structure, but also multifunctional operation in nature. Late-model mobile phones have evolved from telephone devices toward information terminals dealing with multimedia information, involving audio, video as both still and dynamic media, data, radio, digital TV, and internet access. All of this, in addition to telephone voice, requires antennas that are small, compact, built-in, low cost, yet able to provide high-performance facilities to deal with high-data-rate information, and handle multiband communications.

Types of antennas for these applications are necessarily small, compact planar types, generally represented by various printed patch antennas and modified PIFA (Planar Inverted-F Antenna) combined with variously shaped wire elements or printed strips, stubs, slots, and so forth. Antennas are designed to be installed not only in mobile terminals, but also in various small wireless equipment and apparatuses, on which wireless systems are installed, including small portable terminals, personal computers including standard, laptop, and tablet types, USB cards and dongles, and TVs.

Typical wireless systems, including mobile phones, are as follows:

Mobile phones
GSM 850 (850 MHz band),
GSM 900 (900 MHz band),
GSM 1800 (1.8 GHz band),
GSM 1900 (1.9 GHz band),
UMTS (Universal Mobile Telephone Systems) (2 GHz band),
LTE (Long Term Evolution) (700 MHz, 2.3 GHz, and 2.5 GHz bands)
Wireless systems
Bluetooth (2.4 GHz band),
WiFi (IEEE 802 11 a/g/n: 2.4 and 5 GHz bands),
WLAN (IEEE 802.11 b: 2.45/5.2/5.8 GHz),
WiMAX (Fixed: IEEE 802 16 2004: 3.5 and 5.8 GHz bands and
Mobile: IEEE 802 16e: 2.3, 2.5, and 3.5 GHz)
GPS (1.5 GHz band)

Various types of multiband microstrip patch antennas composed with various shapes combined with slots/slits embedded on the patch surface have so far been introduced. In this section, however, antennas for operating at more than two bands will be described.

*8.1.2.1.2.1.1 A printed λ/8 PIFA operating at penta-band*

A PIFA is designed to operate at one-eighth ($\lambda/8$) wavelength as the fundamental resonance mode for applying to WWAN (Wireless Wide Area Network) system [16]. The antenna is installed in a mobile phone, having a simple structure comprised of two radiating strips of length about $\lambda/8$ at 900 MHz. The configuration generates two $\lambda/8$ modes to cover two lower modes for operation of GSM 850/900, and at the same time, two higher-order modes or $\lambda/4$ modes at about 1900 MHz to operate at a wider upper band for GSM 1800/1900/UMTS. The antenna geometry is illustrated in Figure 8.35, in which dimensional parameters are given. The antenna covers penta band, yet occupies only a small printed area of 15 mm $\times$ 31 mm or 465 mm$^2$. The antenna is fed using a coupling feed, by which the ordinarily large input impedance of a traditional $\lambda/8$ mode PIFA is greatly reduced, and successful excitation of the $\lambda/8$ mode for a PIFA is achieved. Measured and simulated return-loss characteristics are depicted in Figure 8.36.

*8.1.2.1.2.1.2 Bent-monopole penta-band antenna*

A metal-wire bent-monopole antenna (BMA) fed by mini-coaxial cable jointly with a thin printed ground line demonstrates that it operates at penta-band: CDMA, GSM, DCS, PCS, and WCDMA bands [17]. The antenna geometry is illustrated in Figure 8.37, which shows (a) antenna geometry and its feed point, (b) front- and back-side views of PCB (Printed Circuit Board). The feed and ground points are connected to a 50-Ω coaxial cable (10 cm long) with an SMA connector. The dimensions of the BMA are $a = 38$, $b = 40$, and $c = 6$ (in mm), and the radius and the length, respectively, of the feeding mini-coaxial cable are 1.13 and 46 mm, which is put tightly on thc BMA to serve as a reactive loading. The effect of the reactive loading is reduction of the electrical

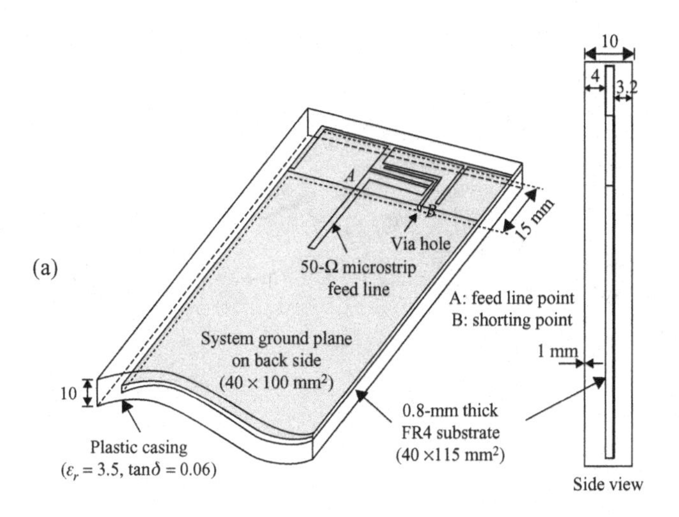


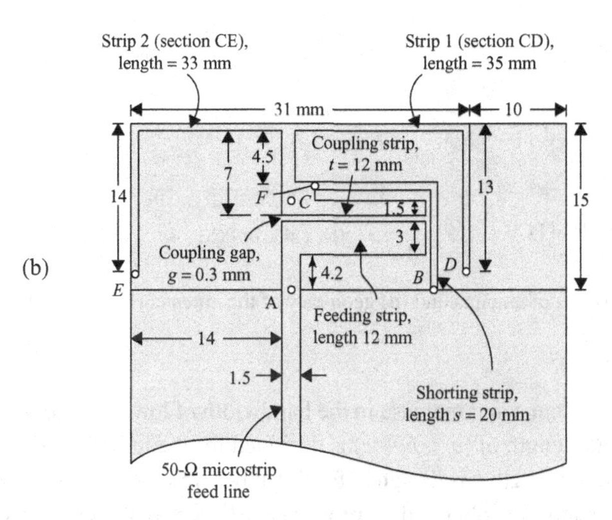


**Figure 8.35** (a) Geometry of printed λ/8-PIFA for penta-band operation and (b) dimensions of PIFA pattern ([16], copyright © 2009 IEEE).

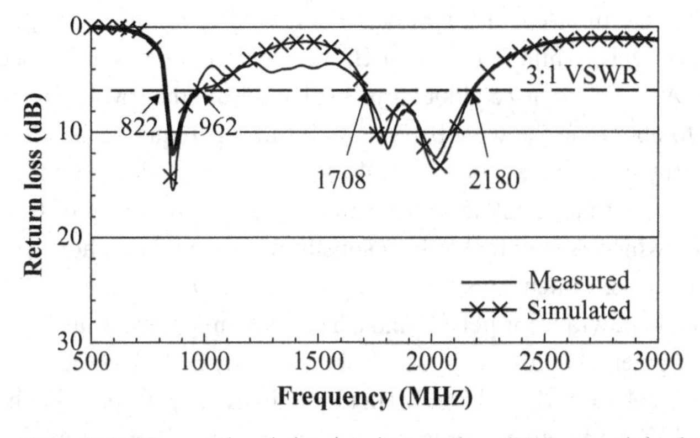


**Figure 8.36** Measured and simulated return loss ([16], copyright © 2009 IEEE).

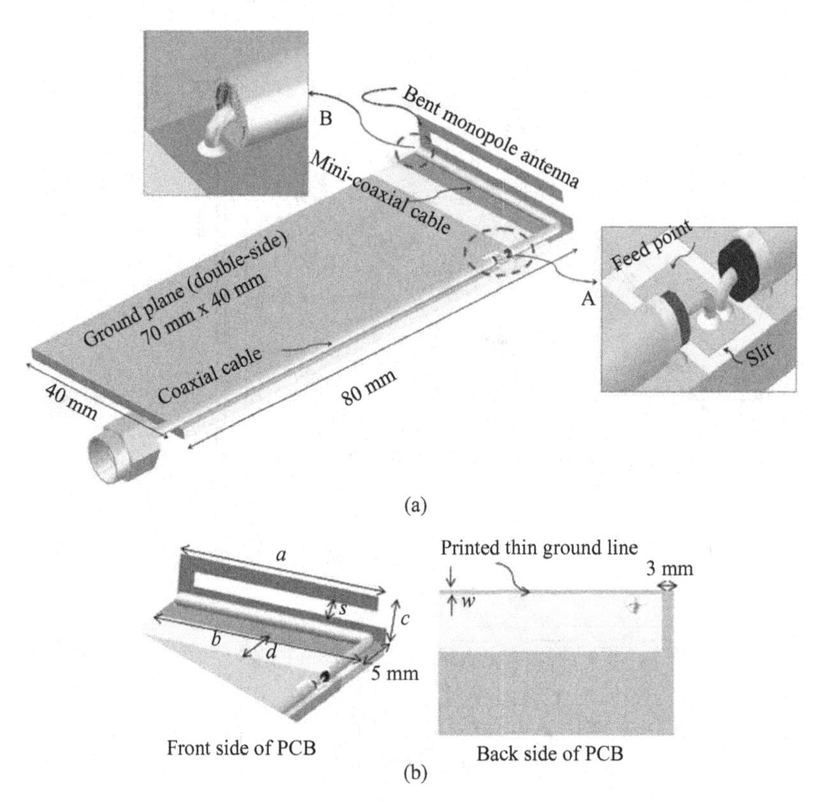


**Figure 8.37** (a) Dimensions of antenna and (b) geometry of the antenna ([17], copyright © 2011 IEEE).

length of monopole antenna and increase in the bandwidth of lower and upper operating frequencies. The total length of $(a + b + c)$ is designed for the lower frequency band of about 892 MHz and the length $b$ is designed for the upper frequency band of 1800 MHz. Good impedance matching is obtained by proper selection of parameters $s$ (slit-space), $w$ (width of ground line), and $d$ (spacing between the ground plane and the BMA). There are two points A and B to connect the mini-coaxial cable. At the point A, the inner and outer conductor of the mini-coaxial cable are connected to the feed and ground point of the PCB, respectively, while at the point B, the inner conductor is connected to the corner of the BMA and the outer conductor is insulated from the BMA. The thin ground line contributes to obtaining the wide bandwidth for lower frequency bands.

In the experiment, an antenna with size of $40 \times 5 \times 6$ mm is placed on the top side of a rectangular FR4 substrate (thickness of 1.5 mm, $\varepsilon_r$ of 4.3 and tan $\delta$ of 0.023) having size $80 \times 40$ mm, which is assumed to be a substitute for a mobile phone platform. By the parametric analysis, dimensions $s = 2$, $w = 0.5$, and $d = 5$ (in mm) are selected to achieve desired bandwidth for penta-bands. The measured and simulated return-loss characteristics are given in Figure 8.38, which shows nearly 200 MHz (22%) in the lower operating band (CDMA/GSM) and 540 MHz (30%) in the upper operating band (DCS, PCS, and WCDMA). Radiation patterns are similar to those of a monopole antenna and maximum gain in the $x$–$y$ plane is obtained as 1.87 dBi and 0.91 dBi for 960 MHz and

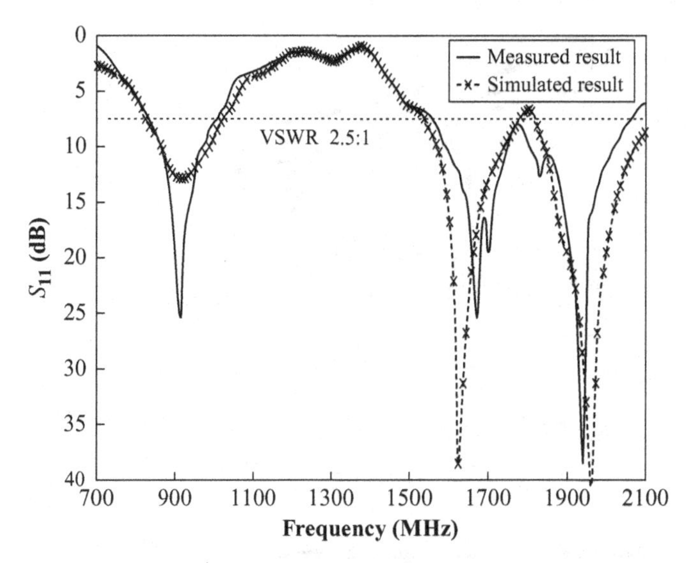


**Figure 8.38** Measured and simulated return-loss characteristics ([17], copyright © 2011 IEEE).

1880 MHz, respectively, and in the *x–z* plane (horizontal plane), 0.7 dBi and 2.07 dBi for 960 MHz and 1880 MHz, respectively.

Addition of a printed thin ground line behind the antenna contributes to increasing the bandwidth at both lower and higher bands, also providing an effect to form a balanced-feed structure like a sleeve balun, because of its length being nearly $\lambda/4$ at 1800 MHz, allowing a short circuit at the base to present an infinite impedance at the top.

The balanced feed to a balanced antenna is useful to reduce the current flow on the ground plane so that the operator's hand effect on the antenna performance can be mitigated.

### *8.1.2.1.2.1.3 Loop antenna with a U-shaped tuning element for hepta-band operation*

The printed loop antenna is designed to cover GSM 850/900/DCS/PCS/UMTS and WiMAX, with a U-shaped tuning element printed on the back side of the circuit board when applied to a laptop computer [18]. The antenna geometry is illustrated in Figure 8.39, which shows (a) 3D view, (b) plan view of the front side, and (c) plan view of the back side. The antenna is printed on an FR4 substrate with thickness of 0.8 mm, $\varepsilon_r = 4.4$, and mounted on the top right corner of a vertical ground plane of size 200 × 160 mm, which is the supporting metal frame of an LCD panel. The antenna measures only 65 × 10 × 0.8 mm, because it is coated on double sided PCB. The U-shape on the back side is a tuning element, which is affixed to the ground plane. Measured and simulated reflection coefficients are shown in Figure 8.40. The results exhibit bandwidth for –6 dB reflection coefficient 140 MHz (820–960 MHz) in the GSM band and 1190 MHz (1710–2900 MHz) in the DCS/PCS/UMTS bands. It also shows that –10 dB bandwidth is sufficient for WLAN and WiMAX applications. Radiation patterns are similar to that of a monopole. Table 8.4 gives average gain, peak gains, and efficiency for five frequency bands.

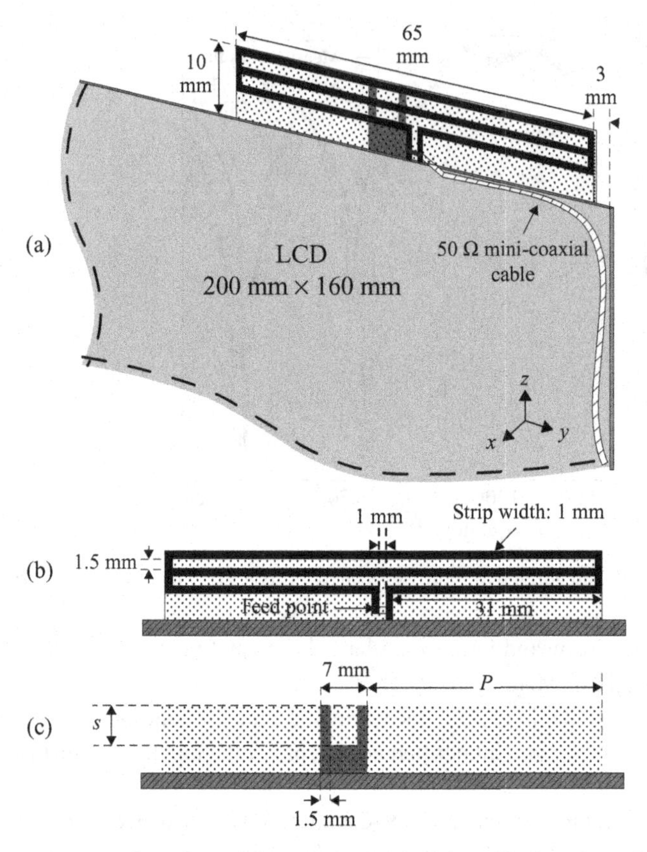


**Figure 8.39** Geometry of the antenna: (a) 3D view, (b) plan view of the front side, and (c) plan view of the back side ([18], copyright © 2010 IEEE).

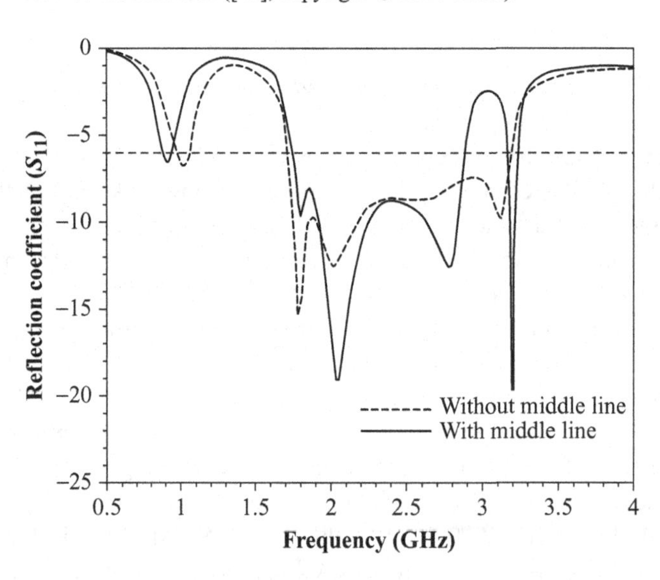


**Figure 8.40** Return loss with and without middle line ([18], copyright © 2010 IEEE).

**Table 8.4** Measured peak and average gains and efficiency ([18], copyright © 2010 IEEE)

| Frequency (GHz) | 0.9 | 1.80 | 2.10 | 2.45 | 2.70 |
|---|---|---|---|---|---|
| Peak Gain (dBi) | 1.12 | 3.13 | 2.26 | 1.82 | 3.21 |
| Average Gain (dBi) | −3.21 | −4.87 | −1.24 | −0.83 | −0.79 |
| Efficiency (%) | 46.4 | 60.5 | 80.6 | 74.8 | 67.6 |

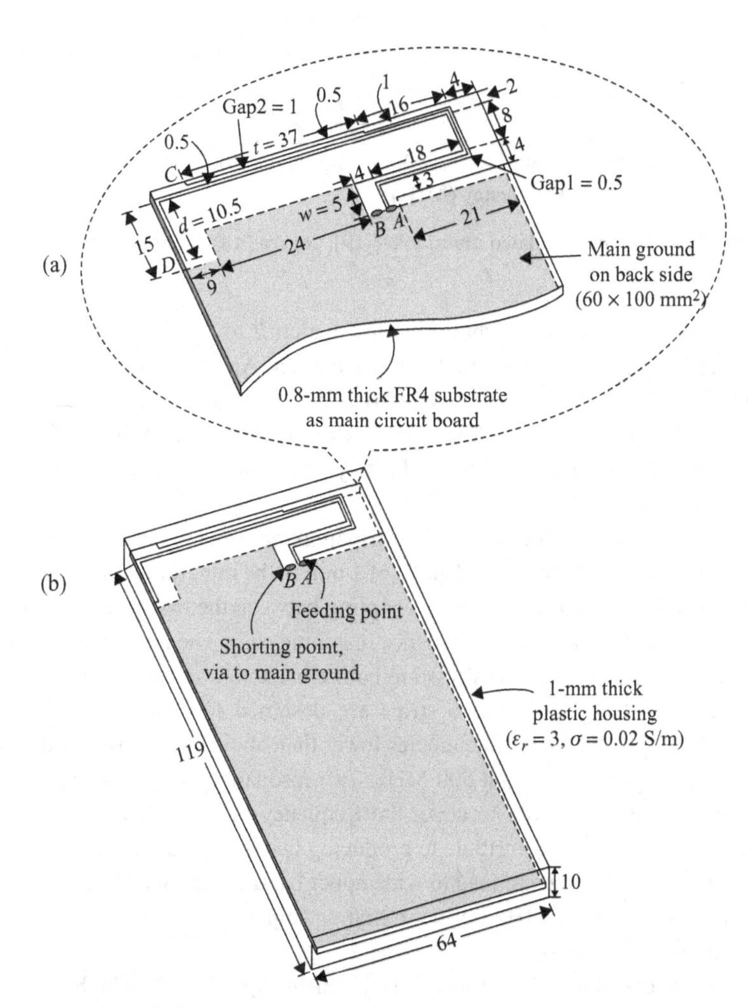


**Figure 8.41** Geometry of the antenna in the mobile phone: (a) antenna structure and (b) antenna embedded on the ground plane ([19], copyright © 2010 IEEE).

*8.1.2.1.2.1.4 Planar printed strip monopole for eight-band operation*

A planar printed strip monopole with closely coupled parasitic shorted strips for eight-band LTE/GSM/UMTS operation to be used in a mobile phone is described in [19]. Geometry of the antenna along with the dimensions are illustrated in Figure 8.41. The antenna has a simple structure comprised of a driven planar monopole (section *A* to *C*

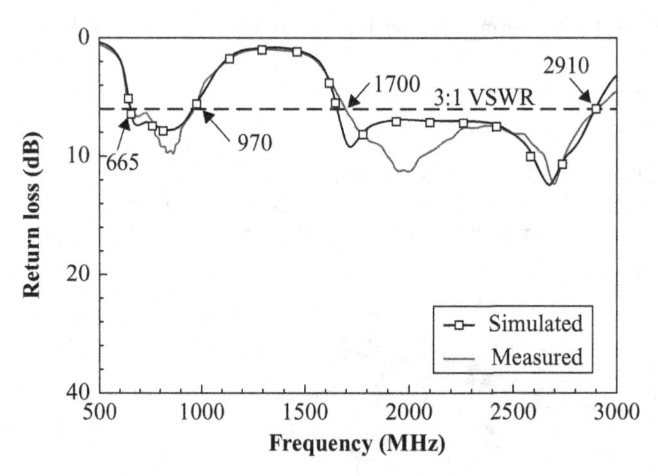


**Figure 8.42** Measured and simulated return loss ([19], copyright © 2010 IEEE).

in Figure 8.41) and a parasitic shorted strip (section *B* to *D* in Figure 8.41), both with comparable length and closely coupled to each other. An FR4 substrate with thickness of 0.8 mm and size of 115 × 60 mm is used as the systems circuit board. Point *A* of the driven strip monopole is the feeding point of the antenna, while point *B* of the parasitic strip is short-circuited to the top edge of the system ground plane through a via-hole in the system circuit board.

There are two gaps between the driven strip monopole and parasitic shorted strip: gap 1 of 0.5 mm in the front section and gap 2 of 1 mm in the remaining section of length $t =$ 37 mm. These gaps provide capacitive coupling between the two strips, and adjustment of the coupling will make good impedance matching easy. The length of the driven strip monopole and parasitic shorted strips are both close to a quarter wavelength, although slightly different lengths. The two strips are designed to contribute to their lowest resonance modes, with one at frequencies lower than about 800 MHz and the other one at frequencies higher than about 800 MHz. Two resonance modes can be incorporated to produce a wide bandwidth to cover the frequency range of 698 to 960 MHz. The parasitic shorted strip can contribute to producing two higher resonance modes of about 1700 MHz and 2700 MHz and lead to wide upper band coverage of the frequency range from 1710 MHz to 2690 MHz by incorporating higher-order resonance modes of the driven strip monopole.

Measured and simulated return loss is given in Figure 8.42, which shows a wide bandwidth of 305 MHz (665–970 MHz), covering LTE 700/GSM 850/900 operations. The upper band shows a further wide bandwidth of 1210 MHz (1700–2910 MHz), which covers GSM 1800/1900/UMTS/LTE 2300/2500 operation.

The antenna exhibits gain of –0.4 to 1.1 dBi for the lower band (698–960 MHz) and 2.7 to 4.4 dBi for the higher band (1710–2690 MHz), and efficiency of 53%–76% and 52%–75% over the lower and higher bands, respectively.

Since the antenna is designed for installing on a mobile phone handset, SAR (Specific Absorption Rate) values pertaining to an operator's head should be taken into

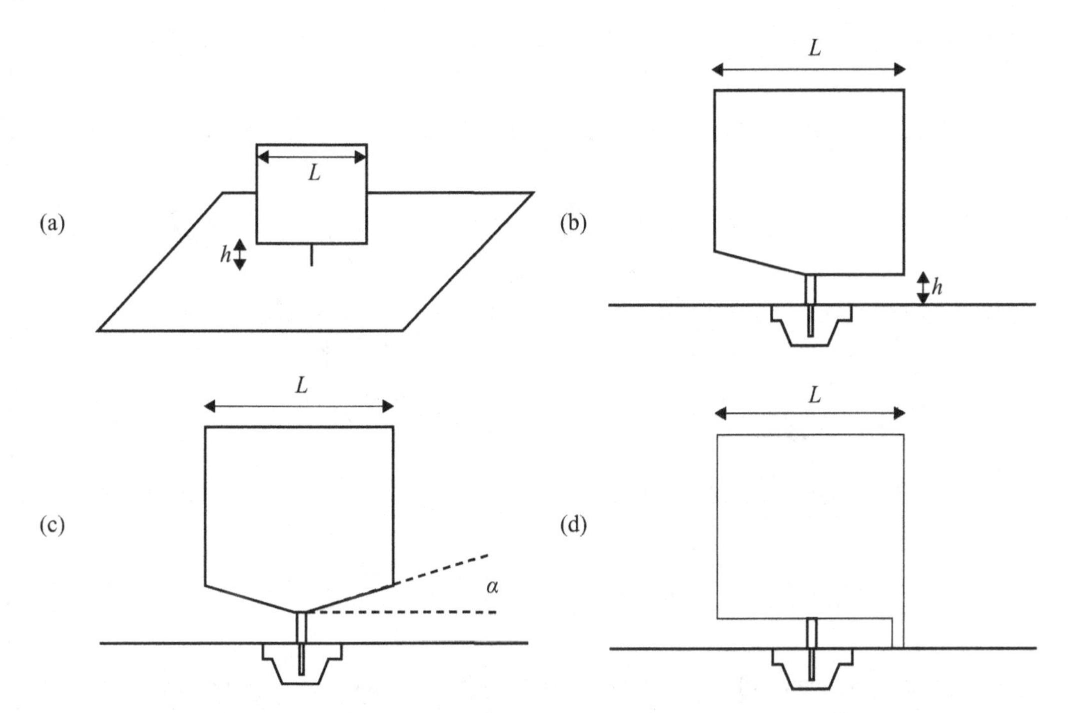


**Figure 8.43** Square planar monopole: (a) on the ground plane, (b) with asymmetric beveling, (c) with symmetrical beveling, and (d) with a shorting post ([20], copyright © 2003 IEEE).

consideration as an important subject when the antenna is designed. The SAR values are calculated by using the simulator SEMCAD for a case where the proposed antenna is mounted at the bottom of the system circuit board. By using the input power of 24 dBm for the GSM 850/900 and 21 dBm for the GSM 1800/1900, UMTS, and LTE operations, the SAR values per gram of head tissue are obtained all well below the SAR value limit of 1.6 W/kg. This result indicates that the proposed antenna can be employed in mobile phone handsets without any concern over effect of radiation from the mobile phone.

*8.1.2.1.2.2 Wideband antennas*

*8.1.2.1.2.2.1 Planar monopole antennas*

A planar quarter-wavelength monopole is a simple wideband antenna, which is generally convenient to match to 50 Ω with unbalanced feed, and does not require a balun. The planar monopole antenna was first described briefly in the literature in 1968 [20] and more detail was given in 1976 in [21], where wide impedance characteristics of a planar quarter-wave monopole were discussed. Subsequently studies on planar monopoles have continued, dealing with various types and shapes of antennas theoretically and experimentally. Among them, an example was that a disk-shaped planar monopole had shown return loss greater than 10 dB over an impedance bandwidth ratio in excess of 10 to 1 [22]. One of the simplest typical planar shapes is a square monopole vertically standing on the ground plane as shown in Figure 8.43, which depicts (a) the

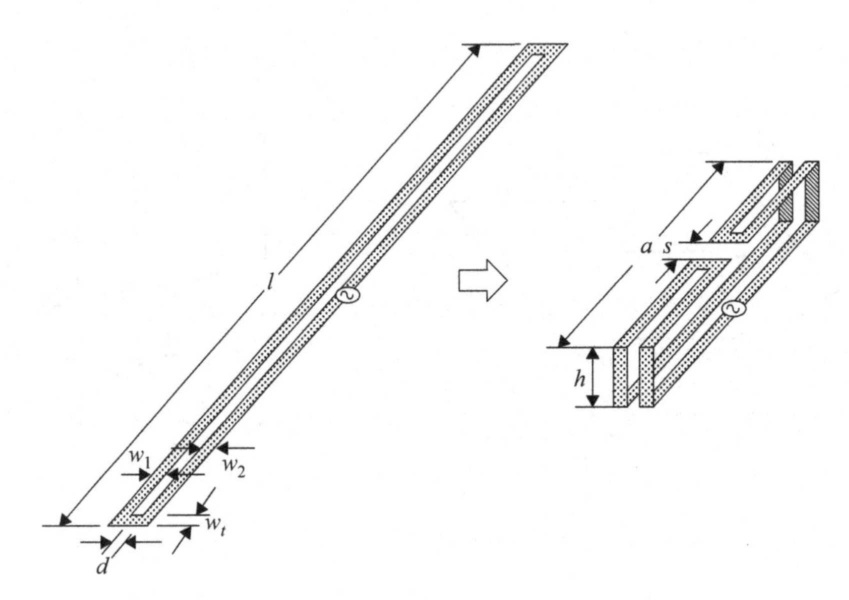


**Figure 8.44** Geometry of a folded loop [27b].

basic geometry, (b) modified versions at its bottom side with asymmetrical beveling, (c) with symmetrical beveling, and (d) with a shorting post. There have been various planar geometries such as rectangle, trapezoid, triangle, and so forth [23–25].

In practical applications, a conventional monopole structure, comprising an antenna standing vertically on the ground plane, is not preferable, because it usually requires an infinite or comparable ground plane, leading to increasing the antenna volume, contrary to the need to make the antenna small and compact. The recent trend is to use planar monopole structures, comprised of the antenna and the ground plane being placed in the same plane. By this means the antenna size can be reduced, granted that the size of ground plane is reduced. To maintain the wide bandwidth with reduced-size ground planes, various techniques have been developed. Antenna profiles can be all sorts of planar geometries; disk, rectangle, triangle, ellipse, and so forth. Some of them will be described later.

*8.1.2.1.2.2.2 Folded loop antenna*

The folded loop antenna was initially designed as an antenna having a balanced structure and fed by a balanced line to suppress the current flowing into the ground. This is important in portable devices in order to mitigate the body effect on the antenna performance, produced by the current on the ground plane, which is perturbed by the operator's hand and head [26].

The antenna is fabricated by folding a long thin rectangular loop of total one wavelength to form a half-wavelength folded dipole, which appears as a folded two-wire transmission line as illustrated in Figure 8.44 [27a]. The antenna is effectively divided into two parts; a dipole structure and a two-wire transmission line. The dipole part essentially contributes to radiation while the two-wire transmission line plays a significant role for determining the antenna impedance [27b] (refer to Appendix I following this

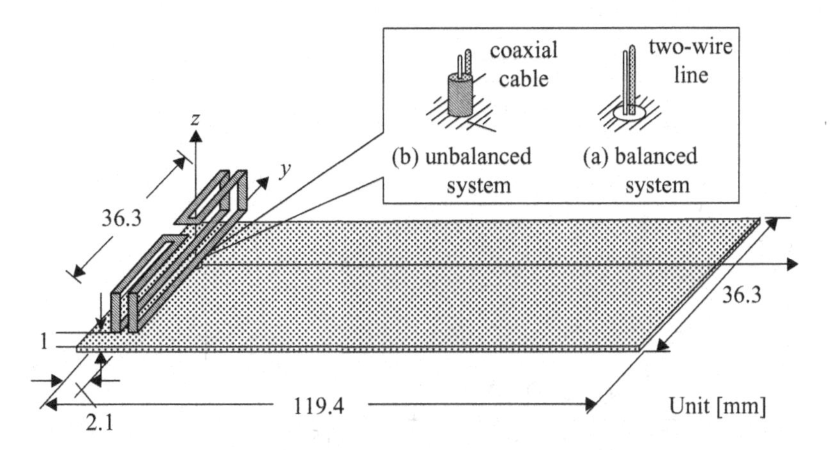


**Figure 8.45** Geometry of the folded loop placed on the ground plane [27b, 28].

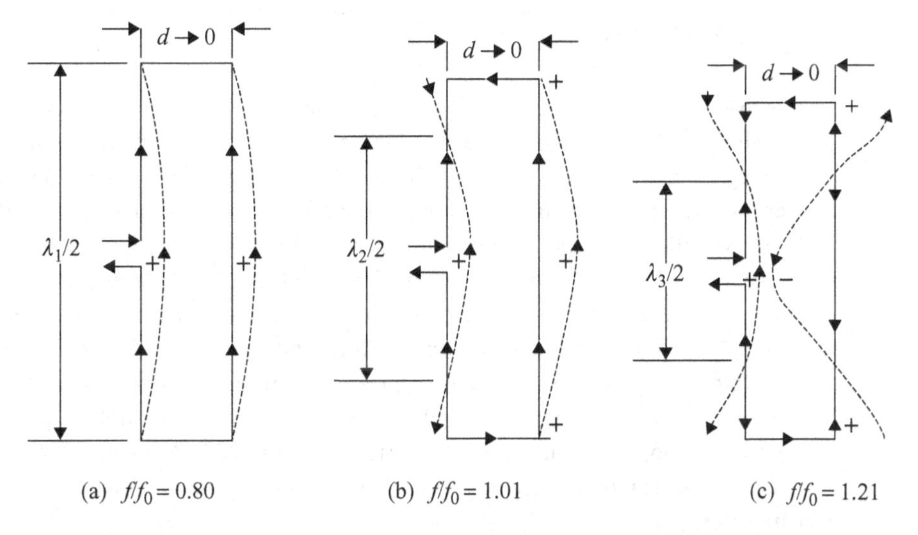


**Figure 8.46** Current distributions for resonance on the folded loop element ([27b, 28], copyright © 2001 IEICE).

section). The antenna is typically mounted on a ground plane in small equipment as shown in Figure 8.45, in which feeding structure is depicted in the inset, and dimensions are given as an example when the antenna is used for a typical mobile phone handset using finite ground plane [27c].

With this antenna structure, resonance occurs at several frequencies as several current modes are observed on the antenna elements, which are depicted in Figure 8.46. Resonances occur at (a) the first frequency $f_1 = 0.80\, f_0$, (b) the second frequency $f_2 = 1.01\, f_0$, and (c) the third frequency $f_3 = 1.21\, f_0$, where $f_0$ is the center frequency of the frequency band [27c]. It was found that a wideband performance was obtained as a result of combined excitation of three closely separated frequencies. Calculated and measured VSWR characteristics when the center frequency $f_0$ is 3470 MHz are shown in Figure 8.47, which indicates a wide bandwidth covering from 0.8 to 1.2 $f/f_0$ for VSWR

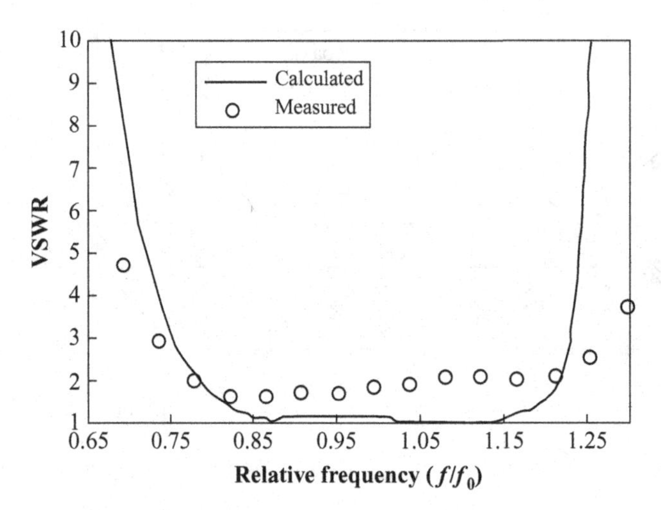


**Figure 8.47** Calculated and measured VSWR ($s = 20$, $h = 8$, $d = 1$, $w_1 = w_3 = 1$, $w_2 = 4$ [mm]) ([28], copyright © 2001 IEICE).

less than two. Current flows on the ground plane of a handset model are also calculated and shown to be very low as shown in Figure 8.48 [27, 28], illustrating almost all areas of the ground plane having about 30 dB lower values compared with that on the antenna element. Reduction of currents on the ground plane is also beneficial to mitigate the SAR (Specific Absorption Rate) value against the human head [29].

Having the balanced structure, another important feature is to bestow the antenna self-balance property (refer to Appendix II following this section), with which a balun is not required and direct feeding with a coaxial line (unbalanced line) is allowable [27b, 30].

A folded loop can be modified to place sideward on the ground plane and further reduce the size in half as Figure 8.49 shows [31]. The figure illustrates a practical half-size folded-loop antenna, hereafter referred to as a BFMA (Built-in Folded Monopole Antenna), being mounted on the ground plane of a mobile phone, along with dimensional parameters [31].

A PIFA and a BFMA, both resonating at the same frequency, are compared in terms of physical volume and bandwidth. It is assumed that a PIFA is mounted on the same size ground plane as a BFMA as Figure 8.50 shows, and the PIFA has the size ($a = b = 19.5$, $d = 5$, and $h = 7$ mm), by which the resonance frequency is the same as the center frequency 2250 MHz of the BFMA. Measured bandwidth for the PIFA and BFMA are 380 MHz and 340 MHZ, respectively, corresponding to 16.0% and 15.1% in the relative bandwidth with respect to the center frequency, and the physical volume of BFMA is estimated to be 42% of that of PIFA [31]. It can be said from this result that BFMA is smaller than PIFA, yet exhibits wider bandwidth.

#### *8.1.2.1.2.3 Ultrawideband antennas*

Ultrawideband (UWB) system is a recent hot topic in both industrial and academic fields. A UWB system is defined as a communication system that has either a bandwidth larger than 500 MHz or a relative bandwidth larger than 0.2 times the center frequency. The FCC allocated the frequency band between 3.1 GHz and 10.6 GHz for unlicensed systems

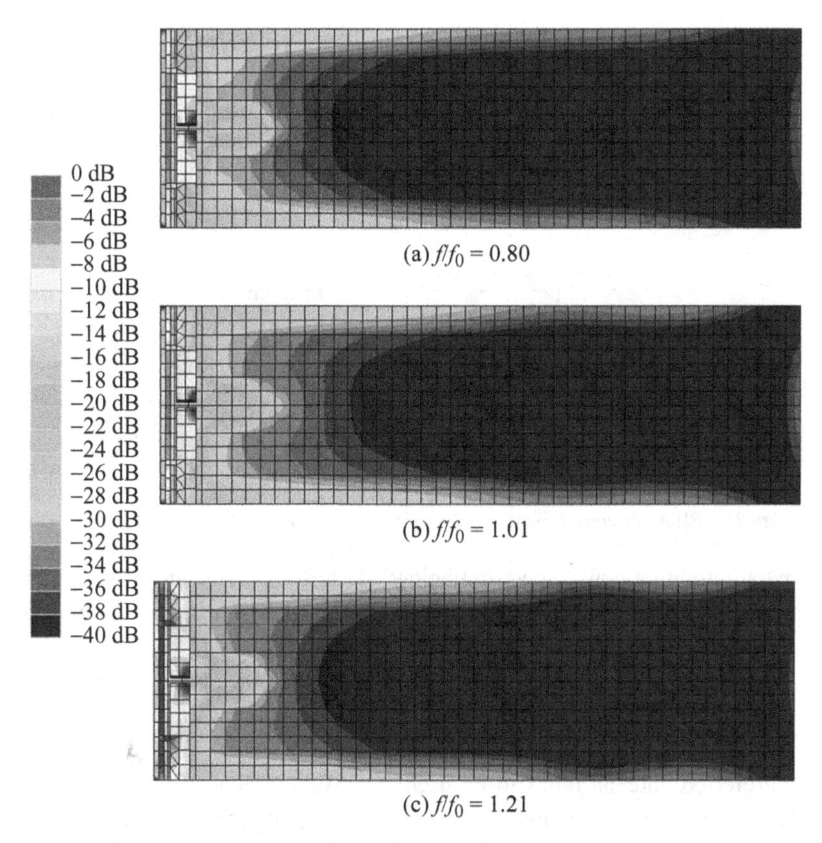


**Figure 8.48** Current distributions on the ground plane: (a) $f/f_0 = 0.80$, (b) $f/f_0 = 1.01$, and (c) $f/f_0 = 1.21$ ([27][28], copyright © 2001 and 2002 IEICE).

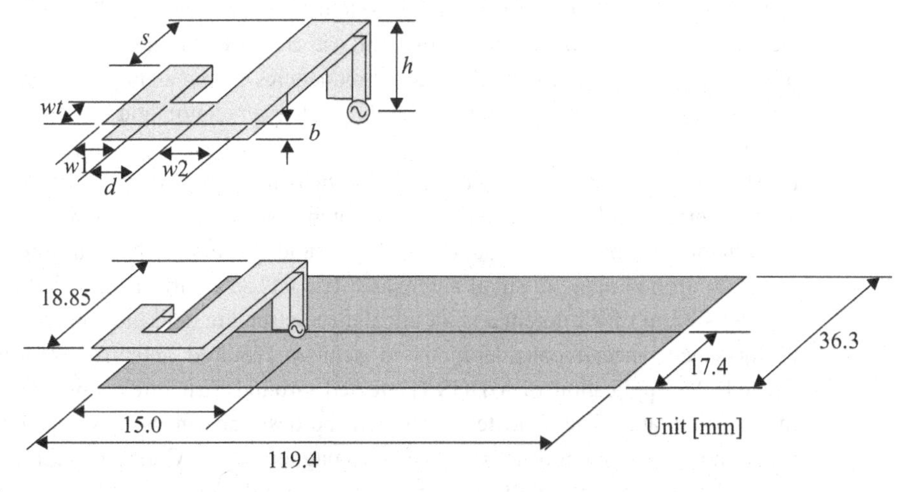


**Figure 8.49** BFMA configuration and the antenna on the ground plane ($w_1 = w_2 = 5$, $w_t = 1.5$, $b = 0.5$, $h = 7$, $s = 12$ [mm]) [27b].

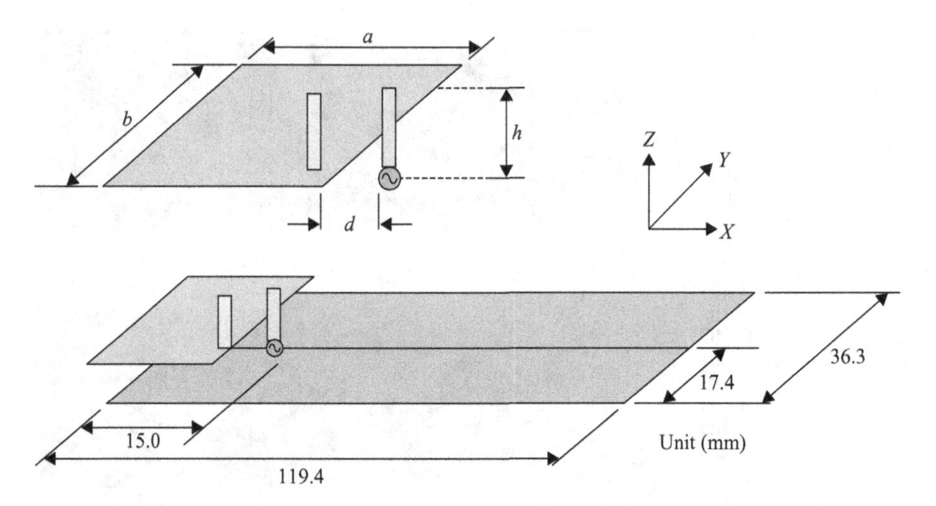


**Figure 8.50** PIFA for comparison ($a = b = 19.5, d = 5, h = 7$ [mm]) [27b].

operating with a power spectrum below –41.6 dBm. Since those systems operate with relatively low power, the communication link is limited to a short range and equipment is generally constructed in small size. Hence, antennas to be installed in such small equipment should accordingly be small. However, since the systems are designed for using signal transmissions that concern high-data-rate information, new challenges are imposed to develop small antennas yet having wideband.

Preferred antenna types to be used for UWB systems are planar ones with small dimensions, yet being simple, compact, low cost, and of course having appropriate performances that satisfy the system operation. In many cases electrically small antennas (ESA) with wideband features are required. Since conventional ESAs have generally narrow bandwidths, development of such small wideband antennas needs additional considerations beyond the techniques used to develop conventional ESAs. The simplest technique to produce wide bandwidth is full use of an area for the radiator, or generation of multiple resonances at closely spaced frequencies within a single antenna structure. For this purpose, a planar structure is fully used as a radiator, and current paths on the antenna surface are modified by modifying antenna shape and introducing slots/slits on the surface of the antenna. How to obtain a wide bandwidth with small-sized antennas is another serious problem. Reduction of the antenna size is attained not only by reduction of radiator size, but also reduction of the ground plane size. It should be noted that since the ground plane in small antennas often acts as a part of the radiator when the feeding system of the antenna is an unbalanced structure, and influences the radiation performance, special considerations to achieve required antenna performance are needed. An application of a DGS (Defected Ground Structure) concept can be one of the candidates. The DGS technique will be described in the next section, because it is a modified version of an EBG (Electromagnetic Band Gap) and exhibits bandgap properties and slow-wave effects that help to reduce the ground-plane size.

Planar structure is advantageous in applying antennas to very small equipment, as it can be easily embedded on a printed circuit board or integrated with other RF electronic

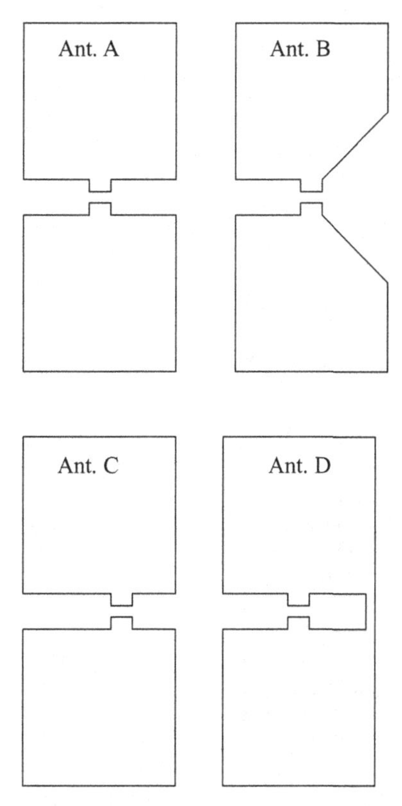


**Figure 8.51** Planar square dipole: (a) original dipole, (b) dipole with beveling at one-side, (c) asymmetrically fed dipole, and (d) dipole with a shorting stub at one side [32].

devices. Also feed and matching structures play an important role for achieving wideband performance.

In UWB antennas, time response as well as frequency response usually is studied, as the UWB system was initially used for impulse communication systems. However, time response is not discussed here, because antennas described here are generally applied to ordinary wideband communication and other wireless wideband systems that are not concerned with such impulse systems.

There have been many papers treating UWB antennas. Various types of UWB antenna have been introduced; dipole types [32, 33], monopole types [34–42], shape modified antennas [35–38, 46, 47], antennas with reduced ground plane [39], slot/slit embedded antennas [40–45], and so forth.

The representative types of UWB antennas will be described here.

*8.1.2.1.2.3.1 Dipole-type UWB antennas*

*8.1.2.1.2.3.1.1 Square patch dipoles*

The simplest dipole antennas are illustrated in Figure 8.51, which shows A: a center-fed square dipole, B: a beveled center-fed square dipole, C: an offset-fed square dipole and D: a center-fed square dipole with a shorting pin [32].

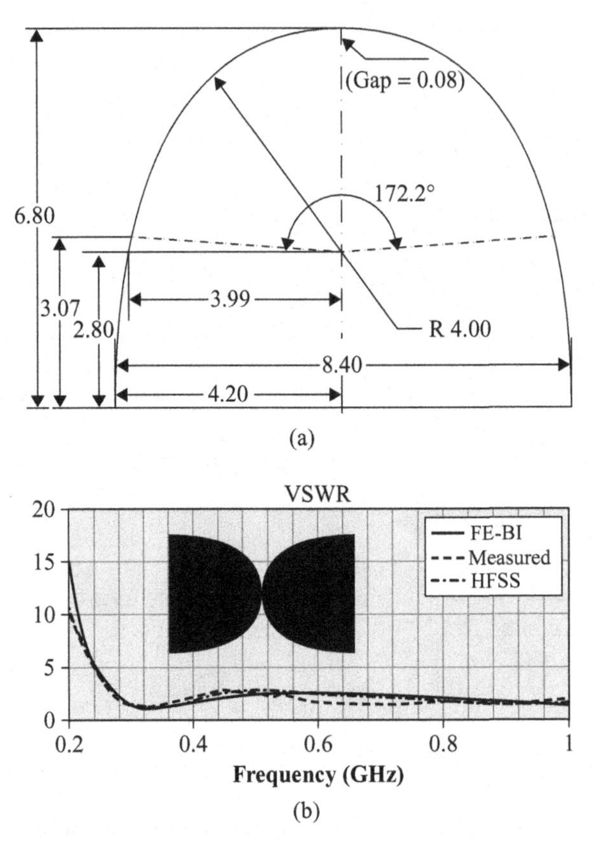


**Figure 8.52** Initial flare dipole (a) Geometry and dimensions (in inches) and (b) shaped dipole and frequency characteristics ([33], copyright © 2009 IEEE).

Impedance bandwidth for $S_{11} < -10$ dB of these antennas are: A: 3.60–8.21, B: 3.40–10.17, C: 3.61–15.30, and D: 2.87–10.04 (in GHz). In terms of bandwidth, antenna C has widest, while antenna A has the narrowest. In addition, antenna C exhibits the most stable impedance response compared with other types. Antenna D introduces an additional resonance at about 2.7 GHz due to the shorting pin, which also provides inductive reactance below 2.7 GHz, while other types are capacitive.

### *8.1.2.1.2.3.1.2 A flare dipole with shape optimized using splines*

A novel planar wideband antenna capable of operating from 190 MHz to 1000 MHz with gains greater than 0 dBi (having size $\lambda/5 \times \lambda/7$ at 190 MHz) is introduced in [33a]. The initial flare dipole has the outer curve which draws a 172.2° arc with 4 inches radius, being connected to a trapezoid as is depicted in Figure 8.52(a), where dimensions in inches are given. Bandwidth characteristics of the initial flare dipole are also shown. The shaped dipole is depicted in Figure 8.52(b), which has dimensions of 13.68 inches long (including the gap) and 8.40 inches wide, corresponding to $\lambda/3 \times \lambda/5$ at 280 MHz. Although the initial flare dipole has wide bandwidth of 3.5:1 (280 to 1000 MHz) as shown in Figure 8.52(b), further wide bandwidth of 5.25:1 (190 to 1000 MHz) is targeted and

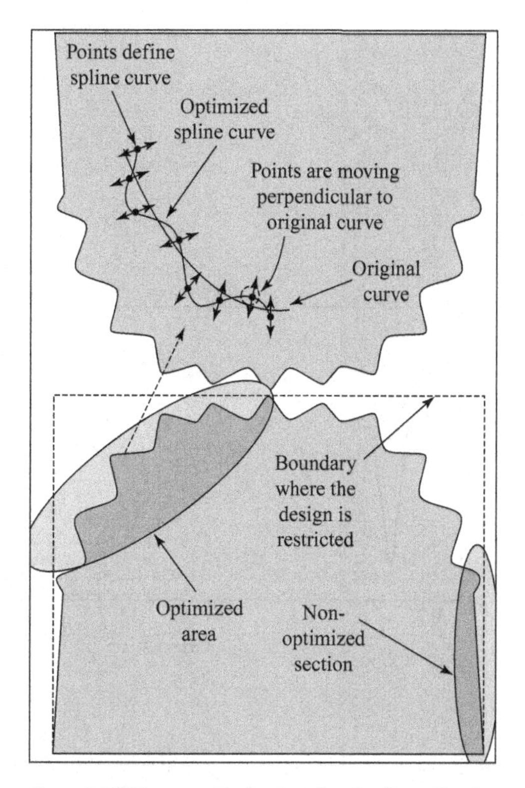


**Figure 8.53** Shape optimization for the flare dipole; a spline is defined at points placed around the perimeter of the flare dipole arc ([33a], copyright © 2009 IEEE and [33b], copyright © 2007 IEEE).

achieved by optimizing the flare shape and employing multiple stage L-C matching circuits at its feeding port. The initial flare dipole has gain greater than 0 dBi beyond 280 MHz, but the same gain at lower frequencies around 190 MHz without affecting bandwidth is achieved by an optimized flare dipole.

To lower the resonance frequency, the outer shape curved portion is modified by optimization of the shape, taking points distributed along the circular arc to define a spline as shown in Figure 8.53, illustrating an approach for shape optimization of the flare dipole; points are placed around the arc perimeter to define a Spline.

A printed shape-optimized dipole is illustrated in Figure 8.54: (a) the photo, (b) simulated and measured return loss, (c) measured gain for the original and optimized dipole, and (d) an enlarged version for the lower frequency regions of (c).

To match the antenna impedance, $N$ number of two port network/stages are designed by applying the optimizing process, which is described in [33b].

More about spline-shaped UWB antennas will be described in the next section.

### *8.1.2.1.2.3.2 Monopole UWB antennas*

The simplest monopole UWB antenna is a half-sized planar square dipole with the ground plane on the same plane. Most monopole-type UWB antennas have modified

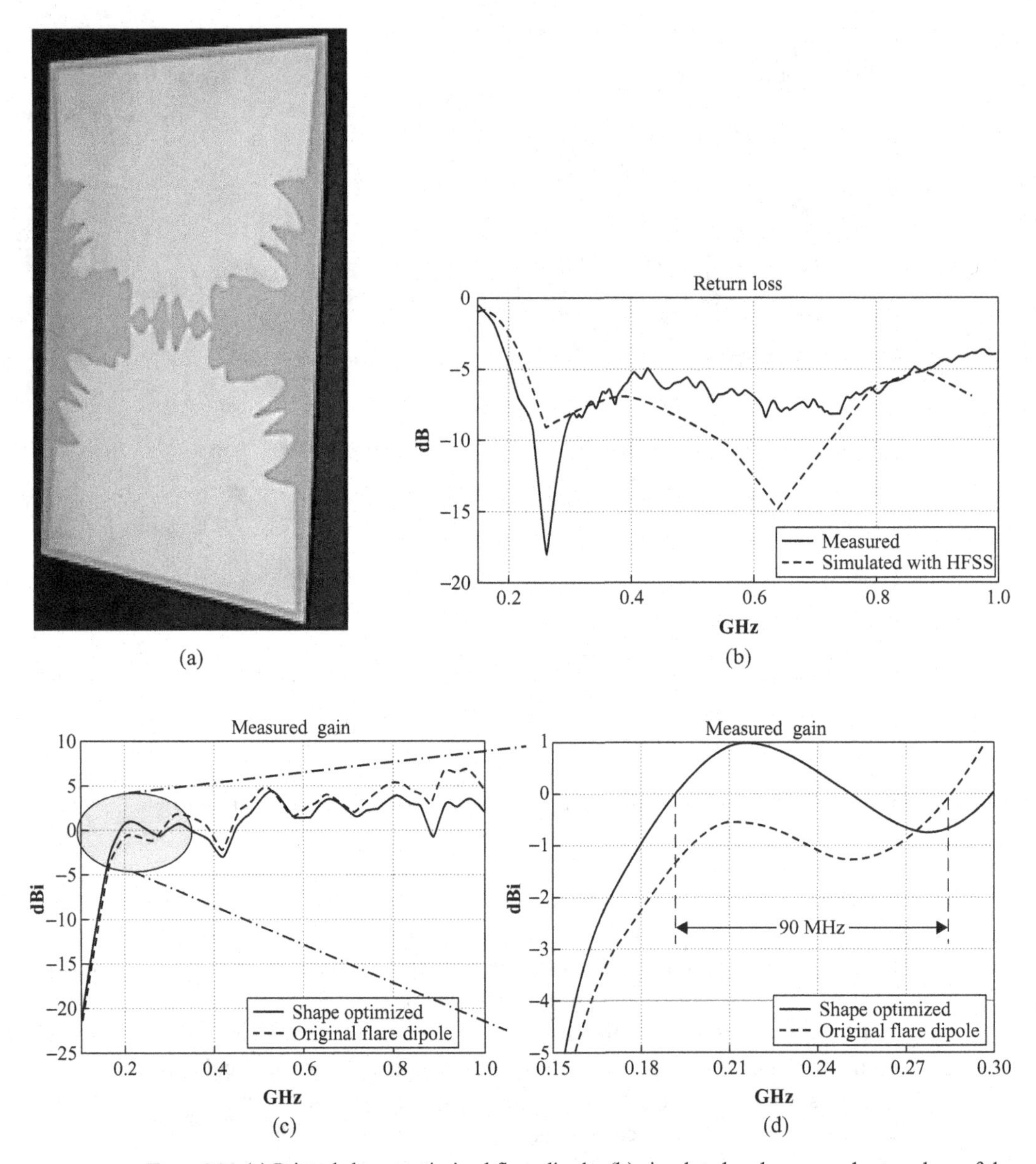


**Figure 8.54** (a) Printed shape optimized flare dipole, (b) simulated and measured return loss of the optimized dipole, (c) measured gain of the original and optimized dipole, and (d) expanded version of (c) around 0.2 GHz. ([33], copyright © 2009 IEEE).

shape, that is, the patch sides are modified to attain longer current paths, and the number of current paths increases so as to achieve multiple resonances beginning from lower frequencies, and consequently produce wideband behavior. The representative ones are illustrated in Figure 8.55, where (a) and (b) are a beveled square patch [34], (c) a patch with binomial curve [35], (d) with staircase-profile [36], and (e) with spline-shape [38].

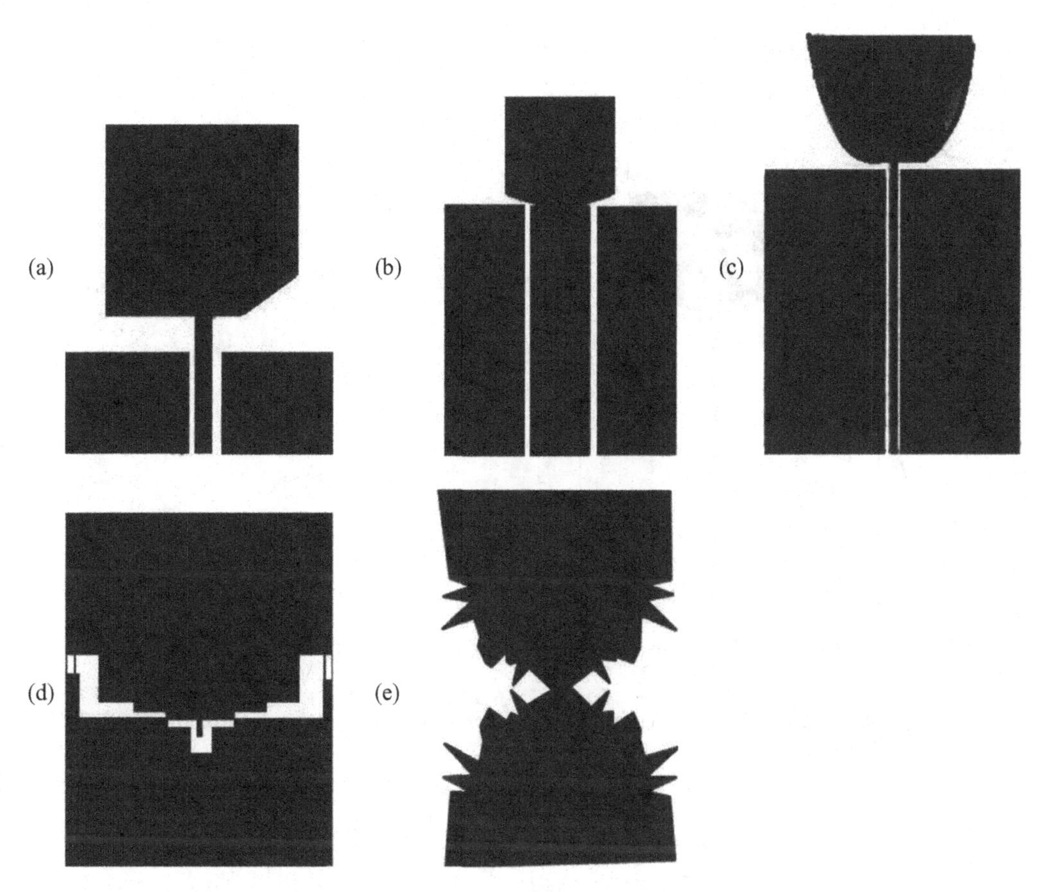


**Figure 8.55** Various planar monopole antennas.

Most monopole UWB antennas have planar structure and are placed on the same plane with the reduced-size ground plane that allows reduction of total size (volume) of the antenna and makes integration of an antenna into electronic devices easy.

*8.1.2.1.2.3.2.1 Binomial-curved patch antenna*

The edge curve of a printed patch modified in shape according to the binomial function is introduced in [35]. Antenna geometry is illustrated along with geometrical parameters in Figure 8.56. The curved boundary as a function of the coordinates $(x, y)$ is expressed by

$$y = f(x) = G + \ell(x/2w)^N \tag{8.1}$$

where $G$ is the gap between the patch and the ground plane, $w$ is the width of the top side of the patch, $\ell$ is the length of the patch, and $N$ is the order of the binomial function. As $N$ increases from unity toward infinity, the shape varies from triangular to nearly rectangular. By parametric analysis, it was found that the optimum values for $G$ and $N$ are 0.45 mm and 4, respectively, with other parameters fixed; $w = 30$ mm,

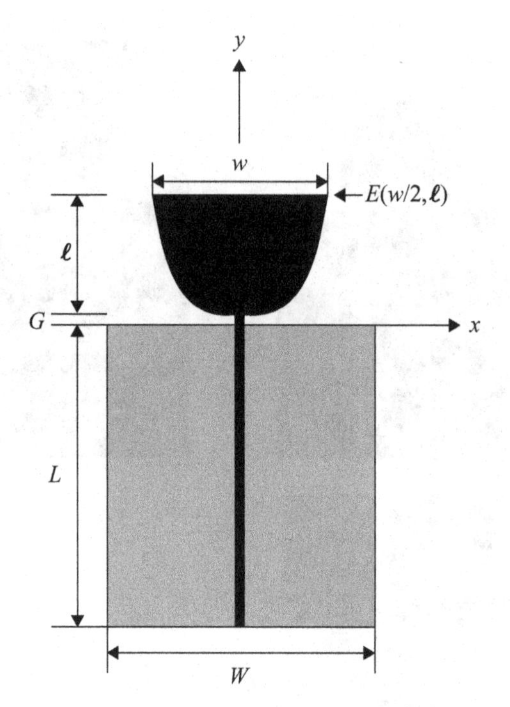


**Figure 8.56** Geometry of the planar binomial-curved monopole antenna ([34], copyright © 2010 IEEE).

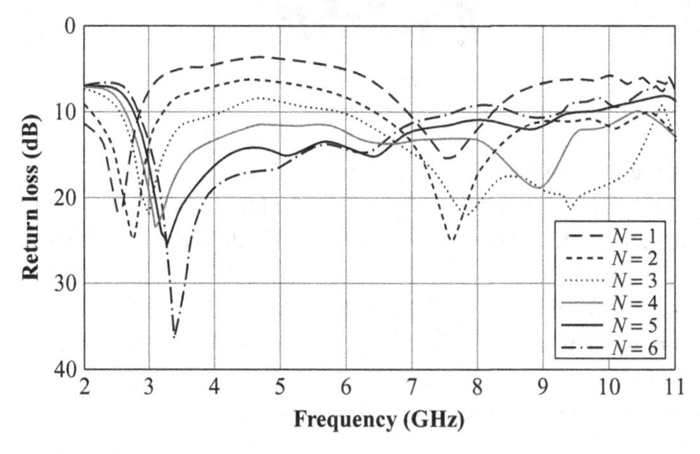


**Figure 8.57** Simulated return loss of the antenna with different $N$ ([35], copyright © 2008 IEEE).

$\ell = 20$ mm, $W = 46$ mm, and $L = 50$ mm. With these parameters, the widest 10-dB return loss bandwidth obtained by simulation was from 2.7 GHz to 11 GHz, while from 2.59 to 10.97 GHz by measurement. Simulated return loss is shown in Figure 8.57. The measured peak (average) gain for the frequencies 3.1, 5.0, and 8.0 GHz, respectively, are –0.88 (–3.65), –2.36 (4.51), and 1.54 (–2.1) (all in dBi).

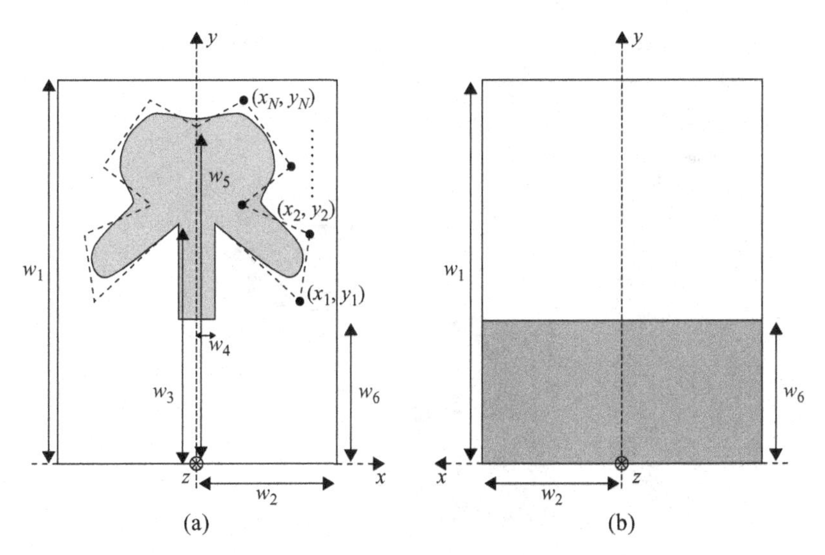


**Figure 8.58** Spline curved planar monopole antenna, indicating points for defining spline; (a) front view and (b) back view ([38a], copyright © 2007 IEEE).

### *8.1.2.1.2.3.2.2 Spline-shaped antenna*

A spline-shaped UWB antenna was synthesized [38a]. An innovative design approach based on the use of a spline description is applied to create a novel UWB antenna geometry. It is also used for formulation of the synthesis in terms of return loss at the input port and coupling properties of a system with identical antennas modeling the UWB communication. A suitable implementation of the PSO (Particle Swarm Optimization) has been integrated with spline-based shape generator and a MoM-based electromagnetic simulator.

The representative parameters to be optimized are shown in Figure 8.58, where the coordinates of the $n$th control points to be determined by the optimization are given as $P_n\ (x_n, y_n)$, taken on the coordinate $(x, y)$. Here $n = 1,2,\ldots N$; $N$ being the total number of the control points used to describe the antenna geometry.

The antenna is printed on the front side of the substrate (thickness 0.78 mm and $\varepsilon_r = 3.38$), having length $w_1 = 69.2$ mm and half-width $w_2 = 10$ mm, and the ground plane of length $w_6 = 51$ mm is printed on the lower part of the back side of the substrate. The antenna geometry is characterized by the array of geometric variables

$$X = \{(x_n, y_n), n = 1, \ldots, N; w_1, w_2, \ldots, w_6\}. \tag{8.2}$$

In the UWB communication system, impedance matching and distortionless conditions for the UWB bandwidth are imposed as the electrical constraints. As for the impedance matching over the UWB bandwidth,

$$|S_{11}(f)| \leq -10 \text{ dB}.$$

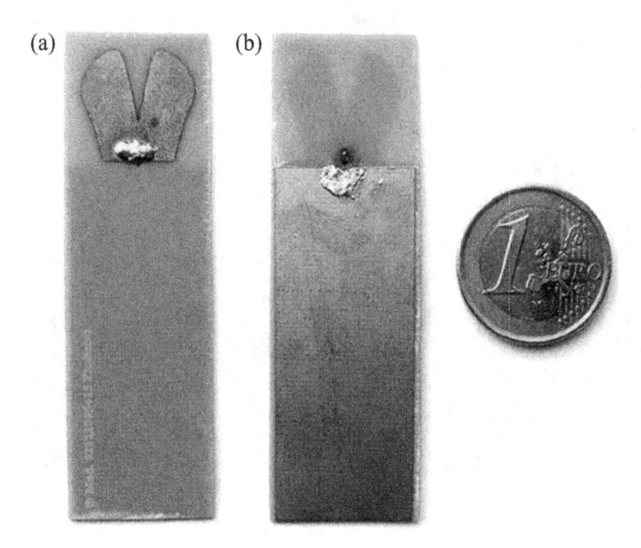


**Figure 8.59** A proto-type dongle antenna: (a) front view and (b) back view ([38a], copyright © 2007 IEEE).

And for the condition of distortionless system, the antenna is required to satisfy a condition pertaining to the magnitude of $S_{21}$, $|S_{21}(f)|$ to be

$$\Delta|S_{21}| \leq 6 \text{ dB}.$$

And the group delay $\tau_g$,

$$\Delta\tau_g \leq 1 \text{ ns}.$$

Here $\Delta|S_{21}|$ and $\Delta\tau_g$, respectively, denote the maximum variation in the whole frequency band of $|S_{21}|$ and $\tau_g$.

The antenna is required to be placed on the platform of the size 100 × 60 (in mm).

By the optimization, coordinates of the control points are; $p_1$ (6.9, 50.6), $p_2$ (9, 55.5), $p_3$ (7, 62.7), $p_4$ (2.7, 66.3), and $p_5$ (1.9, 61.8) all in mm. As for the feeding line, $w_4$ = 5.4, and $w_3$ and $w_5$, which define the range of contour variations along the $y$-axis, are 51.6 and 56, respectively. A prototype antenna is illustrated in Figure 8.59, (a) front view and (b) back view. Simulated and measured return loss is shown in Figure 8.60.

The spline-shaped UWB antenna is applied to integrate in a wireless USB (Universal Serial Bus) dongle [38b]. It has a miniaturized planar structure with maximum extension 39.2 × 19.2, within which the radiator occupies only an area of 16.2 × 19.2 mm. The antenna is printed on a two-sided dielectric substrate ($\varepsilon_r = 3.38$ and thickness 0.78 mm) and the geometry is defined by the set of values of the descriptive parameters; $\varphi_1$ (length of the substrate) = 39.2, $\varphi_2$ (half-width of the substrate) = 9.6, $\varphi_3$ (half width of the feeding point) = 2.1, and $\varphi_4$ (length of the ground plane) = 23.0 (in mm). Antenna geometry is determined by the spline-description, giving control points on the $(x, y)$

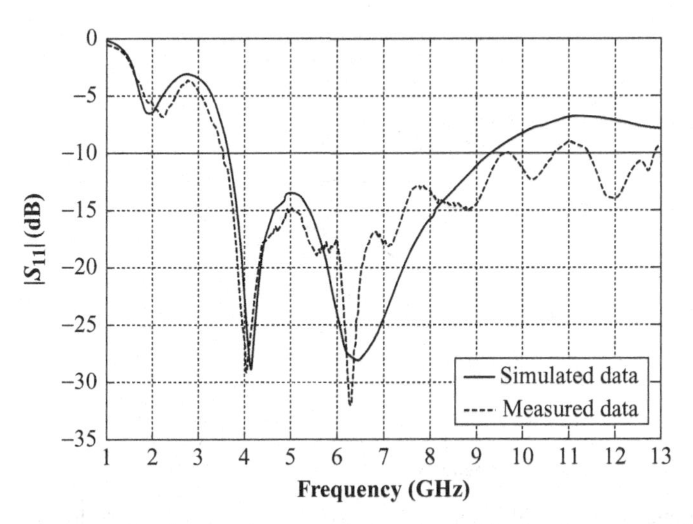


**Figure 8.60** Simulated and measured return loss of the dongle antenna ([38a], copyright © 2007 IEEE).

**Figure 8.61** Geometry of a UWB dongle antenna front view ([38b], copyright © 2008 IEEE).

coordinates as: $p_1 = (2.1, 25.0)$, $p_2 = (6.6, 29.5)$, $p_3 = (8.6, 29.4)$, $p_4 = (7.3, 35.9)$, $p_5 = (6.9, 34.9)$, $p_6 = (2.2, 32.4)$, and $p_7 = (0.0, 33.8)$. These points are given on the antenna geometry illustrated in Figure 8.61.

Measured and simulated return loss is given in Figure 8.62. It shows bandwidth of 2 GHz from 3 GHz to 5 GHz for −10 dB return loss. In terms of $\Delta\ |S_{21}|$, it is 5 dB, which is smaller than the requirement, with average $|S_{21}|$ it is about −23 dB.

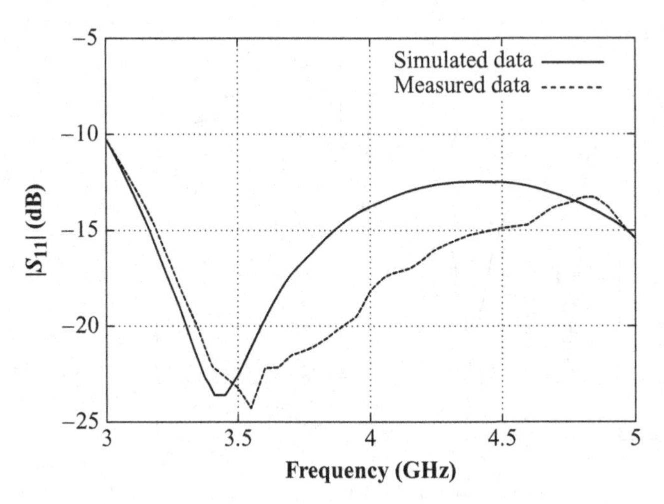


**Figure 8.62** Return loss of a UWB dongle antenna ([38b], copyright © 2008 IEEE).

### *8.1.2.1.2.3.3 UWB antennas with slot/slit embedded on the patch surface*

Slots/slits are embedded on the patch surface in order to lengthen the current paths and increase the number of currents so that multiple resonances occur, leading to production of ultra wide bandwidth. Use of slots/slits has another objective; that is, to produce stop bands within the UWB system band for avoiding interference against other wireless systems, for instance, WLAN (5-GHz bands). For another purpose, a slit is used on the patch to reduce the current on the ground plane so that contribution of the ground plane to radiation is reduced. This leads to reduction in the size of the ground plane at the same time.

Various shapes of slots/slits are applied to patch antennas, depending on the purposes. Examples of such patch antennas with band-notch performance are U-slot on a square patch [39], circular/elliptical patch [40a], circular/ elliptical slot [40b], circular/elliptical slot with U-shaped tuning stub [40c], H-shaped plate and rectangular slots [41], rectangular patch with a notch and a strip [42], pentagon shaped-slot [43], tapered ring slot [44], and octagonal wide slot with square ring [45].

#### *8.1.2.1.2.3.3.1 A beveled square monopole patch with U-slot*

A beveled square monopole patch is introduced in the previous section [20] as one of the useful UWB antennas. In order to avoid interference from other wireless systems operating in the UWB band, a slot/slit (notch) is embedded on the patch surface. Figure 8.63(a) depicts a square beveled monopole patch along with $S_{11}$ characteristics, and the patch, in which a thin U-slot is employed, is shown in Figure 8.63(b), which also provides dimensional parameters [39]. Figure 8.63(c) is the $S_{11}$ characteristics, giving a stop band within the UWB band as a result of a U-slot application. On a beveled square monopole patch antenna, four mode currents $J_0, J_1, J_2$, and $J_3$ flow on the surface as Figure 8.64 illustrates. $J_0$ is loop current, which is a special non-resonant inductive mode, $J_1$ is vertical current flowing along the monopole, associated with resonance at

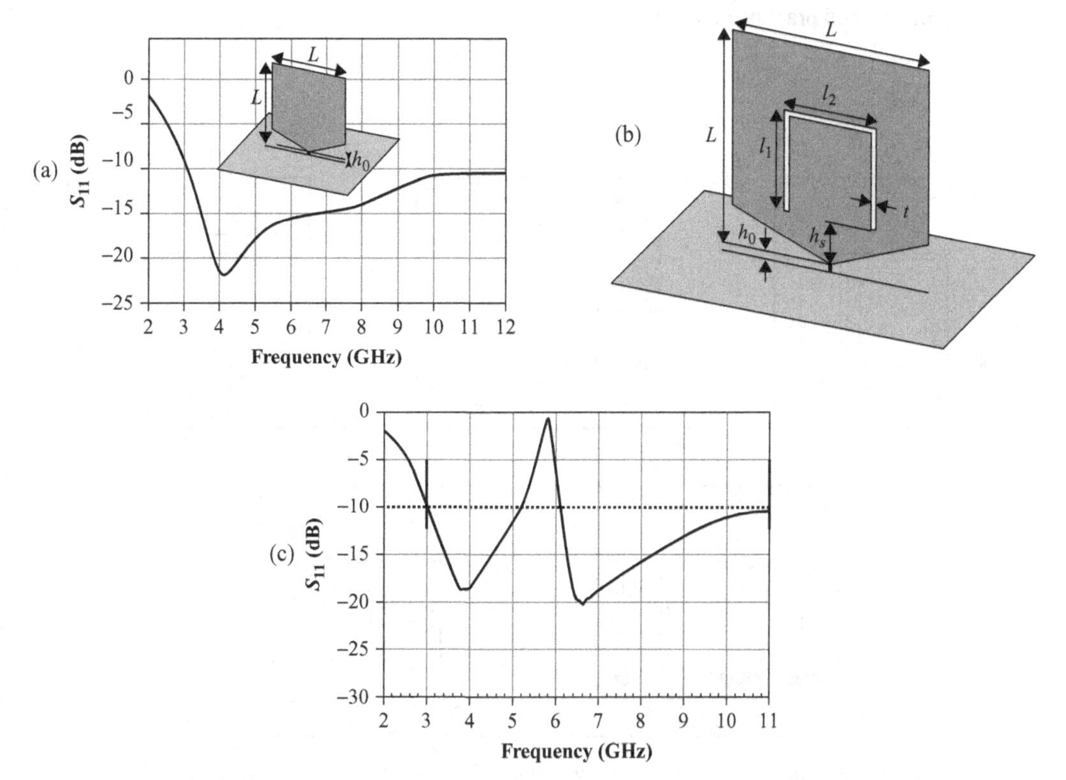


**Figure 8.63** (a) Geometry of a beveled square planar monopole ($L = 19$ mm, $h_0 = 0.2$ mm) with $S_{11}$ characteristics, (b) geometry of a beveled square monopole with a resonant U-slot ($L = 19$, $l_1 = 10$, $l_2 = 8$, $t = 1$, $h_s = 4$, and $h_0 = 0.2$ (in mm)), and (c) reflection coefficient of the antenna ([39], copyright © 2010 IEEE).

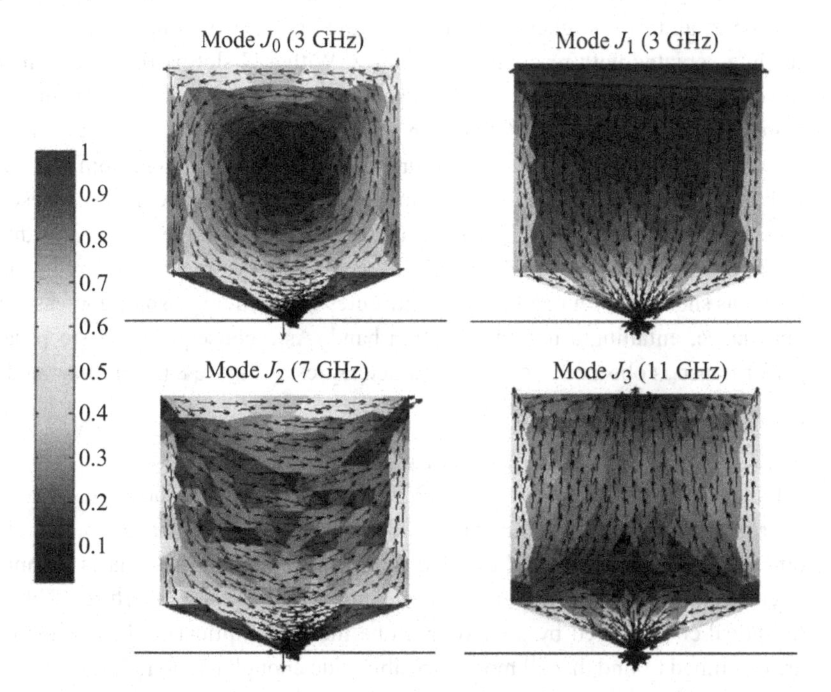


**Figure 8.64** Normalized current distributions on the beveled square monopole antenna for the first four characteristic modes ([39], copyright © 2010 IEEE).

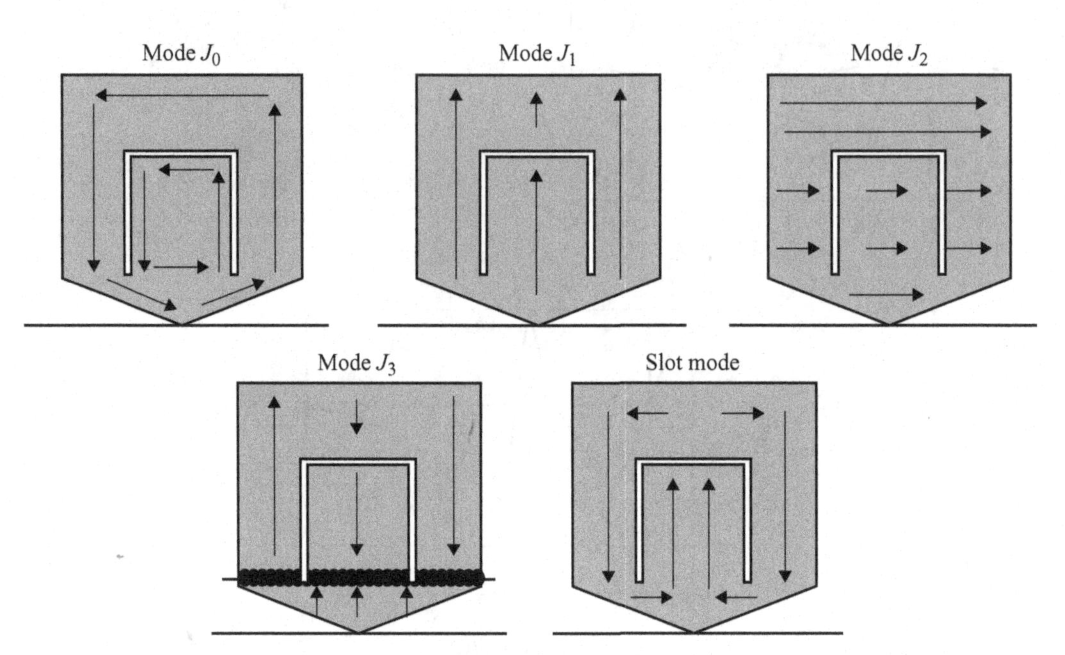


**Figure 8.65** Symbolic patterns of the current distributions on the beveled square monopole antenna loaded with a thin inverted-U-slot for the first four characteristic modes ([39], copyright © 2010 IEEE).

2.75 GHz, $J_2$ is horizontal current parallel to the ground plane, associated with resonance at 7 GHz, and $J_3$ is higher-order-mode vertical current with null near the base of the patch, associated with resonance at 12 GHz. With a U-slot on the patch, an additional current mode is produced, generating an additional resonance in the structure. Graphical representation of the current flows on the beveled square patch with a thin U-slot is illustrated in Figure 8.65, in which current flows of five modes, from $J_0$ to $J_3$ and an additional mode, referred to as a slot mode $J_s$, are shown with arrows. These currents, except $J_s$, flow in paths similar to those on the patch without the U-slot. $J_s$ is divided into two at the center of the patch, and each of them flows symmetrically circling around the U-slot as shown in Figure 8.65, and contributes to producing a sharp increase in antenna impedance, entailing a narrow rejection band. As a consequence, a steep band-reject performance is observed in the $S_{11}$ characteristics as Figure 8.63(c) illustrated.

### *8.1.2.1.2.3.3.2 Circular/Elliptical slot UWB antennas*

Circular/Elliptical CPW-fed slot UWB antennas are introduced in [40a], where a microstrip line-fed UWB antenna is also discussed. Antenna geometry along with dimensional parameters is illustrated in Figure 8.66. The antenna is comprised of a circular/elliptical stub that excites a similar-shaped slot aperture. Three models (elliptical/circular) fed by a CPW and one model (elliptical) fed by a microstrip line are examined to find that all models exhibit wide enough bandwidth for UWB operation with satisfactory radiation efficiency and radiation patterns. The antenna element is

**Table 8.5** Dimensional parameters of four antennas I, II, III and IV ($h = 1.575$ mm and $\varepsilon_r = 3$) ([40a], copyright © 2006 IEEE)

| in (mm) | $L$ | $W$ | $L_1$ | $R_1$ | $L_2$ | $R_2$ | $d$ | $dw$ | Type |
|---|---|---|---|---|---|---|---|---|---|
| Prototype I | 40 | 35 | 6 | 8 | 12 | 16 | 8 | 0.3 | CPW Elliptical |
| Prototype II | 40 | 40 | 7.5 | 7.5 | 15 | 15 | 8 | 0.3 | CPW Circular |
| Prototype III | 90 | 90 | 20 | 20 | 35 | 35 | 12 | 0.3 | CPW Circular |
| Prototype IV | 40 | 35 | 6 | 8 | 12 | 16 | 8 | 0.3 | Microstrip Elliptical |

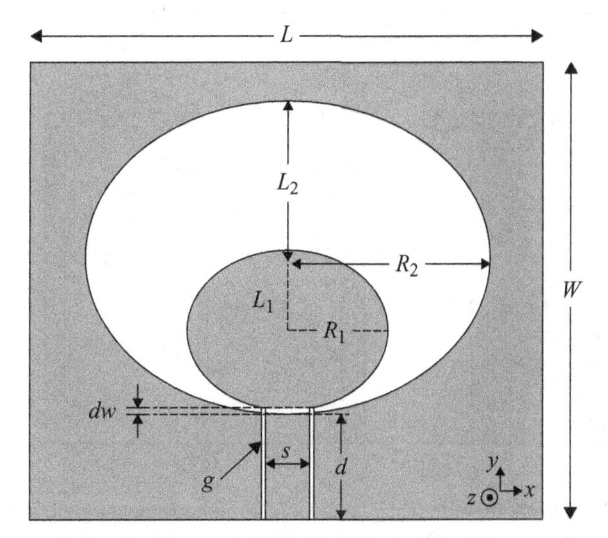


**Figure 8.66** Geometry of antenna and dimensional parameters ([40a], copyright © 2006 IEEE).

etched on the substrate of 1.575 mm thickness with $\varepsilon_r = 3$, and dimensions of the four types of antenna given in Table 8.5 are used.

From the experimental results, bandwidth observed for the Prototype I (elliptical model) is 17.35 GHz (from 2.65 to 20 GHz), or 153%. For the Prototype II (circular model) it is 17.05 GHz (2.95 GHz to 20 GHz), or 148%, and the Prototype III (circular model) demonstrates 175%, beginning from 1.3 GHz that is a lower frequency than that of other models, being at 2 GHz band, as a consequence of the enlarged ground plane. Regarding the radiation patterns, an almost omnidirectional profile is observed in lower frequencies, becoming somewhat directional in higher frequencies. The maximum gain obtained is about 4.5 dBi.

Design of a slightly modified circular/elliptical slot UWB antenna is described in [40b]. Geometries of the elliptical- or circular-shaped monopole antennas are depicted in Figure 8.67(a) and (c), respectively, and the complementary versions of these antennas are also shown in (b) and (d). The antenna structure is formed with planar conducting surfaces of either two ellipses or circles in a two-sided conductor-coated substrate. The primary radiating element and the microstrip feed are on one side of the substrate, while the ground plane is on the other side. The radiating slot is formed by the intersection

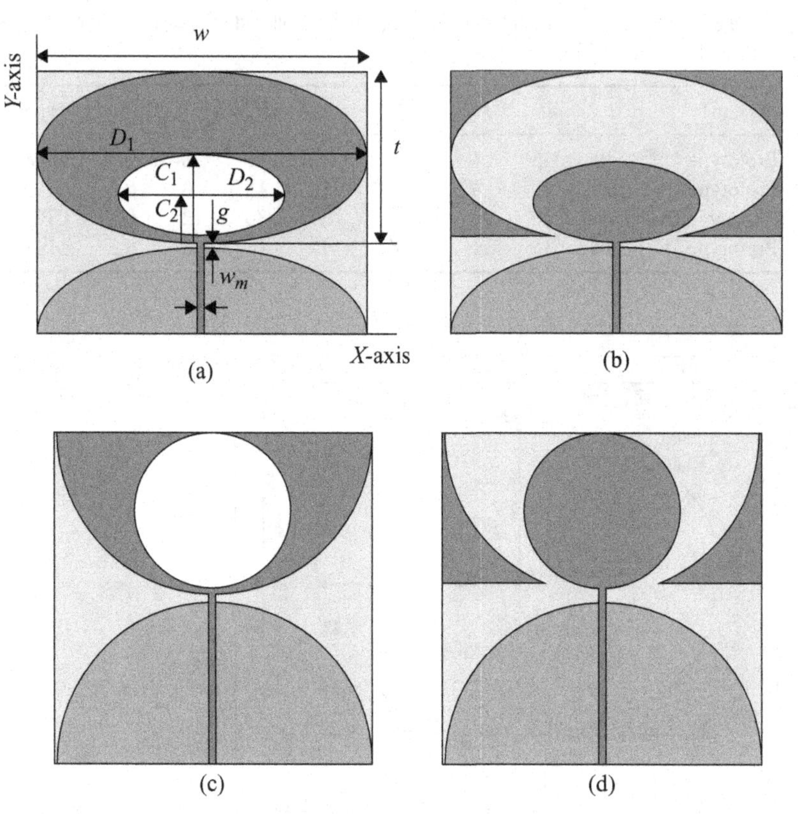


**Figure 8.67** Four UWB antennas: (a) elliptical E-monopole, (b) complementary elliptical E-monopole, (c) circular E-monopole, and (d) complementary circular E-monopole ([40b], copyright © 2009 IEEE).

of two ellipses/circles in the manner shown in Figure 8.66. The size of the slot opening determines the lowest frequency $f_\ell$ of the operation and is selected to be $\lambda_{eff}/2$ ($\lambda_{eff} = c/f_\ell$ : effective wavelength, $\varepsilon_{eff} = (\varepsilon_r + 1)/2$ and $c$ : velocity of light). The dimensions $D_1$ and $D_2$ are chosen as $D_1 = w$ and $D_2 = w/2$, respectively, and $w$ and $\ell$, respectively, are selected to be $\lambda_{eff}/2$ and $\lambda_{eff}/4$. The ground plane has the shape of a half ellipse (circle) for the elliptical (circular) monopole whose dimensions are chosen to be similar to those for the larger ellipse (circle) of the radiating structure. Dimensions of the center of large and small ellipses (circles) $C_1$ and $C_2$, respectively, measured from the end of the feeder are determined by taking ratios $R_1$ and $R_2$ as $C_1 = D_1\ R_1/2$ and $C_2 = D_2\ R_2/2 + w_m$, where $w_m$ is the width of the feed line and $R_1$ and $R_2$ take a value 0.5 for the ellipse and 1.0 for the circle. Here the width $g$ between the radiator and the ground plane is taken as $w_m/2$, which is around half of the feeder width. For the complementary versions, $C_2 = D_2\ R_2/2$.

Four types of antenna models were manufactured, and analysis of antenna performance was carried out by changing the substrate having different $\varepsilon_r$ and thickness. The results exhibit that the proposed design formulas enable the development

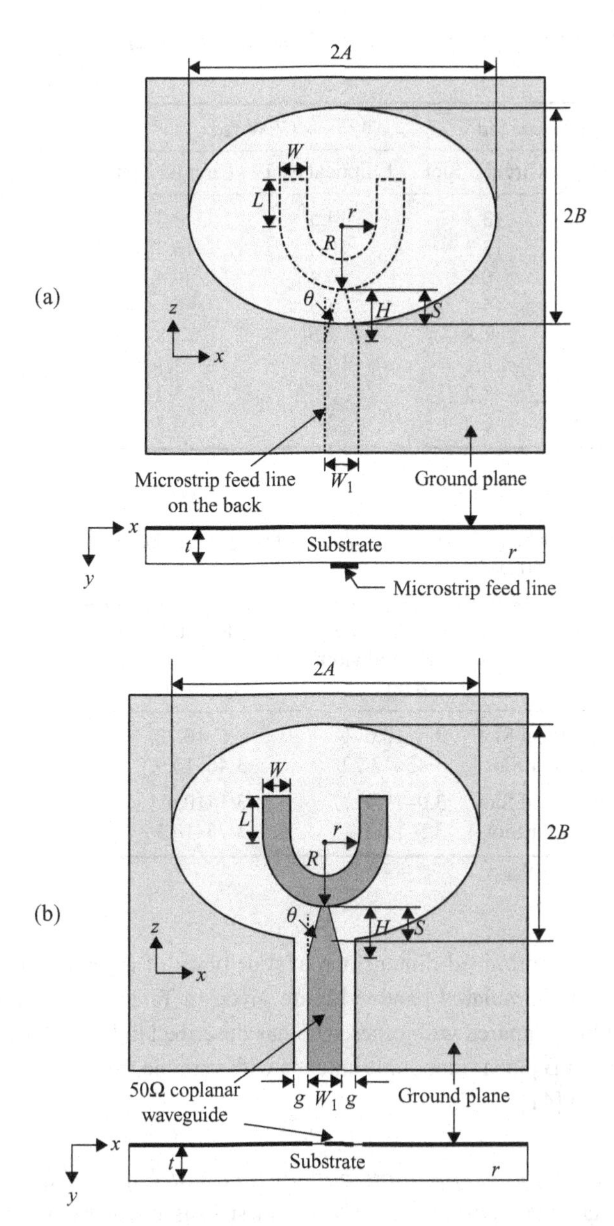


**Figure 8.68** Geometry of printed elliptical slot antenna with U-stub fed by (a) microstrip line and (b) CPW ([40c], copyright © 2006 IEEE).

of UWB antennas with suitable radiation characteristics, covering UWB bandwidth with well-behaved omnidirectional radiation patterns and more than 90% radiation efficiency.

Elliptical/circular slot antennas with U-shaped stub [40c] are studied both theoretically and experimentally and shown to have satisfying UWB characteristics with smaller size. Figure 8.68 illustrates antenna geometries of different feeding types by (a) microstrip

**Table 8.6** Optimized dimensions of printed elliptical/circular slot antenna ([40c], copyright © 2006 IEEE)

| | Microstrip line fed | | CPW fed | |
|---|---|---|---|---|
| | Elliptical Slot | Circular Slot | Elliptical Slot | Circular Slot |
| $A$ (mm) | 16 | 13.3 | 14.5 | 13.3 |
| $B$ (mm) | 11.5 | 13.3 | 10 | 13.3 |
| $S$ (mm) | 0.6 | 0.5 | 0.4 | 0.4 |
| $R$ (mm) | 5.9 | 5 | 5.5 | 5 |
| $r$ (mm) | 2.9 | 1.8 | 2.5 | 1.8 |
| $H$ (mm) | 3.3 | 3.1 | 2.5 | 3.1 |
| $W$ (mm) | 3 | 3.2 | 3 | 3.2 |
| $L$ (mm) | 6 | 6.7 | 3 | 4.3 |

**Table 8.7** Measured and simulated bandwidth of printed elliptical/circular slot antennas ([40c], copyright © 2006 IEEE)

| | | Simulated –10 dB bandwidth (GHz) | Measured –10 dB bandwidth (GHz) |
|---|---|---|---|
| Microstrip line fed | Elliptical Slot | 2.6–10.6 | 2.6–10.22 |
| | Circular Slot | 3.45–13.22 | 3.46–10.9 |
| CPW fed | Elliptical Slot | 3.0–11.4 | 3.1–10.6 |
| | Circular Slot | 3.5–12.3 | 3.75–10.3 |

line and (b) CPW. With optimized dimensions of four types of antennas tabulated in Table 8.6, measured and simulated bandwidth are given in Table 8.7, demonstrating much wider bandwidth compared with other antennas described in [40d–40f].

A novel modified UWB planar monopole antenna with variable frequency band-notch function is described in [41].

### *8.1.2.1.2.3.3.3 A rectangular monopole patch with a notch and a strip*

A small printed rectangular patch with a notch and strip is described in [42a]. The antenna is featured in reduced ground plane effect by cutting a notch from the radiator and attaching a strip asymmetrically to the radiator, while keeping wide bandwidth covering the UWB band. Figure 8.69 depicts antenna geometry, showing (a) a printed rectangular monopole patch and (b) a rectangular monopole patch with a notch and a strip. By slotting the radiator and/or modifying the shape of the radiator as well as the ground plane, the size can be reduced to 30 × 30 or 25 × 25 (in mm) from 40 × 50 mm, a usual size for similar printed antennas [42b, 42c].

In addition, by adding a strip to the top side of the rectangular radiator, the length of the radiator can be reduced to 30 mm as illustrated in Figure 8.69(b). Further reduction of the

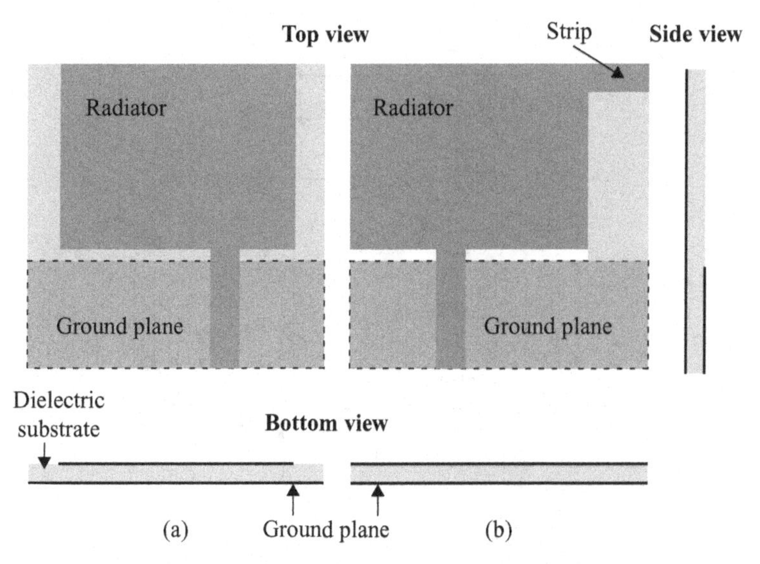


**Figure 8.69** (a) Planar rectangular monopole with the finite ground plane and (b) the antenna with a strip ([42a], copyright © 2007 IEEE).

antenna size can be attained by cutting a notch on the radiator as shown in Figure 8.70(a), where dimensional parameters are also described. Presence of the notch on the radiator leads to concentration of the current distributions on the right portion of the radiator, where the notch is cut, while the currents at the left portion of the radiator as well as the ground plane are very weak. Figure 8.70 illustrates the current distributions on the radiator and the ground plane, comparing (b) with notch and (c) without notch at 3, 5, 6, and 10 GHz. Observing this current distribution, it can be said that the notch plays a significant role to determine the lower operating frequencies, and subsequent impedance matching at around 3 GHz will become more sensitive to the notch dimension than the shape and size of the ground plane. The reason for this is that the currents on the ground plane are much weaker than those on the radiator. Thus it is important to notice that the effects of the ground plane and RF cable on the antenna performance at lower frequencies can be suppressed greatly by the notch [42a]. As the operating frequency increases, current flow becomes stronger on the feeding strip, on the junction of the radiator and the feeding strip, and on the ground plane. Thus impedance matching is greatly affected by the gap $g$ between the patch and ground plane. The lowest frequency $f_\ell$ is determined by the path length $L_\ell$ of the current flow around the notch on the right portion of the patch, which is the sum of the horizontal path from the feeding point, the vertical path from the bottom of the radiator, and the length and the width of the horizontal strip, giving $f_\ell = c/\lambda_\ell(\lambda_\ell = 2L_\ell\sqrt{\varepsilon_r + 1/2})$.

Here $f_\ell = 3.10$ GHz. Simulated and measured return loss is provided in Figure 8.71. The radiation patterns are almost omnidirectional at lower frequencies, while more directional at higher frequencies. The radiation efficiency varies from 79% to 95% across the entire bandwidth 3.1–10.6 GHz.

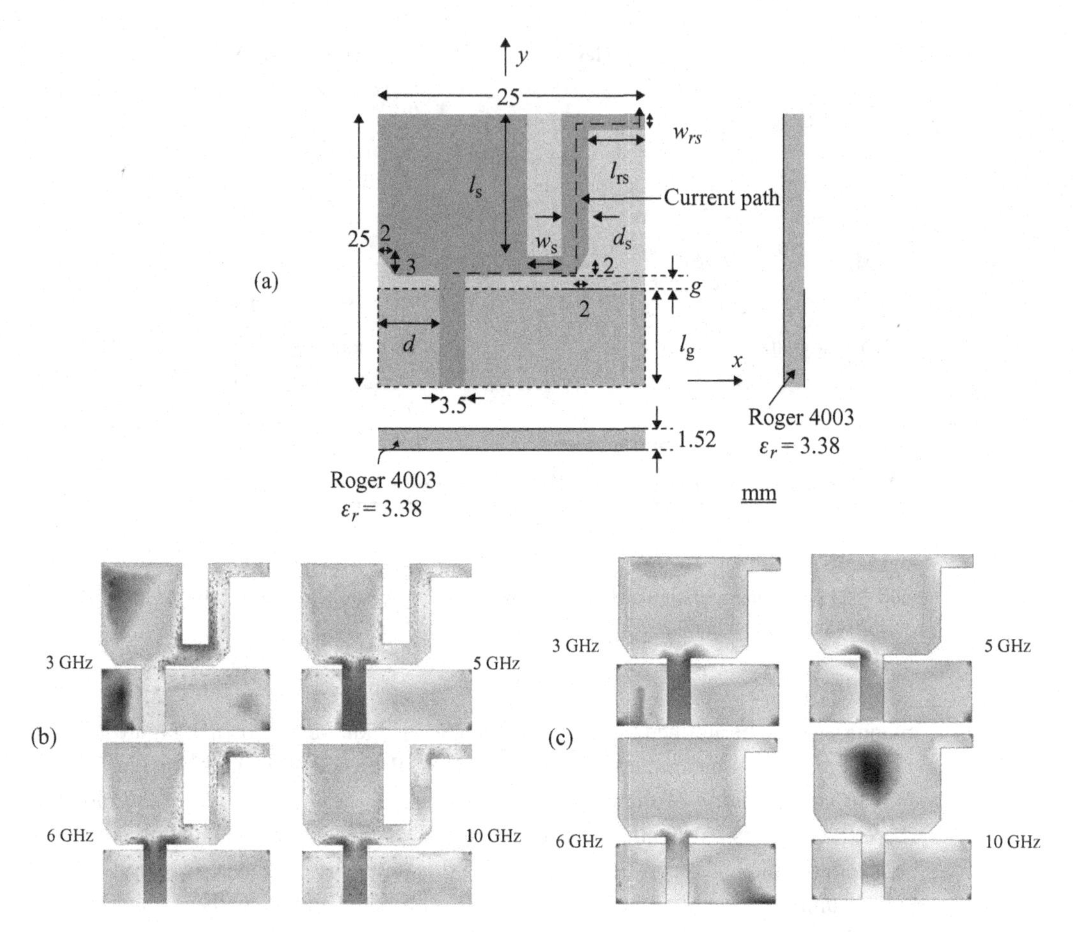


**Figure 8.70** Planar rectangular monopole antenna with a strip and a notch; (a) antenna geometry, (b) current distributions on the antenna with a notch, and (c) the antenna without notch ([42a], copyright © 2007 IEEE).

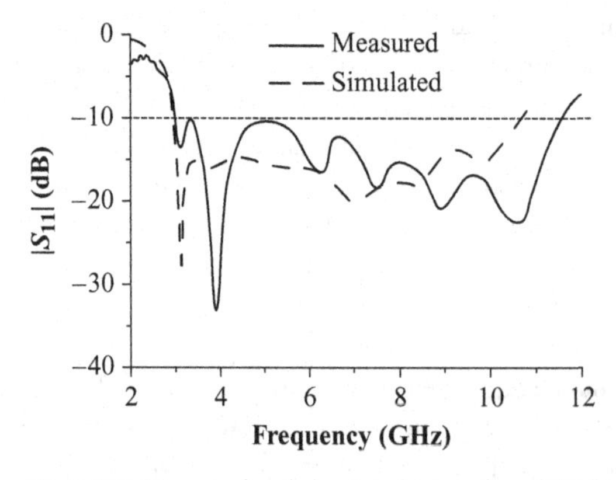


**Figure 8.71** Measured and simulated return loss ([42a], copyright © 2007 IEEE).

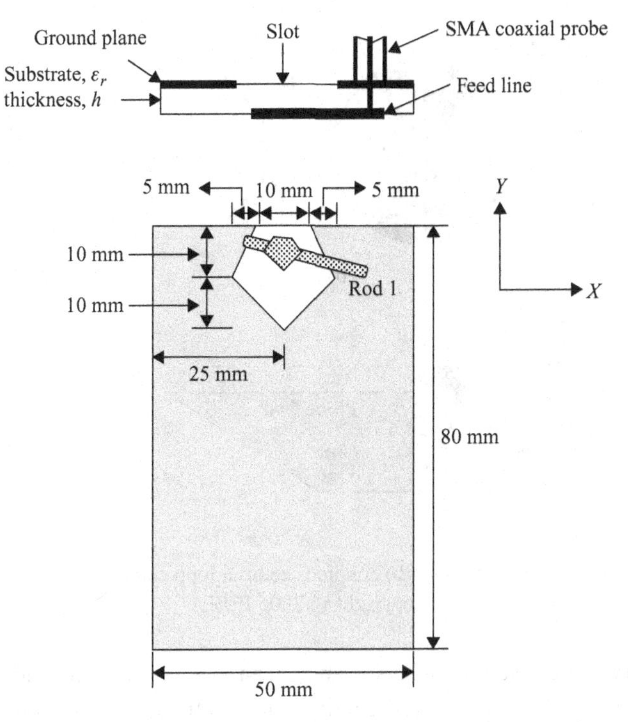


**Figure 8.72** Pentagon-shape microstrip slot antenna fed by a microstrip line ([43], copyright © 2009 IEEE).

### *8.1.2.1.2.3.4 Modified shaped UWB antennas*

Planar antennas with modified shapes are also developed for the UWB applications. Two examples of them are (a) pentagonal shaped-slot antenna, and (b) sectorial loop antenna.

#### *8.1.2.1.2.3.4.1 Pentagon-shape microstrip slot antenna*

The antenna geometry is illustrated in Figure 8.72 [43]. Three models are considered; model A with a straight feed line, model B with tilted feed line and model C with tilted feed line on a different substrate from that of the model A and B. The substrate used for models A and B has $\varepsilon_r = 2.20$ and tan $\delta = 0.0004$, whereas for model C, $\varepsilon_r = 4.50$ and tan $\delta = 0.02$, The thickness of the substrate is 1.58 mm for all the models. The antenna can be designed to mount on the small substrate (ground plane) with the size of 50 mm × 80 mm, which is a similar size to the wireless card used in usual wireless equipment. The antenna will occupy only the top 20 mm or 25% of the ground plane length, leaving enough space available to mount RF devices and circuitry on it. Even with this small size, the impedance bandwidth obtained was maximum 124% (2.65–11.30 GHz), exceeding the UWB bandwidth of 110% (3.10–10.60 GHz), as a result of combination of the pentagon-shaped slot, feed line, and pentagon stub. For models B and C, the feed line is rotated by 15°. In terms of the bandwidth, model A exhibited 106% (2.6–8.4 GHz), model B provided the largest of all 124% and model C obtained 116%

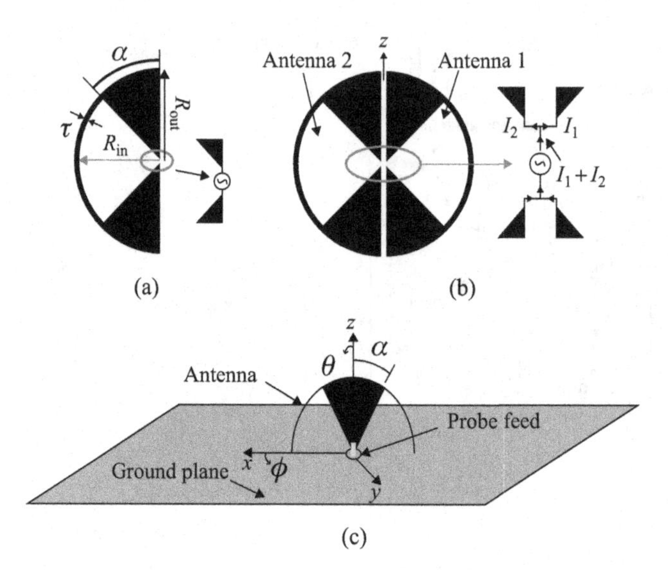


**Figure 8.73** (a) A sectorial loop antenna, (b) coupled sectorial loop antenna (CSLA), and (c) a half of CSLA over the ground plane ([47], copyright © 2005 IEEE).

(2.4–9 GHz). The tilted feed in models B and C can enhance the bandwidth compared with model A. The radiation pattern is nearly omnidirectional, but becomes directional in higher-frequency ranges. The gain variation was 3.25 dBi (3.00–6.25 dBi) over the frequency range between 3.5 GHz and 13.0 GHz.

In order to obtain a directional pattern, a conducting sheet as a reflector is applied on the back of model B. A 50-mm square sheet is placed with spacing $d$, varying from 5 mm to 25 mm and it was found for obtaining the directional patterns within the UWB range that $d = 10$ mm was suitable for the lower-frequency range (3–7GHz) while $d = 5$ mm was suitable for the higher-frequency range (8–11 GHz).

Other types of wideband or multiband slot antenna are a tapered ring slot antenna [44], antenna using square ring resonator [45], and a V-shaped antenna [46].

### *8.1.2.1.2.3.4.2 Sectorial loop antenna (SLA)*

A compact loop antenna, being referred to as a sectorial loop antenna (SLA), composed of an arch and two sectors, is depicted in Figure 8.73, showing (a) topology of the SLA, (b) topology of the coupled SLA, and (c) a half of an SLA placed on the ground plane [47]. The antenna can provide wide enough bandwidth to cover the UWB operation band by combining two identical SLAs in parallel and controlling the mutual coupling between them as shown in Figure 8.73(b). When two identical SLAs are connected in parallel, the antenna system is equivalently expressed in two-port network terms by using relationships between the impedances $Z_{ij}$ and currents $I_{ij}$ ($i = 1$ for one antenna and $j = 2$ for another antenna) as follows:

$$\begin{aligned} V_1 &= Z_{11}I_1 + Z_{12}I_2 \\ V_2 &= Z_{21}I_1 + Z_{22}I_2 \end{aligned}. \tag{8.3}$$

Here $Z_{11}$ and $Z_{22}$ are self-impedance of each antenna, $Z_{ij}$ is the mutual impedance between two antennas, and $V_1$, $V_2$, $I_1$, and $I_2$ are voltages and currents at the input port of the antenna elements as Figure 8.73(b) shows. Since the two antennas are identical and the system is symmetrical, $Z_{11} = Z_{22}$, and $Z_{12} = Z_{21}$. Also $V_1 = V_2$, and $I_1 = I_2 = I$. Thus the input impedance $Z_{in}$ of the antenna is given by

$$Z_{in} = (Z_{11} + Z_{12})/2. \tag{8.4}$$

To obtain wide bandwidth, variation of $Z_{11}$ and $Z_{12}$ with respect to frequency must counteract each other. This can be achieved by optimizing the geometrical parameters of the antenna that are the loop inner and outer radii $R_{in}$ and $R_{out}$ and the sector angle $\alpha$. When $R_{in}$ and $R_{out}$, respectively, are 13 and 14 mm, relatively constant $Z_{in}$ is obtained for $40° < \alpha < 80°$. These three parameters also determine the lowest operating frequency $f_\ell$, which is given by

$$f_\ell = 2c/(\pi - \alpha + 2)(R_{in} + R_{out})\sqrt{\varepsilon_r}. \tag{8.5}$$

By using this, the average radius of the loop $R_{av} = (R_{in} + R_{out})/2$ can be determined. Then optimum $\alpha$ and $\tau = (R_{in} - R_{out})$ need to be determined. Through studies of some experimentally fabricated antennas, the optimum values for these parameters are found. They are $\alpha = 60°$, $R_{av} = 13.5$ mm, and $\tau = 0.4$ mm. This $\tau = 0.4$ mm, which is the smallest value (thinnest loop radius), is chosen, because bandwidth becomes wider as $\tau$ tends to be smaller. With these parameters, an antenna with bandwidth of 3.7 to 11.6 GHz is obtained.

### 8.1.3 Integration of functions into antenna

An antenna, into which active devices or circuits are integrated to enhance the antenna performance or function, is referred to as AIAS (Active Integrated Antenna System) [48]. It has received considerable attention, because the technique will provide surpassing performances or functions to antennas without enlarging the dimensions. There have been quite a few papers and books which have dealt with AIAS [49–51]. The AIAS is not necessarily ESA, but most of them are FSA. However, they have useful features in possibly being manufactured with relatively small size, compact structure, and yet low cost. Representative AIASs are those with enhanced gain [52], operating band [53, 54], and functions of reconfigurable performances such as variation of tuning frequency, switching of operating bands [55, 56], or control of radiation patterns [57–59], and so forth.

IPASs (Integrated Passive Antenna Systems) also play an important role in reducing the antenna size and enhancing the antenna performances; however, because integration of slots/slits into antenna systems has been treated in other sections, it is not mentioned here.

Here in this section, three examples will be described.

#### 8.1.3.1 An oscillator-loaded microstrip antenna

Configuration of the antenna along with the dimensions is shown in Figure 8.74 [60]. The antenna is comprised of two transmission lines, one wide line as a radiator and

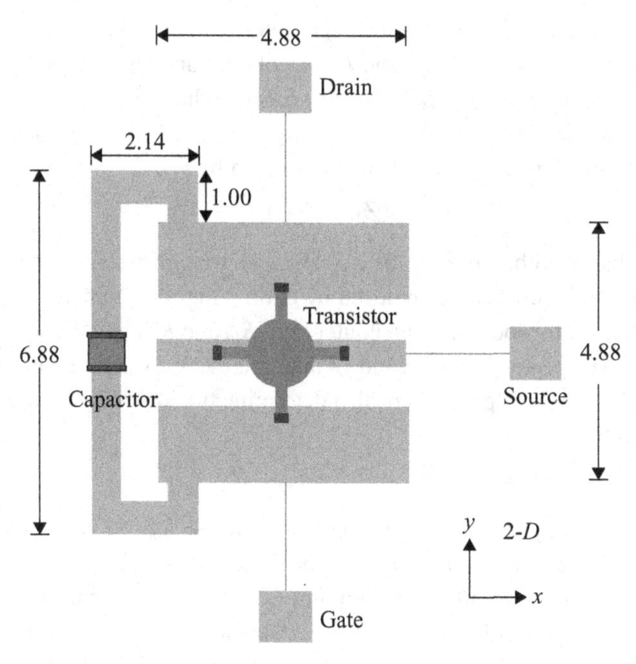


**Figure 8.74** Geometry and dimensions of single-element antenna integrated with an oscillator along with dimensional parameters ([60], copyright © 2008 IEEE).

one narrow line to serve as a feedback loop. A transistor oscillator circuit is mounted on the wider transmission line with two source terminals connected to the central line with narrow width and the drain and gate terminals connected to the two wider lines. The length of the two radiating sections combined with the feedback loop is $3/2\lambda_g$ ($\lambda_g$: guided wavelength). Similar current and charge distributions are created by this circuit arrangement and the radiation patterns typical of a microstrip patch antenna can be obtained. The antenna pattern is etched on the microwave laminate with thickness of 0.635 mm and dielectric constant of 10.2. The transistor is an HF FET (High Frequency Field Effect Transistor) having super low-noise characteristics and gain of 8.5–9 dB over the frequency range between 6 and 12 GHz. A 1.2 pF capacitor is placed on the radiator line to serve as DC isolation between the drain and gate and at the same time provide RF feedback from the drain to gate. The transistor circuit is biased using a single 1.5 V battery between the source and drain terminals with the gate terminal remaining open.

This antenna is fabricated for testing and the performances are measured. The gain is 10 dBi at 8.5 GHz and the EIRP is 11.2 dBm. The phase noise is –87.5 dBc/Hz at 100 kHz offset, that is attributed primarily to the transmission feedback circuit in addition to low-noise characteristics of the HF FET. Figure 8.75 shows measured transmission data $S_{21}$ with $V_{drain} = 1.2$ V and $V_{gate} = 0$, and for the no-transistor case. The inset in the figure depicts radiated power.

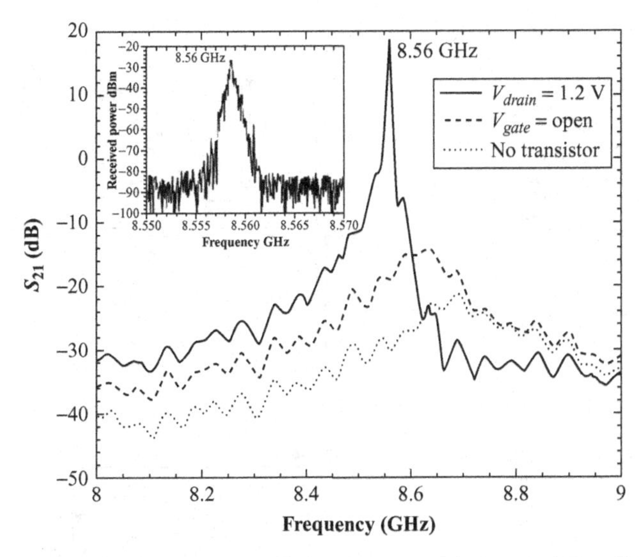


**Figure 8.75** Measured transmission data $S_{21}$ with $V_{drain}$ and $V_{gate}$ ([60], copyright © 2008 IEEE).

### 8.1.3.2 A reconfigurable PIFA with integrated PIN-diode and varactor

By integrating a PIN-Diode for switching and a varactor for fine-tuning into a PIFA structure, an antenna capable of multiband operation for several mobile communication systems is proposed [61]. The geometry of the antenna and its dimensional parameters are illustrated in Figure 8.76, showing (a) 3D view, (b) top view, (c) side view, and (d) front view. By varying capacitance of the varactor on an impedance-matching short-line, fine-tuning of operating frequencies can easily be achieved and by switching the radiating elements by means of the PIN-diode status (on and off), operating frequency bands largely separated can be selected. The antenna can cover four bands; USPCS (1.85–1.99 GHz), WCDMA (1.92–2.18 GHz), m-WiMAX (3.4–3.6 GHz), and WLAN (5.15–5.825 GHz).

### 8.1.3.3 Pattern reconfigurable cubic antenna

A unique single-feed cubic antenna capable of pattern reconfigurable performance is introduced in [59]. The antenna is a metallic cubic cavity with a slot radiator on each of its six surfaces and can radiate in a $4\pi$ steradian range to receive incident waves with any polarization. The pattern reconfiguration is achieved by using a PIN diode which opens or shorts at the center of the slot on the cube, thus producing change of the radiation pattern. The operating frequency is 5 GHz. Schematic drawing of the antenna is shown in Figure 8.77. The performances of this structure are described by two aspects; resonance modes of the cube which radiate through the slot and the resonance of the slot. The cube dimension $a$ can be determined by taking these two effects into account and considering the first fundamental modes $TE_{011}$, $TE_{101}$, and $TE_{110}$. With cube dimension $a = 37.5$ mm, slot length $l_s = 27$ mm, and probe length $l_p = 27$ mm, the resonance frequencies of the cube and the slot are 5.8 GHz (bandwidth is about 2.5%) and 5 GHz

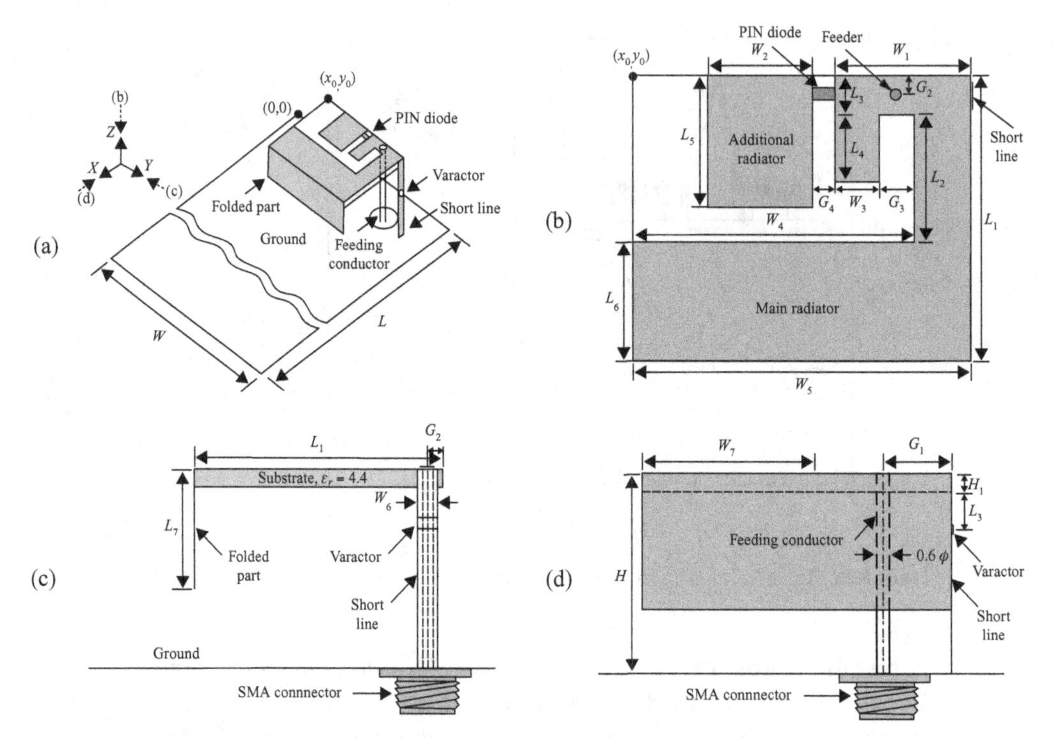


**Figure 8.76** Geometry of a PIFA loaded with a PIN-diode: (a) 3D view, (b) top view, (c) side view, and (d) front view ([61], copyright © 2010 IEEE).

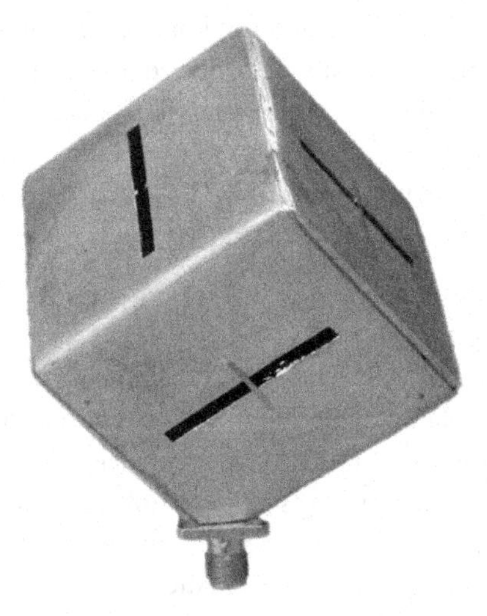

**Figure 8.77** 3D view of a cubic reconfigurable antenna ([59], copyright © 2009 IEEE).

(bandwidth is about 6.5%), respectively. To have two frequencies close in order to obtain a wider bandwidth, the cube length and probe length are changed to $l_s = 39$ mm and $l_p = 38$ mm, respectively and achieved the bandwidth of 11.3% at 5.2 GHz with the resonance frequencies of the cube at 5.05 GHz and the slot at 5.4 GHz, respectively. The

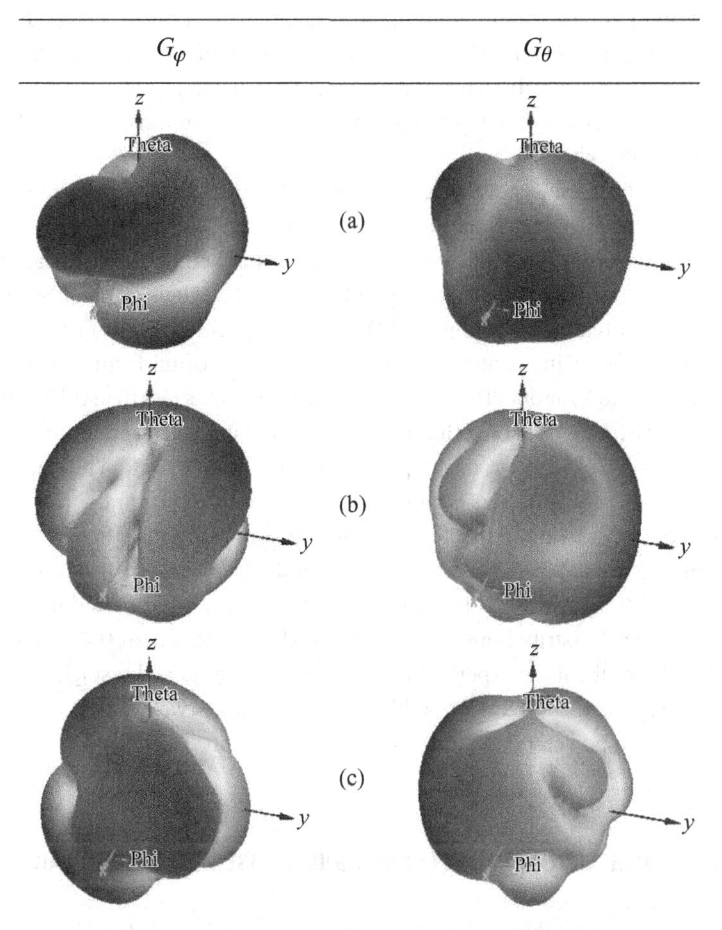


**Figure 8.78** Simulated radiation patterns along $\varphi$ (on left) and $\theta$ (on right) for (a) configuration 1, (b) configuration 2, and (c) configuration 3 ([59], copyright © 2009 IEEE).

radiation patterns are varied by short-circuiting or open-circuiting the center of the slot, resulting in cancelling or producing radiation. This action can be performed by using a PIN diode which switches the state of on or off to cause short- or open-circuiting the slot. Reconfiguration of the pattern is realized by selecting which slots are short-circuited. Three configurations 1, 2, and 3 are considered; each one contains two short-circuited slots on the lower sides and one on the upper side on the cube. Figure 8.78(a), (b), and (c), respectively, illustrate the simulated radiation patterns at 5.4 GHz for the configurations 1, 2, and 3, for $G_\varphi$ (gain pattern along $\varphi$) on the left and $G_\theta$ (gain pattern along $\theta$) on the right. Switching the configuration is equivalent to rotating the cube around the probe on an angle of 120°. As can be noted in Figure 8.78 the antenna radiates in a $4\pi$ steradian range with a maximum gain toward a certain direction, and hence by switching the configuration the maximum gain direction changes. The maximum gain is evaluated as approximately 3.7 dBi. In addition to the variation in the maximum gain direction, the radiated powers, in other words $G_\varphi$ and $G_\theta$, in a given direction are not identical depending upon in which configuration the antenna is set. That is, a ratio $G_\varphi/G_\theta$ can be higher or lower than unity. By taking advantage of this pattern performance, the antenna can achieve pattern diversity in terms of polarization. It can be also said that the antenna is applicable to diversity of not only polarization, but also power and phase by means of the configuration switching.

Diversity performance is evaluated both theoretically and experimentally. A measure of the diversity performance is the correlation coefficient $\rho$, which determines the quality of the communication channel in diversity and is defined as the correlation between incident wave envelopes of different polarizations [62]. It is a function of complex incident fields of two polarizations, XPD, a ratio of incident wave powers, and arrival angle distributions of incident waves. With assumption of uniform distribution in the angle of arrival (AoA) and XPD $< -10$ dB, the simulated $\rho$ is less than 0.12, but with non-uniform distributions, $\rho$ takes different values depending on the configuration, because incident waves have different envelopes depending on the AoA. Thus three envelope correlations, $\rho_{12}$, $\rho_{13}$, and $\rho_{23}$, because of three available pattern configurations, must be taken into consideration. As for the non-uniform AoA distributions, a Laplacian distribution is assumed [59]. With assumption of the Laplacian distribution and XPD $=$ 0 (indoor condition) and 6 dB (urban field condition), $\rho$ is evaluated higher than that in case of uniform distributions, but is estimated to be less than 0.47, which is still a practical and useful value. Experiments verified these simulated results.

It can be said that the reconfigurable cubic antenna is well suited to apply pattern diversity and also power, phase, and polarization diversity.

## 8.2 Design and practice of PCSA (Physically Constrained Small Antennas)

A PCSA is an antenna that is not necessarily ESA, but has portions sized similarly to ESA. Typical PCSAs would be low-profile antennas, having the height comparable with the size of ESA. They are microstrip antennas, PIFAs, and printed antennas.

### 8.2.1 Low-profile structure

Low-profile structure can be comprised of an antenna element and a ground plane, which provides an image effect. When the ground plane is a PEC (Perfect Electric Conductor), the image effect is negative for an electric source placed parallel to the ground plane, whereas it is a positive effect for a magnetic source. The PEC ground plane can be replaced by a high impedance surface (HIS) or a surface that exhibits an electromagnetic bandgap (EBG) property that can produce a positive image of the antenna, reducing the antenna profile as well as improving the antenna performance. Use of EBG is quite beneficial not only for miniaturizing the dimensions, increasing gain, and bandwidth, but also for reducing the mutual coupling between two closely mounted antennas. Recently, much interest has focused on applications of DGS (Defected Ground Surface) associated with low-profile antennas and development of small antennas.

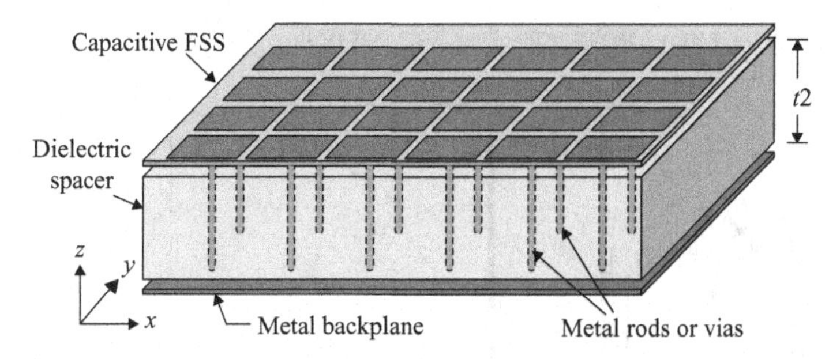


**Figure 8.79** 3D view of a mushroom-like HIS structure ([64], copyright © 2003 IEEE).

### 8.2.2 Application of HIS (High Impedance Surface)

The concept of HIS was first introduced in 1999 [63], where a mushroom-like HIS surface was treated. The mushroom-like surface has periodic structure, consisting of a lattice of metal plates connected to a solid metal sheet by vertical conducting metal vias. Figure 8.79 shows a sketch of the typical mushroom-like structure [64]. The structure can be visualized as mushrooms protruding from the back-plane surface. Since the protrusions are small compared to the operating wavelength, their electromagnetic properties can be described by using lumped capacitors and inductors, behaving as a network of parallel resonant L-C circuits, which acts as a two-dimensional filter to block the current flow along the surface. The surface impedance is modeled as a parallel resonant L-C circuit, which exhibits high impedance over a predetermined tuned frequency band. The surface acts equivalently as a frequency selective surface (FSS). The HIS can be considered as a kind of two-dimensional photonic crystal that prevents the propagation of radio frequency surface currents within the bandgap. The surface does not support propagating surface waves and its image currents are not in phase reversal, but in-phase, allowing a radiating element to lie directly adjacent to the surface, while still radiating efficiently. Figure 8.80(a) shows an example of the transmission coefficient $S_{21}$ of a surface with suppression bandwidth from 0.98 to 1.35 GHz and (b) gives the phase of the reflection coefficient of a surface with an operational bandwidth from 0.88 to 1.35 GHz [64].

The HIS has proven to be useful as an antenna ground plane on which the surface wave is suppressed, resulting in less radiation in the backward direction. The reflection phase is unusual, allowing an antenna to lie directly adjacent to the ground plane (HIS) without being shorted out, by which antenna gain is enhanced [63].

Design methodology for the mushroom-like HIS was described in [64].

An HIS can be constituted by surfaces other than the mushroom-like one. Use of periodically corrugated reflectors shown in Figure 8.81 is an example, where a planar hexagonal dipole is placed close to the surface with its axis parallel to the grooves, which in this case is referred to as a system of *H*-type corrugation surface [65]. When a dipole is placed with its axis perpendicular to the grooves, it is referred to as a system of *E*-type corrugation surface. By combining *E*-type and *H*-type corrugated reflectors, and by applying a UWB dipole with only 21-mm profile on the surface, a very wide

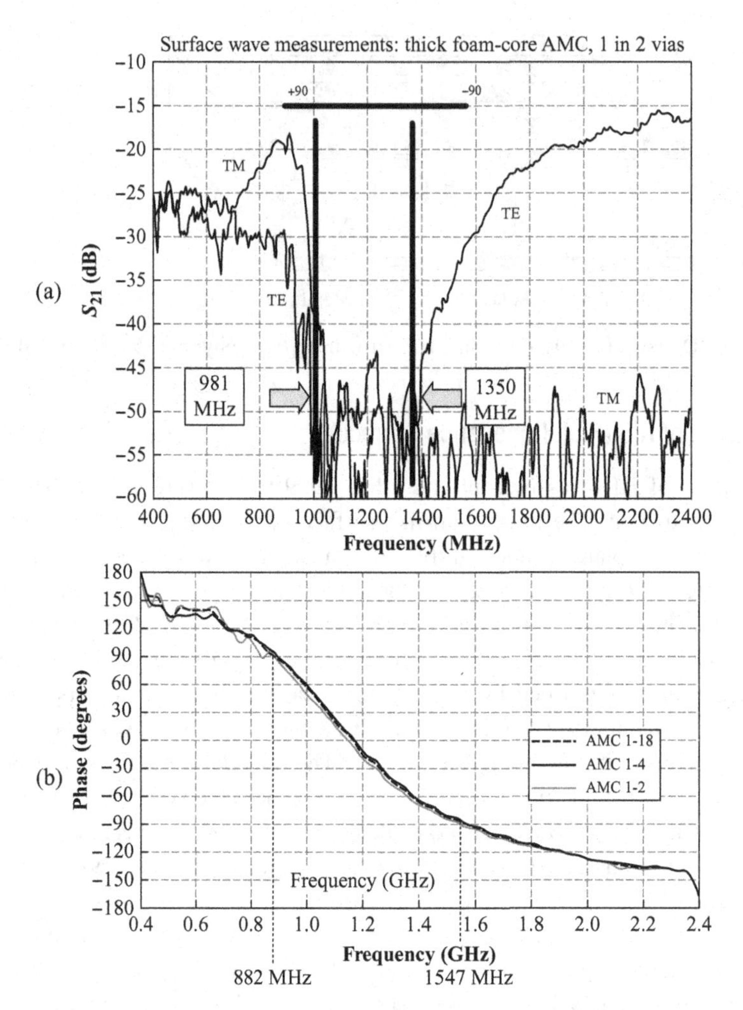


**Figure 8.80** Measured (a) surface wave transmission coefficient and (b) phase for HIS ([64], copyright © 2003 IEEE).

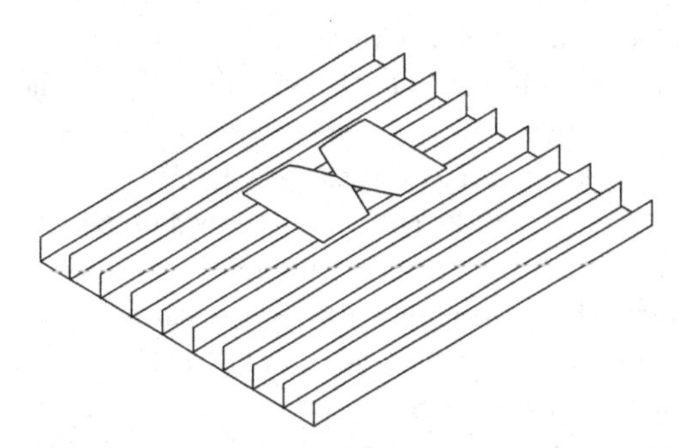

**Figure 8.81** A source dipole antenna over the corrugated ground plane ([65], copyright © 2010 IEEE).

**Figure 8.82** The FSS unit cell ([66], copyright © 2010 IEEE).

impedance bandwidth was obtained, for VSWR < 2, from 2.75 to 8.35 GHz. The radiation pattern was stable and unidirectional, and gain over 6 dB within the frequency band of 2.75 to 6.0 GHz was obtained.

Another example is an FSS with a periodical array of cells, having a pattern that is a combination of a Jerusalem cross and a three-step fractal patch shown in Figure 8.82 [66]. The low-profile monopole antenna on the HIS ground plane with the substrate thickness of 0.07λ was able to obtain an enhanced gain of 7.73 dB from 5.35 dB.

### 8.2.3 Applications of EBG (Electromagnetic Band Gap)

#### 8.2.3.1 Miniaturization

A probe-fed patch antenna backed by a mushroom-type EBG substrate is designed, and it demonstrates miniaturization that achieved 66.83% [67]. Miniaturization is achieved by employing the property of EBG structures, which supports slow wave propagation as its first propagating mode, having a longer effective wavelength than those in free space and dielectrics. The EBG structure, over which a patch antenna will be embedded, is designed to operate in its slow wave region in order to achieve a lower frequency operation of the patch antenna for the size reduction. Here the operating frequency of 2.4 GHz is used, to which the size of a conventional patch antenna is designed to be 31.9 mm in length and 40.9 mm in width. A proposed patch antenna embedded on the mushroom-type EBG surface is illustrated in Figure 8.83, in which a unit cell is shown by its top and side views along with dimensional parameters. The number of cells is designed to have slightly larger area than the patch, so a square surface with 4 × 4 cells to cover the patch is selected. In the initial design, the unit cell size $a$ is chosen to be 5.5 mm, based on the wavelength of slow wave mode $\lambda_{slow} = 34.5$ mm. Other parameters are $D_{via} = 0.8$, $g = 0.25$, $h_1 = 1.524$, and $h_2 = 0.762$. The dielectric constant of the substrate is 3.66. By doing optimization, starting from the initial values with $\lambda_{slow}/2$, the patch size, $L$ (length) × $W$ (width), respectively, is determined to be 18.3 × 17.4 mm. The reduction of the patch area from the area of a conventional patch is estimated to be 66.83% in its ratio, and 49.58% even when the area of the EBG surface is included.

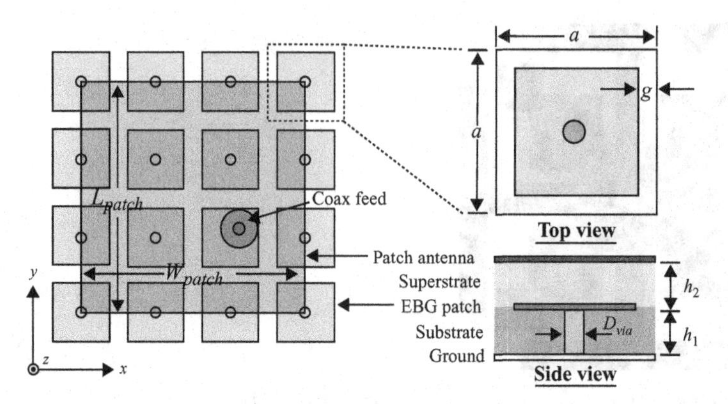


**Figure 8.83** Miniaturized patch antenna on a mushroom type EBG substrate ([67], copyright © 2010 IEEE).

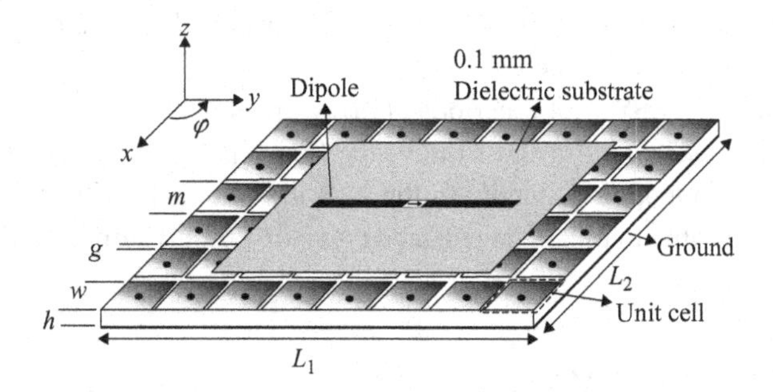


**Figure 8.84** A dipole closely placed over an EBG surface ([69], copyright © 2008 IEEE).

The antenna exhibits rather narrow bandwidth and low efficiency because of the thin substrate.

In the paper [67], the effect of EBG on the coupling between two antennas is investigated, showing that *H*-plane coupling between two units of this type of EBG patch antenna is much lower than that of conventional patches.

### 8.2.3.2 Enhancement of gain

It is natural that gain can be enhanced by a patch antenna when embedded on the EBG surface, because the surface wave on the ground plane is suppressed. Subsequently current flow into the backside of the ground plane can be reduced, leading to increase of radiation into forward directions. A simple example is a dipole antenna embedded on the mushroom-type EBG surface described in [63, 68, 69]. A wideband dipole embedded on the EBG surface is described in [69]. Figure 8.84 illustrates a schematic of the EBG with dipole, giving also dimensional parameters. The antenna is designed to obtain wide bandwidth as well as higher gain based on the optimization of the interaction between impedances of a primary antenna and its imaginary dipoles with the reflection phase of a mushroom-type EBG structure. The peak gain obtained by calculation is between 5.5 and 8.3 dBi over the frequency region from

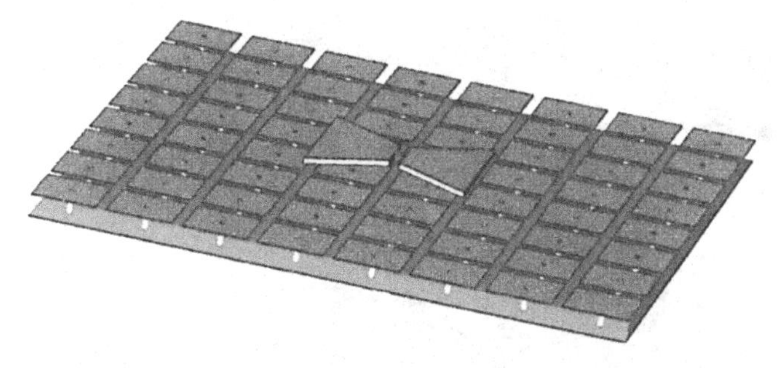

**Figure 8.85** Planer bow-tie antenna closely placed over the EBG ground plane ([71], copyright © 2008 IEEE).

1.7 GHz to 2.5 GHz, while measured and simulated bandwidths are 38% and 41%, respectively. Since the antenna is a balanced type, a balun is necessary for avoiding interference of unbalanced current produced on the feeding cable without a balun. By using a wideband balun, the antenna was shown to obtain wider bandwidth as compared with a case without a balun. The dipole is made of two metal strips with 1-mm width and 45.7-mm length, and is printed on the substrate with $\varepsilon_r = 4.5$, having the size of 94 × 94 mm. The EBG structure is composed with 6 × 8 cells, each cell (mushroom) having the size of 12.4 mm in width, and 6.0 mm in height, being arrayed with gaps of 0.4 mm. The bandwidth of 1.6–2.5 GHz covers frequency bands of several wireless systems such as DCS (Digital Communication Systems: 1.71–1.88 GHz), GSM (Global Systems for Mobile Communication: 1.85–1.99 GHz), PCS (Personal Communication Systems: 1.85–1.99 GHz), UMTS (Universal Mobile Telecommunication System: 1.92–2.17 GHz), and WLAN (Wireless Local Network: 2.4–2.485 GHz).

### 8.2.3.3 Enhancement of bandwidth

With the close spacing of a thin dipole and its image element, currents of these elements are in phase, leading to decrease in the reactive energy surrounding the dipole. Consequentially, the operating bandwidth increases. It was shown that the EBG ground plane requires a reflection phase in the range of 90°± 45° for a low-profile straight wire dipole antenna to exhibit a good return loss [70] and the design of the EBG ground plane follows to meet this requirement at a desired operating frequency. A dipole is designed to operate at the same frequency, with consideration to placing it closely over the EBG surface.

A dipole antenna introduced previously in [69] was designed by following the above design concept and it demonstrated a fairly wideband performance covering 1.7 GHz–2.5 GHz with center frequency of about 2.0 GHz.

A dipole antenna embedded over the EBG ground plane designed to exhibit about 1.4:1 impedance bandwidth is shown in [71]. In the same way, a bow-tie dipole antenna over the EBG ground plane is described to show much greater bandwidth. Figure 8.85 depicts the antenna placed over the EBG ground plane. The bow-tie dipole antenna has the thickness of 0.01λ, an overall length of 0.30λ, and an overall width of 0.26λ at 300 MHz. It is located 0.02λ over the EBG ground plane, which is composed of

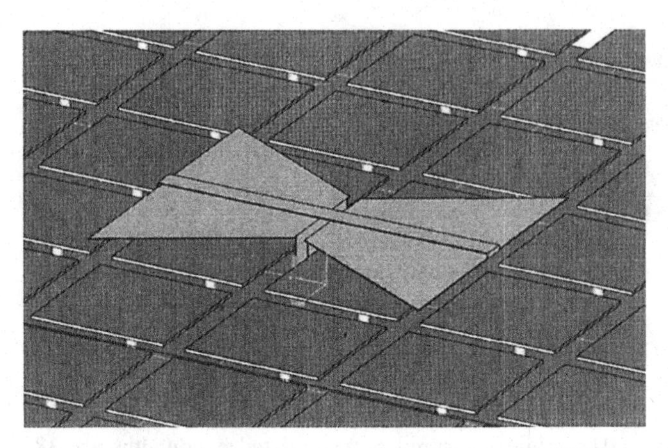

**Figure 8.86** A folded bow-tie dipole over the EBG surface ([71], copyright © 2008 IEEE).

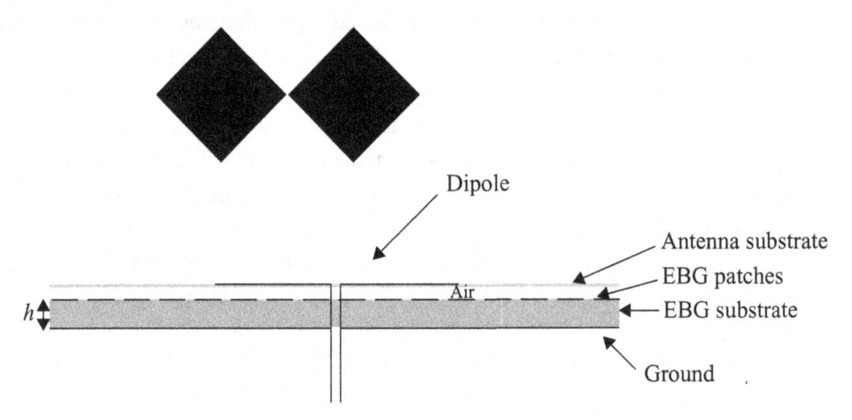


**Figure 8.87** A diamond dipole backed by the EBG ([72], copyright © 2007 IEEE).

8 × 8 cells (mushrooms), with each cell having the size of 0.12λ, via radius of 0.005λ, gap between cells of 0.02λ, and the substrate ($\varepsilon_r = 2.2$) thickness of 0.04λ at a frequency of 300 MHz. To obtain good matching for a wide bandwidth, the antenna is modified to a folded structure, with both edges of the bow-tie element folded by a narrow strip over the bow-tie element as shown in Figure 8.86. With this folded bow-tie structure, the –10 dB return loss bandwidth spanned a frequency range from 306 MHz through 419.5 MHz.

Other types of antennas than straight wire and bow-tie can be useful. A square-patch dipole, called a diamond dipole, shown in Figure 8.87 [72] is embedded on the EBG surface and achieved a wide return-loss bandwidth of 1.4:1 (33%) taking the radiation pattern into consideration. Also a sleeve dipole over the EBG is treated similarly and shown to have a bandwidth of 1.28:1 (26%) [72].

### 8.2.3.4 Reduction of mutual coupling

Since the EBG structure has a feature of suppressing the surface wave propagation on the EBG surface because of its bandgap property, the mutual coupling between two

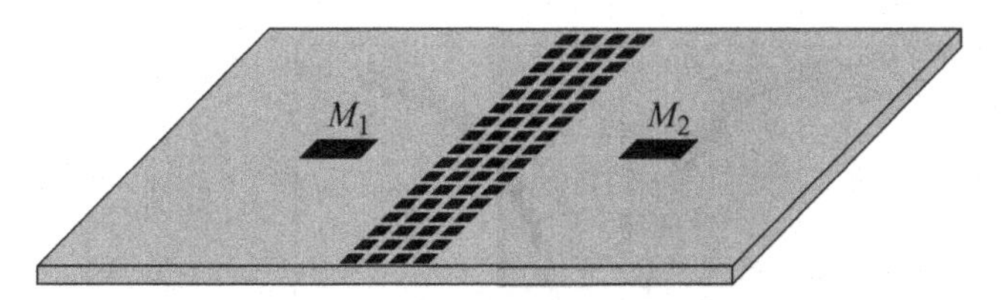


**Figure 8.88** Microstrip antennas separated by the mushroom-type EBG ([68a], copyright © 2007 IEEE).

antennas can be suppressed by locating them over the EBG structure, if the antennas operate within the EBG bandgap frequency band [68a]. For an array of two antennas, for instance, the EBG substrate is placed between two antennas to allow arraying them with close separation as shown in Figure 8.88 [68a]. The EBG structure, in this example consisting of four columns of cells (patches), is inserted between two patches. The two patches have the same size of 7 × 4 mm and are placed at a distance of 38.8 mm (0.75λ at 5.8 GHz). The substrate thickness of the system is 2 mm, and $\varepsilon_r$ is 10.2. Here, the cell (mushroom) size of the EBG structure is considered for three cases, 2, 3, and 4 mm, the gap between cells is 0.5 mm, and the via radius is 0.3 mm. Figure 8.89(a) provides the return loss $S_{11}$ and (b) gives $S_{21}$, corresponding to the mutual coupling. In each figure, comparisons between cases with and without the EBG structure are shown. All the antennas resonate around 5.8 GHz and show better than –10 dB matches, although the EBG substrate may somewhat affect the impedance match of the antennas. Without the EBG structure, the antennas show strong mutual coupling of –16.15 dB, whereas the mutual coupling level changes with the EBG structure's presence. When the cell size is 2 mm, since the EBG bandgap is higher than the resonance frequency, the mutual coupling is not reduced, leaving still the high level of –15.85 dB. In case of 3-mm cell size, the resonance frequency 5.8 GHz falls inside the EBG bandgap so that the surface waves are suppressed and the mutual coupling is greatly reduced to the level of –25.03 dB. When the cell size is increased to 4 mm, the EBG bandgap becomes lower than the resonance frequency. Thus the mutual coupling is not improved, leaving the level as strong as –16.27 dB.

### 8.2.4 Application of DGS (Defected Ground Surface)

The concept of DGS emerged from studies of PBG structure in electromagnetics, now being referred to as the EBG structure [73]. An EBG substrate is implemented on the ground plane surface to achieve the stopband property over a frequency range and thus a "defected" ground plane surface (DGS) is created. The DGS structure is constituted of modified EBG structures, having some compact geometries such as a unit cell as a single defect, or in periodic configuration with a small period number on the ground plane, which provides stopbands and slow-wave nature over a frequency range with a different manner of EBG. Hence, the DGS can be referred to as a unit cell

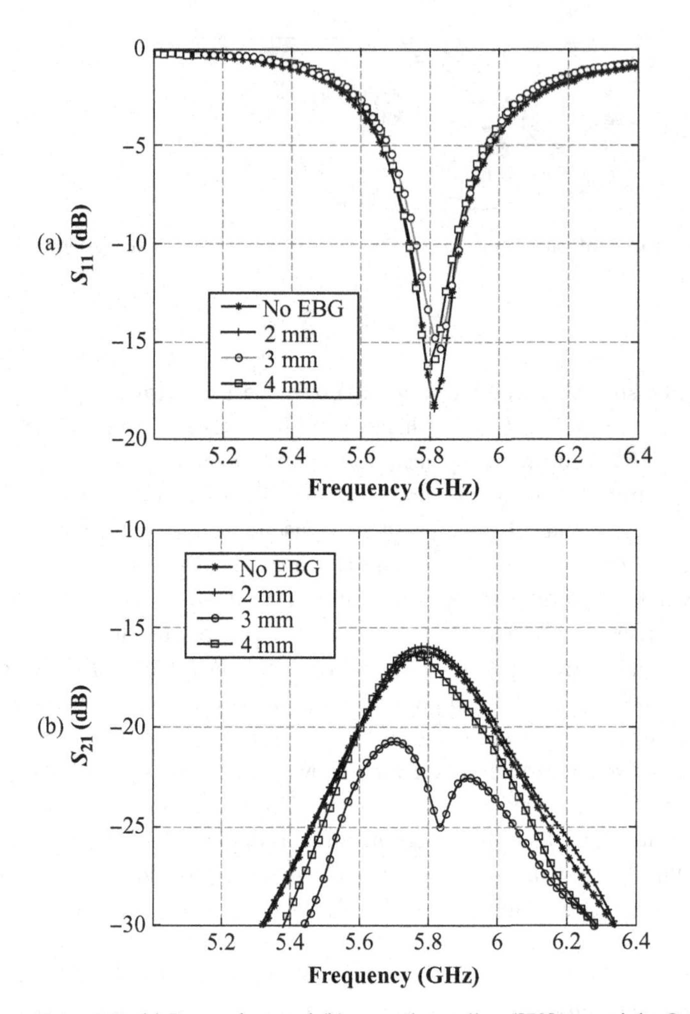


**Figure 8.89** (a) Return loss and (b) mutual coupling ([72], copyright © 2007 IEEE).

EBG, or as an EBG having limited number of cells and a period of repetition. In a practical DGS, the defects can be created by etching periodic or single patterns on the ground plane substrate. Various shapes and sizes have been explored so far, including ring, meander line, triangle, dumbbells, spirals, H-shape, U-shape, and their modifications. Figure 8.90 depicts some geometries of the DGS unit. The DGS structure can be applied to antennas, microwave filters, power amplifiers, oscillators, and so forth. Shapes and sizes of the structure depend on the desired operating frequency and the required performances.

#### 8.2.4.1 Ring-shaped DGS

Figure 8.91 illustrates a concentric-ring DGS [74]. Two configurations with DGS structure etched on the ground plane are shown in the figure; (a) is #1 and (b) is #2, which

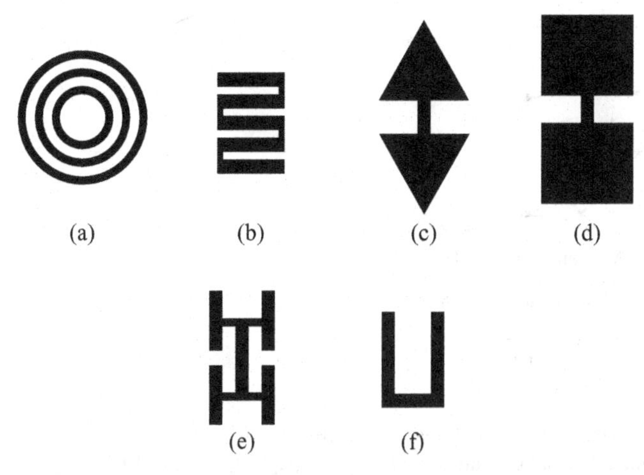


**Figure 8.90** Some examples of DGS geometries: (a) ring (b) meander line (c) triangle (d) dumbbell (e) spiral H-shape and (f) U-shape.

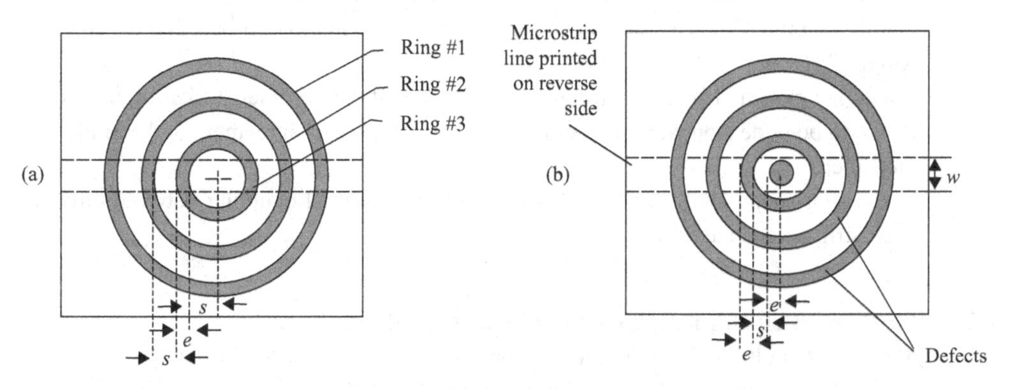


**Figure 8.91** Concentric ring (CR) DGS: (a) configuration no. 1 and (b) configuration no. 2 ([74], copyright © 2006 IEEE).

is similar to #1 but with a small circle defect at the center. The DGS circles are etched on the ground plane and are shown by dark color in the figure. A pair of dotted lines represents a microstrip line etched on the reverse side of the substrate. The substrate $\varepsilon_r$ is 2.33 and the thickness is 1.575 mm. The dimensions of the DGS are: $2e = 0.1\lambda_g = 2$ mm, $h = 0.08\lambda$, and $s = 3e$, where the operating frequency $f_g = 10$ GHz. The width of the microstrip line $w = 4.7$ mm. Figure 8.92 shows measured and simulated transmission characteristics $S_{21}$ of DGS #1, comparing measured $S_{21}$ with and without DGS. It shows the stopband effect of DGS by $S_{21} < -10$ dB over the frequency range of 9 to 11 GHz. With the DGS #2, an almost identical result was obtained.

As a modification of these DGSs, a case where the DGS structure is backed by a secondary ground plane, a metal plate, is studied, and the performance measured. In

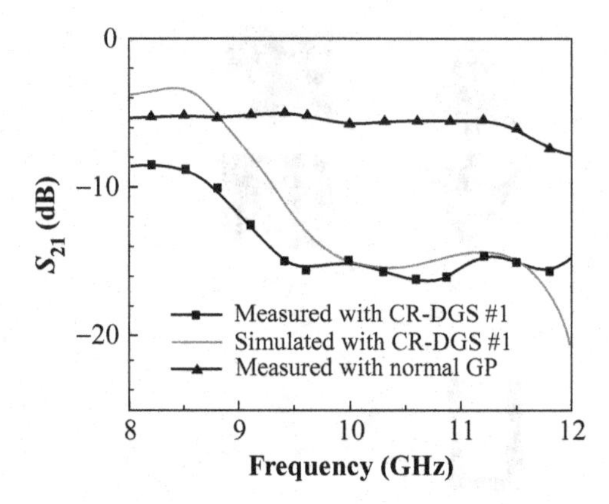


**Figure 8.92** Measured and simulated transmission characteristics of a microstrip etched on CR-DGS no. 1 and a normal ground plane ([74], copyright © 2006 IEEE).

this case, a deeper stopband was observed over the frequency range of 10.6 to 11.4 GHz compared with the case without the second ground plane. This type of DGS is useful for suppressing interference, phase noise, and harmonics in microstrip-based active antenna design [74].

Another modified DGS is a structure having the half-ring geometry. Measured data are shown to be wide stopband operation over 8 to 10 GHz and agreement with simulation was noted.

The ring-shape DGS can be applied to suppress mutual coupling between circular microstrip antenna array elements.

#### 8.2.4.2 Multiband circular disk monopole patch antenna

A compact multiband patch antenna consisting of a circular disk monopole with an L-shape slot cut on the ground plane, forming a DGS surface is proposed [75]. The slot creates two orthogonal current paths on the ground plane, producing two additional resonances in the impedance of the antenna. The geometry of the antenna is depicted in Figure 8.93, where (a) shows top view and (b) gives perspective view. Dimensional parameters are also provided in the figure. The antenna structure is low profile, constructed in completely uniplanar form, since it does not require use of any via as in the mushroom structure or lumped-element components. The antenna is designed on a low-cost substrate having $\varepsilon_r = 4.34$, $\tan\delta = 0.016$, and height $h_{sub} = 1.59$ mm. Other dimensional parameters are the length of the monopole $L_m = \lambda_g/4$, and the radius of the disk $r = 8$ mm, which is designed to produce the low resonance frequency around 3.1 GHz. The ground plane size is determined to be $L_g \times W_g = 12 \times 24$ mm taking as granted that the antenna will be employed for a card-size module used in MIMO (Multi-Input Multi-Output) systems, to which mounting at least three antennas on the size of 75 × 34 × 5 mm is possible. Resonance frequencies are 5.6 and 8 GHz and the return-loss performance below –10 dB is from 3.7 to 9.3 GHz.

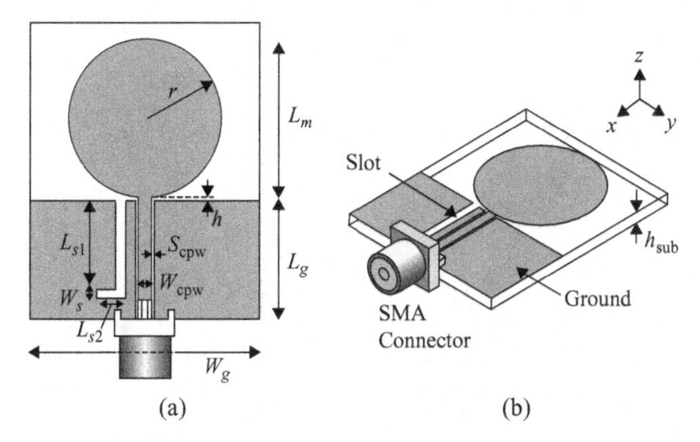


**Figure 8.93** Multiband printed monopole antenna with a DGS ([75], copyright © 2008 IEEE).

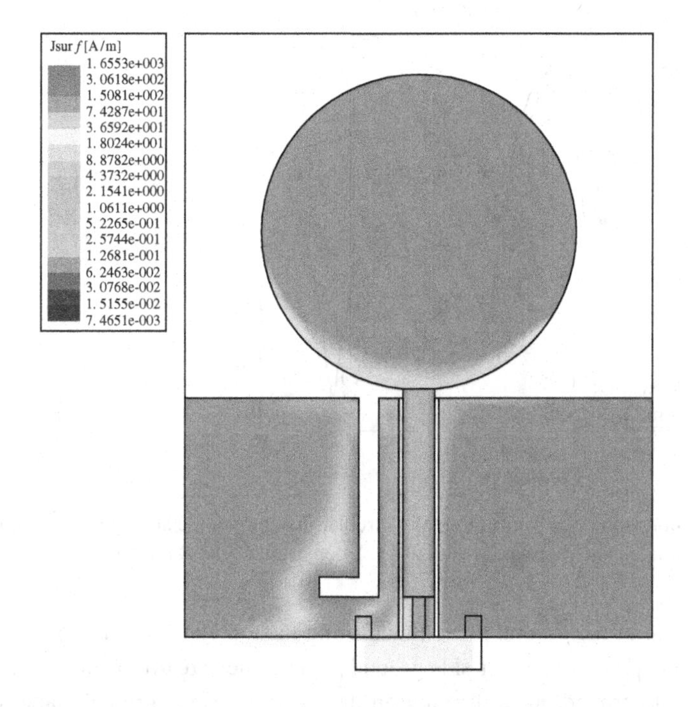


**Figure 8.94** Simulated surface current on the conductors of the multiband monopole antenna on the DGS at 2.7 GHz ([75], copyright © 2008 IEEE).

Employing the DGS structure by inserting an L-slot on the ground plane, two additional resonances around 3 GHz result, while retaining the original high-frequency resonances around 5.6 and 8 GHz. Change in the surface current distributions on the ground plane due to the L-shaped slot can be observed in Figure 8.94. The L-slot forces current on the left-side of the ground plane to wrap around the slot, while the current

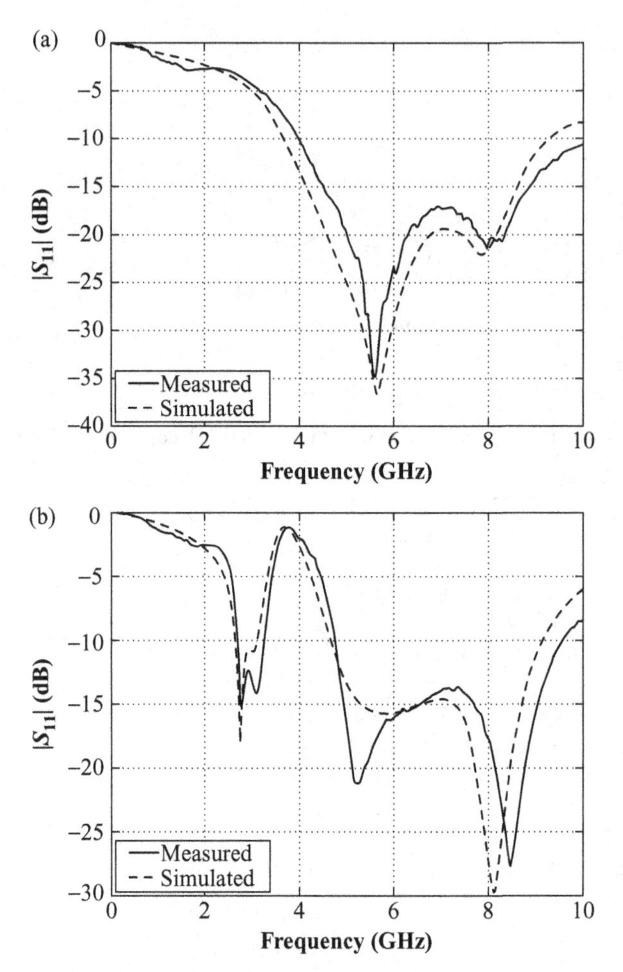


**Figure 8.95** Comparison of the return loss of a circular disk monopole antenna (a) without and (b) with DGS (simulated and measured results) ([75], copyright © 2008 IEEE).

on the right-side of the ground plane flows along the top edge. The L-slot thus creates alternate current paths on the left-side ground plane, which are orthogonal to the current on the right-side ground plane, flowing on the top edge of the ground plane. Because these two currents have slightly different path lengths, two resonance frequencies result. Also these two currents generate orthogonal polarization, allowing two distinct adjacent resonances to merge, which contributes to extending the bandwidth around 3 GHz. Figure 8.95 illustrates measured and simulated return loss for (a) without DGS (no L-slot on the ground plane) and (b) with DGS. From the figure, it is observed that the simulation shows –10 dB $S_{11}$ bandwidth of 520 MHz for the low band (2.62–3.14 GHz) and 4.51 GHz for high-band (4.66–9.17 GHz), and slightly different but still similar results by measurement.

There have been many papers treating application of DGS. Some of them are suppression of cross polarization [76], reduction of harmonics [77], and reconfigurable

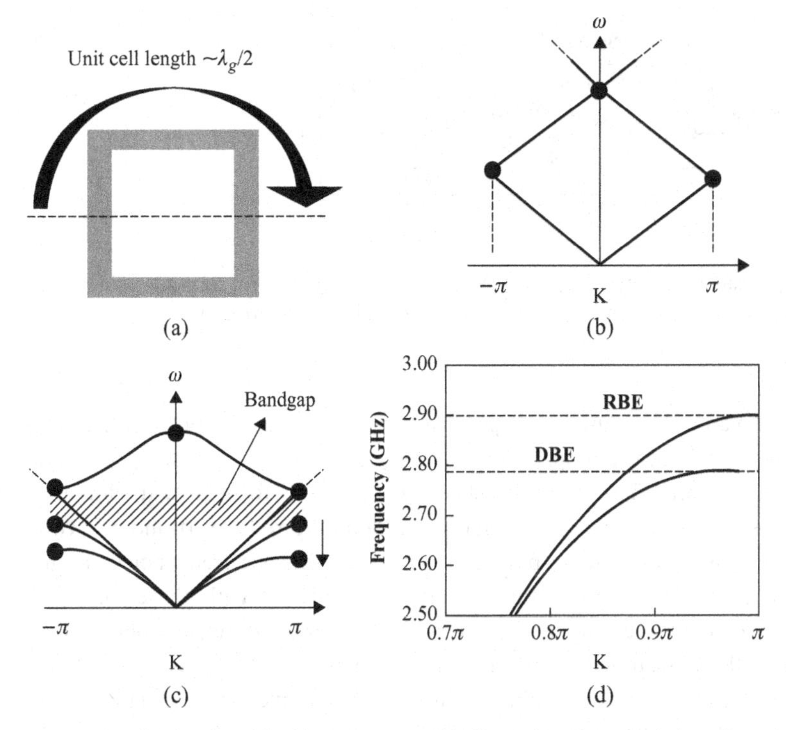


**Figure 8.96** (a) Simple printed loop antenna, (b) dispersion characteristics of a unit cell forming the rectangular loop antenna, (c) bending $k$–$\omega$ characteristics to shift resonance to lower frequencies, and (d) magnified view of the dispersion characteristics around the band edge ([83], copyright © 2009 IEEE).

DGS [78], which has a potential of a number of applications, where the reconfiguration performance is required.

## 8.2.5 Application of DBE (Degenerated Band Edge) structure

EBG structures have unique properties of frequency bandgap and group delay diminishing ($\partial\omega/\partial k = 0$) near the bandgap. This property has been employed to improve antenna performance in various ways as was described in previous sections. In order to realize miniaturization of antennas, the typical common techniques are lowering the resonance frequency and reducing the wave velocity in the antenna structure. Reducing the band edge frequency of the EBG structure satisfies this concept. However, if lowering the band edge frequency can be realized, further downsizing of the antenna will be possible. The DBE structure is a device that can be harnessed.

The general properties and applications of DBE structure have been discussed in detail by J. L. Volakis [79–83]. The concept of emulating the DBE structure, not using materials like a crystal, but a microstrip substrate is demonstrated by a printed loop structure, for example, as depicted in Figure 8.96(a) [83]. Lowering the band edge frequency can

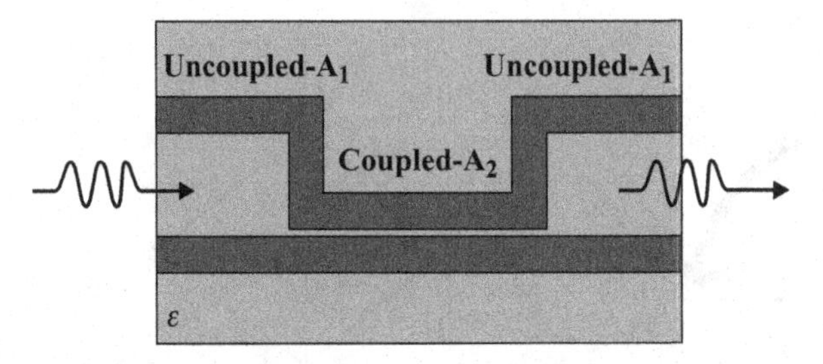


**Figure 8.97** Concept of creating a DBE surface on a microstrip substrate using a pair of coupled and uncoupled transmission lines ([83], copyright © 2009 IEEE).

be explained by using the dispersion diagram of Figure 8.96(b), which is generated by using a half of the loop as the unit cells to form the periodic structure shown in the Figure 8.96(a). The loop consists of two unit cells formed by the top and bottom half loops. Two traces on the dispersion diagram (Figure 8.96(b)) indicate two propagating modes of either positive or negative $k$, and resonance frequencies are given at points $k = \pm\,\pi$ and 0, because these points are associated with phases of matching for two propagating waves on the loop. To achieve lower resonance frequency, it is necessary to lower the $k$–$\omega$ curve at $k = \pi$ as shown in Figure 8.96(c). This can be achieved by using DBE structure, which exhibits a 4th-order $k$–$\omega$ curve, descending more sharply around the band edge than conventional EBG structure having a 2nd-order $k$–$\omega$ curve as shown in Figure 8.96(d).

A practical structure to replace a volumetric DBE material (crystal) composed of mis-aligned anisotropic layers can be equivalently implemented by coupled and uncoupled transmission line pairs printed on a uniform substrate as shown in Figure 8.97 [83]. A microstrip (MS) DBE antenna composed of two-circularly cascaded unit cells designed for resonance at 2.59 GHz is illustrated in Figure 8.98. The antenna has broadside gain of 6.9 dB at 2.59 GHz with 0.8% bandwidth for $|S_{11}| < -10$ dB. A physical footprint of the antenna on the Duroid substrate ($\varepsilon_r = 2.2$) is $\lambda_0/4.3 \times \lambda_0/5$. The radiated field is linearly polarized and the radiation efficiency is over 95%. The narrow bandwidth can be improved by using a thicker substrate. By changing the substrate thickness from 250 mils to 500 mils, the bandwidth increased to 8.7%, but with lowered resonance frequency to 2.3 GHz and reduced gain to 6.2 dB. As the antenna features a highly concentrated field at the center of the antenna, it is suited to an array that can be tightly arranged. Figure 8.99 depicts an example of such an array, which is a 4 × 4 array, constructed in the size of $1.2\lambda_0 \times 1.2\lambda_0$, with separation of each element $\lambda_0/15$ at the center frequency $f_0 = 2.55$ GHz. The directivity is 13 dB and the bandwidth is 2%.

A DBE antenna with reduced size and yet wide bandwidth is designed. The antenna depicted in Figure 8.100(a) has reduced footprint of 0.85 × 0.88 (in inches) ($\lambda_0/9.6 \times \lambda_0/9.3 \times \lambda_0/16$ at 1.45 GHz) etched on 2 × 2 inches alumina substrate ($\varepsilon_r = 9.6$, $\tan\delta = 3 \times 10^{-4}$) with thickness of 500 mils, having a metal plate as the ground plane on the back surface. On the alumina substrate, the resonance frequency is down to 1.455 GHz

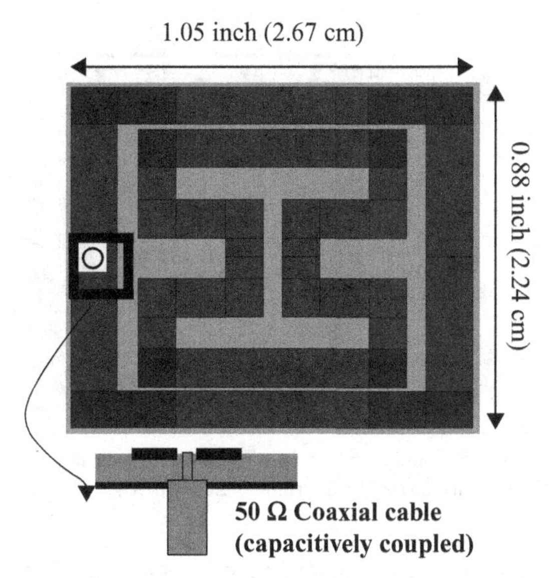


**Figure 8.98** Microstrip (MS) DBE layout on a 2 × 2 (in inches) substrate ([83], copyright © 2009 IEEE).

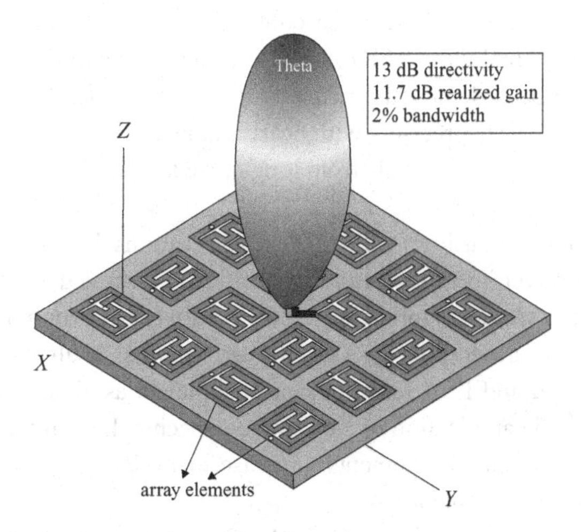


**Figure 8.99** Radiation pattern of a 4 × 4 MS-DBE array using element in Fig. 8.98 ([83], copyright © 2009 IEEE).

and the bandwidth is increased to 3.5%. Tangential electric field magnitude on the top surface around the antenna element is shown in Figure 8.100(b).

## 8.3 Design and practice of PSA (Physically Small Antennas)

A PSA is an antenna that is not necessarily ESA, but simply is a physically small antenna, including ESA. Various recently emerged wireless systems, for not only

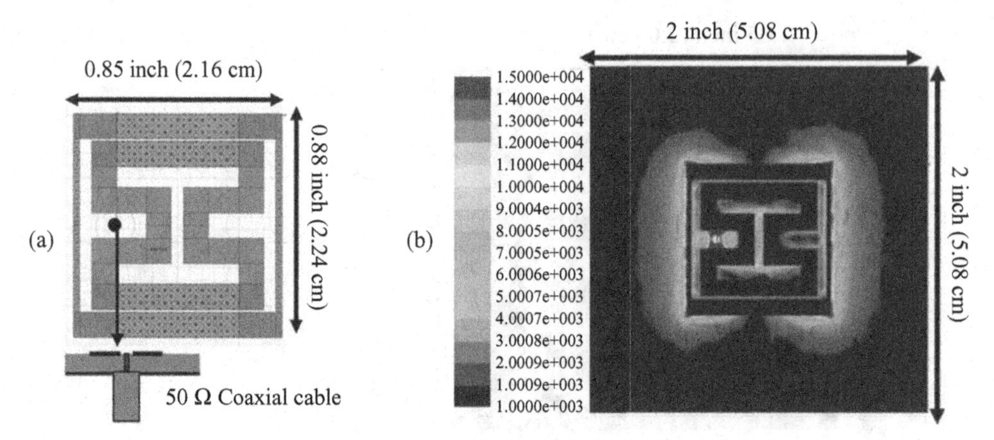


**Figure 8.100** (a) Small-sized MS-DBE layout on a 2 × 2 (in inches) substrate and (b) tangential electric field magnitude on the top surface ([83], copyright © 2009 IEEE).

communications, but also for control, sensor, data transmission, identification, remote sensing, wireless power transmission, and so forth, require small antennas to fit into the units of small pieces of equipment for those systems. Typical wireless applications are such systems as various short-range communications, for instance NFC (Near Field Communications), including RFID (Radio Frequency Identification), where numerous applications have been deployed practically in the recent decade, and radio watches/clocks, which operate very precisely with nearly standard time, being synchronized automatically by receiving time signals from long-wave standard-time broadcasting stations.

Types of antennas are not necessarily specific to these applications, but generally they are conventional types. However, most of them are specifically designed to fit into the small equipment used in the various applications and yet satisfy the requirements for each system. Antennas are not necessarily ESAs; however, antennas in almost all these cases need to be miniaturized and ESA techniques are employed as almost inevitable means. The techniques include application of slow-wave concepts, lowering resonance frequency, filling space with radiation elements, increase of radiation modes, and so forth, as covered previously in this chapter.

As the antenna dimensions become smaller, evaluation of antenna performances tends to become harder to obtain correct results. Special techniques to evaluate antenna performances often are required. However, the appropriate electromagnetic simulations may gain greater importance for replacing the measurements when evaluating antenna performances. The simulation can be used even for design of small antennas.

Another important problem is impedance matching of antenna to the load, conventionally a resistance of 50 Ω. In cases of very small RFID equipment, for example, the antenna is often directly connected to the RF circuits, which has impedance not of 50 Ω, but usually a much higher impedance. Further, connection of a type of balanced antenna with unbalanced circuits or vice versa may often be encountered. Without good

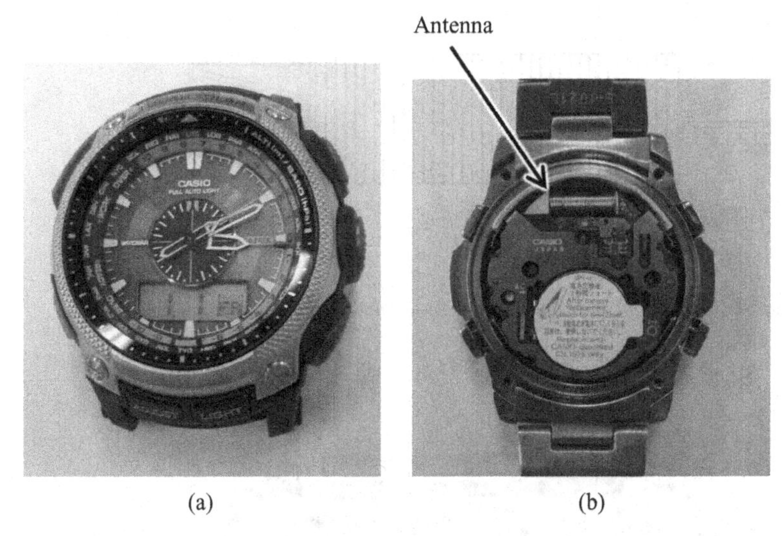


**Figure 8.101** (a) A wristwatch front view and (b) inside view to show a small coil antenna installed wristwatch ([84], copyright © 2007 IEEE, and [85], copyright © 2006 IEEE).

matching between the antenna and the load, the best system performance cannot be realized, meaning the desired operation range in the RFID system, for instance, might never be attained. Unfortunately, it has so far been recognized that there have been many systems operating in mismatched condition without careful considerations on the design. It may not be easy to obtain the perfect matching conditions, especially when the system operates over a wide frequency band; however, since it is an indispensable requirement, a means must be contrived to achieve suitably good matching. The method is not very specific, but conventional ways of matching can be applied.

## 8.3.1 Small antennas for radio watch/clock systems

Standard-time signals are broadcast from numerous radio stations of the world. In Japan, there are two long-wave broadcasting stations for JST (Japan Standard Time), which provide accuracy of $\pm 1 \times 10^{-12}$. A watch/clock, into which a very small receiver with a small magnetic-core loop antenna is integrated, receives the JST signal through the broadcasting of either 40 or 60 kHz from one of the two stations and automatically adjusts the time display to agree closely with the JST. A view of a wristwatch, in which a small receiver with a coil antenna is installed, is illustrated in Figure 8.101; (a) the front view and (b) the inside view, where the antenna is indicated with an arrow [84]. An example of a coil antenna is shown in Figure 8.102 [84, 85]. In the case of amorphous metal material, the core is composed of multilayered very thin film materials as shown in Figure 8.102. The amorphous material has permeability $\mu_r$ of around 8800, the thickness of a film is 0.16 μm, and a core is comprised of 40 films, half of which are bent slightly upward at the edge of the core to improve the sensitivity, Figure 8.103 shows a fabricated antenna. The antenna is designed to operate at 40 and/or 60 kHz, with length

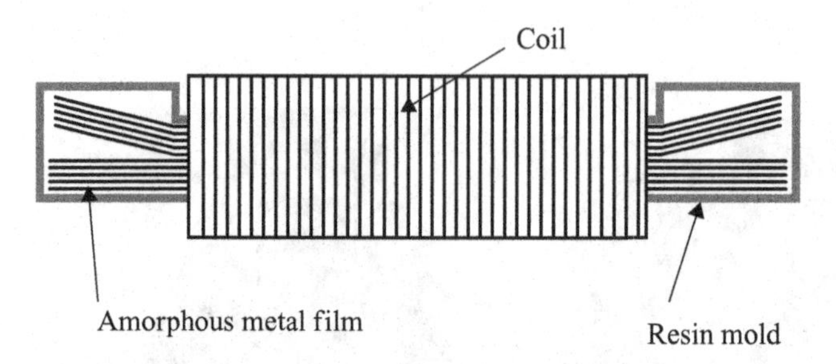


**Figure 8.102** A small amorphous metal laminated core loop antenna [84–85].

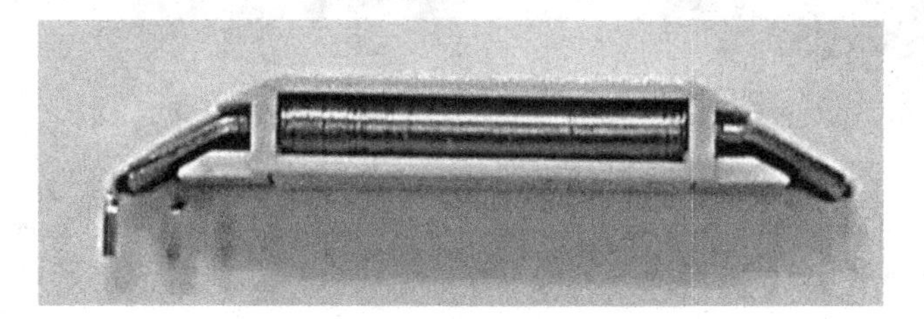

**Figure 8.103** Fabricated amorphous metal core loop antenna [84–85].

of 16 mm and number of turns about 1200. It was found that permalloy material can also be used for the core of the antenna instead of amorphous metal materials.

### 8.3.2 Small antennas for RFID

The main function of the RFID systems is to retrieve data or information about specific objects which were previously inserted. The systems include a transmitter and receiver, which are generally referred to as interrogator and transponder, respectively, and the operating frequency ranges cover from LF to UHF bands, dependent on the purpose of the system, and the communication ranges differ from almost zero distance to far-field distances as defined by radio-wave propagation. There are two types of systems, passive and active types. In the passive type, the transponder does not have an internal power source, whereas the active-type transponder is powered. The passive transponder operates with the received power from the electromagnetic field generated by the interrogator. Hence, the operating range of the passive system is very limited to almost zero distance in the range of tens of centimeters by using frequencies of LF and HF bands, to which electromagnetic coupling or an inductive communication system is usually employed to make an air link. By contrast, the active system can perform long-range operation, using higher frequencies such as VHF and UHF. The systems operating at higher frequencies than UHF regions are referred to as Microwave RFID systems.

Antenna types depend on the type of the systems. For the passive RFID systems, where the magnetic coupling or inductive communication system is employed, loop or coil type antennas are generally used. The devices that carry data or information are

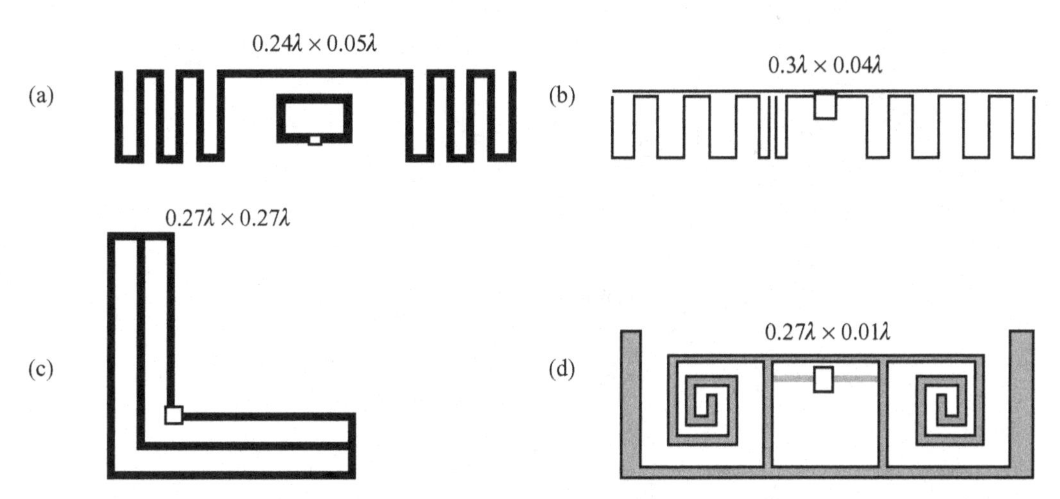


**Figure 8.104** Examples of small wire-type RFID antennas ([87], copyright © 2006 IEEE, and [90], copyright © 2006 IEEE).

often called a tag, instead of transponder, because of their physical shape, and they can also function as visual tags for pallets or cases of goods. Other types of antennas can be used in the tag. Antennas in tags, which have generally small, compact, or thin structures, are designed by applying miniaturization techniques.

For the active RFID systems, transponders have an internal power source, typically a battery, by which devices in the interrogator can be triggered for sending signals back to the interrogator. The interrogators are also called readers, because in earlier RFID systems, they were only capable of reading data or information sent from transponders. Varieties of antennas are employed for both transponders and interrogators in the active systems, although small antennas are still preferred. Numerous types of antennas have been developed and employed in practice for various types of RFID equipment. In this section, some typical RFID antennas are described.

#### 8.3.2.1 Dipole and monopole types

To reduce the antenna size, meander line structures are widely employed in various types of RFID antennas. Some examples are shown in Figure 8.104 [86]. In the figure, (a) shows an antenna ($f$ = 915 MHz) with a loop to couple with the meander line [87]; (b) depicts a meander line antenna ($f$ = 920 MHz) loaded with a bar, by which antenna impedance can be adjusted along with the meander line length [88]; (c) is a doubly folded L-shaped dipole ($f$ = 915 MHz), useful to fit the corner of a box and being visible from almost every angle [89]; and (d) provides a multi-conductor antenna ($f$ = 900 MHz) having double T-match scheme and spiral folding to achieve the required inductance [90]. There are many other meander line RFID antennas; however, here only a few among them are introduced.

A unique-configuration antenna, named forked shape monopole antenna is proposed in [91]. The antenna is designed for a compact USB dongle application. The geometry of

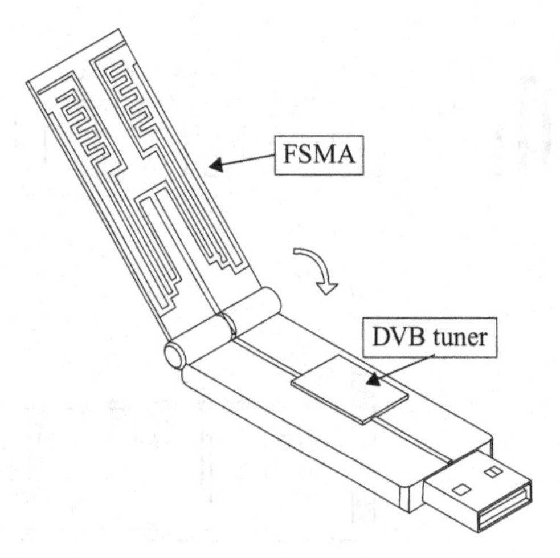


**Figure 8.105** A FSMA installed USB dongle ([91], copyright © 2010 IEEE).

the antenna is illustrated in Figure 8.105. The fork line is used to create a capacitive coupling effect to reduce the antenna size and enhance the bandwidth. The antenna is designed to operate in a very wide frequency band, 470 to 860 MHz, so as to receive digital video broadcasting (DVB). Furthermore, the upper frequency band of the antenna can be extended to 1142 MHz, leading to use in the GSM (Global System for Mobile Communications) band.

A wired bow-tie dipole RFID tag antenna recessed in a cavity is introduced in [92]. Figure 8.106 illustrates antenna geometry with top view in (a) and cut view in (b). The antenna is a cavity-backed type, allowing easy installation in a metallic container, vehicle, aircraft, and so forth. The antenna having the total length of 89 mm and the wing width of 38 mm is designed to operate at 915 MHz and is placed on the surface of the cavity, having the size of 140 × 80 × 50 mm. Through experiment, it was confirmed that maximum reading distance by using the antenna with optimized dimensions was about 3.2 times greater when compared with a commercial label-type RFID tag.

A planar dipole antenna ($f = 915$ MHz) designed to allow direct mounting on materials often encountered in practical applications is described in [93]. The outer shape of the planar element is an inverse exponential curve, which is illustrated in Figure 8.107. An inductive coil is attached to the input of the antenna to ease matching to an RFID chip that is mounted at the center of the coil. The size of the antenna is 100 × 25 mm. The reflection coefficient was lower than –5 dB from 860 to 960 MHz. The range performance of the antenna was examined when placed on materials such as cardboard, a paper block, plastic, wood, a bottle of tap water, and a glass bottle, in terms of material relative permittivity and the material thickness. It was found that if the tag is placed on materials such as cardboard, plastic, wood or in free space, having lower permittivity and less thickness, the tag range was around 8 m. However, materials of higher permittivity

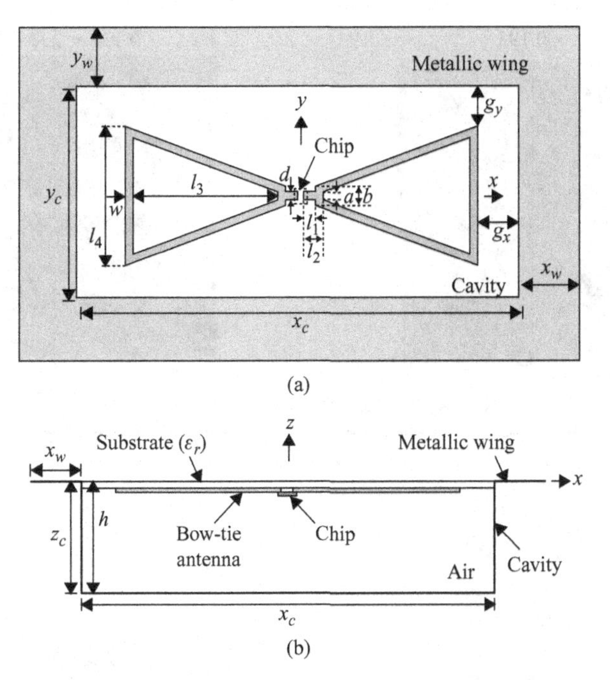


**Figure 8.106** Geometry of wired bow-tie RFID tag antenna: (a) top view and (b) cut view ([93], copyright © 2010 IEEE).

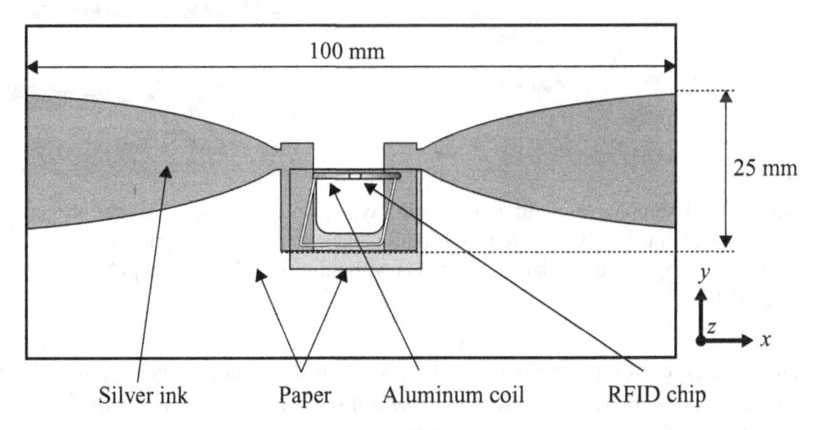


**Figure 8.107** A flared planar dipole antenna loaded with an RFID reader ([93], copyright © 2010 IEEE).

and greater thickness such as thick paper block, bottle of tap water, glass bottle either empty or filled with water, affect the tag range and decrease it to about 2 m.

#### 8.3.2.2 Inverted-F configuration antennas

To reduce the length of a monopole antenna the element is folded, and to ease the impedance matching, a shorting pin is added to form an Inverted-F (IF) shape. Various

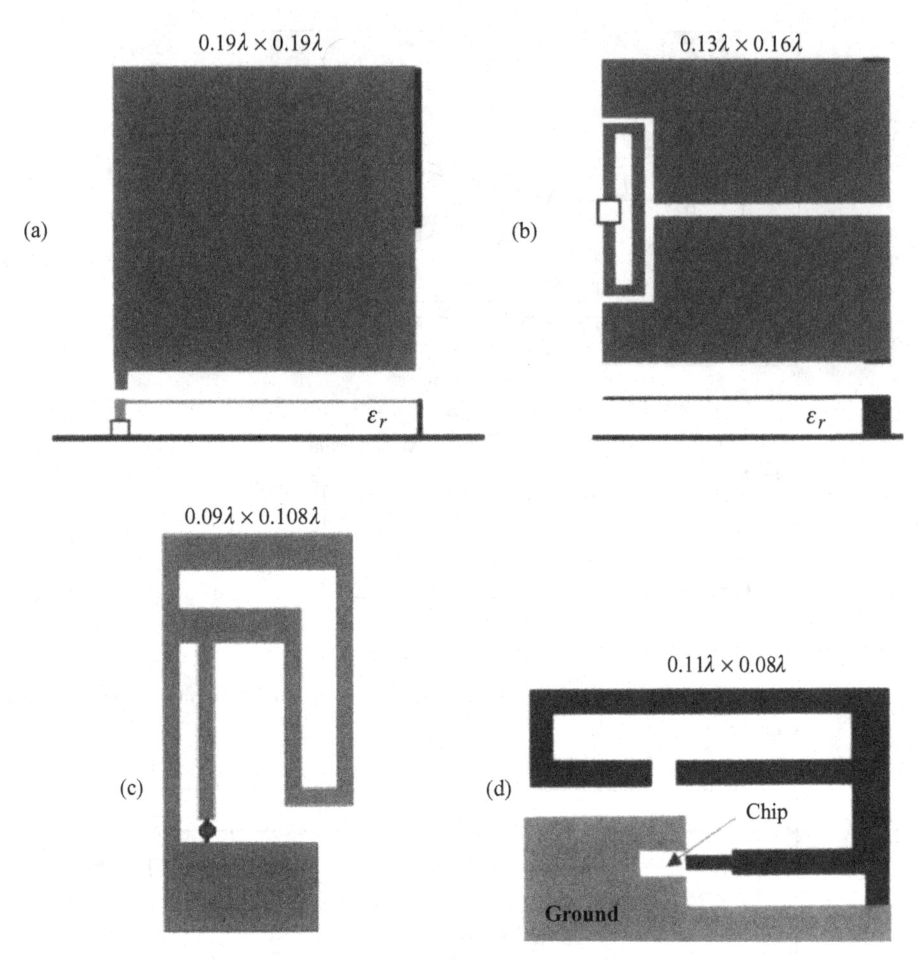


**Figure 8.108** Examples of modified Inverted-F type planar antennas: (a) conventional type, (b) two-layer double-PIFA tag fed by a small rectangular loop, (c) coplanar I-F antenna with an additional stub, and (d) coplanar IF antenna with multiple folded lines ([94], [95–96], copyright © 2006 IEEE).

modified IF configurations are applied to form RFID antennas. Figure 8.108(a) shows a PIFA (Planar IF Antenna) ($f$ = 870 MHz) with a square element [86, 94]; (b) depicts a double-PIFA tag ($f$ = 900 MHz) fed by a small rectangular loop [95]; (c) illustrates a PIFA ($f$ = 870 MHz) with an additional horizontal stub [96]; and (d) a dual-band PIFA ($f$ = 2400 MHz and 5300 MHz) with multiple folded elements [97].

Another type of antenna to which IF configuration is applied is a miniaturized IFA loaded with an element having a Hilbert trace at the end of the IFA element as shown in Figure 8.109 [98]. The antenna size is 35 × 6 × 1.6 mm including the ground plane. The center frequency is 2.45 GHz, return loss is 29 dB, and bandwidth for less than 10 dB return loss is 230 MHz (2.33–2.56 GHz), corresponding to 9.4% in terms of the relative bandwidth. The operating frequency can be shifted by adjusting the length of the Hilbert trace, which also leads to increasing the bandwidth. The gain can be adjusted

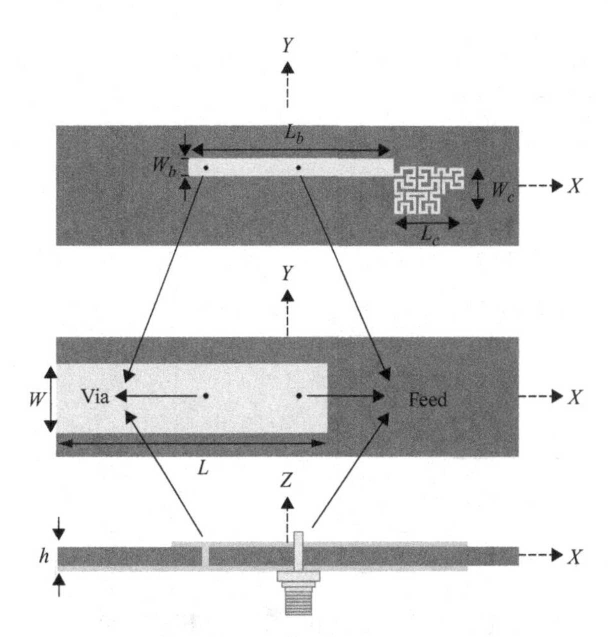


**Figure 8.109** Geometry of an IFA top loaded with Hilbert trace [97].

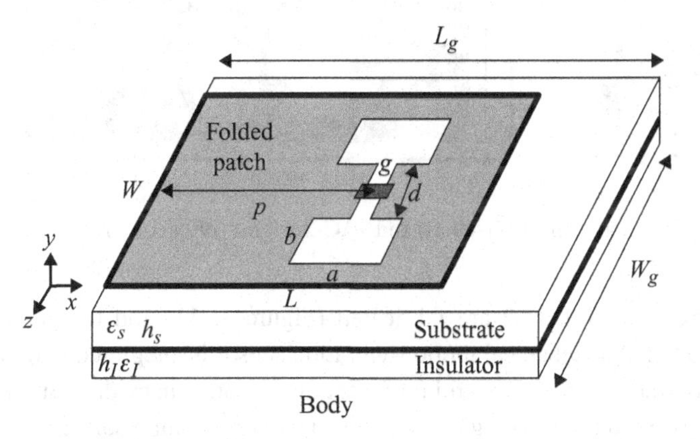


**Figure 8.110** A folded patch with H-shaped slot loaded with a chip ([99], copyright © 2010 IEEE).

by appropriately designing the size and the shape of the ground plane. With this antenna configuration, antenna performances can be improved, while achieving miniaturization of the antenna.

#### 8.3.2.3 Slot type antennas

A slot embedded folded rectangular patch is designed to apply to a wearable UHF RFID sensor tag [99]. Layout of the tag is illustrated in Figure 8.110, where dimensional parameters are also provided. The antenna is essentially a series-fed L-shaped patch and radiates from the open edge of the patch and the slot embedded on the planar element. A sensor chip is located in the middle of the gap between the two slots. Two prototype

**Table 8.8** Parameters of two tags [mm] ([99], copyright © 2010 IEEE)

| Parameter | TAG-1 | TAG-2 |
|---|---|---|
| $W_g$ | 60 | 90 |
| $W$ | 60 | 80 |
| $L_g$ | 49 | 60 |
| $L$ | 49 | 49 |
| $p$ | $L/2$ | 15 |
| $a$ | 8 | 18 |
| $b$ | 10 | 9 |
| $d$ | 10 | 10 |
| $g$ | 3 | 4 |
| $hs$ | 4 | 4 |

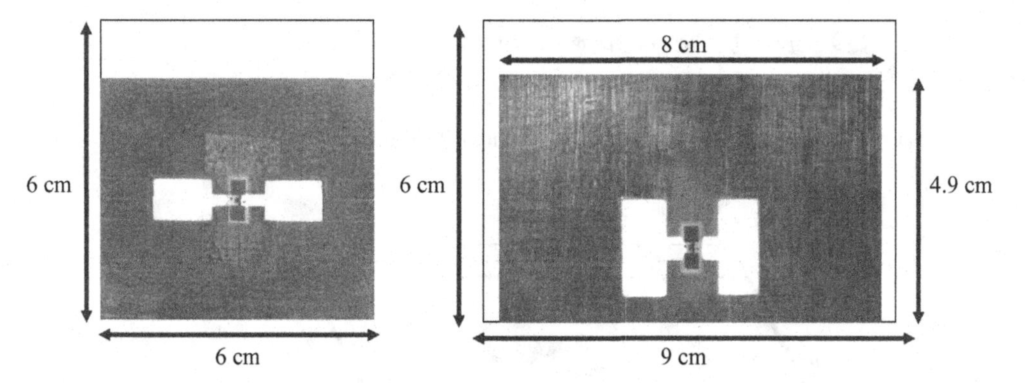


**Figure 8.111** Fabricated antenna TAG-1(left) and TAG-2 (right) ([99], copyright © 2010 IEEE).

RFID tags TAG-1 and TAG-2 are fabricated (Figure 8.111) and the parameters are given in Table 8.8. A mechanical motion-vibration sensor is integrated into the antenna structure. This kind of sensor is useful for detection of motion in medical applications, to assist diagnosis of some neurological diseases, involving compulsory arm movements in domestic healthcare, for example to track the behavior of aged persons [99]. The maximum reading range of TAG-1 is 1.5 m by using a short-range reader that emits not more than 0.5 W EIRP at the tag's microchip power of –30 dBm, while 2.1 m for TAG-2. When a long-range reader (emitting up to 3.2 W EIRP) is used, it increases to 4 m for the TAG-1 and 5.5 m for the TAG-2.

The antenna has a size comparable to a credit card and can be applied to any part of the body. Improvement of the tag is concerned with realization of flexible conformal tags by applying textile technology.

A compact antenna using series OCSRR (Open Complementary Split Ring Resonator) is designed for the passive RFID tags [100]. The OCSRR unit is composed of a circular slot and a conductor loop inserted inside the slot as shown in Figure 8.112(a). Figure 8.112(b) gives the equivalent circuit of the OCSRR unit. The inductance $L_{OCSRR}$

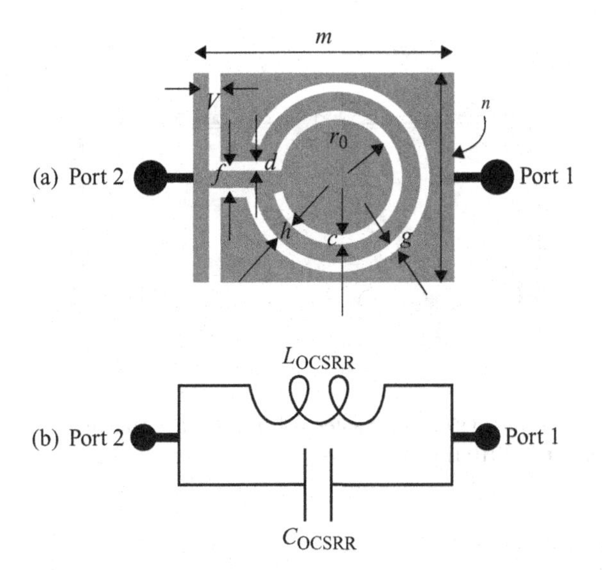


**Figure 8.112** (a) An OCSRR unit and its dimensional parameters and (b) the equivalent circuit ([100], copyright © 2010 IEEE).

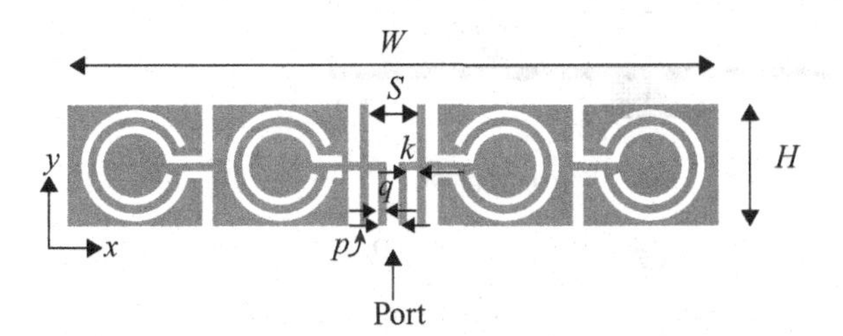


**Figure 8.113** Proposed UHF RFID antenna comprised of series connected OCSRR units ([100], copyright © 2010 IEEE).

is created by the conducting loop between the two inner and outer ring slots and the capacitance $C_{OCSRR}$ is between the inner conducting disk and the surrounding outer conducting planes. By connecting OCSRR units in series, and adjusting the size, input resistance and bandwidth can be increased, which is advantageous for matching of small tag antennas that generally have small resistance and high reactance components. Particularly with several units, the inductance becomes four times that of the conventional CSRR and makes for a feasible power harvesting circuit in the passive IC on the RFID tag. The antenna is comprised of four units of OCSRR as shown in Figure 8.113.

The antenna operates at 920 MHz and obtains a maximum reading range of 5.48 m with dimensions of $0.036\lambda_0 \times 0.17\lambda_0$ ($\lambda_0$ is the free space wavelength at 920 MHz).

Other examples of RFID tags using slots are a thin L-shaped slot dipole loaded with a meander line at the end of the dipole as shown in Figure 8.114 [101], and a patch antenna on which circular slots are embedded to produce circular polarization as shown in Figure 8.115 [102].

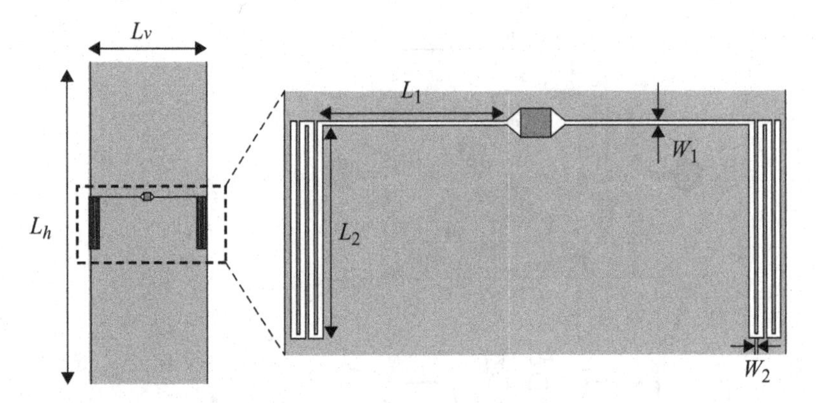


**Figure 8.114** Geometry of a proposed slot dipole antenna loaded with meandered line having RFID chip ([101], copyright © 2010 IEEE).

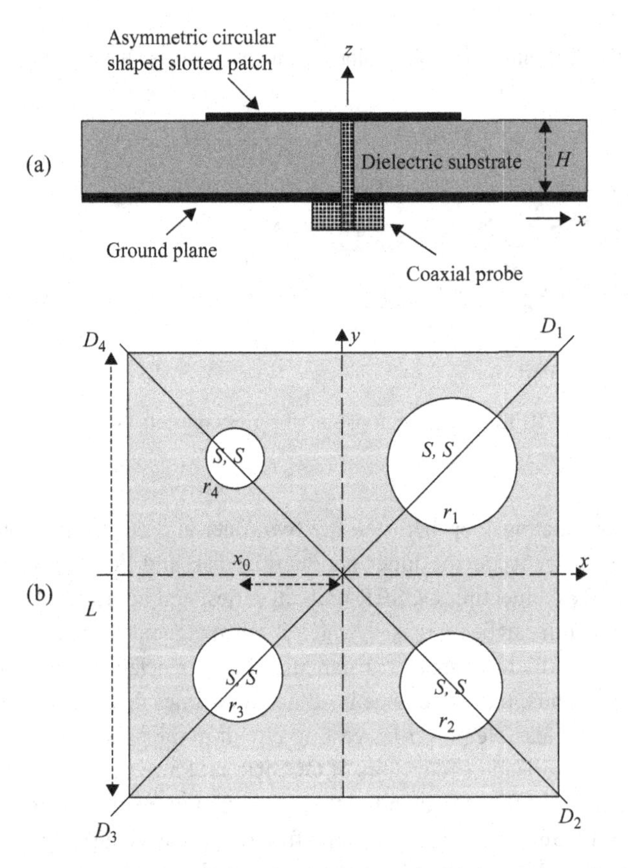


**Figure 8.115** Unequal size circular slots embedded square patch: (a) cross sectional view and (b) top view ([102], copyright © 2010 IEEE).

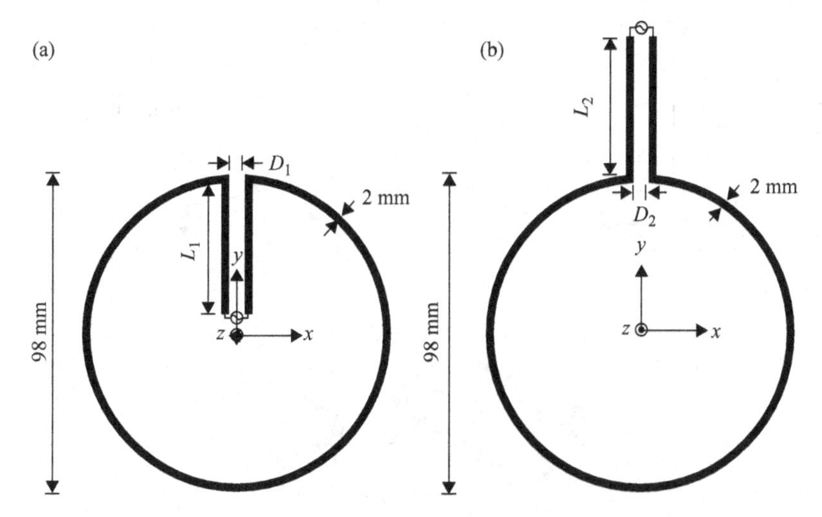


**Figure 8.116** Circular loop antenna with extended line fed by (a) internal line and (b) external line ([103], copyright © 2006 IEEE).

### 8.3.2.4 Loop antenna

A small circular loop antenna (CLA) operating at 911.25 MHz is designed for application to an RFID tag [103]. A one-wavelength and a half-wavelength CLA with a short stub are fabricated as shown in Figure 8.116, where (a) depicts an antenna with an internal stub, while (b) illustrates an antenna with an outer stub. The antenna dimensions are given in the figure. The stub increases the loop length so that the current path is increased and the diameter of the loop is reduced. The ratios of size reduction compared with a general CLA are 83% and 92%, respectively, for the one-wavelength CLA and the half-wavelength CLA. The return loss, –10 dB bandwidth, and gain are –11.9 dB, 12 MHz (1.3%), and 1.18 dBi, for the one wavelength CLA and –16.5 dB, 48 MHz (5%) and –0.58 dBi, respectively.

Another example is a square loop, but a part of the element is taken out, leaving three corners, as shown in Figure 8.117(a) [104]. Two elements are used to form a quasi-Yagi type antenna, with one element as a driver and another as a reflector, in order to produce a directional pattern as depicted in Figure 8.117(b), and consequent higher gain. The antenna is designed to install in a hand-held RFID reader as illustrated in Figure 8.117(c) for its operation in the near-field environment. The total dimensions of the antenna are 115 × 115 × 13.8 mm for operation of 433.92 MHz. The 10-dB bandwidth is 2.31 MHz (0.53%), covering the required bandwidth of 433.67–434.17 MHz for the ISO/IEC standards. When the reader is held by an operator, the radiation pattern deviates slightly from the desired direction, the orientation of the reader, because the operators typically hold the reader at a slant. Then, the gain is 2.5 dBi at angle $\theta = 20°$ (azimuth) at $\varphi = 0°$ (elevation), while 2.1 dBi at $\theta = 0°$.

The antenna radiates circular polarization and the bandwidth in terms of both impedance and for axial ratio less than 3, is 0.58% (433.23–435.7 MHz) and 1.01% (430.5–434.9 MHz) at $\theta = 0°$ and 20°, respectively.

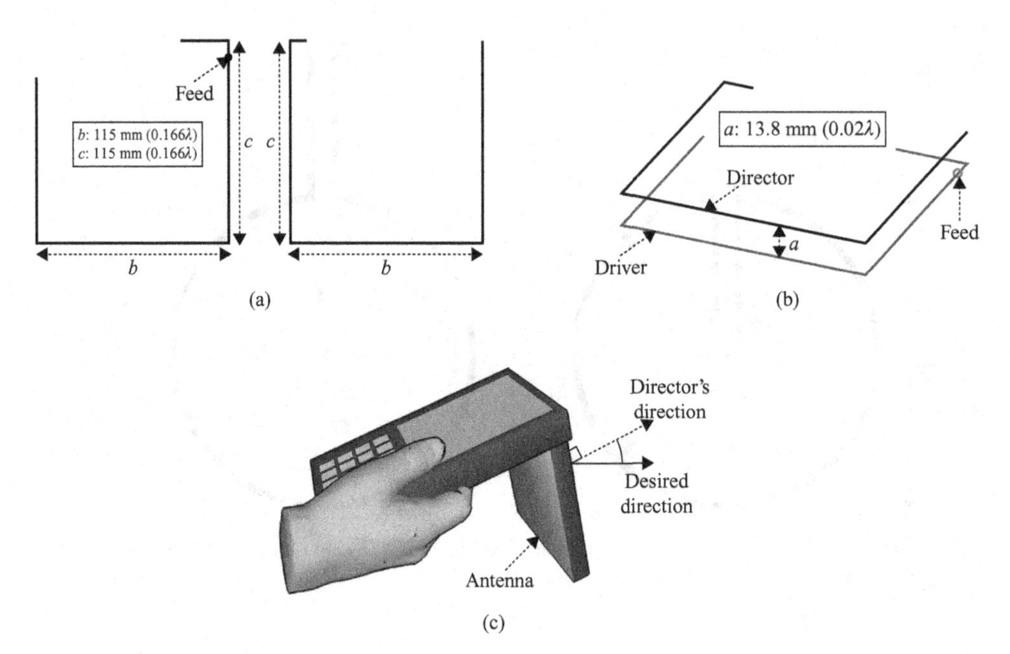


**Figure 8.117** A hand-held RFID device loaded with a two-wired broken square loop antenna ([104], copyright © 2010 IEEE).

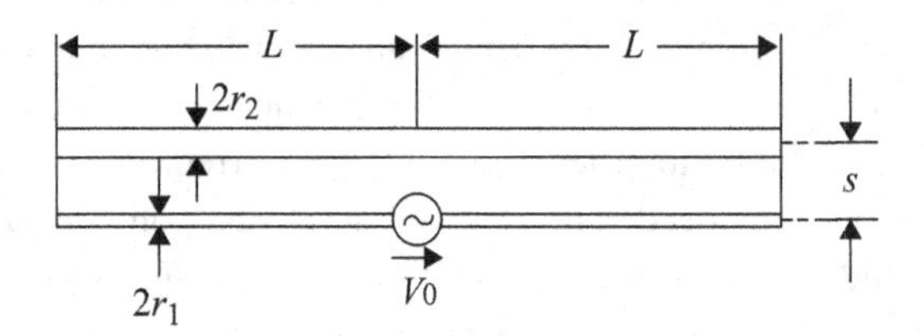


**Figure 8.A1** A folded dipole [27b].

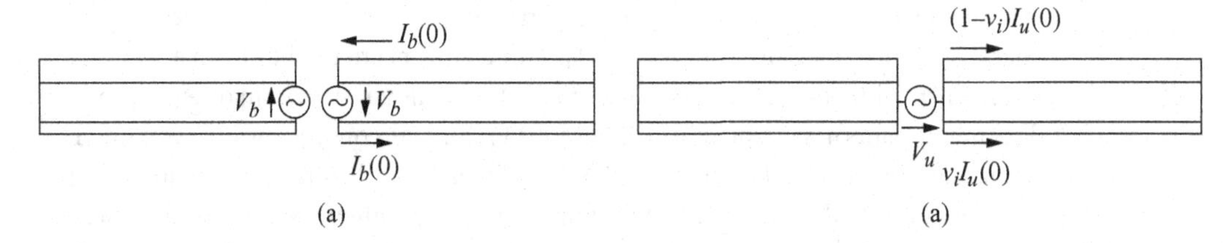


**Figure 8.A2** Equivalent expression of a folded dipole: (a) balanced mode and (b) unbalanced mode [27b].

## *Appendix I*

By folding an antenna structure, the input impedance depends on the ratio $v_i$ of the current distributions on the antenna elements. A folded antenna of length $2L$ is modeled as depicted in Figure 8A.1 [27b], in which current flows on each element having radius of $r_1$ and $r_2$ are denoted as $I_1$ and $I_2$, respectively, and the driven voltage is $V_0$. This model is equivalently divided into two parts as shown in Figure 8A.2; one is a balanced

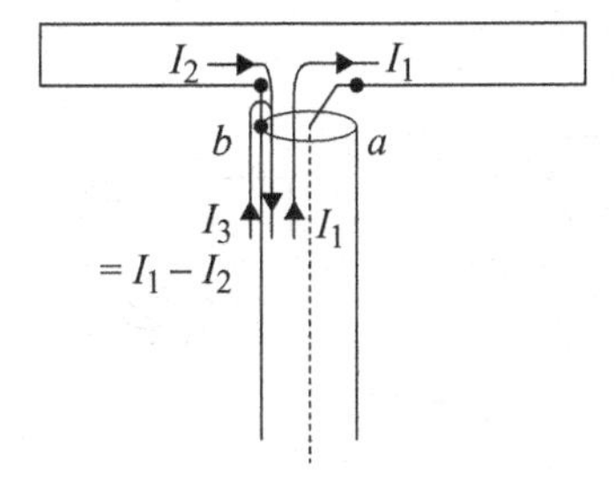


**Figure 8.A3** Current distributions at the feed point of the folded dipole connected with a coaxial cable [27b].

mode shown in (a), on which the currents $I_b(0)$ flow on both elements [$I(0)$ : current at the driven terminals] and the driven voltage is $V_b$. The other part is an unbalanced mode shown in (b), on which current flow on one element is $I_{u1} = (1 - \nu_i)\, I(0)$ and on other element is $I_{u2} = \nu_i\, I(0)$, respectively, and the driven voltage is $V_u$. Relationships between these voltages are given by

$$\begin{aligned} 2(1 - \nu_i)V_b + V_u &= V_0 \\ -2\nu_i V_b + V_u &= 0 \end{aligned} . \tag{A8.1}$$

From this

$$\begin{aligned} V_b &= V_0/2 \\ V_u &= \nu_i V_0 \end{aligned} . \tag{A8.2}$$

Then, input impedance $Z_{in}$ of the antenna is given by

$$\begin{aligned} Z_{in} &= V_0/[\nu_i I_u(0) + I_b(0)] \\ &= 1/[(Z_u/\nu_i) - 1 + 2Z_b - 1] \end{aligned} \tag{A8.3}$$

where $Z_u = V_u/I_u\,(0)$ and $Z_b = V_b/I_b$.

When two elements have the same radius, $\nu_i = 1/2$.

*Appendix II*

When a coaxial cable is directly connected to a folded dipole as shown in Figure 8A.3, current $I_1$ flowing into the dipole from the inner conductor of the coax returns to the coax as the sum of the current $I_2$, current at the opposite terminal of the dipole, and $I_3$, current on the outer conductor of the coax, which is equal to the difference of currents $I_1$ and $I_2$ [27b]. This is equivalently rewritten as a model shown in Figure 8A.4(a). This model is further divided into two modes; A and B, respectively, as Figure 8A.4(b) and (c) show. In each mode, the following relationships exist, by

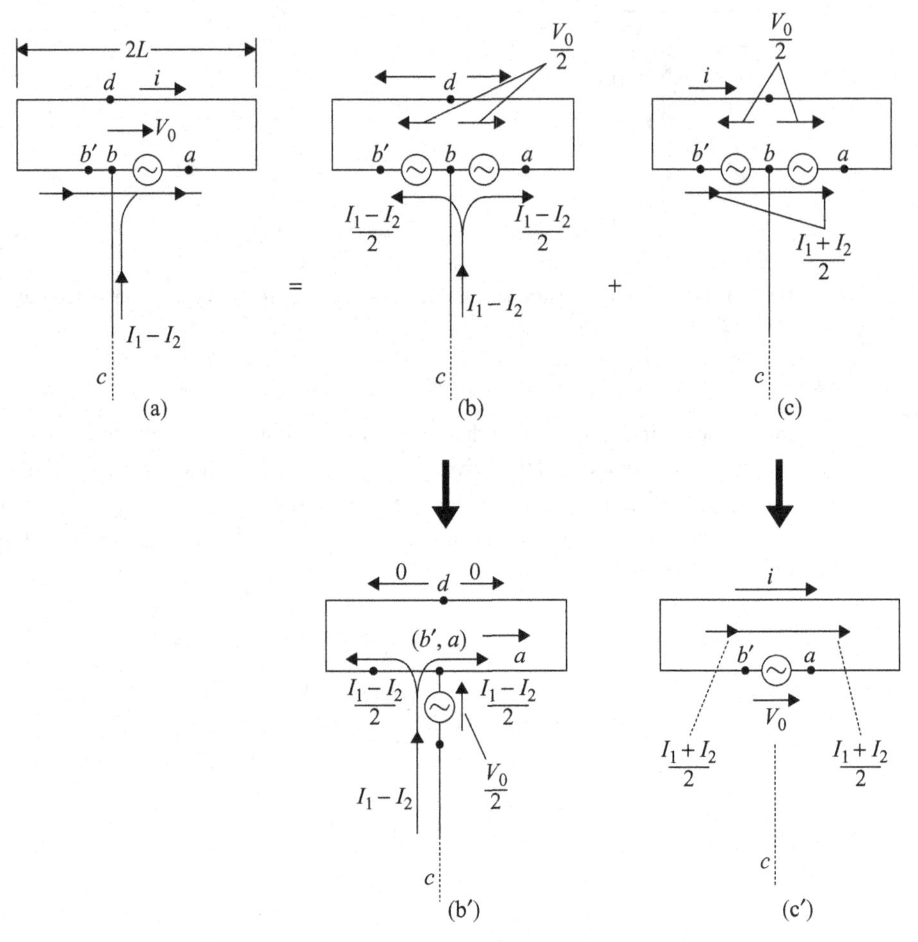


**Figure 8.A4** Current and voltage at the feed point of a folded dipole (a) original mode, (b) and (b′) A mode, and (c) and (c′) B mode [27b].

using $Z_a$ and $Z_b$, impedances of A mode and B mode, respectively, seen at the terminals *b–d*,

$$\begin{aligned} V_0/2 &= (I_1 - I_2)/Z_a \text{ for A mode}, \\ V_0 &= (I_1 + I_2)Z_b/2 \text{ for B mode}. \end{aligned} \tag{A8.4}$$

From these,

$$\begin{aligned} (I_1 - I_2) &= V_0/(2Z_a) \\ (I_1 + I_2) &= V_0/(Z_b/2) \end{aligned}. \tag{A8.5}$$

When the length of antenna $L$ is a half wavelength, that is, one side is a quarter wavelength, $Z_a$ becomes infinite, then $I_1 = I_2 = I$, and $(I_1 + I_2)/2 = I$. This means that the current on the outer conductor of the coax $(I_1 - I_2)$ disappears, and the same current $I$ flows in and out from the coax, implying that a balun is not required. This is the

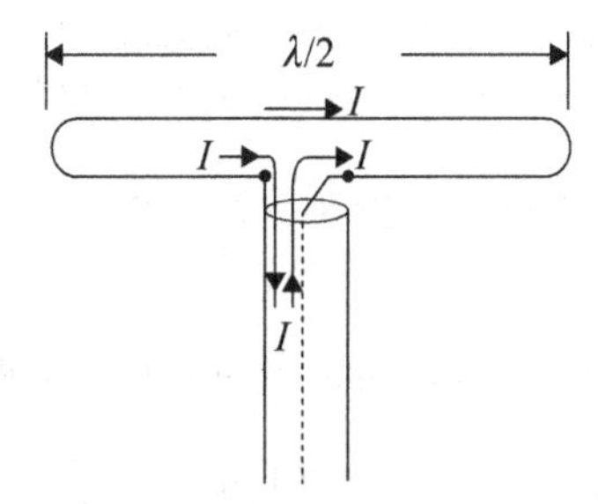


**Figure 8.A5** Direct connection of a coaxial cable to a half-wave folded dipole [27b].

self-balanced property of a folded dipole. With this property, the coax can be directly connected to the folded dipole, as Figure 8A.5 depicts [27b, 30].

## References

[1] F. Yang, X-X Zhang, and Y. Rahmat-Samii, Wide-Band E-Shaped Patch Antennas for Wireless Communications, *IEEE Transactions on Antennas and Propagation*, vol. 49, 2001, no. 7, pp. 1094–1100.

[2] Y. Chen, S. Yang, and Z. Nie, Bandwidth Enhancement Method for Low Profile E-Shaped Microstrip Patch Antennas, *IEEE Transactions on Antennas and Propagation*, vol. 58, 2010, no. 7, pp. 2442–2447.

[3] A. Khidre, K. F. Lee, F. Yang, and A. Elsherbeni, Wideband Circularly Polarized E-Shaped Patch Antenna for Wireless Applications, *IEEE Antennas and Propagation Magazine*, vol. 52, 2010, no. 5, pp. 219–229.

[4a] S. K. Sharma and L. Shafai, Performance of a Novel $\psi$-shape Microstrip Patch Antenna with Wide Bandwidth, *IEEE Antennas and Wireless Propagation Letters*, vol. 8, 2009, pp. 468–471.

[4b] S. K. Sharma and L. Shafai, Investigation of a Novel $\psi$-shape Microstrip Patch Antenna With Wide Impedance Bandwidth, *IEEE APS International Symposium* June 2007, Digest vol. 45, pp. 881–884.

[5] K. L Chung, A Wideband Circularly Polarized H-Shaped Patch Antenna, *IEEE Transactions on Antennas and Propagation*, vol. 58, 2010, pp. 3379–3383.

[6] Z. N. C. Nasimuddin and X. Qing, Dual-Band Circularly Polarized S-Shaped Slotted Patch Antenna with a Small Frequency Ratio, *IEEE Transactions on Antennas and Propagation*, vol. 58, 2010, pp. 2112–2225.

[7a] S. I. Latif, L. Shafai, and S. K. Sharma, Bandwidth Enhancement and Size Reduction of Microstrip Slot Antennas, *IEEE Transactions on Antennas and Propagation*, vol. 53, 2005, no. 3, pp. 994–1003.

[7b] Y-Shin Wang and S-J Chung, A Short Open-End Antenna with Equivalent Circuit Analysis, *IEEE Transactions on Antennas and Propagation*, vol. 58, 2010, no. 5, pp. 1771–1775.

[8a] T. Huynh and K. F. Lee, Single-Layer Single-Patch Wideband Microstrip Antenna, *Electronics Letters*, vol. 31, 1997, no. 16, pp. 1310–1312.

[8b] A. K. Shackelford, K. F. Lee, and K. M. Luk, Design of Small-Size Wide-Bandwidth Microstrip-Patch Antennas, *IEEE Antennas and Propagation Magazine*, vol. 45, 2003, no. 1, pp. 75–83.

[8c] C. L. Mak, K. M. Luk, K. F. Lee, and Y. L. Chow, Experimental Study of a Microstrip Patch Antenna with an L-shaped Probe, *IEEE Transactions on Antennas and Propagation*, vol. 48, 2000, no. 5, pp. 777–783.

[8d] K. F. Lee, Y. X. Guo, J. A. Hawkins, R. Chiar, and K. M. Luk, Theory and Experiment on Microstrip Patch Antennas with Shorting Wall, *IEE Proceedings, Microwaves, Antennas & Propagation*, vol. 147, 2000, pp. 521–525.

[8e] A. K. Shackelford, K. F. Lee, K. M. Luk, and R. Chair, U-Slot Patch Antenna with Shorting Pin, *Electronics Letters*, vol. 37, 2001, no. 12, pp. 729–730.

[8f] R. Chair *et al.*, Miniature Wide-Band Half U-Slot and Half E-Shaped Patch Antennas, *IEEE Transactions on Antennas and Propagation*, vol. 53, 2005, no. 8, pp. 2645–2652.

[8g] C. L. Mak, R. Chair, K. F. Lee, K. M. Luk, and A. A. Kishk, Half-U-slot Patch Antenna with Shorting Wall, *International Symposium USNC/CNC/URSI National Radio Science Meeting*, vol. 2, 2003, pp. 876–879.

[9] K. F. Lee *et al.*, The Versatile U-Slot Antenna, *IEEE Antennas and Propagation Magazine*, vol. 52, 2010, no. 1, pp. 71–88, and p. 86 in vol. 52, 2010, no. 2.

[10] K. F. Lee, S. L. Steven, and A. Kishk, Dual- and Multiband U-Slot Patch Antennas, *IEEE Antennas and Wireless Propagation Letters*, vol. 2, 2008, p. 64.

[11] K. F. Tong and T. P. Wong, Circular Polarized U-Slot Antenna, *IEEE Transactions on Antennas and Propagation*, vol. 55, 2007, no. 8, pp. 2382–2385.

[12] S. Weigand *et al.*, Analysis and Design of Broad-Band Single-Layer Rectangular U-Slot Microstrip Patch Antennas, *IEEE Transactions on Antennas and Propagation*, vol. 51, 2003, no. 3, pp. 457–468.

[13] K.-L. Wong, *Compact and Broadband Microstrip Antennas*, John Wiley & Sons, 2002, p. 239.

[14] J.-Y. Sze and K.-L. Wong, Bandwidth Enhancement of a Microstrip-Line-Fed Printed Wide-Slot Antenna, *IEEE Transactions on Antennas and Propagation*, vol. 49, 2001, no. 7, pp. 1020–1024.

[15] J.-Y. Jan and J.-W. Su, Bandwidth Enhancement of a Printed Wide-Slot Antenna With a Rotated Square, *IEEE Transactions on Antennas and Propagation*, vol. 53, 2005, no. 6, pp. 2111–2114.

[16] C.-H. Chang and K.-L. Wong, Printed λ/8-PIFA for Penta-Band WWAN Operation in the Mobile Phone, *IEEE Transactions on Antennas and Propagation*, vol. 57, 2009, no. 5, pp. 1373–1381.

[17] C.-M. Peng *et al.*, Bandwidth Enhancement of Internal Antenna by Using Reactive Loading for Penta-Band Mobile Handset Application, *IEEE Transactions on Antennas and Propagation*, vol. 59, 2011, no. 5, pp. 1728–1733.

[18] C.-W. Chiu and Y.-J. Chi, Printed Loop Antenna with a U-Shaped Tuning Element for Hepta-Band Laptop Applications, *IEEE Transactions on Antennas and Propagation*, vol. 58, 2010, no. 11, pp. 3464–3470.

[19] F.-H. Chu and K.-Lu Wong, Planar Printed Strip Monopole with a Closely-Coupled Parasitic Shorted Strip for Eight-Band LTE/GSM/UMTS Mobile Phone, *IEEE Transactions on Antennas and Propagation*, vol. 58, 2010, no. 10, pp. 3426–3431.

[20] M. J. Ammann and Z. N. Chen, Wideband Monopole Antennas for Multi-Band Wireless Systems, *IEEE Antennas and Propagation Magazine*, vol. 45, 2003, no. 2, pp. 146–150.

[21] G. Dubost and S. Zisler, *Antennas a Large Bande*, Masson, 1976, pp. 128–129.

[22] S. Honda *et al.*, On a Broadband Disk Monopole Antenna, *Technical Report of Television Society Japan*, ROFT 91–55 (1991–10) 1991.

[23] J. A. Evans and M. J. Ammann, Planar Trapezoidal and Pentagonal Monopole with Impedance Bandwidth in Excess of 10:1, *IEEE APS International Symposium Digest*, vol. 3, Orlando, 1999, pp. 1558–1561.

[24] Z. N. Chen and M. Y. W. Chia, Impedance Characteristics of EMC Triangular Planar Monopoles, *Electronics Letters*, vol. 37, 2001, no. 21, pp. 1271–1272.

[25] Z. N. Chen, Impedance Characteristics of Planar Bow-Tie Like Monopole Antennas, *Electronics Letters*, vol. 36, 2000, no. 13, pp. 1100–1101.

[26] H. Morishita, H. Furuuchi, and K. Fujimoto, Performance of Balanced-fed Antenna System for Handsets in the Vicinity of a Human Head or Hand, *IEE Proceedings, Microwaves, Antennas & Propagation*, vol. 149, 1999, no. 2, pp. 85–91.

[27a] H. Morishita *et al.*, A Balance-Fed Loop Antenna System for Handset, *IEICE Transactions on Fundamentals of Electronics, Communications and Computer Sciences*, vol. E82-A, 1999, no. 7, pp. 1138–1143.

[27b] S. Hayashida, Analysis and Design of Folded Loop Antennas, Doctoral Dissertation, National Defense Academy, Japan, 2006. (in Japanese).

[27c] S. Hayashida, H. Morishita, and K. Fujimoto, A Wideband Folded Loop Antenna for Handsets, *IEICE*, vol. J86-B, 2003, no. 9, pp. 1799–1805 (in Japanese).

[28] Y. Kim *et al.*, A Folded Loop Antenna system for Handsets Developed and Based on the Advanced Design Concept, *IEICE Transactions on Communications*, vol. E84-B, 2001, no. 9, pp. 2468–2475.

[29] H. Morishita, Y. Kim, and K. Fujimoto, Analysis of Handset Antennas in the Vicinity of the Human Body by the Electromagnetic Simulator, *IEICE Transactions on Electronics*, vol. E84-C, 2001, no. 7, pp. 937–947.

[30] S. Hayashida, H. Morishita, and K. Fujimoto, Self-Balanced Wideband Folded Loop Antenna, *IEE Proceedings, Microwaves, Antennas & Propagation*, vol. 153, 2006, no. 1, pp. 7–12.

[31] S. Hayashida *et al.*, Characteristics of Built-in Folded Monopole Antenna for Handsets, *IEICE Transactions on Communications*, vol. E88-B, 2005, no. 6, pp. 2275–2283.

[32] X. H. Wu and Z. N. Chen, Comparison of Planar Dipoles in UWB Applications, *IEEE Transactions on Antennas and Propagation*, vol. 53, 2005, no. 6, pp. 1973–1983.

[33a] S. Koulouridis and J. L. Volakis, A Novel Planar Conformal Antenna Designed with Splines, *IEEE Antennas and Wireless Propagation Letters*, vol. 8, 2009, pp. 34–36.

[33b] S. Koulouridis and J. L. Volakis, Miniaturization of Flare Dipole via Shape Optimization and Matching Circuits, *IEEE APS International Symposium* 2007, pp. 4785–4788.

[34] W. Dullaert and H. Rogier, Novel Compact Model for the Radiation Pattern of UWB Antennas Using Vector Spherical and Slepian, *IEEE Transactions on Antennas and Propagation*, vol. 58, 2010, no. 2, pp. 287–298.

[35] C.-W. Ling *et al.*, Planar Binomial Curved Monopole Antennas for Ultra-wideband Communication, *IEEE Transactions on Antennas and Propagation*, vol. 55, 2007, no. 9, pp. 2622–2624.

[36] D. Valderas *et al.*, UWB Staircase-Profile Printed Monopole Design, *IEEE Antennas and Wireless Propagation Letters*, vol. 7, 2008, pp. 255–259.

[37] M. Sun, Y. P. Zhang, and Y. Lu, Miniaturization of Planar Monopole Antenna for Ultra-wideband Radios, *IEEE Transactions on Antennas and Propagation*, vol. 58, 2010, no. 7, pp. 2420–2425.

[38a] L. Lizzi *et al.*, Optimization of a Spline-Shaped UWB Antenna by PSO, *IEEE Antennas and Wireless Propagation Letters*, vol. 6, 2007, pp. 182–185.

[38b] F. Viani *et al.*, A Miniaturized UWB Antenna for Wireless Dongle Devices, *IEEE Antennas and Wireless Propagation Letters*, vol. 7, 2008, pp. 714–717.

[39] E. A-Daviu *et al.*, Modal Analysis and Design of Band-Notched UWB Planar Monopole Antennas, *IEEE Transactions on Antennas and Propagation*, vol. 58, 2010, no. 5, pp. 1457–1467.

[40a] E. S. Angelopoulos *et al.*, Circular and Elliptical CPE-Fed Slot and Microstrip-Fed Antennas for Ultrawideband Applications, *IEEE Antennas and Wireless Propagation Letters*, vol. 5, 2006, pp. 294–297.

[40b] A. M. Abbosh and E. Bialkowsky, Design of Ultrawideband Planar Monopole Antenna of Circular and Elliptical Shape, *IEEE Antennas and Wireless Propagation Letters*, vol. 7, 2009, pp. 17–23.

[40c] P. Li and X. Chen, Study of Printed Elliptical/Circular Slot Antennas for Ultrawideband Applications, *IEEE Transactions on Antennas and Propagation*, vol. 54, 2006, no. 6. pp. 1670–1675.

[40d] J. Y. Chion, J. Y. See, and K. L. Wong, A Broadband CPW-fed Strip Loaded Square Slot Antenna, *IEEE Transactions on Antennas and Propagation*, vol. 51, 2003, no. 4, pp. 719–721.

[40e] H. D. Chen, Broadband CPW-fed Square Slot Antennas with a Wideband Tuning Stub, *IEEE Transactions on Antennas and Propagation*, vol. 51, 2003, no. 8, pp. 1982–1986.

[40f] R. Chair, A. A. Kishk, and K. F. Lee, Ultrawide-band Coplanar Waveguide-fed Rectangular Slot Antenna, *IEEE Antennas and Wireless Propagation Letters*, vol. 3, 2004, no. 12, pp. 227–229.

[41] R. Zaker and J. Nourinia, Novel Modified UWB Planar Monopole Antenna with Variable Frequency Band-Notch Function, *IEEE Antennas and Wireless Propagation Letters*, vol. 7, 2008, pp. 112–116.

[42a] Z. N. Chen, T. S. P. See, and X. Qing, Small Printed Ultrawideband Antenna with Reduced Ground Plane, *IEEE Transactions on Antennas and Propagation*, vol. 55, 2007, no. 2, pp. 383–388.

[42b] H. S. Choi *et al.*, A New Ultra-wideband Antenna for UWB Applications, *Microwave Optical Technology Letters*, vol. 40, 2004, no. 5, pp. 399–401.

[42c] K. Chung, H. Park, and J. Choi, Wideband Microstrip-fed Monopole Antenna with a Narrow Slit, *Microwave Optical Technology Letters*, vol. 47, 2005, no. 4, pp. 400–402.

[43] S. K. Rajgopal and S. K. Sharma, Investigation on Ultrawideband Pentagon Shape Microstrip Slot Antenna for Wireless Communications, *IEEE Transactions on Antennas and Propagation*, vol. 57, 2009, no. 5, pp. 1353–1359.

[44] T.-G. Ma and C.-H. Tseng, An Ultrawideband Coplanar Waveguide-Fed Tapered Ring Slot Antenna, *IEEE Transactions on Antennas and Propagation*, vol. 54, 2006, no. 4, pp. 1105–1110.

[45] W.-J. Lui, C.-H. Cheng, and H.-B. Zhu, Improved Frequency Notched Ultrawide-band Slot Antenna Using Square Ring Resonator, *IEEE Transactions on Antennas and Propagation*, vol. 55, 2007, no. 9, pp. 2445–2450.

[46] H. Elsadek and D. Nashaat, Multiband and UWB V-Shaped Antenna Configuration for Wireless Communications Applications, *IEEE Antennas and Wireless Propagation Letters*, vol. 7, 2008, pp. 89–91.

[47] N. Behdad and K. Sarabandi, A Compact Antenna for Ultrawide-Band Applications, *IEEE Transactions on Antennas and Propagation*, vol. 53, 2005, no. 7, pp. 2185–2191.

[48] K. Fujimoto, Integrated Antenna Systems, in K. Chang (ed.) *Encyclopedia of RF and Microwave Engineering*, vol. 3, John Wiley and Sons, 2005, pp. 2113–2147.

[49] A. Mortazawi, T. Itoh, and J. Harvey, *Active Antennas and Quasi-Optical Arrays*, IEEE Press, 1999.

[50] J. A. Navarrow and K. Chang, *Integrated Active Antennas and Spatial Power Combining*, John Wiley and Sons, 1996.

[51] K. Chang, R. A. York, P. S. Hall, and T. Itoh, Active Integrated Antennas, *IEEE Transactions on Antennas and Propagation*, vol. 50, 2002, no. 3, pp. 937–943.

[52] G. Elli and S. Liw, Active Planar Inverted-F Antenna for Wireless Applications, *IEEE Transactions on Antennas and Propagation*, vol. 57, 2009, no. 10, pp. 2899–2906.

[53] C. R. White and G. M. Rebeiz, Single- and Dual-Polarized Tunable Slot-Ring Antennas, *IEEE Transactions on Antennas and Propagation*, vol. 51, 2003, no. 1, pp. 19–26.

[54] L. M. Feldner *et al.*, Electrically Small Frequency-Agile PIFA-as-a Package for Portable Wireless Devices, *IEEE Transactions on Antennas and Propagation*, vol. 55, 2007, no. 11, pp. 3310–3319.

[55] S.-L. S. Yang and A. A. Kishk, Frequency Reconfigurable U-Slot Microstrip Patch Antenna, *IEEE Antennas and Wireless Propagation Letters*, vol. 7, 2008, pp. 127–129.

[56] D. E. Anagnostou and A. A. Gheethan, A Coplanar Reconfigurable Folded Slot Antenna Without Bias Network for WLAN Applications, *IEEE Antennas and Wireless Propagation Letters*, vol. 8, 2009, pp. 1057–1060.

[57] C. Zhang *et al.*, A Low-Profile Branched Monopole Loop Reconfigurable Multiband Antenna for Wireless Applications, *IEEE Antennas and Wireless Propagation Letters*, vol. 8, 2009, pp. 216–219.

[58] A. Edalati and T. A. Denidni, Reconfigurable Beamwidth Antenna Based on Active Partially Reflective Surfaces, *IEEE Antennas and Wireless Propagation Letters*, vol. 8, 2009, pp. 1087–1090.

[59] J. Sarrazin, Y. Mahe, and S. Avrillon, Pattern Reconfigurable Cubic Antenna, *IEEE Transactions on Antennas and Propagation*, vol. 57, 2009, no. 2, pp. 310–317.

[60] C. H Mueller *et al.*, Small-Size X-Band Active Integrated Antenna with Feedback Loop, *IEEE Transactions on Antennas and Propagation*, vol. 56, 2008, no. 5, pp. 1236–1241.

[61] J-H Lim *et al.*, A Reconfigurable PIFA Using a Switchable PIN-Diode and a Fine-Tuning Varactor for USPCS/WCDMA/m-WiMAX/WLAN, *IEEE Transactions on Antennas and Propagation*, vol. 58, 2010, no. 7, pp. 2404–2411.

[62] K. Fujimoto (ed.) *Antenna Systems for Mobile Communications*, Artech House, 2009, pp. 247–248.

[63] D. Sievenpiper *et al.*, High-Impedance Electromagnetic Surfaces with a Forbidden Frequency Band, *IEEE Transactions on Microwave Theory and Techniques*, vol. 47, 1999, no. 11, pp. 2059–2074.

[64] S. Clavijo, R. E. Diaz, and W. E. Mckinzie, Design Methodology for Sievenpiper High-Impedance Surfaces: An Artificial Magnetic Conductor for Positive Gain Electrically Small Antennas, *IEEE Transactions on Antennas and Propagation*, vol. 51, 2003, no. 10, pp. 2678–2690.

[65] Q. Wu, J. Geng, and D. Su, On the Performance of Printed Dipole Antenna with Novel Composite Corrugated-Reflectors for Low-Profile Ultrawideband Applications, *IEEE Transactions on Antennas and Propagation*, vol. 58, 2010, no. 12, pp. 3829–3846.

[66] G. Gampala, R. Sammeta, and C. J. Reddy, A Thin, Low-Profile Antenna Using a Novel High Impedance Ground Plane, *Microwave Journal*, July 2010, pp. 70–80.

[67] A. Suntives and R. Abhari, Design of a Compact Miniaturized Probe-Fed Patch Antennas Using Electromagnetic Bandgap Structure, *IEEE APS International Symposium*, 2010,

[68a] F. Yang and Y. Rahmat-Samii, Microwave Antennas Integrated with Electromagnetic Band-Gap (EBG) Structure: A Low Mutual Coupling Design for Array Applications, *IEEE Transactions on Antennas and Propagation*, vol. 51, 2003, no. 10, pp. 2936–2946.

[68b] Z. Li and Y. Rahmat-Samii, PBC, PMC, and PEC Ground Plane: A Case Study for Dipole Antenna, *IEEE APS International Symposium 2000*, vol. 4, pp. 2258–2261.

[69] M. Z. Azad and M. Ali, Novel Wideband Directional Dipole Antenna on a Mushroom Like EBG Structure, *IEEE Transactions on Antennas and Propagation*, vol. 56, 2008, no. 5, pp. 1242–1250.

[70] F. Yang and Y. Rahmat Samii, Reflection Phase Characterizations of the EBG Ground Plane for Low Profile Wire Antenna Applications, *IEEE Transactions on Antennas and Propagation*, vol. 51, 2003, no. 10, pp. 2691–2703.

[71] S. R. Best and D. L. Hanna, Design of a Broadband Dipole in Proximity to an EBG Ground Plane, *IEEE Antennas and Propagation Magazine*, vol. 50, 2008, no. 6, pp. 52–64.

[72] L. Akhoondzadeh-Asl, *et al.*, Wideband Dipoles on Electromagnetic Bandgap Ground Plane, *IEEE Transactions on Antennas and Propagation*, vol. 55, 2007, no. 9, pp. 2426–2434.

[73] D. Guha, S. Biswas, and Y. M. M. Antar, Defected Ground Structure for Microstrip Antennas, in D. Guha and Y. M. M. Antar (eds.) *Microstrip and Printed Antennas*, John Wiley and Sons, 2011, Chapter 12.

[74] D. Guha, *et al.*, Concentric Ring-shaped Defected Ground Structures for Microstrip Applications, *IEEE Antennas and Wireless Propagation Letters*, vol. 5, 2006, pp. 402–405.

[75] M. A. Antoniades and G. V. Eleftheriades, A Compact Multiband Monopole Antenna with a Defected Ground Plane, *IEEE Antenna and Wireless Propagation Letters*, vol. 7, 2008, pp. 652–655.

[76] D. Guha, M. Biswas, and Y. M.M. Antar, Microstrip Patch Antenna with Defected Ground Structure for Cross Polarization Suppression, *IEEE Antennas and Wireless Propagation Letters*, vol. 4, 2005, pp. 455–458.

[77] Y. J. Sung, M. Kim, and Y. S. Kim, Harmonic Reduction with Defected Ground Structure for a Microstrip Patch Antenna, *IEEE Antennas and Wireless Propagation Letters*, vol. 2, 2003, pp. 111–113.

[78] H. B. El-Shaarawy *et al.*, Novel Reconfigurable Defected Ground Structure Resonator on Coplanar Waveguide, *IEEE Transactions on Antennas and Propagation*, vol. 58, 2010, no. 11, pp. 3622–3628.

[79a] J. L. Volakis *et al.*, Antenna Miniaturization Using Magnetic Photonic and Degenerated Band-Edge Crystals, *IEEE Antennas and Propagation Magazine*, vol. 48, 2008, no. 5, pp. 12–27.

[79b] J. L Volakis, C. Chen, and K. Fujimoto, *Small Antennas: Miniaturization Technique & Applications*, McGraw-Hill, 2010, chapter 7.

[80] S. Yarga, K. Sertel, and J. L. Volakis, Degenerate Band Edge Crystals for Directive Antennas, *IEEE Transactions on Antennas and Propagation*, vol. 56, 2008, no. 1, pp. 119–126.

[81] S. Yarga, K. Sertel, and J. L. Volakis, A Directive Resonator Antenna Using Degenerate Band Edge Crystals, *IEEE Transactions on Antennas and Propagation*, vol. 57, 2009, no. 3, pp. 799–803.

[82] S. Yarga, K. Sertel, and J. L. Volakis, Multilayer Directive Resonator Antenna Operating at Degenerate Band Edge Modes, *IEEE Antennas and Wireless Propagation Letters*, vol. 8, 2009, pp. 287–290.

[83] G. Mumcu, K. Sertel, and J. L. Volakis, Miniature Antenna Using Printed Coupled Lines Emulating Degenerate Band Edge Crystals, *IEEE Transactions on Antennas and Propagation*, vol. 57, 2009, no. 6, pp. 1618–1623.

[84] K. Abe, Analysis and Design of Very Small Antennas for Receiving Long Wave Signals, Doctoral Dissertation, Tokyo Institute of Technology, 2007.

[85] K. Abe and J. Takada, Performance Evaluation of a Very Small Magnetic Core Loop antenna for an LF Receiver, *Proceedings of Asia-Pacific Microwave Conference*, 2006, TH3C-4, pp. 935–938.

[86] G. Marrocco, The Art of UHF RFID Antenna Design: Impedance-Matching and Size-Reduction Techniques, *IEEE Antennas and Propagation Magazine*, vol. 50, 2008, no. 1 pp. 66–79.

[87] W. Choi *et al.*, RFID Tag Antenna with a Meandered Dipole and Inductively Coupled Feed, *IEEE APS International Symposium, 2006*, pp. 1347–1350.

[88] A. Toccafondi and P. Braconi, Compact Load-Bars Meander Line Antenna for UHF RDID Transponder, *First European Conference on Antennas and Propagation*, Nice France, 2006, p. 804.

[89] S. A. Delichatsios *et al.*, Albano Multidimensional UHD Passive RFID Tag Antenna Design, *International Journal of Radio Frequency Identification Technology and Applications*, 1.1 January 2006, pp. 24–40.

[90] C. Cho, H. Cho, and I. Park, Design of Novel RFID Tag Antenna for Metallic Objects, *IEEE APS International Symposium, 2006*, pp. 3245–3248.

[91] C.-K. Hsu and S.-J. Chung, A Wideband DVB Forked Shape Monopole Antenna with Coupling Effect for USB Dongle Application, *IEEE Transactions on Antennas and Propagation*, vol. 58, 2010, no. 9, pp. 2029–3036.

[92] D. Kim and J. Yeo, A Passive RFID Tag Antenna Installed in a Recessed Cavity in a Metallic Platform, *IEEE Transactions on Antennas and Propagation*, vol. 58, 2010, no. 12, pp. 3814–3820.

[93] T. Deleruyelle *et al.*, An RFID Tag Antenna Tolerant to Mounted Materials, *IEEE Antennas and Propagation Magazine*, vol. 52, 2010, no. 4, pp. 14–19.

[94] H. Hirvonenen *et al.*, Planar Inverted-F Antenna for Radio Frequency Identification, *Electronics Letters*, vol. 40, 2004, no. 4, pp. 848–850.

[95] B. Yu *et al.*, Balanced RFID Tag Antenna Mountable on Metallic Plates, *IEEE APS International Symposium 2006*, pp. 3237–3240.

[96] C. H. Cheng and R. D. Murch, Asymmetric RFID Tag Antenna, *IEEE APS International Symposium 2006*, digest, pp. 1363–1366.

[97] I. Y. Chen *et al.*, Folded Dual-Band (2.4/5.2 GHz) Antenna Fabricated on Silicon Suspended Parylene Membrane, *Asia Pacific Microwave Conference Proceedings*, December 2005, Proc. no. 4, pp. 4–8.

[98] J.-T. Huang, J.-H. Shiao, and J.-M. Wu, A Miniaturized Hilbert Inverted-F Antenna for Wireless Sensor Network Applications, *IEEE Transactions on Antennas and Propagation*, vol. 58, 2010, no. 9, pp. 3100–3103.

[99] C. Occhuzzi, S. Cippitelli, and G. Marrocco, Modeling, Design and Experimentation of Wearable RFID Sensor Tag, *IEEE Transactions on Antennas and Propagation*, vol. 58, 2010, no. 8, pp. 2490–2498.

[100] B. D. Braaten, A Novel Compact UHF RFID Tag Antenna Designed with Series Connected Open Complementary Split Ring Resonator (OCSRR) Particles, *IEEE Transactions on Antennas and Propagation*, vol. 58, 2010, no. 11, pp. 3728–3733.

[101] J. Kim *et al.*, Design of a Meandered Slot Antenna for UHF RFID Applications, *IEEE APS International Symposium 2010*, 206.5

[102] Z. N. C. Nasimuddin and X. Qing, Asymmetric-Circular Shaped Slotted Microstrip Antennas for Circular Polarization and RFID Applications, *IEEE Transactions on Antennas and Propagation*, vol. 58, 2010, no. 12, pp. 3821–3828.

[103] H.-K. Ryu and J.-M. Woo, Small Circular Loop Antenna for RFID Tag, *International Symposium on Antennas and Propagation*, 2006, a341, r315.

[104] J. J. Yu and S. Lim, A Miniaturized Circularly Polarized Antenna for an Active 433.92 MHz RFID Handheld Reader, *IEEE APS International Symposium 2010*, 206.1.

# 9 Evaluation of small antenna performance

## 9.1 General

Generally, evaluation of small antenna performance tends to be difficult, as the size of an antenna becomes very small. The difficulty of small-antenna measurement originates in small dimensions, asymmetry of antenna structure, and the antenna characteristics sensitive to the influence of its environment. Determination of correct or precise antenna performance will become difficult as the antenna dimension becomes smaller. Evaluation of antenna performance is performed by means of EM simulation and measurement. In this chapter, measurement of small antenna performances is described, whereas simulation is dealt with in Chapter 10. There are no essential differences between small antenna measurement and ordinary antenna measurement. However, as the size of an antenna becomes smaller, special considerations are required in order to assure the reliable evaluation of small antenna performances. Typical attentions are paid for cases:

(1) When the size of an antenna is very small compared with its wavelength or the size of a nearby conductor.
(2) When antenna structure is asymmetric and complicated.
(3) When an antenna is located in a complicated environment.

Although particular considerations are given to the measurement of small antennas, there may be some cases where no effective method is found for the accurate determination of the antenna characteristics. For example, when the size of the antenna is extremely small, the resistance becomes too small to be determined. Another case is where the antenna structure is so complicated that precise measurement is very difficult to perform.

To measure very small input resistance of a small antenna, for example, disturbance due to the connection of the instrument, which might be much larger than the size of the input terminals of the antenna, would give rise to significant errors in the measurement, etc.

When an antenna structure is symmetric, the antenna characteristics equivalent to those of the full structure can be measured by the image theory using a ground plane with a size of several wavelengths. If a ground plane can be used, it will become easier to perform exact evaluation of antenna performances as compared with the measurement without using a ground plane. Therefore, when an antenna to be measured has a simple and symmetrical structure, the antenna structure may be halved by placing it over a ground plane, and the use of a coaxial cable for feeding the antenna becomes easy. In measurement by using a coaxial cable and ground plane, a balun (Balanced-to-Unbalanced Transformer) becomes unnecessary. Moreover, the measurement becomes stable, since noise and interference are isolatable from the feed of the antenna. Furthermore, it is also an advantage that there is no unbalanced current, which would likely disturb measurements, by feeding an antenna with a coaxial cable from beneath a ground plane. As, for example, with portable mobile terminals having an unbalanced mode terminal as an antenna connection terminal, a coaxial cable can be used in measurement. However, in actual measurement, since a ground plane is not infinitely large, unwanted unbalanced currents arise on a ground plane, and these currents also flow around to the back side of a ground plane. As a consequence, exact measurement of the input impedance becomes difficult, and radiation patterns will be distorted. For this reason, in order to measure the antenna characteristics in an environment as near as possible to free-space conditions, a ground plane with the size of a few wavelengths is usually used. However, in radiation pattern measurement, even if the ground plane has any kind of size, the influence by a finite size of a ground plane is not avoided, owing to the scattered wave from the edge of a ground plane and radiation produced by the currents flowing to the back side of the ground plane.

Since the characteristics of a small antenna installed in small wireless equipment are strongly influenced by nearby materials including electronic components and equipment case, special considerations are required for measurement. The coaxial cable connected to an antenna for measurement becomes especially a problem. Since a coaxial cable is a metallic line, it needs to be arranged so that current may not flow on the outside conductor of the coaxial cable when it is connected to the antenna, and it is most important how this is realized for the small antenna measurement on a portable mobile terminal. Furthermore, since the antenna characteristics are greatly dependent on the spatial relationship of a human body and an antenna, it is also important to perform measurements with consideration of the antenna characteristics under their environment.

There is a serious problem in measurement when an antenna has a balanced terminal at the feed point and a coaxial cable is connected to the terminal, and unbalanced currents are induced on the outside surface of the feed cable. A similar situation is encountered when a small antenna has asymmetry, irregular, or complicated structure. In that case, a balun should be used, and careful means for avoiding coupling, proximity influence, etc. must be taken. Thus, since the image theory by PEC (perfect electric conductor) GP ground plane cannot be used in case of measurements under conditions close to actual use, a special device is needed. The method for reducing undesired current on the outside surface of a coaxial cable in the unbalanced-fed antenna is different from that in the balanced-fed antenna. In the case of unbalanced-fed antennas, ferrite beads may be attached to a coaxial cable, or a coaxial cable feed is arranged near the antenna to attain measurement error as low as possible. For example, a coaxial cable might be oriented so that it may intersect perpendicularly with linear polarization when radiation patterns are measured. A simple way to check the influence of the coaxial cable is to touch the cable by hand and observe variation of the input impedance characteristics on the network analyzer. Balanced-fed antennas need baluns when they are fed by a coaxial cable.

Such is a case often observed when a balanced-fed antenna is directly connected with a coaxial cable, where undesired currents may flow on the outside surface of the coaxial cable, greatly influencing measurements. If such unwanted current on the outside surface of a coaxial cable can be prevented from flowing by the above-mentioned measure, the input impedance characteristic of an antenna can be measured. However, radiation pattern measurement needs special attention around the nulls of the radiation pattern, since the original antenna characteristic may change due to the existence of a coaxial cable between the antennas. Moreover, there is the method of connecting a small oscillator directly to an antenna, without using a coaxial cable. As a small oscillator is installed inside a mobile portable terminal and is connected with an antenna directly or through a balun, measurement of a radiation pattern is possible without the influence of a coaxial cable, but input impedance cannot be measured this way. Therefore, to obtain complete results, it is necessary to measure input impedance characteristics by another method.

There is a method of using an optical system for the measuring method that conquers the above-mentioned fault. This method transforms radio frequency signals into light signals and transmits the light signals by an optical fiber. After that, light signals are transformed back into radio frequency signals. By using the optical cable made of dielectric instead of a coaxial cable, the influence of currents flowing on the outside of a coaxial cable can be removed. Moreover, in this method, input impedance characteristics can also be measured, and balanced-to-unbalanced transformation also made unnecessary by miniaturizing the area connected with an antenna. Therefore, this is a promising measuring method for the future.

## 9.2 Practical method of measurement

Because of the great variety of antenna configurations, the method of measurement should be selected in such a way that the proper measurement can be performed, depending on the antenna configurations and mounting conditions, to achieve accurate results. When an antenna under test has a configuration that is not simple and symmetric, there are many factors that must be taken into account to achieve accurate results from the measurement.

The small antennas installed on portable mobile terminals can be classified into the unbalanced-fed antenna types shown in Figure 9.1 and the balanced-fed antenna types shown in Figure 9.2 according to the feeding structure. As examples of unbalanced-fed antennas, there are a monopole antenna on a case, an inverted-F antenna, a microstrip antenna and so on. As a balanced-fed antenna, there is a loop antenna etc. by return.

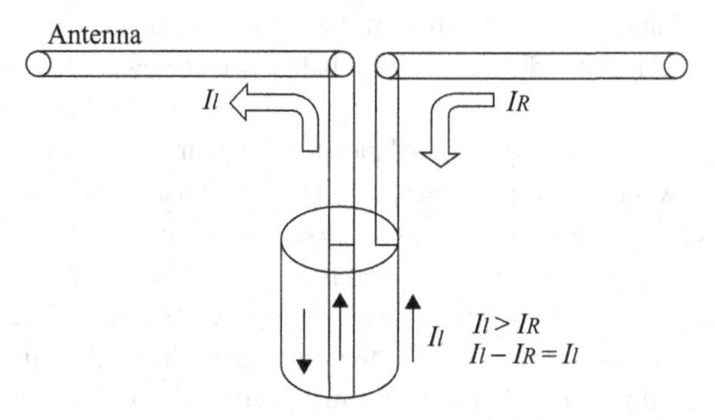


Figure 9.1 Unbalanced-fed antenna.

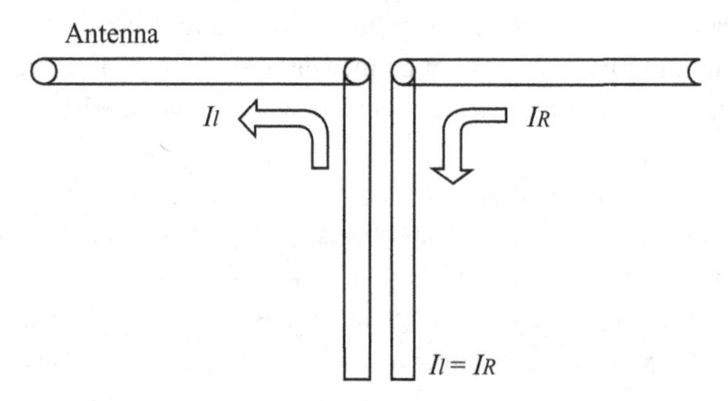


Figure 9.2 Balanced-fed antenna.

## 9.2.1 Measurement by using a coaxial cable

### 9.2.1.1 Measurement of unbalanced-fed antennas

It is possible to measure the input impedance characteristics in the tip of a coaxial cable by connecting the outside conductor of a coaxial cable to the case of a portable terminal and connecting a central conductor to the antenna for the unbalanced-fed antenna. Moreover, if the case structure is designed so that a coaxial connector may be attached in a case and the central conductor of the coaxial cable may serve as a feed line to an antenna, that is convenient for measurement. In this case, by short-circuiting the center pin and case of a coaxial connector linked to the antenna side, the electrical contribution from the measuring side at the tip of a coaxial cable can be known and compensated. Besides, it is necessary to choose carefully the position and direction for alignment of the coaxial cable connecting to the case so as to not disturb the current flowing on the case, and to align the coaxial cable in a direction that does not affect the radiation pattern [1]. Moreover, the balun method of preventing unwanted currents on the outside of the coaxial cable, including the dual-band balun of the same structure as the Spertopf, or bazooka balun of a sleeve antenna is also effective [2][3]. If the optical fiber explained in Section 9.2.3.1 is used instead of a coaxial cable, ideal measurements will be attained even when current flows on a whole metal case.

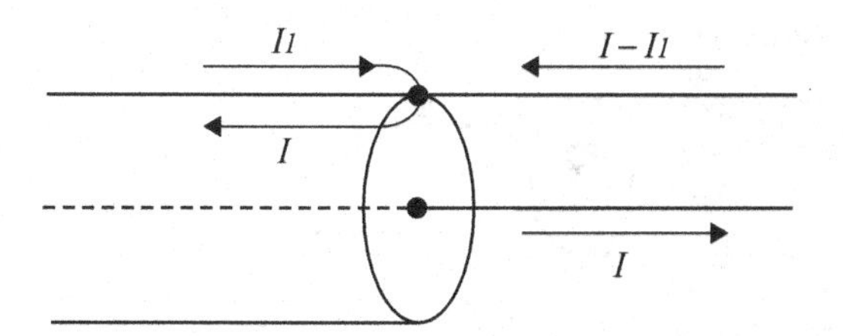


**Figure 9.3** Current in the connection between the balanced antenna and the coaxial cable.

#### 9.2.1.2 Measurement of balanced-fed antenna

In measurement of balanced-fed antennas, a balun is needed between the antenna and the coaxial cable linked to a measuring instrument. Figure 9.3 shows a branch of the current in the connection between the balanced antenna and the coaxial cable. The effect which reduces the current flowing on the outside conductor of the coaxial cable is provided by the balun. As the undesired current is suppressed by using a balun, the unwanted radiation by the current on the surface of a cable disappears. Moreover, since a balanced antenna is also excited in balanced mode, the measurement of the original antenna characteristics can be attained.

Figure 9.4 shows some typical balun configurations which have been used. There are a strip type, a Bazooka type, a quarter-wavelength coaxial type, and a half-wavelength coaxial type, and all of these are simple to produce. As each of these baluns has about 10% relative bandwidth, it is sufficient to manufacture a single balun for the center frequency in antenna measurement of mobile terminals. However, the above-mentioned baluns have a large fault in antenna measurement of mobile terminals, because they have sizes comparable to a wavelength. There is another balun, however, which consists of a 180-degree hybrid to use in a higher-frequency band [4]. The characteristic of this balun is improved by putting in some resistance. It is possible to make this balun easily on a printed circuit board and various sizes and frequencies can be chosen according to the purpose. Such are the advantages of this balun, shown in Figure 9.5.

### 9.2.2 Method of measurement by using small oscillator

For radiation pattern measurement, use of a small, integrated oscillator which removes the influence of an external coaxial cable is very effective [5]. Although the measurement using a small oscillator is inexpensive and easy, it cannot measure phase information. The small oscillator, which operates from a battery in the metal case, is shown in Figure 9.6 [5]. Under the present circumstances, it is useful to insert a small fixed attenuation (6 dB for example) between an antenna and the small oscillator. Thereby, influence on the oscillator resulting from a mismatch of the antenna is minimized.

The system of measurement using a small oscillator is shown in Figure 9.7. Since measurement over a frequency band may require tuning the oscillator throughout a frequency band, it is useful to connect a spectrum analyzer to the receiving probe antenna. Quantitative measurements then are taken by comparing the spectrum analyzer

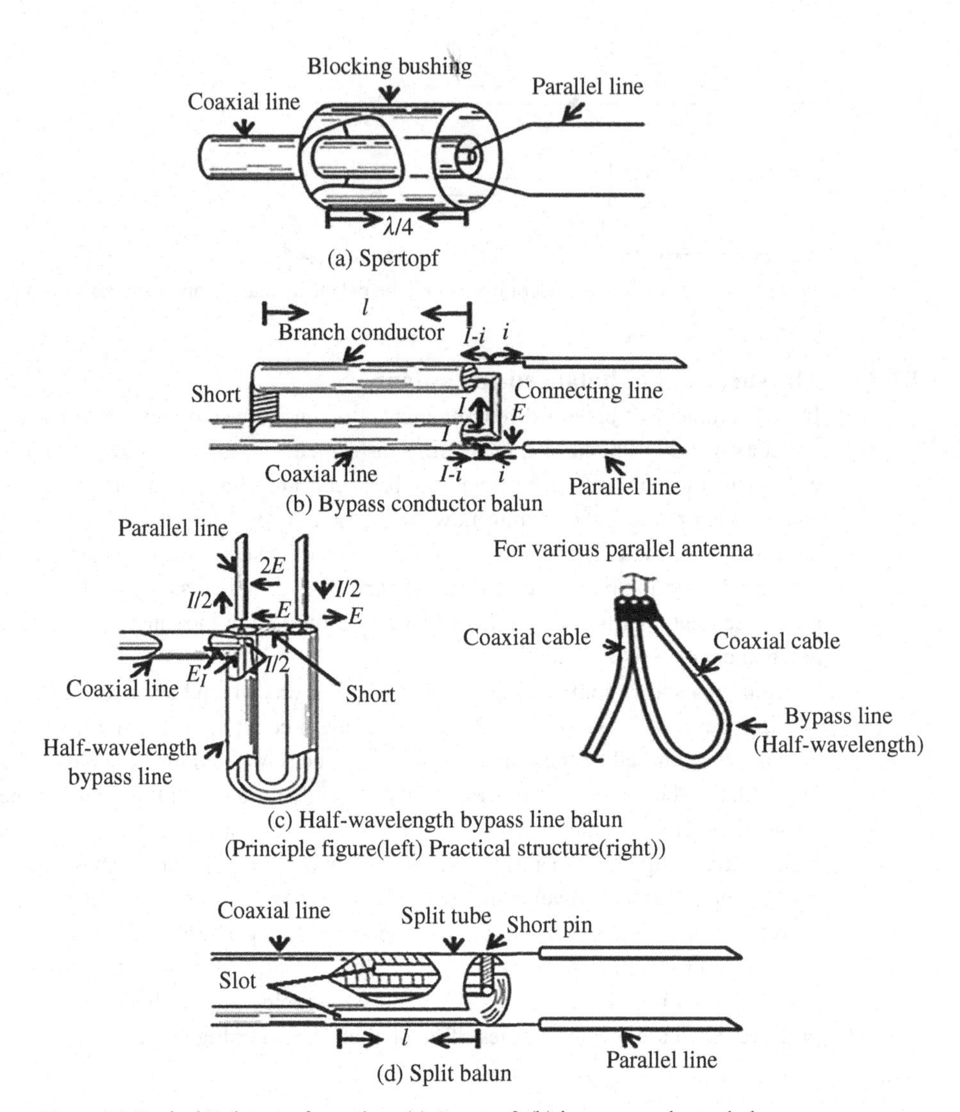


**Figure 9.4** Typical Balun configuration: (a) Spertopf, (b) bypass conductor balun, (c) half-wavelength bypass line balun, and (d) split balun.

readings when the oscillator is connected to a standard dipole antenna, to readings obtained when the same oscillator is connected to the unit under test.

The distance between the antenna of a personal mobile terminal in the presence of a human body at the time of measurement, and a receiving antenna is needed to be at least $2D^2/\lambda$ [6]. $D$ is the size of the portion considered to contribute to radiation of the antenna containing a human body, and $\lambda$ is the operating wavelength. For example, when the spherical model with a diameter of 20 cm is used as a human body head model and there is 10 cm distance between a spherical model and an antenna, $D$ is 30 cm.

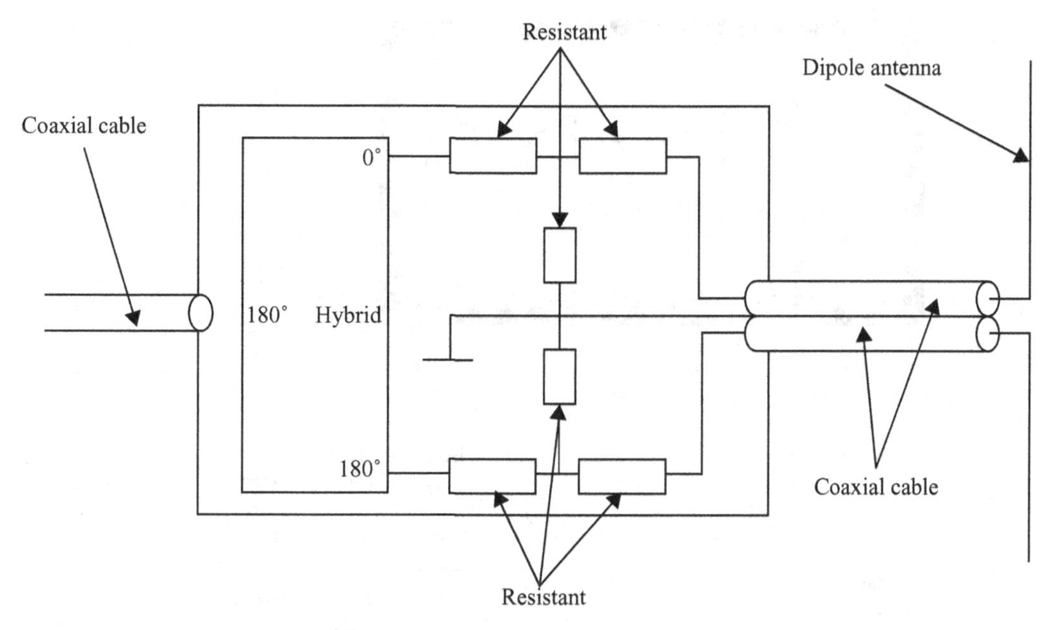


Figure 9.5 Balun with 180° hybrid.

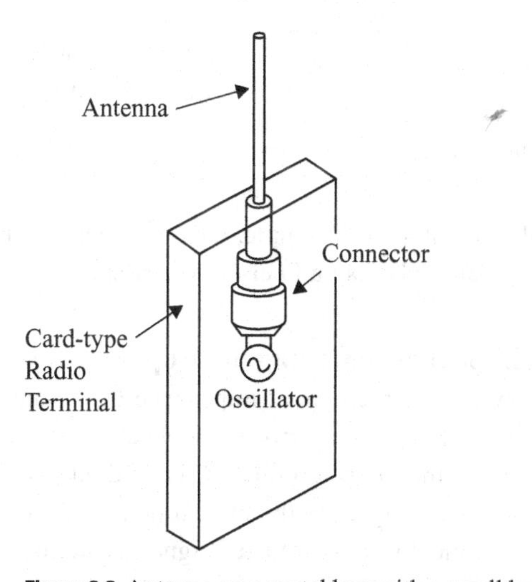


Figure 9.6 Antenna on a metal box with a small built-in oscillator.

### 9.2.3 Method of measurement by using optical system

Fiber-optics for feed instead of coaxial cables has been introduced in order to remove the undesired influence of coaxial cables [1]. Although conventional methods using a small oscillator can remove the influence of the coaxial cable for radiation patterns, much effort is required to measure at multiple frequencies because it can measure only one frequency for one measurement. Further, phase pattern and impedance characteristics

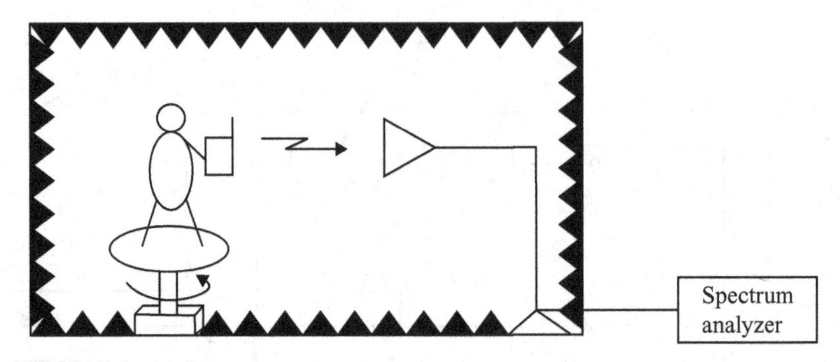


**Figure 9.7** System of measurement using a small oscillator.

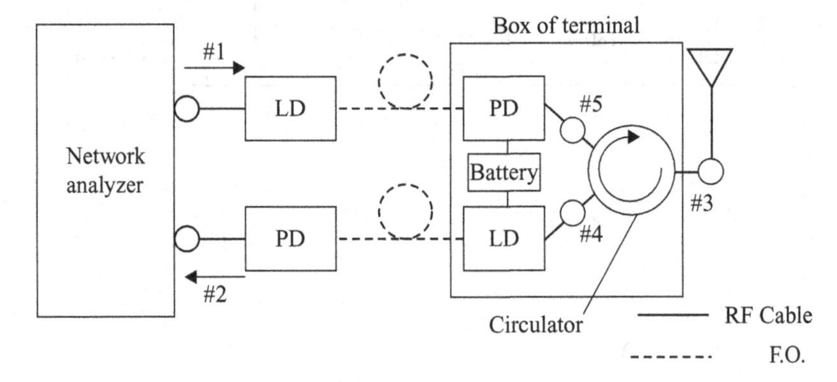


**Figure 9.8** Measurement system for impedance.

cannot be obtained using a small oscillator. For example, measurement systems for the input impedance and the radiation pattern by using fiber-optics are shown.

### 9.2.3.1 Measurement for the input impedance by using fiber-optics

Figure 9.8 shows measurement systems for the input impedance of the antenna on a radio terminal by using fiber-optics. Fiber-optics transmit optical signals modulated by RF (radio frequency) to and from the radio terminal. A laser diode module (LD) with direct laser modulation converts RF signal to the optical domain. A photo diode module (PD) then re-converts the optical signal to an RF signal. Batteries supply the LD and PD modules with electric power in the radio terminal. In these measurement systems, no metallic cable is connected to the radio terminal. Therefore, the influence of measurement cables is reduced remarkably. The SOLT (Short, Open, Load, Through) method can be used to calibrate the measurement results of the impedance.

Figure 9.9 shows the measurement model. A quarter-wavelength monopole antenna is mounted on the top of a rectangular box, which represents a radio terminal. The length of the box is $L$ as a parameter. This model is measured by both RF cable and fiber-optics for comparison. When we use RF cable, it is connected at either position A or B. In case

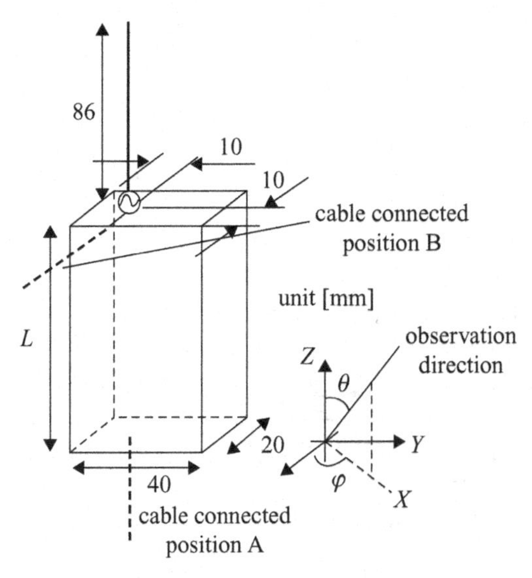


**Figure 9.9** Measurement model.

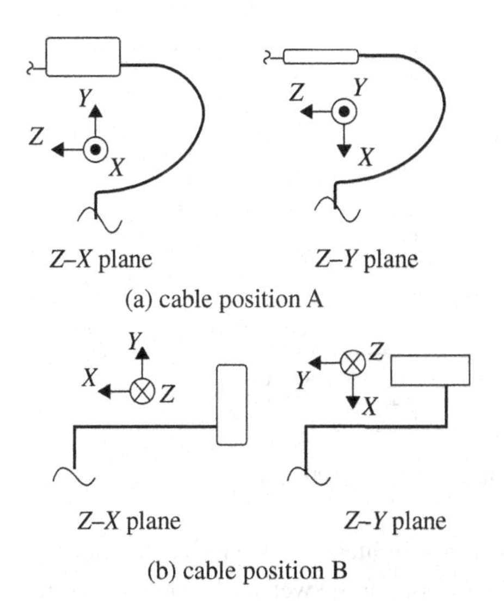


**Figure 9.10** Cable setting for radiation pattern measurement.

of fiber-optics, the cable is connected at position A. For radiation pattern measurement, the RF cable is set up as Figure 9.10. Every measurement is performed in the 900 MHz band.

The measured input impedance characteristics obtained by using RF cable and fiber-optics are shown in Figures 9.11 and 9.12, where calculated results using FDTD are also shown. The length of the radio terminal $L$ is different in Figures 9.11 and 9.12. In Figure 9.11 with $L = 80$ mm, results with the RF cable at position A are much

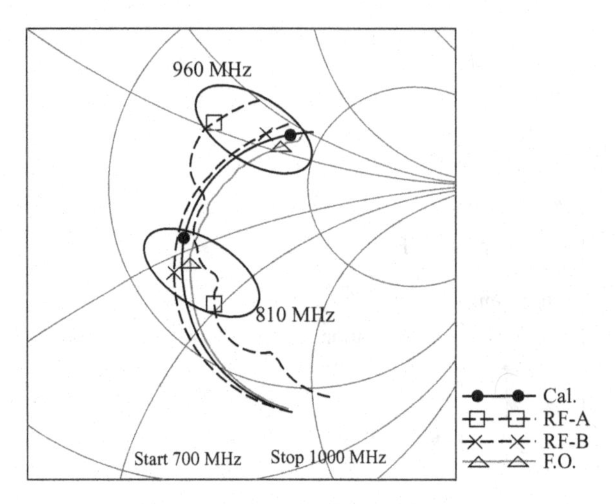


**Figure 9.11** Measurement results for impedance ($L = 80$ mm).

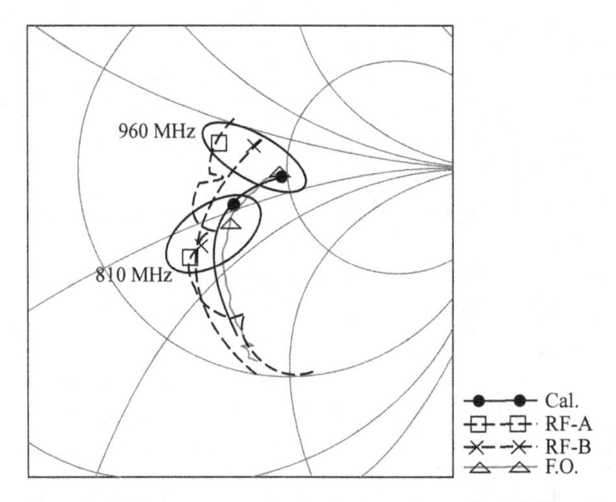


**Figure 9.12** Measurement results for impedance ($L = 100$ mm).

different from calculations because of the influence of the RF cable. Measurements with RF cable at position B and with fiber-optics agree well with calculations in this case. In Figure 9.12 with $L = 100$ mm, results with RF cable at position A and B disagree with calculations. Only the result of fiber-optics agrees well with calculations in this case. It can be said from these facts that a measurement system using fiber-optics can achieve accurate results without dependence on the measurement model.

#### 9.2.3.2 Measurement of the radiation pattern by using fiber-optics

Figure 9.13 shows measurement systems for the radiation pattern. This system is based on the radiation pattern measurement system using a network analyzer. Only one cable feeding the radio terminal is replaced by optical fiber. For radiation pattern measurement,

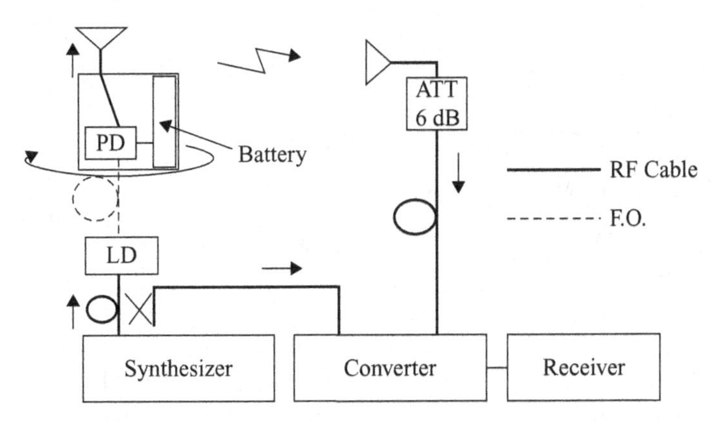


**Figure 9.13** Measurement system for the radiation pattern.

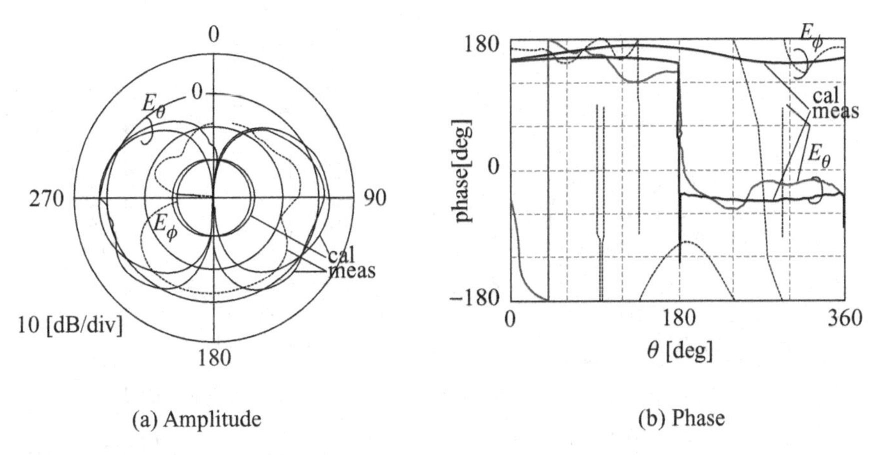


**Figure 9.14** Radiation pattern measured by the RF cable at position A.

only the PD and battery to drive the PD should be mounted on the radio terminal. Using this system, phase patterns can be obtained. Further, multiple frequency measurement is available for one measurement setup. Therefore, the effort required to measure multiple frequency characteristics can be reduced compared with conventional methods using a small oscillator.

The radiation patterns with $L = 80$ mm using RF cable and the fiber-optics are shown in Figures 9.14–9.16. In Figure 9.14, RF cable connected at position A influences results of amplitude and phase patterns. Measured results do not agree with calculations. In the case of RF cable at position B as shown in Figure 9.15, accuracy of measurement is improved, but there is still error between measured and calculated results. In Figure 9.16, measured results obtained by using fiber-optics can achieve good agreement with calculated results. These facts show the advantages of the measurement systems using fiber-optics.

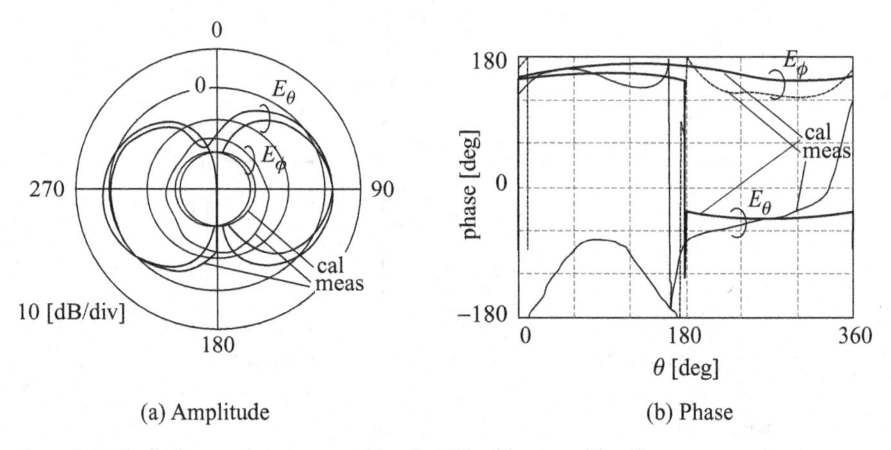


**Figure 9.15** Radiation pattern measured by the RF cable at position B.

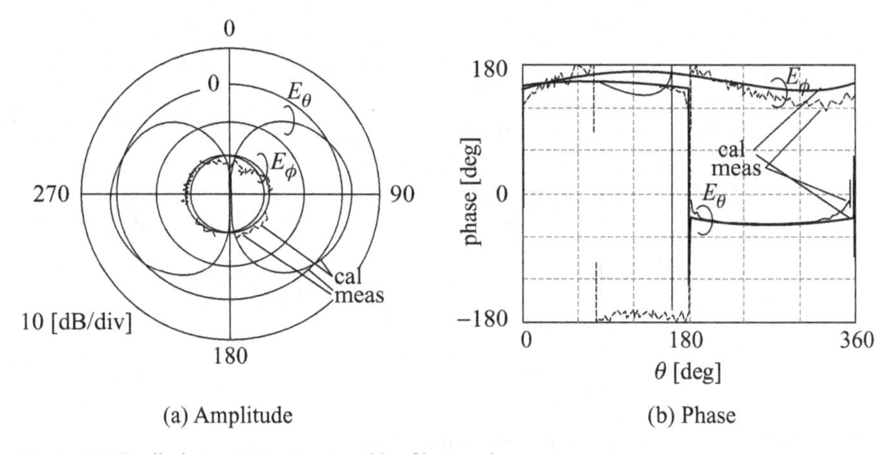


**Figure 9.16** Radiation pattern measured by fiber-optics.

## 9.3 Practice of measurement

### 9.3.1 Input impedance and bandwidth

Small antennas have some characteristics that distinguish them from ordinary antennas. A very small antenna can be considered essentially as an electric dipole or a magnetic dipole and the reactive impedance is highly capacitive or inductive, whereas the resistive impedance is very small. With this impedance characteristic, particular care must be taken for small antennas. If the length of a dipole is shorter than $0.1\lambda$, perfect match to the load of 50 $\Omega$ is very difficult.

If the antenna under test has such small size that its resistance is, for example, several ohms or less, the instrument to be used should have high enough accuracy and preciseness to measure such a low resistance value. A typical case is where the efficiency

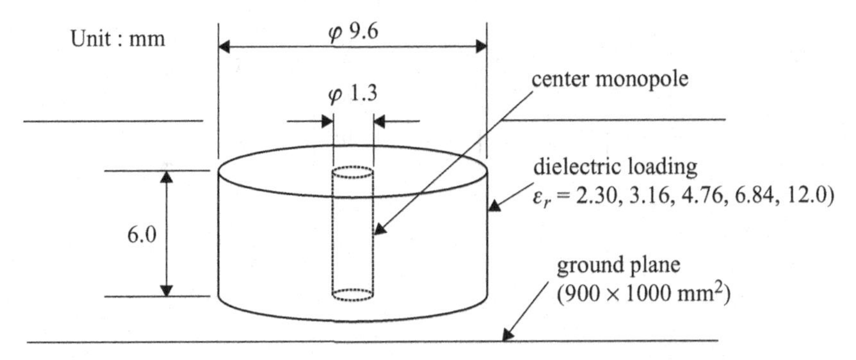


**Figure 9.17** Dielectric-material loaded dipole.

of an Electrically Small Antenna (ESA) that has dimensions on the order of $0.1\lambda$ or less is determined. When an antenna resistance component in the VHF–UHF or higher ranges is less than 0.5 Ω, for example, there may be no way to determine the true impedance because instruments with high enough accuracy to determine such a small value may not be available.

One possible way to overcome this problem is to use a scale model, which is made with $n$-times enlarged dimensions of the original antenna, and use lower frequency, $1/n$ of the operating frequency for the antenna under test. This $n$ is referred to as the scale factor, being greater than unity. With the enlarged model, the measurement is easily performed and error in the measurement can be made smaller. This method is based on the similarity relation, in which the Maxwell equation can be equally applied to both an original antenna and an enlarged antenna, which has similar shape, but $n$-times scaled dimensions. The typical electromagnetic parameters in this case, where $\omega_1 = n\omega_2$, are denoted by permeability $\varepsilon_1 = \varepsilon_2$, permittivity $\mu_1 = \mu_2$, conductivity $\sigma_1 = n\sigma_2$, impedance $Z_1 = Z_2$ and directivity $D_1 = D_2$, where $\omega_1$ and $\omega_2$, respectively, are the operating frequency of the original antenna and the frequency used in measurement.

The features of using a scale model are: (1) evaluation is easily performed by using a convenient size and in a less disturbed environment, (2) parameters can be flexibly selected in the measurement, and (3) cost necessary for instrument and fabrication of model are low. The scale factor can be smaller than unity, when an antenna under test is very large or heavy, and evaluation can be performed by a smaller antenna with the concept similar to the case when $n$ is greater than unity.

Meanwhile, a successful measurement for very small impedance with reasonable accuracy was reported in [7], in which very low impedances of dielectric-material loaded dipoles (Figure 9.17) were introduced. The results are shown in Figure 9.18, where small values of radiation resistance and inverse of input reactance with respect to the relative permittivity of loaded dielectric material are shown. Efficiency–bandwidth ratio, which corresponds to Wheeler's radiation power factor, is used to evaluate small values of radiation resistance along with the Wheeler-cap method for efficiency measurement.

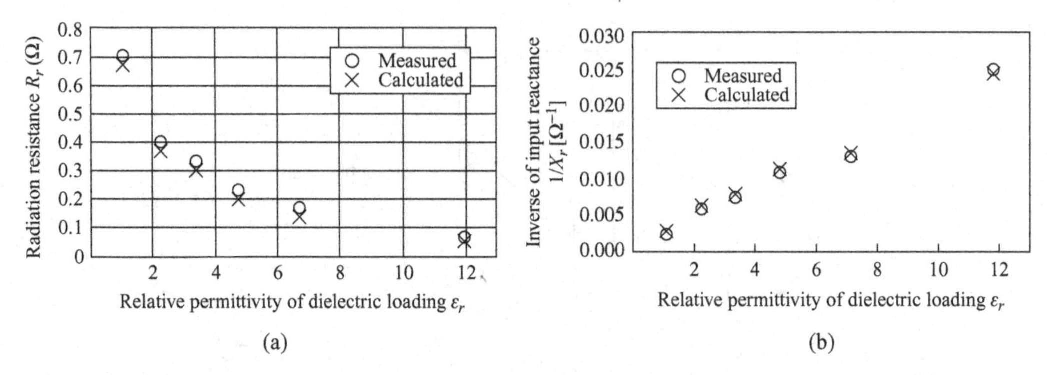


**Figure 9.18** Radiation resistance and inverse of input reactance.

Stability of the instrument is also important because, if the instrument is not stable enough, errors due to fluctuation in the measurement system become involved in the measured results. For instance, when a small impedance of an ESA, e.g., about 1 Ω is measured, stability of the instrument must be high enough to determine one-tenth of 1 Ω or so. The results obtained by the measurement with an unstable instrument are meaningless.

The bandwidth $\Delta f$ is the range of frequencies on either side of a center frequency, at which the antenna characteristics such as impedance, gain, radiation efficiency, and so forth, are within an acceptable range of those at the center frequency. For narrow band antennas like small antennas, a ratio of $\Delta f/f_0 = B$ is defined as

$$B = \Delta f/f_0 = (f_1 - f_2)/f_0 \tag{9.1}$$

where $f_1$ and $f_2$, respectively, are the higher and the lower frequencies, at which the power declines 3 dB from the maximum and $f_0$ is the center frequency of the bandwidth $\Delta f$. This $B$ is referred to as the relative bandwidth and is practically used instead of the bandwidth $\Delta f$. The center frequency $f_0$ is given by

$$f_0 = (f_1 + f_2)/2. \tag{9.2}$$

Quality factor $Q$ can be used instead of the relative bandwidth $B$ when the bandwidth $\Delta f$ is narrow ($B < 10$ %),

$$B = 1/Q. \tag{9.3}$$

## 9.3.2 Radiation patterns and gain

The radiation field for an antenna consists of a near-field region close to the antenna and a far-field region further away from the antenna. Here, measurement for the far-field radiation pattern is presented. The far-field region can be defined by calculating the phase error due to the finite size of the antenna. Figure 9.19 shows the difference in the path length from the observation point to the center of the antenna system, and to the end of the antenna. This path length difference can be translated into a phase difference

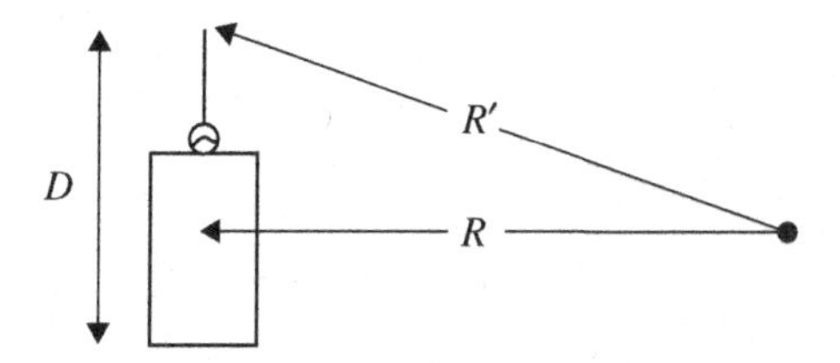


**Figure 9.19** Definition of the far field.

as indicated by the equation:

$$\Delta\varphi = \frac{2\pi}{\lambda}\left(R' - R\right) = \frac{2\pi}{\lambda}\left\{\sqrt{R^2 + \left(\frac{D}{2}\right)^2} - R\right\}$$
$$= \frac{2\pi}{\lambda}\frac{1}{R}\left(\frac{D}{2}\right)^2 = \frac{\pi D^2}{4\lambda R}. \tag{9.4}$$

For the phase error to be less than $\pi/8$, the distance $R$ from the antenna must be such that:

$$R > 2D^2/\lambda. \tag{9.5}$$

The above equation is often used to define the far-field distance for antennas.

### 9.3.3 Efficiency measurement

There are three typical methods to determine the antenna efficiency: pattern integration method, $Q$-factor method, and Wheeler method [8].

The pattern integral method determines $P_{rad}$ from the integration of the radiation pattern intensity, which is obtained from the far-field measurement. While pattern integration is considered the most accurate method available, especially when the absolute efficiency of an antenna is desired, it usually takes a rather long time, because it needs to measure radiation patterns and to perform integration of the radiation patterns over a spherical surface completely enclosing the antenna.

On the contrary, $Q$-factor and Wheeler methods have the advantage of being quick and easy to employ, as compared to the pattern integration method. These two methods relate the antenna efficiency to the input impedance rather than the far-field pattern integration. These two methods have the advantage that the measurement does not require any particular system, but can employ ordinary methods, and also can be performed in ordinary laboratory rooms. However, there is a trade-off between the accuracy and ease of the measurement because the accuracy of the antenna efficiency obtained by these two methods is relatively lower than by the pattern integration method.

The $Q$-factor method is based on a measurement of two antenna $Q$ values, one is the $Q_{RL}$ of the test antenna and another is the $Q_R$ of an ideal antenna. $Q_{RL}$ can be determined by measuring the impedance of the test antenna. $Q_R$ can be determined from calculation based on Chu's theory. If the $Q$ of an antenna is high, it can be interpreted as the reciprocal of the relative bandwidth $B$. The $B$ can be known from the impedance–frequency characteristics. $B$ is defined as the bandwidth for the half-power

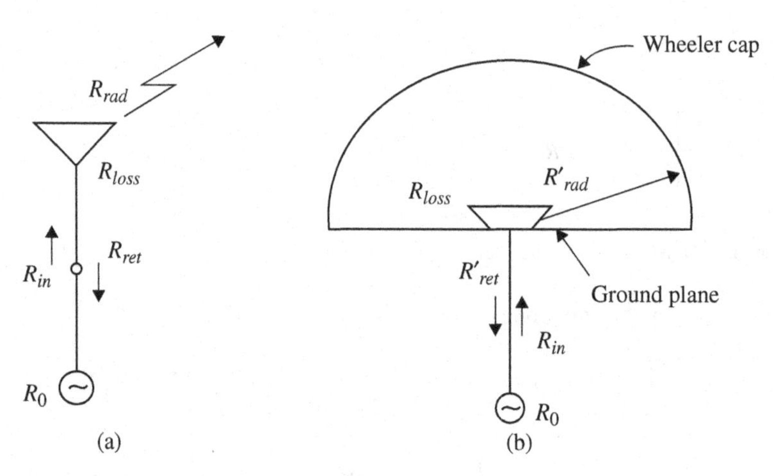


**Figure 9.20** Wheeler method.

frequencies that occur for a power reflection coefficient of 0.5 (VSWR = 5.83). Because of the inherent difficulty in determining $Q_R$, the Wheeler method is recommended because it is simpler in practice.

In the Wheeler method (Figure 9.20), the antenna resistance is measured in two ways: $R_{in}$ of an antenna reradiating into free space and $R_{loss}$ of an antenna placed within a Wheeler cap, which eliminates radiation from the antenna. By using these two resistances $R_{loss}$ and $R_{in}$, $R_{rad}$ can be found, since $R_{in} = R_{rad} + R_{loss}$. Then the efficiency $\eta$ is calculated from

$$\eta = R_{rad}/(R_{rad} + R_{loss}). \tag{9.6}$$

The dimension of a cap is recommended to be about $\lambda/2\pi$, although there have been no exact theories to prove this idea. By experimental studies, it is understood that there seems to be no severe limitation in the cap size unless it greatly exceeds $\lambda/2\pi$. It was also confirmed that the radius of the Wheeler cap can be as large as 10 times a radian length, granted that resonance and interference of the Wheeler cap on the frequency characteristics of the testing antenna can be avoided [9]. For the shape of a cap, other types than a sphere such as a cube, a rectangular solid, can be used. The shape somehow depends on the size and type of test antennas. For a thin, short monopole, both hemispheric and cubic shapes can be used. A cubic shape seems to be better for antennas such as an inverted-F antenna, because it has an element parallel to the sides of the cap, and the distance between the antenna element and the cap can be made nearly equal; thereby, variation in the antenna current is made small after the cap covers the antenna.

In any case, care must be observed for the resonance or anti-resonance of the antenna, cap, or both. Note that the principle of the Wheeler method is based on the measurement of the series resistance or the parallel conductance. At or near resonance or anti-resonance frequencies, resistance or conductance becomes very low or approaches an infinitely large value and no consistency in the measured values can be expected.

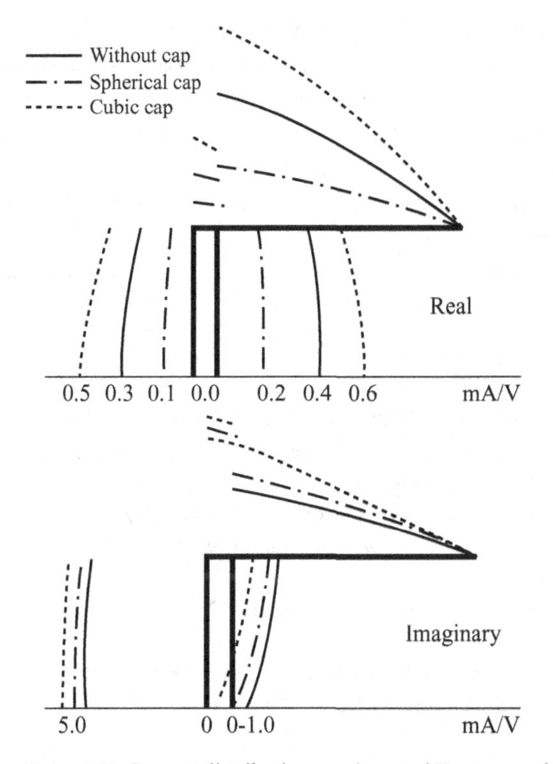


**Figure 9.21** Current distributions on inverted F antenna element (800 MHz).

The inner surface of a cap should be plated to reduce the ohmic loss. Usually, the interaction of the cap surface current with the antenna current is ignored because it can be assumed to be very small. However, it may become an appreciable value when the antenna element approaches the inner surface of the cap because the cap surface current may interact with antenna current. Then the cap surface loss becomes included in the antenna loss. Plating the cap surface with gold is best.

Another problem in the Wheeler method lies in an assumption that the antenna current is not changed by the presence of the Wheeler cap. This is not really true. Figure 9.21 shows that the antenna current of an inverted-F antenna varies due to a Wheeler cap [10]. Variation of the current depends on the shape and the size of the Wheeler cap as well as the type and the size of the antenna. Errors in the measurement are greatly increased, especially at the frequency of resonance or anti-resonance, which depend on the type and size of the antenna as well as the shape and size of the Wheeler cap.

Radiation efficiency of a small antenna located in some complicated environment can be evaluated in the reverberation chamber [11]. The radiation efficiency, for instance, of a small built-in antenna installed in a small handset, which is held by an operator, differs from that of a small antenna located in free space, because the radiated power from the handset is varied by the environmental materials. In addition, in practical operation, especially in an urban field, the incident or generated field by the antenna is not uniform, but rather random, including cross-polarized radio waves, due to the environmental conditions in the wave propagation path. In the reverberation chamber, radiation efficiency can be evaluated in an environment, where non-uniform field, which equivalently corresponds to the field around the antenna, can be easily produced. In the evaluation of the radiation efficiency, the radiation power from the handset is measured in the reverberation chamber and the efficiency is determined by taking a ratio of the radiated power to the input power to the antenna [12].

Input impedance of a very small antenna located near lossy materials can also be evaluated in the reverberation chamber [13]. An example is again a small built-in antenna installed in a small handset and a lossy human operator holds the handset.

## References

[1] T. Fukasawa, K. Shimomura, and M. Ohtsuka, Accurate Measurement Method for Characteristics of an Antenna on a Portable Telephone (in Japanese), *IEICE Transactions on Communications*, vol. J86-B, September 2003, no. 9, pp. 1895–1905.

[2] C. Icheln, J. Krogerus, and P. Vainikainen, Use of Balun Chokes in Small-Antenna Radiation Measurements, *IEEE Transactions on Instrumentation and Measurement*, vol. 53, April 2004, no. 2, pp. 498–506.

[3] C. A. Balanis, *Antenna Theory: Analysis and Design*, 2nd edn., John Wiley and Sons, 1997.

[4] H. Garn, M. Buchmayr, and W. Mullner, Tracing Antenna Factors of Precision Dipoles to Basic Quantities, *IEEE Transactions on Electromagnetic Compatibility*, vol. 40, November 1998, no. 4, pp. 297–310.

[5] H. Saito, I. Nagano, S. Yagitani, and H. Haruki, Radiation Characteristics of Antenna for a Small Radio Terminal in Vicinity of a Human Body (in Japanese), *IEICE Transactions on Communications*, vol. J83-B, October 2000, no. 10, pp. 1437–1445.

[6] T. Uno and S. Adachi, Range Distance Requirements for Large Antenna Measurements, *IEEE Transactions on Antennas and Propagation*, vol. 37, June 1989, no. 6, pp. 707–720.

[7] I. Ida, T. Sekizawa, H. Yoshimura, and K. Ito, Dependence of the efficiency-bandwidth product of small dielectric loaded antennas on the permittivity, *Proceedings of ISAP 2000*, vol. 1, pp. 61–64.

[8] E. D. Newman, P. Bohley, and C. H. Walter, Two methods for measurement of antenna efficiency, *IEEE Transactions on Antennas and Propagation*, vol. AP-23, 1975, no. 3, pp. 457–461.

[9] M. Muramoto, N. Ishii, and K. Itoh, Studies of Radiation Efficiency Measurement by Using Wheeler Cap, *IEICE Transactions*, vol. J-78-B-II, 1995, no. 6. pp. 454–460.

[10] R. Y. Chao, K. Hirasawa, and K. Fujimoto, Wire Antenna Current Distributions within a Wheeler Cap, *IEICE Transactions*, vol. J71-B, 1988, no. 11, pp. 1370–1372.

[11] K. Rosengren *et al.*, Characterization of Antennas for Mobile and Wireless Terminals in Reverberation Chambers, Improved Accuracy by Platform Stirring, *Microwave and Optical Technology Letters*, vol. 30, 2001, pp. 391–397.

[12] T. Sugiyama, T. Shinozuka, and K. Iwasaki, Estimation of Radiated Power of Radio Transmitter Using a Reverberation Chamber, *IEICE Transactons on Communications*, vol. E88-B, 2005, no. 8, pp. 3158–3163.

[13] P. S. Kildal and R. K. Rosengren, Measurement of Free Space Impedances of Small Antennas in Reverberation Chamber, *Microwave and Optical Technology Letters*, vol. 32, 2002, pp. 112–115.

# 10 Electromagnetic simulation

## 10.1 Concept of electromagnetic simulation

Generally, there are some analytical methods to obtain antenna characteristics. However, it can be difficult to obtain the characteristics of small antennas, because they are very limited in dimension and influenced quite sensitively by environmental conditions around them. In many cases, it is hard to verify the performance of small antennas by experiments. For example, when the size of the antenna is extremely small, the radiation resistance becomes too small to be determined accurately. Another case is where the antenna structure is so complicated that precise measurement becomes very difficult to perform.

Recently, antenna engineers have become able to rely on highly specialized electromagnetic (EM) simulators to develop and optimize antenna design [1, 2]. Computer-aided analysis and optimization have replaced the design process of iterative experimental modification of the initial design. Therefore, the EM simulators become of paramount importance in the efficient research and development of new small antennas. An accurate computer-aided design (CAD) could support a clear physical interpretation for characteristics of small antennas.

The operation of an EM simulator is based on the numerical solution of Maxwell's equations in differential or integral form.

Antennas for small mobile terminals are typical small antennas in practice and the EM simulators are often used to design them. Here, the concept of electromagnetic simulation is described along with antennas for small mobile terminals.

The EM simulations including antenna and propagation problems are widely used nowadays as the capabilities of personal computers and workstations increase rapidly. On the other hand, mobile communication systems, such as mobile phones recently deployed under various wireless systems having multiple functions including broadband and near-field communications, and also Radio Frequency Identification applications (RFID), have become common. Small and hand-held equipment used in those systems demand reduction in size and weight. Antennas used for such handsets must also follow downsizing of the handset unit, yet must keep the antenna performance unchanged or even improved. Especially, built-in antennas are required for handsets for the sake of convenience and durability. Because the environmental factors such as the handset material itself and the nearby human body affect the sensitivity and performance of the antenna mounted on the handset, it is not easy to realize an optimum antenna system.

To obtain the characteristics of the installed antenna, considered as one unified system composed of an antenna element and a conducting handset case, analysis by using an EM simulator is very effective [2, 3]. However, as the simulators are not perfect for analyzing antenna problems, designers need to have adequate experience using the simulator and be acquainted with the strengths and weaknesses of the various simulation methods. Representative applications of many analyses of small antennas used in mobile handsets by using EM simulators on the basis of method of moments [4, 5] and FDTD (Finite-Difference Time-Domain) method [3, 6] have been reported so far. In the method of moment (MoM), both the antenna element and the conducting handset case are simulated by a wire-grid model and they are treated as one unified system with the entire system made up of wire elements only [4, 5]. By using FDTD method, the analysis of the handset antenna including the effects of the surrounding objects such as the handset unit and a human body can be easily performed instead of applying approximate methods [3, 6]. However, the EM simulators generally are proprietary in nature and the details of their internal workings are not widely available to general users.

## 10.2 Typical electromagnetic simulators for small antennas

The electromagnetic simulator is based on the numerical solution of Maxwell's equations in differential or integral form. Integral-equation methods make use of Maxwell's equations in integral-equation form to formulate the electromagnetic problem. The kernel of the integrals is formed by a Green's function tensor.

Differential-equation methods are derived directly from Maxwell's curl equations or the Helmholtz wave equations. The most popular differential-equation-based methods are the Finite Element Method (FEM), utilized in Ansoft's *HFSS* software package, and the Finite-Difference Time-Domain Method (FDTD), which is employed by CST's time domain transient solver.

FEM, like FDTD, is a method based on solving a partial differential equation, rather than an integral equation, as in the case for MoM. Both the differential-equation methods are particularly suitable for modeling full three-dimensional volumes that have complex geometrical details.

Here, four types of EM simulators are explained. The IE3D, FIDELITY, HFSS, and Microwave Studio EM simulators, which are commercial software products, are based on MoM, FDTD (Finite-Difference Time-Domain), FEM (Finite Element Method), and FIT(Finite Integration Technique), respectively.

The IE3D simulator is the electromagnetic simulator based on the method of moments, and the primary formulation of IE3D is an integral equation obtained through the use of Green's function. Primary features are: (a) modeling of three-dimensional metallic and dielectric structures, (b) automatic generation of non-uniform meshing with rectangular and triangular cells, (c) automatic creation of small cells on the edges, and so forth. There is no limitation on the shape and orientation of the metallic structures. In the calculation, the IE3D simulator employs a three-dimensional non-uniform triangular and rectangular mixed meshing scheme automatically, and the shape of the antenna and conductor consists of a combination of polygons. In the IE3D simulator, the source program is not open to the public and only a brief description of the method of moments is given in the IE3D User's Manual [8].

The FIDELITY simulator is the electromagnetic simulator based on the FDTD (Finite-Difference Time-Domain) method and its basic principle is to use a finite difference to represent the differential in Maxwell's equations. The basis is to combine the electric and magnetic fields by employing Yee's algorithm [9]. Primary features are: (a) modeling both planar and three-dimensional structures with isotropic or non-isotropic dielectric material, (b) three-dimensional electric field and magnetic field display with slicing capability, (c) automatic non-uniform meshing and meshing-independent geometry, and so on. In the FIDELITY simulator, the source program is also not open to the public and only a brief description of the FDTD method is given in the FIDELITY User's Manual [10].

The HFSS simulator is the electromagnetic simulator based on the full-wave FEM (Finite Element Method). This simulator can analyze the characteristics of a small antenna automatically by solving Maxwell's equations. Primary features are: (a) modeling three-dimensional structures with CAD and dynamic geometry rotation, (b) automatic adaptive mesh generation and refinement, (c) comprehensive materials database containing permittivity, permeability, electric and magnetic loss tangents for common substances, and so on.

Microwave Studio is based on the Finite Integration Technique (FIT) and it allows choosing the time-domain as well as the frequency-domain approach. The Finite Integration Technique first describes Maxwell's equations on a grid space, maintaining properties such as energy conservation, and then forms the specific differential equations, such as the Poisson or wave equations. The Finite Integration Technique can be formulated on different kinds of grids, e.g., Cartesian or general non-orthogonal grids. In the time domain, which is our field of interest, the resulting discrete grid equations of the Finite Integration Technique are at least in "some" cases identical to the discrete equations derived with the classical Finite-Difference Time-Domain (FDTD) method. FDTD was introduced in the mid 1960s, and uses a coordinate-based staggered-grid system and the famous Yee cell. In contrast to the Finite Integration Technique, which is applied to the integral form of the field equations, FDTD (as a subset of the Finite Integration Method) is applied to the differential form of the governing Maxwell curl equations.

## 10.3 Example (balanced antennas for mobile handsets)

This section describes an example of simulation for performance of a small antenna located in a complicated environment. A typical example is a built-in antenna installed in a small mobile terminal, where the antenna is surrounded by various materials, electronic components, and devices that seriously affect antenna performances, and furthermore influenced by an operator who holds the terminal unit. Mobile phone handsets and other wireless terminals in the recent decade have made remarkable progress, and their size and weights have been dramatically reduced [11]. Accordingly, antennas used for such terminals must have followed their downsizing trends, for which some issues are taken up on the design of antenna systems.

Since antennas for mobile handsets cannot be designed in isolation from their host equipment, it is desirable to think of an antenna as a system. Moreover, the antenna performances are significantly influenced by the effects of the surrounding environments such as the human body which is considered as a big lossy medium existing near the antenna. The influence by the human body cannot be neglected as it disturbs the radiation, varies in the input impedance, and causes mismatching loss [11]. Accordingly, a system designer should consider some factors in designing antenna systems for mobile handsets as follows:

- The antenna performances should be maintained or enhanced although the antenna was miniaturized.
- The degradation of the antenna performances caused by a human body should be mitigated.
- The effect on the human body from the antenna radiation should be reduced.
- The antenna performances should be maintained under the multi-path radio environments.

The essential concept in designing antenna systems for handsets is that the antennas should be (1) small size and compact, (2) lightweight, (3) low-profile or built-in type, and have structure to (4) mitigate antenna performance degradation due to body effect, particularly, operator's hand and head, and to (5) reduce the SAR value (Specific Absorption Rate) in the human head. The factors (4) and (5) are required for the latest handset design and need the new technology for realizing antenna systems.

In addition to the above issues, with new deployment of various wireless systems, there is an increasing demand for mobile terminals to enhance functions such as wide-band or multiband operation, yet maintaining small dimensions. Thus, additional design considerations to meet such requirements have become an urgent issue.

The design concept so far applied to most built-in handset antennas is to use the GP (Ground Plane over which an antenna element is placed) as a part of the radiator. This is particularly true because antennas should have small size, and the assistance of the GP is needed in order to achieve enough gain and particularly wide-enough bandwidth. Regarding (1) to (3), e.g., small size and low-profile, antennas having magnetic current as the radiation source placed in parallel to the GP is not only to realize low-profile structure, but also to use its image of the antenna so that the twofold field-strength enhancement is obtained as a result of superposition of the field produced by the image, as shown in Figure 10.1 [12]. In order to realize (4) and (5), antennas having balanced terminals and being fed by a balanced line have been proposed. The effectiveness of using such antennas has been shown in some previous papers [13–16].

So far the planar inverted F antenna (PIFA) has been used in many handsets, as it matches the needs (1), (2), and (3) mentioned above [17]. However, in practical use, gain degradation, which is serious sometimes, has been observed when an operator holds the handset. This is caused by the variation in the current distributions on the GP, which

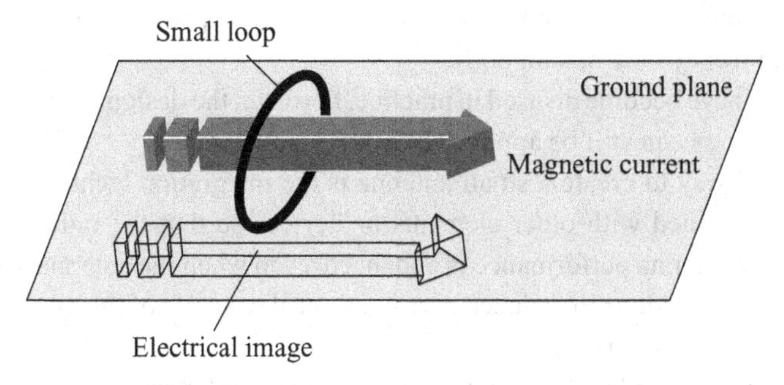


**Figure 10.1** Small loop antenna on a ground plane and magnetic current.

are produced by the excitation of the built-in PIFA element. These current distributions on the GP actually assist the antenna performance, contributing to improve gain and also bandwidth. In turn when the body effect exists, particularly when the operator's hand grasps the handset, these currents vary and as a result the antenna performance degrades; sometimes serious gain reduction and frequency change are observed. It has been commonly observed in handsets where small antennas, not only PIFA but also other types of antennas like ceramic chips, are built in. Thus in order to avoid such antenna-performance degradation, reduction of the currents on the GP has been desired and antennas having balanced structure and being fed by a balanced line have been developed.

There has been a report that has shown another way to reduce the currents on the GP. It is the use of a half-wave monopole applied to a rectangular conducting box that simulates the handset body. The analysis of it has shown almost no current flow on the handset body with use of a half-wave monopole [18]. In practice; however, a 3/8 wavelength or a 5/8 wavelength monopole has been used for the handsets, as they have appropriate input impedance to match the load, and yet cause only a small current flow on the GP. The results of this analysis [8] have contributed to establishing the design concept for the antennas used in conventional PDC handsets. However, antennas designed with this concept do not satisfy the requirements (1) to (3) and (5).

Now it is rather natural that antennas for the latest handsets have been reconsidered and have progressed to be designed with advanced concepts. At the same time, the recent trends that antennas should be small in size, yet have enough performance and enhanced functions are consistently kept in mind and embraced in the design concept. In practice, in order to realize antenna systems which can satisfy the requirements (4) and (5) in addition to (1) to (3), new design concepts should be introduced.

For the purpose of realization of (4) and (5), various types of antennas having balanced structure have been taken into consideration [13–16]. Regarding small-sized and low-profile antennas, use of antennas having a magnetic current source was introduced. One of these was a small loop antenna which was regarded as a magnetic dipole normal to the surface of the loop [12] and it was successfully applied practically to the box-type pager's antenna [19]. A small loop antenna built into a pager was placed in the pager body in such a way that the loop surface was perpendicular to the human body instead of to the GP so that the field produced by the image loop was superposed to obtain twofold field strength in front of the human body.

Pager systems have become disused in practice; however, the design concept employed for the pager antenna can still be applied to realize small antennas.

Another useful way to create a small antenna is the integration technique, by which an antenna is combined with other elements or devices so that the radiation mode is increased and the antenna performance is enhanced, even when the antenna size is made small. A folded loop antenna introduced here is a typical example of this type of antenna, which also has balanced structure.

The design concept for handset antennas thus favors a balanced structure in order to achieve small antennas and yet maintain performance that does not degrade in the vicinity of the human hand and head. In designing the latest handset antennas, other significant considerations are included; these are on the realization of very small antennas, feasibility of impedance matching for wide bandwidth, and so forth.

### 10.3.1 Balanced-fed folded loop antenna

In the previous sections, it has been shown that antennas having balanced structures and being fed with a balanced line are very effective to mitigate the antenna performance degradation due to the body effect and also reduce the radiation toward the human head [13–16]. As Figure 10.2 shows, various types of antennas, such as rectangular loop and L-type loop, which have balanced terminals and are fed with a balanced line, have been introduced. The principle of using a balanced structure is to reduce the current flow on the GP excited by the antenna element placed on the GP.

In turn by applying this concept to the handset antenna design, a further serious issue has been raised. That is how to realize an antenna which has enough gain and bandwidth, even when the size has been made small and the assistance of the GP as a part of the radiator is removed. Problems are to design an antenna which has enough gain and bandwidth with a small-sized structure. The proper impedance matching is also a problem, because there is some difficulty in obtaining desired impedance by conventional antennas, as shown in Table 10.1. Use of a balun in order to feed antennas with balanced terminals provides a sort of complexity for impedance matching and usually makes the bandwidth narrow.

Then a folded loop antenna is taken into consideration as one of the candidates which solves these problems and satisfies the requirements in the design concepts. This antenna has an integrated structure, which is composed of a radiating element and a reactive element; these being constituted by using a two-wire transmission line, and folded at a quarter-wavelength to form a folded half-wave dipole equivalently. The equivalent folded dipole acts mainly as the radiator, and the two-wire line is used for adjusting the antenna impedance [20–22]. The antenna has a one-wavelength loop structure so that no unbalanced current may be produced on the feed line; that is to say, this antenna has a self-balanced structure [21], which is useful to reduce the current flow on the GP. In addition, since this antenna can be built in a small volume by means of its folded structure, small size and low profile can be achieved. Furthermore, use of a two-wire transmission line can

**Table 10.1** Input impedance of balanced-fed antennas ($f_0 = 186$ GHz)

| | Input impedance [Ω] |
|---|---|
| Rectangular loop Type A | 74.31 + j 1393 |
| Rectangular loop Type B | 119.7 + j 1494 |
| Helical dipole | 0.934 + j 80.47 |
| L-type loop | 32.81 + j 997.5 |

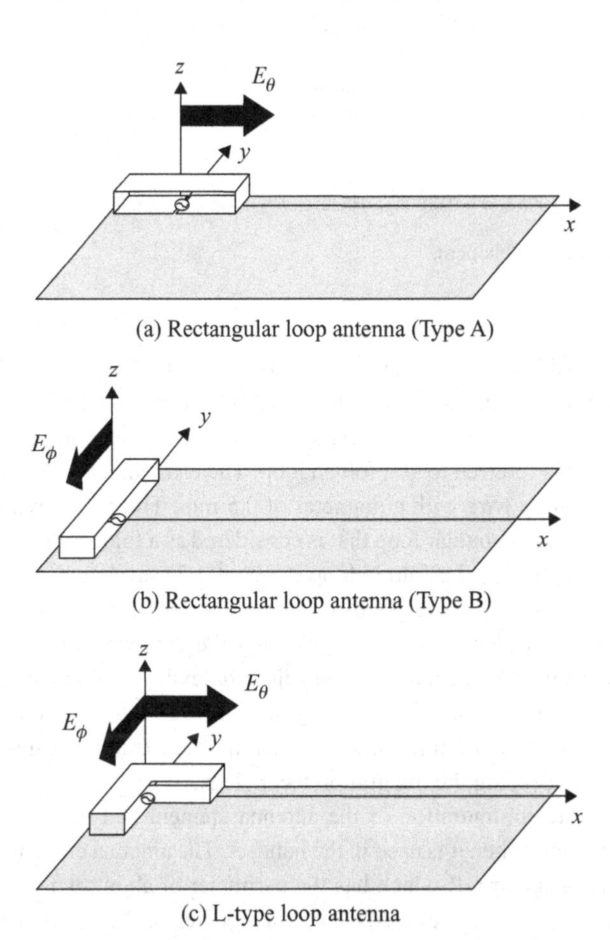


**Figure 10.2** Balanced-fed antennas: (a) rectangular loop antenna (Type A), (b) rectangular loop antenna (Type B), and (c) L-type loop antenna.

make possible flexible adjustment of antenna impedance by changing parameters such as the length and the width of wires, and the distance between the two wires, and no balun is necessary. This antenna can be designed to have enough gain and bandwidth to be applied to handsets presently used in practice. In order to achieve wide bandwidth, the wire should be replaced by a ribbon-shaped element with wider width. A parasitic element placed along with the folded loop element is effective to achieve further wide bandwidth.

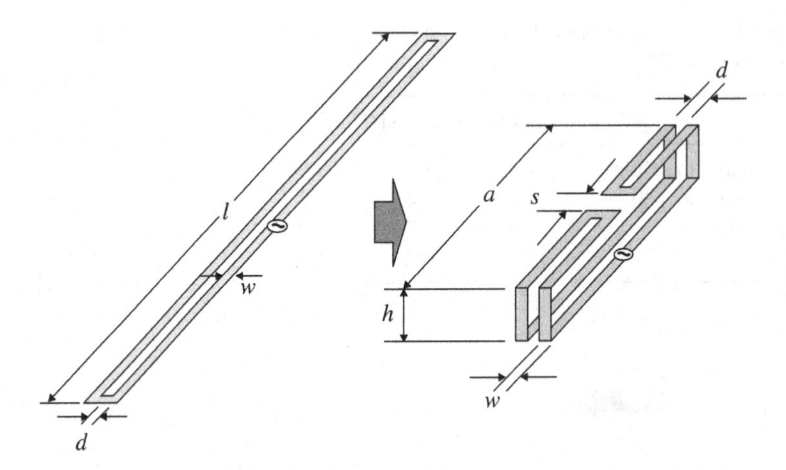


$\lambda = 161.3$ mm, $l = 0.44\,\lambda$, $d = 0.006\,\lambda$, $w = 0.003\,\lambda$, $a = 0.225\,\lambda$, $h = 0.056\,\lambda$, $s = 0.006\,\lambda$

**Figure 10.3** Folded loop antenna element.

## 10.3.2 Antenna structure

A folded loop introduced here is the antenna previously described in 8.1.2 (2)-2b, which is essentially a two-wire transmission line, folded at about a quarter-wavelength to form an equivalent half-wave folded dipole, and yet appears as a one-wavelength loop antenna, from which the antenna is referred to as a folded loop. The element taken as an example here is made of thin copper wire with a diameter of 0.5 mm. The folded wire element forms a very thin ($d \ll \lambda$) rectangular loop that is considered as a folded dipole as shown in Figure 10.3. The length $l$ of the folded loop is about 1/2 wavelength at the center frequency $f_0$ which equivalently corresponds to a half-wavelength folded dipole and at the same time to a one-wavelength loop. By selecting the two-wire transmission line parameters, the antenna input impedance can be adjusted flexibly, since the reactance of the transmission line can be adjusted by selecting the length and the width of the wires and the distance between the two lines. This is the important feature of this antenna, which is constituted by applying the integration technology.

Figure 10.4 shows the configuration of the antenna element and the finite GP that represents a shielding plate when it is used in the handset. The antenna element is placed very closely to the rectangular GP, which has the perimeter of about two wavelengths. Both unbalanced and balanced type of feed are considered here as Figure 10.4(a) and (b) show. The unbalanced feed is tested only to confirm the effectiveness of the unbalanced feed for the folded loop, as the folded loop of a half wavelength has performance similar to the balanced system. The antenna element is fed by either a coaxial cable (Figure 10.4(a)) or a parallel line (Figure 10.4(b)). In the experiment, a semi-rigid coaxial cable with a diameter of $0.011\lambda$ is used and the load of a 50 Ω : 50 Ω chip balun is used for the balanced system. The center frequency $f_0$ is 1860 MHz.

In the numerical analysis, various types of electromagnetic simulators are applied based on the method of moments (MoM) [23], finite-difference time-domain (FDTD) method [24], finite element method (FEM) [25], and finite integration method (FIM) [26].

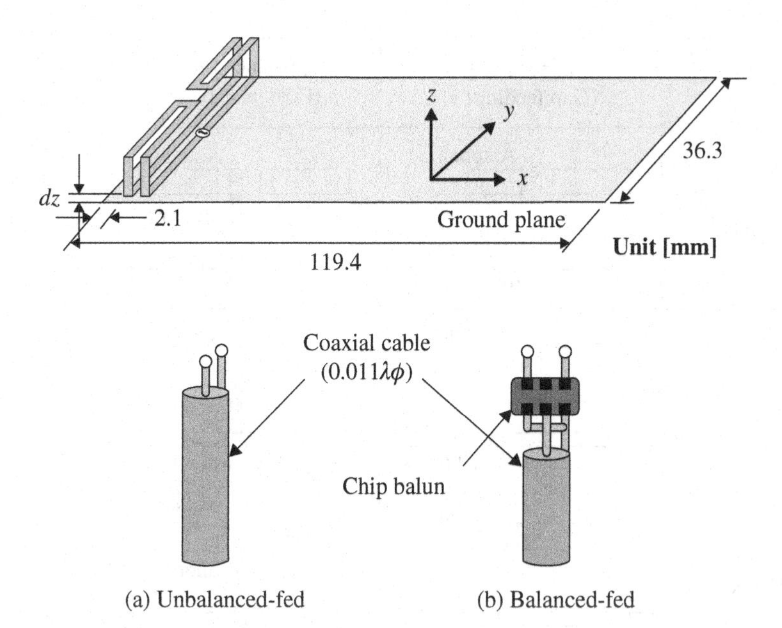


**Figure 10.4** Antenna structure: (a) unbalanced feed and (b) balanced feed.

The electromagnetic simulators including antenna and propagation problems are widely used nowadays as the computational capacities of personal computers and workstations have become adequate. However, as the simulators are not perfect for analyzing antenna problems, users need to have experience with the simulators and be acquainted with their individual characteristics.

Figure 10.5 shows the feeding method of each electromagnetic simulator. In the MoM simulation, since the models are divided into small cells automatically, the standard frequency of the division employs 2500 MHz, higher than the required frequency. To confirm the convergence, the results of four models which have divisions from 10 to 30 cells per wavelength (i.e., size ranging from 0.03λ to 0.1λ) are compared and discussed. When the number of cells per wavelength is 20, the numbers of total cells of unbalanced-fed and balanced-fed models are 509 and 458, respectively. The feed type used in simulation is horizontal port which is called "Extension for MMIC" (Monolithic Microwave Integrated Circuit). A length of the edge cell is one tenth of the feed line's width empirically.

In the FDTD simulator, models are divided into non-uniform meshing of 0.5 to 4 mm. Around the feed point the structure is divided into minimum cells of 0.5 mm, while the part where there is no influence in the calculation is divided into larger cells to reduce the total number of cells. The total number of cells is 49 140 and all parts of models use Perfect Matching Layer, which is referred to as PML, as absorbing boundary conditions. The feed type used in the simulation is "Coaxial Port" and "Gap Port."

In the FEM simulator, models are divided by using automatic adaptive mesh. Therefore, the model of the handset, which is used here as an example, is divided into small cells effectively. The total number of cells is 389 160. The large models such as human

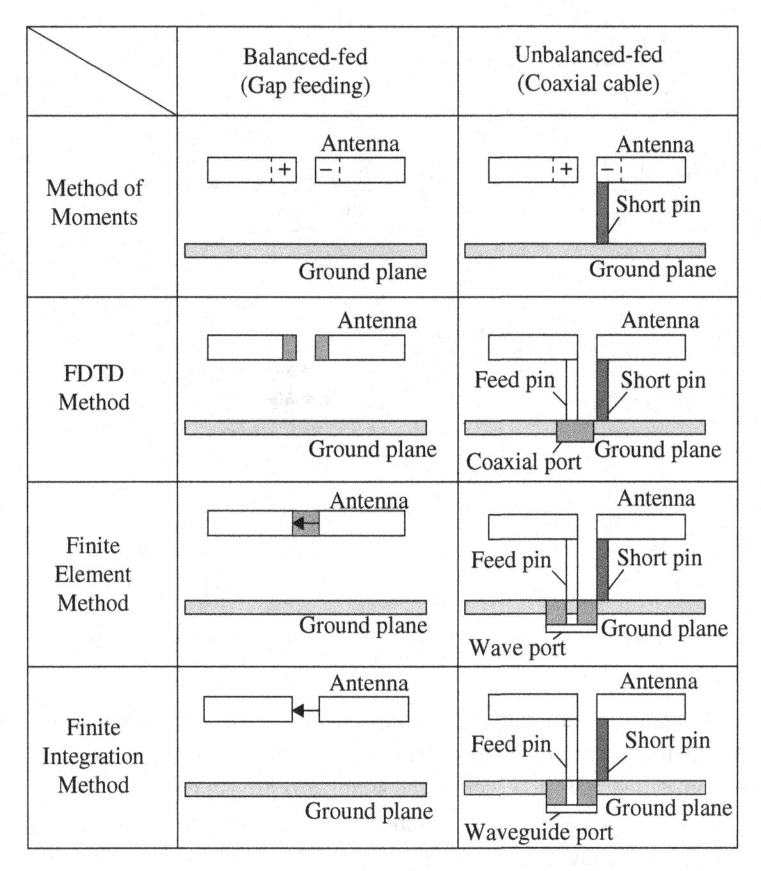


**Figure 10.5** Feeding method of each electromagnetic simulator.

head or hand model are divided into the larger cells by using manual mesh to reduce the number of total cells. The feed type used in the simulation is a coaxial port.

### 10.3.3 Analytical results

Figure 10.6 shows the simulated current distribution on the GP at the center frequency by using the MoM simulator, where (a) shows for the case of an unbalanced system and (b) shows for the case of a balanced system. Since the antenna element is placed on the GP so as to make the system structure symmetrical with respect to the center of the GP (the $x$–$z$ plane) the current distributions on the GP obviously become symmetrical. The figure illustrates the current distributions on the GP only, not including those of the antenna element, and the amplitude of currents is shown with gray shading. In the figure, the fainter shading indicates greater current flow while the darker color part indicates smaller current flow. The antenna element is located on the left side of the GP.

As shown in the figure, while only a slight difference is seen around the feed point between the current distributions in both unbalanced and balanced system, at other points they have almost the same amplitudes. For the unbalanced system, the current

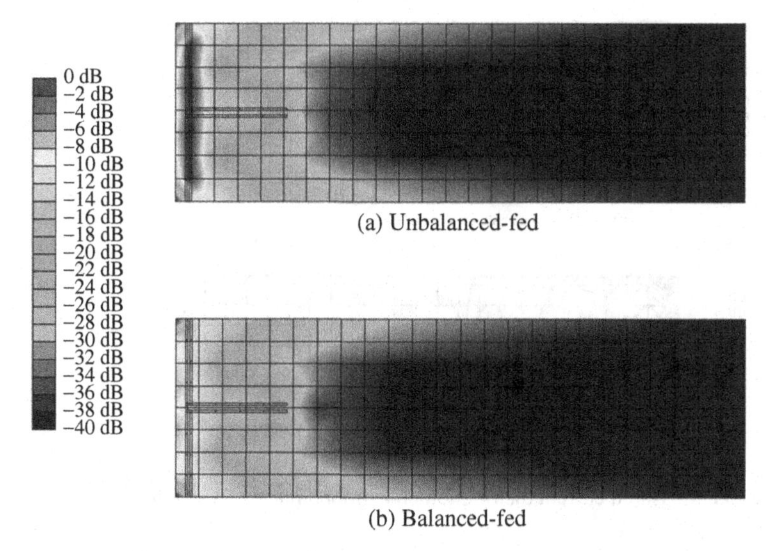


**Figure 10.6** Simulated current distribution on a ground plane (Method of Moments): (a) unbalanced fed and (b) balanced fed.

distributions on the GP become almost symmetric with respect to the $x$-axis as well as the balanced system, even though it is fed by an unbalanced feed line.

The measured current distribution at the center frequency is obtained by using Schmid & Partner Engineering AG's DASY-3 package for near-field evaluation [27]. The magnetic field above the antenna system is detected in phase and amplitude by a probe antenna scanning a plane surface that is about $\lambda/10$ from the GP of the antenna structure. The surface current density $J$ is related to the magnetic field $H$ through the equation $n \times H = J$ ($n$ is a unit vector pointing in the outward direction from the surface). The current distribution is measured in each $3/100\ \lambda$ span on a rectangular plane surface above the antenna structure. The measured current distribution on the GP in the two-dimensional plane is shown in Figure 10.7. In the experimental result, the brightest shading part in the dense contour lines indicates greatest current flow and diminishing in the current flow is shown by the variation of contour lines that become gradually coarse. The current distributions are almost the same for both unbalanced and balanced systems, while there is a slight difference around the folded loop element, resulting from inclusion of current distributions of not only on the GP but also on the antenna element, and they are symmetric with respect to the $x$-axis in both unbalanced and balanced systems. These tendencies are very similar to those of calculated results.

From those results, it can be seen as was expected that a folded loop element with a length of half-wavelength has a self-balance property and hence the unbalanced current does not flow on the feed line nor on the GP, even with a folded loop element fed via the unbalanced line. The self-balance property is described in Appendix II in 8.1.2 (2)-2b.

Figure 10.8 shows the current distribution on the GP simulated by each simulator for the unbalanced-fed model, where (a), (b), and (c) are the results using the FDTD, FEM,

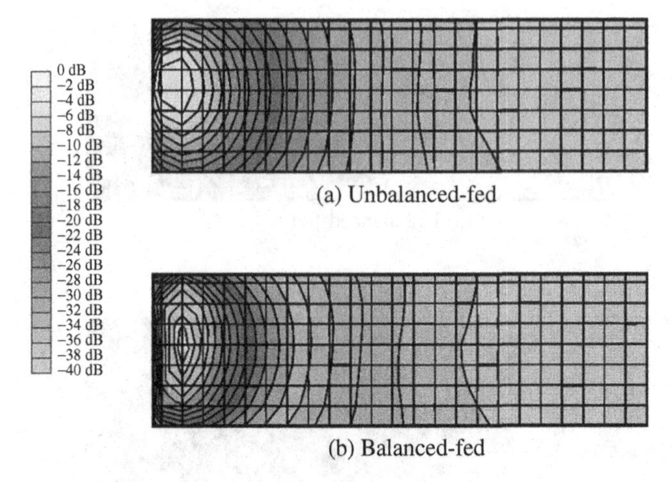


**Figure 10.7** Measured current distribution on a ground plane: (a) unbalanced feed and (b) balanced feed.

and FIM simulators, respectively. In all cases, the current on the GP is reduced by the balanced feeding method, which is the same result as the MoM simulator.

Figure 10.9 shows the measured and simulated return-loss characteristics. The simulated result of each simulator is similar to the measured results. For the results obtained by the MoM simulator, the frequency at the lowest return loss is low in the case of the number of cells per wavelength of 10, compared to other cases. For the results obtained by the FDTD simulator, it is apparent that the calculated results show wider bandwidth than the measured results. For the results of the FEM and FIM simulators, the frequency at the lowest return loss becomes higher as the number of the total cells increases, but remains lower than the measured one.

Figure 10.10 shows the simulated return-loss characteristics of unbalanced and balanced-fed models by using the optimized conditions in all the simulators as well as the measured results. All the simulated results agree very well with the measured results in both models.

Figure 10.11 shows the simulated and measured radiation patterns. All the simulated results agree very well with the measured results, and the patterns of the unbalanced-fed model are almost the same as those of the balanced-fed model. This is because a folded loop antenna has the self-balance effect and the radiation from the GP is almost suppressed. The difference in the maximum gain is below 1 dB as shown in Table 10.2.

### 10.3.4 Simulation for characteristics of a folded loop antenna in the vicinity of human head and hand

In this section, the antenna performances in the vicinity of the human head and hand will be investigated by means of simulation, in which the human head and hand are simulated in consideration of the practical situation of an operator using a handset, and equivalent phantom models are used for the verification.

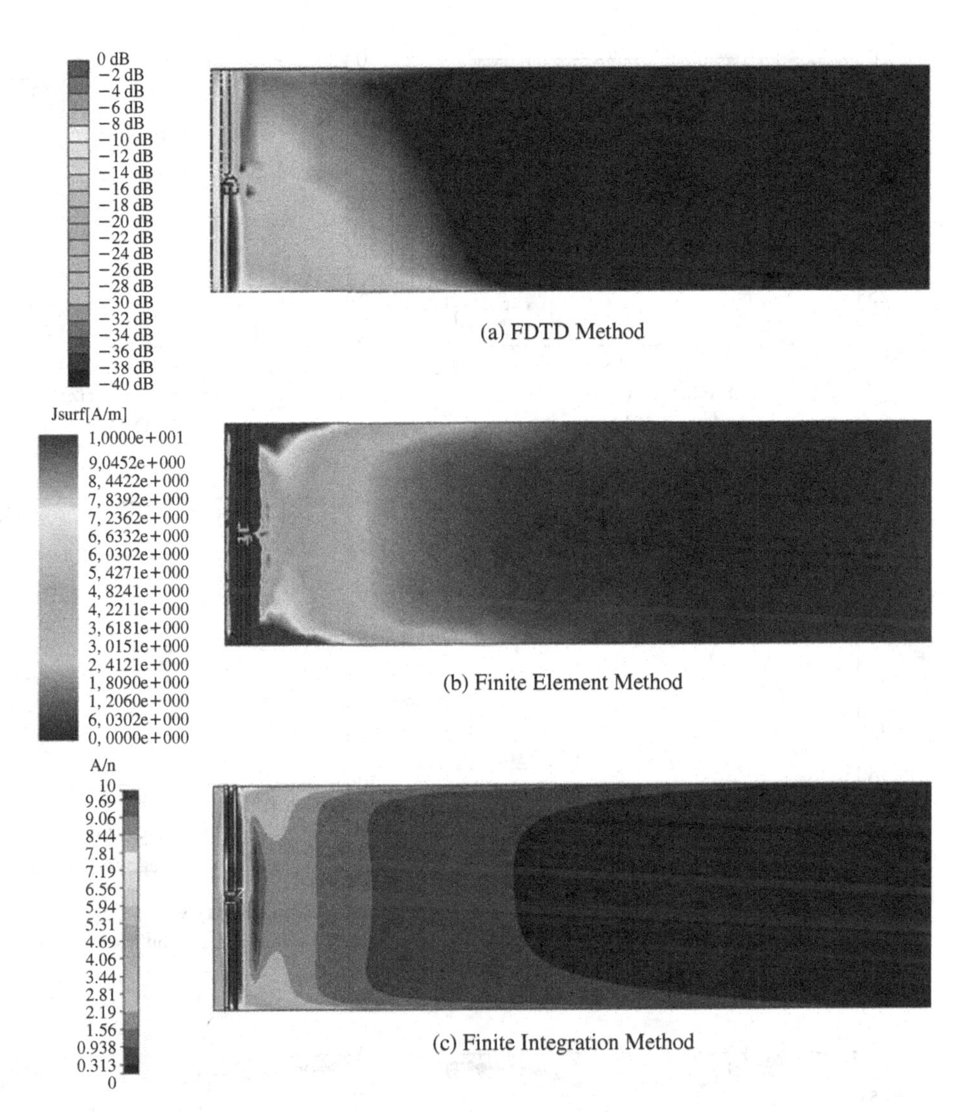


**Figure 10.8** Current distribution on a ground plane simulated by each electromagnetic simulator (unbalanced-fed model): (a) FDTD Method, (b) Finite Element Method, and (c) Finite Integration Method.

#### 10.3.4.1 Structure of human head and hand

Figure 10.12 illustrates a folded loop antenna in the vicinity of the human body, where (a) is the spherical head model, and (b) are the spherical head and hand models. The spherical head model has dielectric properties of relative permittivity of 43.37 and conductivity of 1.204 S/m at 1.9 GHz. The diameter of the head is 200 mm and the distance between the human head model and the antenna is 10 mm. The human hand model has dielectric properties of relative permittivity of 54 and conductivity of 1.45 S/m at 1.9 GHz. There is 4 mm spacing between the inner surface of hand model

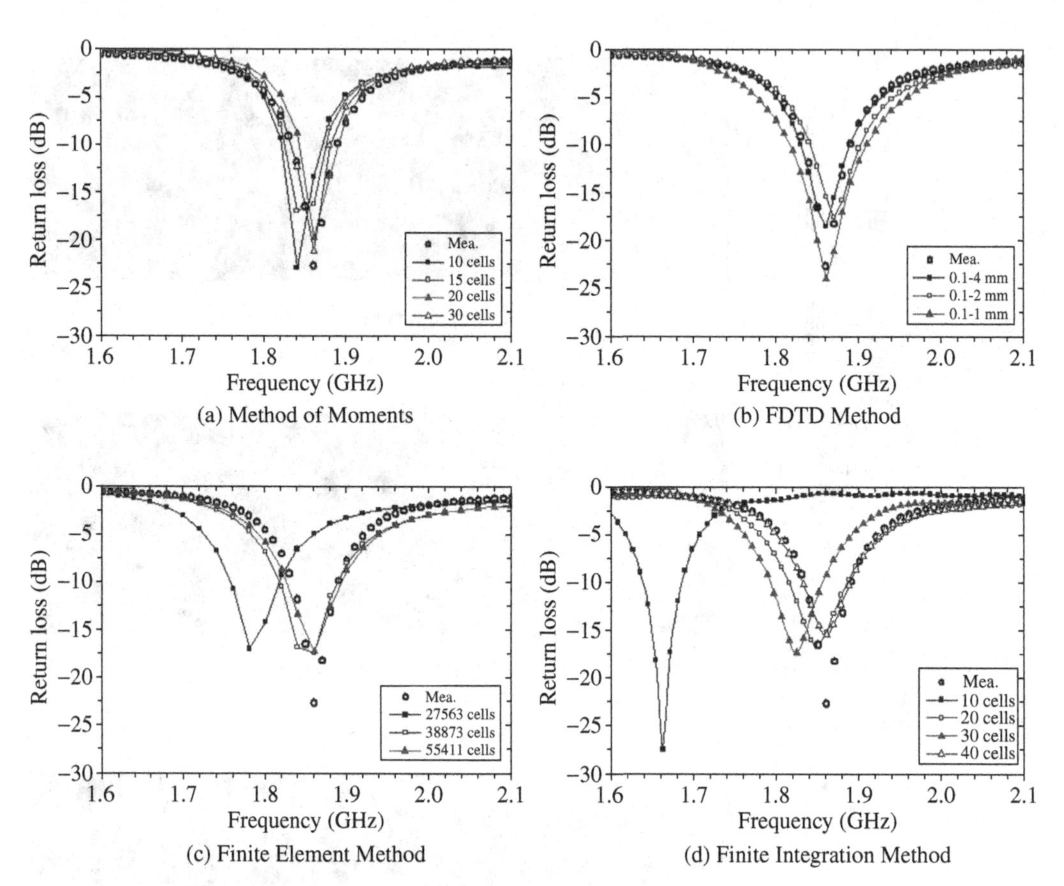


**Figure 10.9** Return-loss characteristics (unbalanced-fed model): (a) Method of Moments, (b) FDTD Method (c) Finite Element Method, and (c) Finite Integration Method.

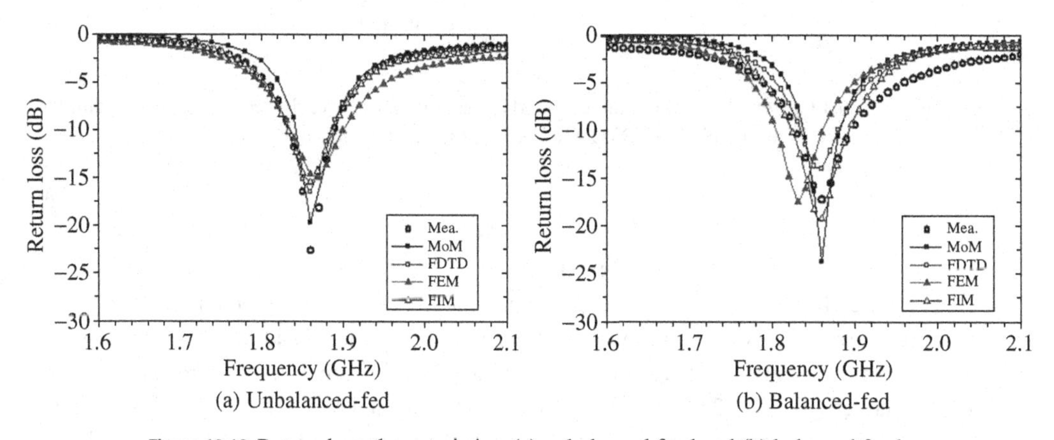


**Figure 10.10** Return-loss characteristics: (a) unbalanced feed and (b) balanced feed.

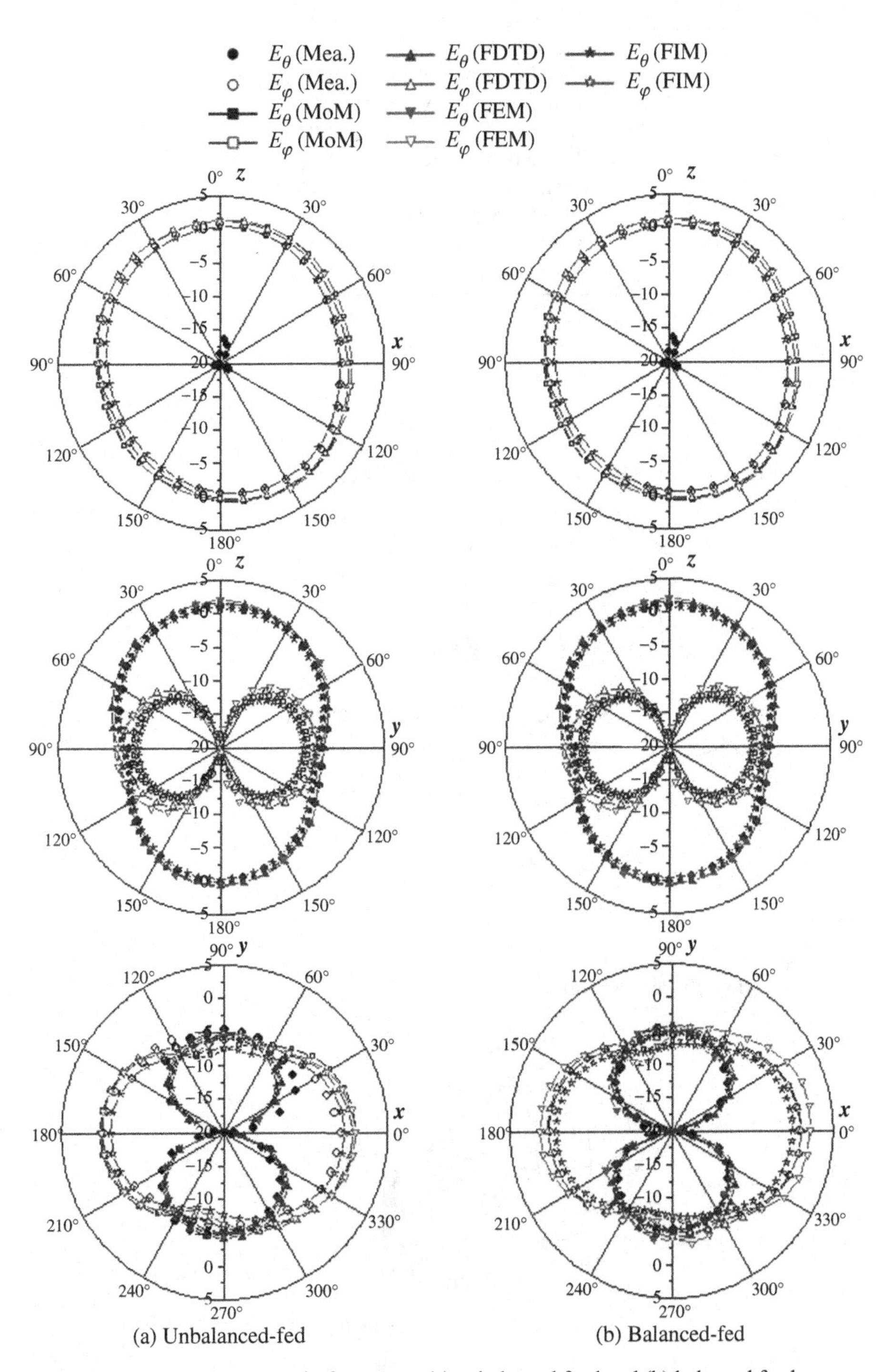


**Figure 10.11** Radiation patterns in free space: (a) unbalanced feed and (b) balanced feed.

**Table 10.2** Gain of a folded loop antenna ($f_0 = 1860$ MHz)

| | | Gain [dBi] |
|---|---|---|
| | Method of Moments | 1.43 |
| | FDTD Method | 1.36 |
| Unbalanced-fed | Finite Element Method | 1.37 |
| | Finite Integration Method | 2.2 |
| | Measured gain | 1.02 |
| | Method of Moments | 1.34 |
| | FDTD Method | 1.76 |
| Balanced-fed | Finite Element Method | 2.05 |
| | Finite Integration Method | 2.2 |
| | Measured gain | 1.11 |

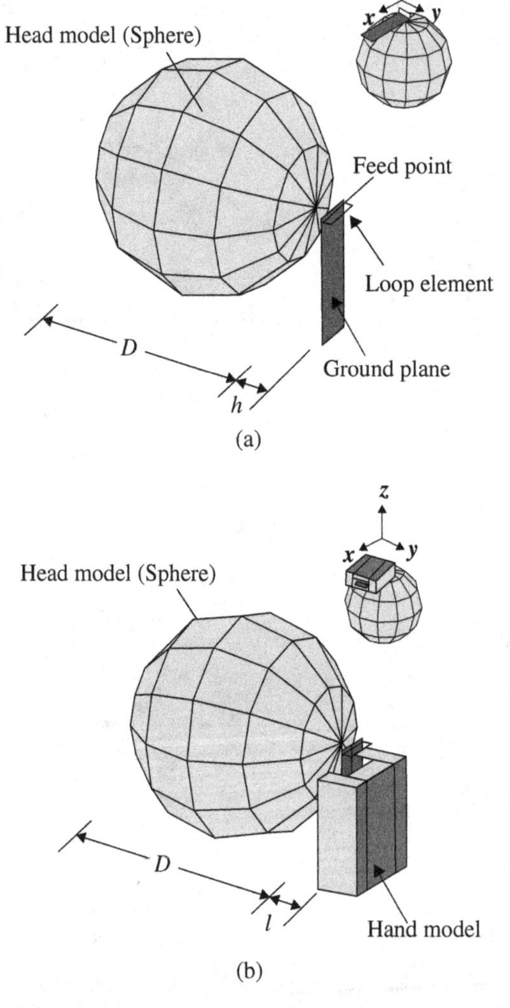


**Figure 10.12** Antenna in the vicinity of a human model: (a) antenna in the vicinity of a spherical head model and (b) antenna in the vicinity of a spherical head model and hand model.

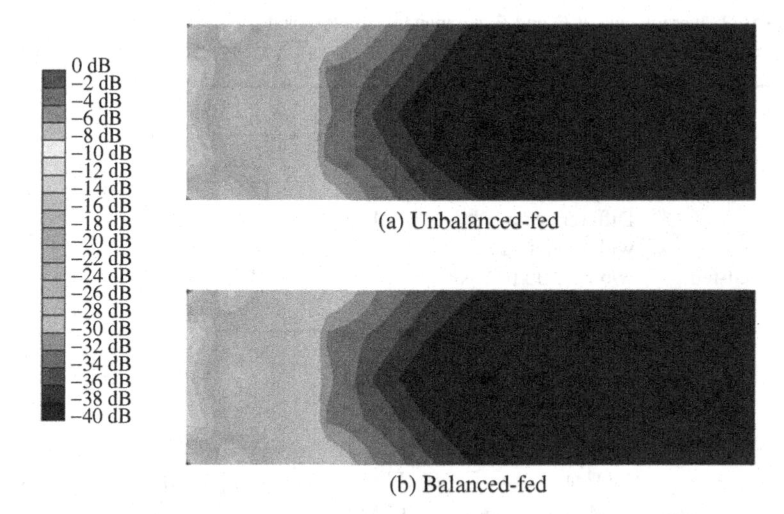


**Figure 10.13** Current distribution of a folded loop antenna in the vicinity of a spherical head model ($f_0 = 1860$ MHz): (a) unbalanced feed and (b) balanced feed.

and the boundary of the handset, because the antenna is assumed to be surrounded by a housing for avoiding direct contact between the user's hand and the interior elements.

In the MoM simulator, the number of cells per wavelength is 10 and the numbers of total cells of unbalanced-fed and balanced-fed models are 2864 and 2918, respectively. In the FDTD simulator, models are divided into non-uniform meshing of 0.5 to 4 mm and the numbers of total cells of unbalanced-fed and balanced-fed models are 367 275 and 313 880, respectively. In the FEM simulator, the numbers of total cells of unbalanced-fed and balanced-fed models are 18 345 and 12 253, respectively. In the FIM simulator, the total number of cells of unbalanced-fed and balanced-fed models are 9 660 189.

Figure 10.13 shows the simulated current distribution on the GP in the vicinity of the head model by using the MoM simulator, where (a) and (b) are the currents of unbalanced and balanced-fed models. As in the case of free space, the current on the GP is reduced even in the vicinity of the head model. Moreover, the result of an unbalanced-fed model agrees very well with that of a balanced-fed model. This shows that the self-balance effect is still maintained even though a folded loop is located in the vicinity of the human head.

#### 10.3.4.2 Analytical results

Figure 10.14 shows the radiation patterns in the vicinity of the head model, where (a) and (b) are the radiation patterns of unbalanced and balanced-fed models. In the figure, the difference of about 4 dB between the simulated and measured results is seen in the vicinity of a head model, while they are almost the same in the free space. It is considered that the calculated gains by each simulator include the different loss of the impedance mismatch and are different from each other, because the radiation patterns by each simulator are calculated at the center frequency of 1860 MHz. Furthermore, users need

**Table 10.3** Average gain of $E_\theta$ and $E_\phi$ components in the vicinity of a head model

| | Head model | $E_\theta$ [dBi] | $E_\varphi$ [dBi] |
|---|---|---|---|
| | with model : a | –1.00 | –9.04 |
| Unbalanced-fed | w/o model : b | –5.82 | –17.81 |
| | Difference : c = b – a | –4.82 | –8.41 |
| | with model : a | –1.11 | –9.40 |
| Balanced-fed | w/o model : b | –5.73 | –18.53 |
| | Difference : c = b – a | –4.62 | –9.13 |

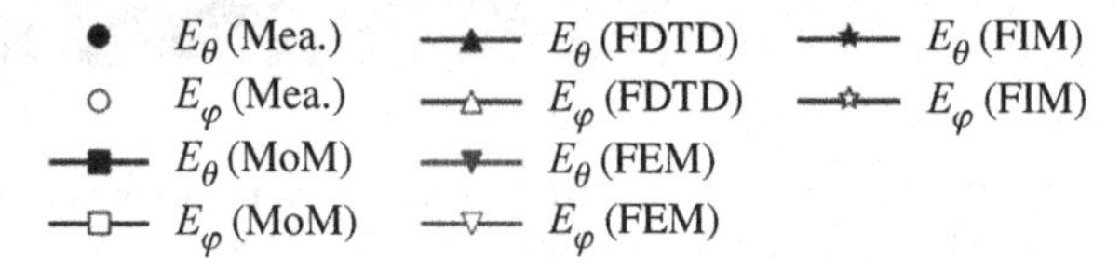


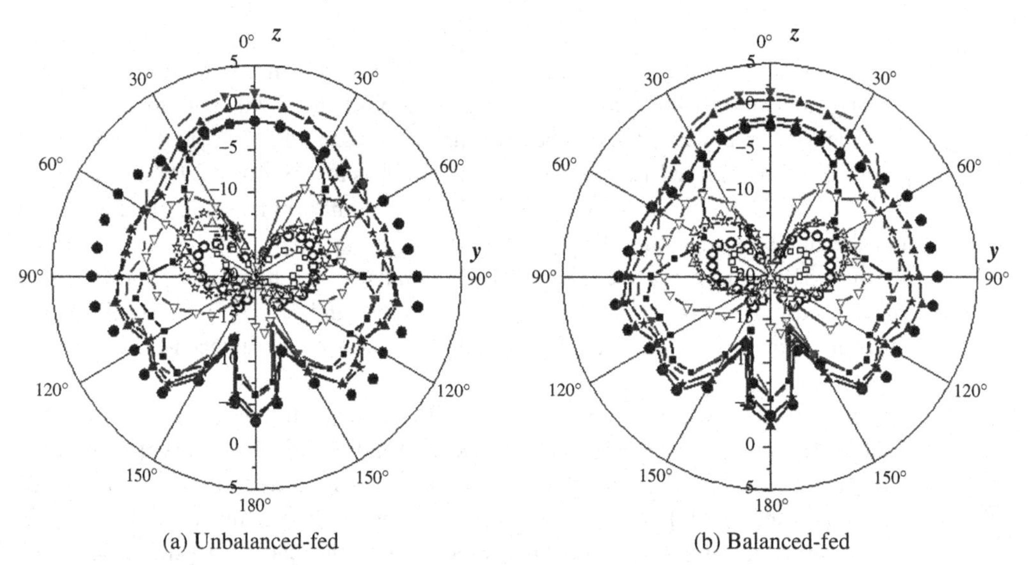


(a) Unbalanced-fed (b) Balanced-fed

**Figure 10.14** Radiation pattern in the vicinity of a spherical head model ($f_0$ = 1860 MHz): (a) unbalanced feed and (b) balanced feed.

to have more familiarity with each simulator in order to design the optimum geometrical configuration, and they must be acquainted with the application of the simulator. The radiation toward the head is decreased while the gain toward $+z$ direction becomes higher. The same as in free space, the radiation pattern of the unbalanced-fed model is almost the same as that of the balanced-fed model, and this shows that the self-balance effect is still maintained even though a folded loop is located in the vicinity of the human head. Table 10.3 shows the average gain of $\theta$- and $\phi$-components of the electrical field

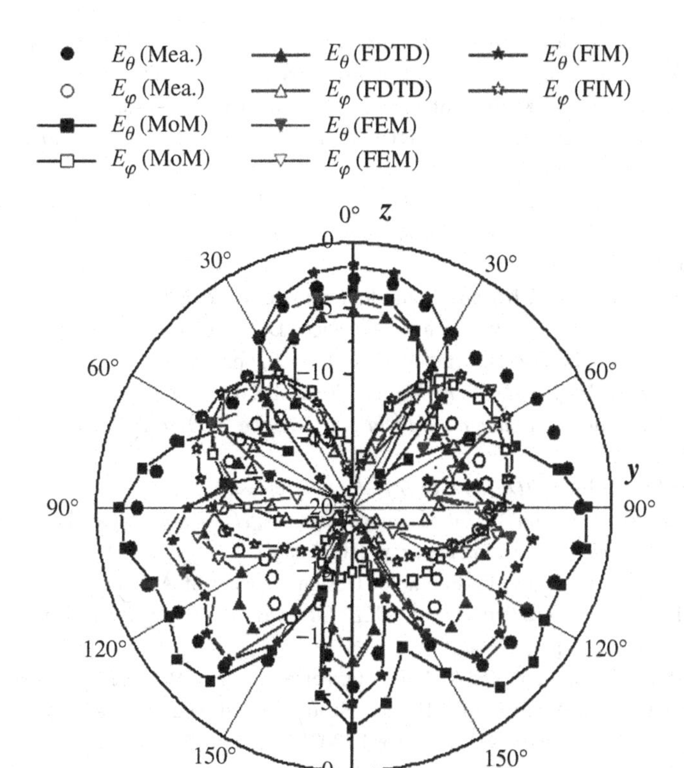


**Figure 10.15** Radiation pattern of an unbalanced-fed model in the vicinity of head and hand models ($f_0 = 1860$ MHz).

in the vicinity of a head model. There is almost no difference between the results of unbalanced and balanced-fed models.

Figure 10.15 shows the radiation pattern of a folded loop antenna in the vicinity of head and hand models at 1860 MHz. By the presence of a hand model, the maximum gain was reduced in the $y$–$z$ plane.

## References


[1] A. Vasylchenko, Y. Schols, W. De Raedt, and G. A. E. Vandenbosch, Quality Assessment of Computational Techniques and Software Tools for Planar-Antenna Analysis, *IEEE Antennas and Propagation Magazine*, vol. 51, February 2009, no. 1, pp. 23–38.

[2] M. Taguchi, On Electromagnetic Simulators for Antenna Analysis, *Proceedings IEICE General Conference*, 1998, C1 TB-1-1, p. 736.

[3] K. Sato, K. Nishikawa, N. Suzuki, and A. Ogawa, Analysis of Antennas Mounted on Portable Equipment Near Human Body, *IEICE Transactions (Japan)*, vol. J79-B-II, November 1996, no. 11, pp. 892–900.

[4] K. Hirasawa and K. Fujimoto, Characteristics of Wire antenna on a Rectangular Conducting Body, *IECE Transactions (Japan)*, vol. J65-B, April 1982, pp. 382–389.
[5] K. Sato, K. Matsumoto, K. Fujimoto, and K. Hirasawa, Characteristics of a Planar Inverted-F Antenna on a Rectangular Conducting Body, *IECE Transactions (Japan)*, vol. J71-B, November 1988, pp. 1237–1243.
[6] K. D. Katsibas, C. A. Balanis, P. A. Tirkas, and C. R. Birtcher, Folded Loop Antenna for Mobile Hand-held Units, *IEEE Transactions on Antennas and Propagation*, vol. 46, 1998, no. 2, pp. 260–266.
[7] Y. Kim, H. Morishita, Y. Koyanagi, and K. Fujimoto, A Folded Loop Antenna System for Handset Development and Based on the Advanced Design Concept, *IEICE Transactions on Communications*, vol. E84-B, September 2001, no. 9, pp. 2468–2475.
[8] Zeland Software, Inc., *IE3D User's Manual*, Release 8.0, Fremont, 2001.
[9] T. Uno, *FDTD for Electromagnetic Field and Antenna Analyses*, Corona Publishing Co., Ltd., 1998, p. 2.
[10] Zeland Software, Inc., *FIDELITY User's Manual*, release 1.5.
[11] K. Fujimoto and J. R. James, *Mobile Antenna Systems Handbook*, Artech House, 1994.
[12] J. D. Kraus, *Antennas*, 2nd edn., McGraw-Hill, 1988.
[13] H. Morishita, Y. Kim, and K. Fujimoto, Design Concept of Antennas for Small Mobile Terminals and the Future Perspective, *IEEE Antennas and Propagation Magazine*, vol. 44, October 2002, no. 5, pp. 30–43.
[14] H. Morishita, H. Furuuchi, and K. Fujimoto, Performance of Balance-fed Antenna System for Handsets in the Vicinity of a Human Head or Hand, *IEE Proceedings: Microwaves, Antennas and Propagation*, vol. 149, April 2002, no. 2, pp. 85–91.
[15] Y. Kim, H. Furuuchi, H. Morishita, and K. Fujimoto, Characteristics of Balance-fed L-type Loop Antenna System for Handsets in Vicinity of Human Head, *Proceedings 2000 International Symposium on Antennas and Propagation*, vol. 3, August 2000, pp. 1203–1206.
[16] H. Morishita, Y. Kim, and K. Fujimoto, Analysis of Handset Antennas in the Vicinity of the Human body by the Electromagnetic Simulator, *IEICE Transactions on Electronics*, vol. E84-C, July 2001, no. 7, pp. 937–947.
[17] T. Taga and K. Tsunekawa, Performance Analysis of a Built-in Planar Inverted F Antenna for 800 MHz Band Portable Radio Units, *IEEE Journal on Selected Areas in Communications*, vol. SAC-5, June 1987, no. 5, pp. 921–929.
[18] K. Hirasawa and K. Fujimoto, Characteristics of Wire Antenna on a Rectangular Conducting Body, *IEICE Transactions on Communications (Japanese)*, vol. J65-B, April 1982, pp. 382–389.
[19] K. Fujimoto, A. Henderson, K. Hirasawa, and J. R. James, *Small Antennas*, Research Studies Press, UK, 1987.
[20] C. A. Balanis, *Antenna Theory: Analysis and Design*, 2nd edn., John Wiley and Sons, 1997.
[21] H. Uchida and Y. Mushiake, *VHF-Antenna*, Corona Publishing (in Japanese), 1961.
[22] R. C. Johnson, *Antenna Engineering Handbook*, 3rd edn., McGraw-Hill. Inc., 1961, Chapter 4.
[23] Zeland Software, Inc., *IE3D* ver. 7.0: www.zeland.com/IE3D.htm
[24] Zeland Software, Inc., *Fidelity* ver. 3.0: www.zeland.com/Fidelity.htm
[25] Ansoft, HFSS 8.0, The Maxwell Online Help System, 2001.
[26] Computer Simulation Technology AG, *CST Microwave Studio* ver. 2006: www.cst.com
[27] Schmid & Partner Engineering AG, www.speag.com/

# 11 Glossary

## 11.1 Catalog of small antennas

Small antennas here are treated in a wider sense than generally used (which concerns only ESA), because of its significance in the antenna and communication community. However, in consideration of the latest trends and requirements for small antennas, the variety of wireless systems including wireless broadband systems and short range radio systems, demands for a variety of antennas having physically constrained dimensions as well as electrically small dimensions and enhanced functions have become urgent. Thus, describing only ESA in the modern book is considered insufficient, whereas introduction of other types of antennas such as PCSA, FSA, and PSA should be preferred, as they have been widely employed in recently deployed wireless systems. The book describes principles of small sizing for antennas, design techniques, including miniaturization, and many antenna examples. The latest design technologies include application of metamaterials, EBG (Electromagnetic Band Gap), HIS (High Impedance Surface), DGS (Defect Ground Surface), and so forth.

This chapter contains a catalog of antennas listed in earlier book chapters, providing the original figure which can be referred by numbers used in the text and figures not appeared in the text with number with A as Fig. Ax. (x: a serial number in the catalog.). The main feature of the chapter is to provides readers with some useful information for designing small antennas and assistance in selecting suitable antennas for systems.

The list gives brief view, main features and some examples of applications. In addition, for the reader's convenience, references and locations, indicating the section in the text where readers can refer, are provided.

## 11.2 List of small antennas

| Brief view | Antenna type | Main features | Applications | References and location in the text |
|---|---|---|---|---|
| | **Short dipole** $\left(2a \le \frac{\lambda}{2\pi},\ d \ll \lambda\right.$ | Low radiation resistance, high capacitive reactance, narrowband<br>Low radiation efficiency (with high loss matching circuit) small, thin, lightweight, low cost | Small mobile terminals (Handset)<br>RFID | [1–4]<br>Chapter 3<br>Figure 3.1 |
| | **Monopole** (on the ground plane) $\left(a \le \frac{\lambda}{4\pi}\right.$ | Low radiation resistance, high capacitive reactance, narrowband<br>Low radiation efficiency (with high loss matching circuit) small, thin, lightweight, low cost | Small mobile terminals (Handset)<br>RFID | [5, 6]*<br>*not necessarily ESA |
| | **Inverted-L (ILA)** (on the ground plane) $L + h \le \frac{1}{4}\lambda$ | Low-profile, thin, lightweight, low cost<br>Low radiation resistance, high capacitive reactance, vertical polarization (with low $h$), horizontal polarization (possible with higher $h$) | HF band communications<br>vehicles, ships | [7, 8]<br>Figure 1.6(c) |
| | **Inverted-F (IFA)** (on the ground plane) $\left(L + h \le \frac{1}{4}\lambda,\ d \ll \lambda\right.$ | Low-profile, thin, lightweight, low cost<br>Easily matching to 50 Ω (with adjustment of $h$ and $d$) | VHF/UHF band communications vehicles<br>Small mobile terminals (Handset) | [7, 9]<br>Figure 1.6(d) |
| | **Small loop** (circular) $ka \ll 1$ | Low radiation resistance, high inductive reactance small, thin, lightweight, compact | Small portable equipment<br>NFC systems (including card type RFID) | [10–12]<br>Figure 3.3 |
| | (rectangular) | Small, thin, lightweight, compact | Small portable equipment<br>NFC systems (including card type RFID) | [13, 14]<br>Figure 3.5 |

| | | | | |
|---|---|---|---|---|
| 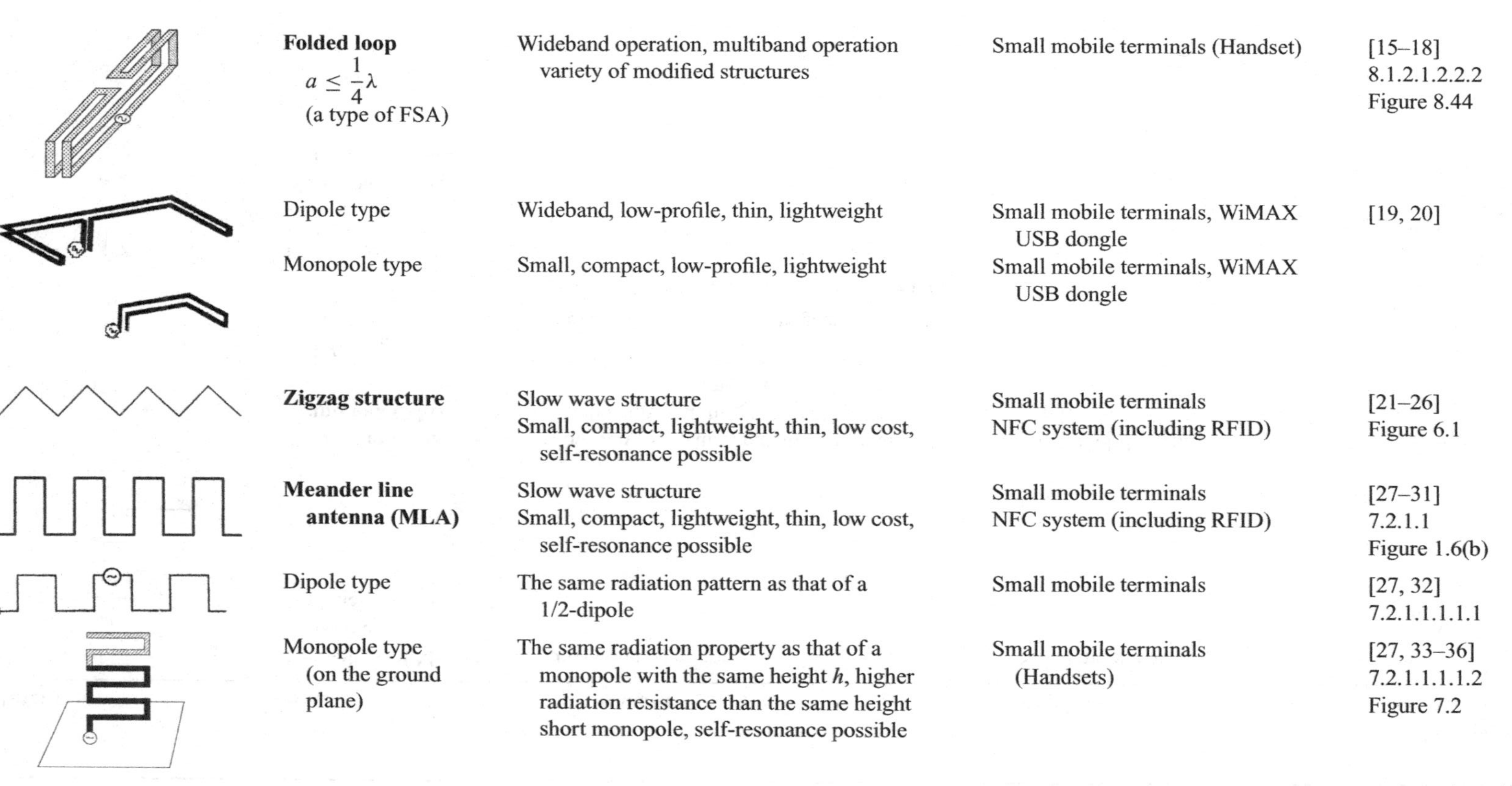 | **Folded loop** $a \leq \frac{1}{4}\lambda$ (a type of FSA) | Wideband operation, multiband operation variety of modified structures | Small mobile terminals (Handset) | [15–18] 8.1.2.1.2.2.2 Figure 8.44 |
| | Dipole type | Wideband, low-profile, thin, lightweight | Small mobile terminals, WiMAX USB dongle | [19, 20] |
| | Monopole type | Small, compact, low-profile, lightweight | Small mobile terminals, WiMAX USB dongle | |
| | **Zigzag structure** | Slow wave structure<br>Small, compact, lightweight, thin, low cost, self-resonance possible | Small mobile terminals<br>NFC system (including RFID) | [21–26]<br>Figure 6.1 |
| | **Meander line antenna (MLA)** | Slow wave structure<br>Small, compact, lightweight, thin, low cost, self-resonance possible | Small mobile terminals<br>NFC system (including RFID) | [27–31]<br>7.2.1.1<br>Figure 1.6(b) |
| | Dipole type | The same radiation pattern as that of a 1/2-dipole | Small mobile terminals | [27, 32]<br>7.2.1.1.1.1.1 |
| | Monopole type (on the ground plane) | The same radiation property as that of a monopole with the same height $h$, higher radiation resistance than the same height short monopole, self-resonance possible | Small mobile terminals (Handsets) | [27, 33–36]<br>7.2.1.1.1.1.2<br>Figure 7.2 |

(*cont.*)

| Brief view | Antenna type | Main features | Applications | References and location in the text |
|---|---|---|---|---|
| | Dual-element MLA (on the ground plane) | Dual-band operation | Small mobile terminals | [37] Figure 7.9 |
| L, p, 2a | **Normal mode helical antenna (NMHA)** $\left(\begin{array}{l}2a \ll \lambda \\ L = np < \frac{1}{4}\lambda\end{array}\right.$ | Slow-wave structure, radiation normal to the helix axis with the same pattern as a short dipole of the same axial length, Self-resonance possible with increase in $n$, even though $L \ll \frac{1}{4}\lambda$, Higher efficiency than that of a dipole with the same length | Small mobile terminals (Handsets) | [38–41] Figure 6.15 |
| | Dipole type | | | [42, 43] Figure 7.57 |
| | Monopole type (on the ground plane) | | | [44, 45] Figure 7.70 |

|  |  |  |  |  |
| --- | --- | --- | --- | --- |
| 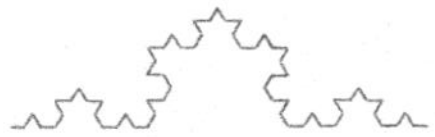  | Wired fractal structure | Slow-wave structure, the same radiation pattern as that of the same length short dipole, higher radiation resistance than the same length dipole, multiband operation, depending on the whole length of the wire | Wideband wireless systems, multiband wireless systems | [46] Figure 6.2 |
| 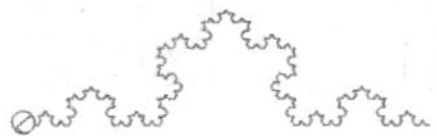  | Monopole type (on the ground plane) |  |  | Figure 7.88 |
|  | **Planar antenna** (planar metal in space or etched on a dielectric substrate) | Small, compact, lightweight, low cost<br>The space and the size of the planar patch mainly determine the lower bound of the impedance bandwidth<br>Wider bandwidth than wire radiator | Portable devices, including RFID tags, mobile phones, sensors, laptops, USB dongles, etc. | [47, 51] Chapter 7 |
| 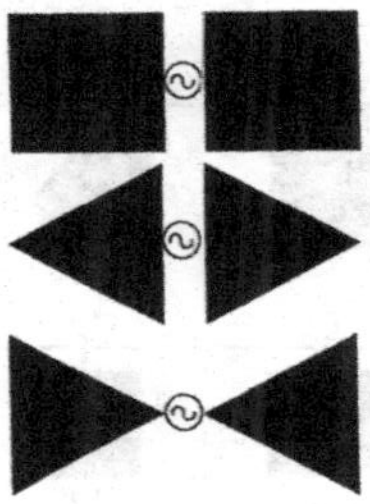  | Dipole type (square, triangular, rectangular, circular, elliptical, trapezoidal, etc. and their variations) |  |  |  |
| 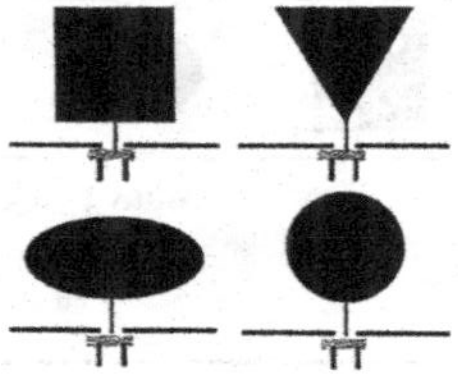  | Monopole type (planar metal in space, or etched on a dielectric substrate) (on a ground plane) |  |  |  |

(*cont.*)

| Brief view | Antenna type | Main features | Applications | References and location in the text |
|---|---|---|---|---|
| | With slotted and/or notched planar radiator of various shapes, a stub attached to the planar radiator | Lowering resonance frequency by slot, lowering the height by stub and notch | | |
| | With stub/loop for impedance matching | Feed gap, feed structure and the bottom shape of the radiator determine impedance matching | | |
| | Planar meander line dipole | Meandered wire dipole with triangular, rectangular, etc, shapes. Size reduction compared to a dipole with the same resonance frequency | RFID tags, mobile phone wireless systems (WLAN, WPAN, etc.) | [52–57]<br>7.2.2<br>Figure 7.85 |
| | Ultra small planar meander line dipole<br>Folded ultra small planar meander line antenna | Miniaturized size (possibly occupying area of $0.05\lambda \times 0.05\lambda$ at UHF bands, with gain −8 dBi, efficiency −9 dB), Folded structure increases radiation resistance and gain, use of dielectric substrate ($\varepsilon_r = 10$) further improves efficiency (up to −2.5 dB) and gain (up to −3.6 dBi) | RFID tag (card type), small wireless equipment | [58, 59]<br>7.2.1.1.1.1.5<br>Figure 7.30 |

| | | | | |
|---|---|---|---|---|
| | Planar meander line monopole embedded on a thin dielectric substrate | Planar densely wound meander line monopole antenna, another meander line monopole element wound in different direction on the rear side for dual polarization, wideband operation covering 400–700 MHz bands | Digital TV Broadcasting, Wideband receiver | [60]<br>7.2.1.1.1.1.2<br>Figure 7.6 |
| | Meandered wire dipole with meandered wire element | Reduced length by using meandered structure on the meandered wire dipole elements | RFID tags, laptop, USB dongle, etc. | [61]<br>7.2.2.1.1 |
| | **Antenna with space-Filling geometries**<br>Peano curve<br>Hilbert curve | Increase in the antenna length within a small area lowers resonance frequency, the antenna length increases as the order of iteration increases, increasing space-filling density, and lowering the resonance frequency, radiation properties are essentially identical, independent of the geometry, related to the occupying area | Design of small antenna and frequency selective surface (FSS)<br>FSS can be applied to thin absorbers and metamaterial bulk media | [62–66]<br>7.2.2<br>Figure 7.81 |
| | **With Fractal Geometry** | | | [55]<br>7.2.2.1.3.2 |
| 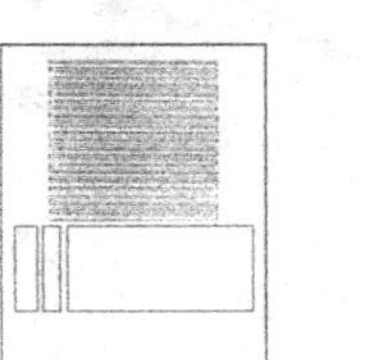 | Koch type | Increase in the wire length decreases radiation resistance and efficiency, as well as the resonance frequency, and increases $Q$ | | 7.103 |

(*cont.*)

| Brief view | Antenna type | Main features | Applications | References and location in the text |
|---|---|---|---|---|
| 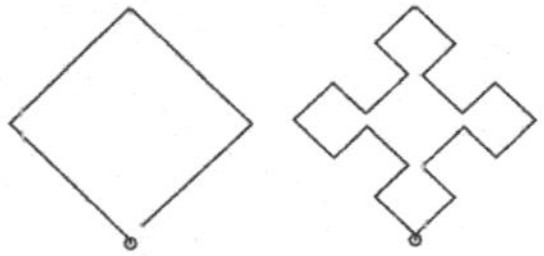 | Minkowsky type | | | Figure 7.103 |
|  | Sierpinsky type | Multiband operation, number of multiple bands are adjusted by control of scale factor to generate Sierpinsky geometry | Multiband systems | [67, 68]<br>7.2.2<br>7.2.2.1.3.3<br>Figure 7.106 |
| | **Monopole type** with: | | | |
| 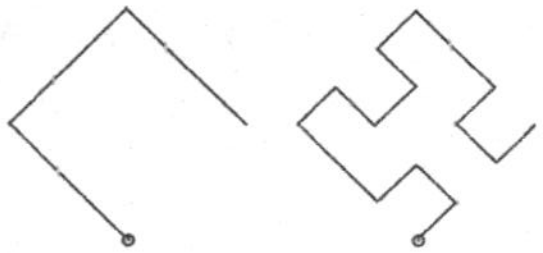 | Hilbert curve | | | Figure 7.103 |
| 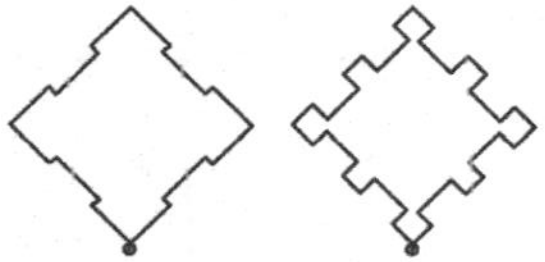 | Meander line, Fractal geometries, etc. | | | [69, 70]<br>7.2.2.1.3.2<br>Figure 7.103 |
|  | PIFA with planar Peano elements | Dual-band operation by using two different Peano wire elements as horizontal element in PIFA structure | Mobile phone, dual band | [71]<br>7.2.2.1.1.1<br>Figure 7.97 |

| | Spiral antenna | | | |
|---|---|---|---|---|
|  | Equiangular spiral<br>$r = r_0 e^{a\varphi}$<br>$\begin{pmatrix} r_1 = r_0 e^{a\varphi} \\ r_2 = r_0 e^{a(\varphi-\delta)} \\ r_3 = r_0 e^{a(\varphi-\pi)} \\ r_4 = r_0 e^{a(\varphi-\pi-\delta)} \end{pmatrix}$<br>$r$ : radius of spiral<br>$r_0$ : initial radius ($\varphi = 0$)<br>$\varphi$ : angle of rotation<br>$a$ : rate of growth<br>$\delta$ : initial angle of $r_2$ | Wideband operation in terms of impedance as well as radiation properties, bi-directional radiation normal to the surface of spiral with circular polarization | Satellite communications, GPS, RFID tags, etc. | [72–75]<br>7.2.2.1.4.1<br>Figure 7.112 |
| | Archimedean spiral<br>$r = r_0 + a\varphi$ | Dense winding of spiral arm in smaller space than equiangular type, more useful in practical applications. Loosely wound spiral arm brings frequency-dependent behavior | | [76]<br>7.2.2.1.4.2<br>Figure 7.113 |
| spiral<br>dielectric | Miniaturized spiral<br>• with inductive loading<br>a. meander line<br>b. zigzag line<br>• with capacitive loading<br>a. dielectric substrate<br>b. dielectric superstrate | Reduced size with integration of inductive/capacitive elements or lumped component on the spiral arms. Reduction of the spiral structure with either dielectric superstrate, substrate or both | | [77]<br>7.2.2.1.4.6<br>[78]<br>7.2.2.1.4.6 |
| | Cavity-backed spiral | Small and thin antenna structure, unidirectional radiation | Vehicle GPS terminals | [79]<br>7.2.2.1.4.5 |

(*cont.*)

| Brief view | Antenna type | Main features | Applications | References and location in the text |
|---|---|---|---|---|
| | Square spiral | Simpler construction than round structure | | [80]<br>7.2.2.1.4<br>Figure 7.133 |
| | **Complementary Structure** | | | |
| | Monopole-slot | Small, compact, planar complementary, monopole-slot antenna, wideband, finite ground plane | RFID tags, USB dongle, etc. | [81–83]<br>6.2.4.1<br>7.2.4.1 |
| | L-shape monopole and dual slot | Small, compact, planar complementary, L-shaped monopole-dual slot antenna, wide bandwidth, finite ground plane | | [81–83]<br>7.2.4.1.1 |
| | T-shaped monopole and dual slot (H-shaped) | Small, compact, planar complementary, T-shaped monopole and dual slot | | [84]<br>7.2.4.1.2 |

| | | | | |
|---|---|---|---|---|
| | Half-circle monopole and dual slot | Electrically small (6×25 mm), compact, complementary, half-circle dual slot antenna, wide bandwidth (2.8∼10.7 GHz for −10 dBi $S_{11}$), gain 2∼5 dB over the entire bandwidth | UWB systems | [85]<br>7.2.4.1.3<br>Figure 7.186 |
|  | Self-complementary spiral | Frequency independent property with infinite structure. Finite bandwidth with truncation of the platform, but still useful bandwidth for practical applications | | [85]<br>6.2.4.1<br>Figure 6.35 |
|  | Self-complementary with a compound space form | | | [85]<br>6.2.4.1 |
| 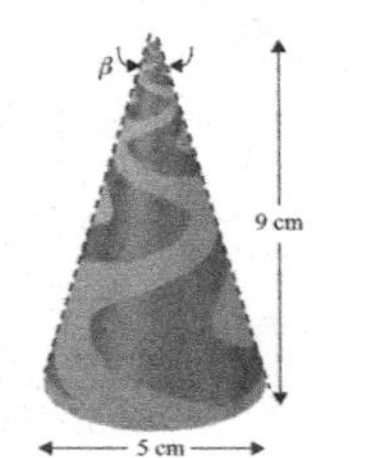 | 3D two-arm conical sinuous antenna | Small dimensions (9 cm height and 5 cm bottom diameter), wide bandwidth (9:1 for VSWR $< 5$), covering five frequency bands (3 to 17 GHz), gain Δ∼5 dBic | UWB systems | [86]<br>7.2.4.1.4<br>Figure 7.190 |

*(cont.)*

| Brief view | Antenna type | Main features | Applications | References and location in the text |
|---|---|---|---|---|
| | **Top-loaded Antenna** | | | |
| | Circular disk | Electrically small size (height), uniform current distribution on the monopole element | | [87, 88]<br>7.2.3.1.1<br>Figure 7.168 |
| | Wire-grid disk dipole, monopole | Self-resonant, resonance frequency lowered by increasing disk diameter, increasing capacitance, a shunt stack point used for matching | | [74]<br>7.2.3.1.1<br>Figure 7.169 |
| | Spiral disk | Self-resonance, resonance frequency lowered by varying spiral wire length, increasing both capacitance and inductance | | [75]<br>7.2.3.1.1<br>Figure 7.169 |
| | Disk-loaded helix | Self-resonance, resonance frequency lowered by both top-loading and helical winding, increasing capacitance and inductance | | [74]<br>7.2.3.1.1<br>Figure 7.169 |

| | | | |
|---|---|---|---|
| 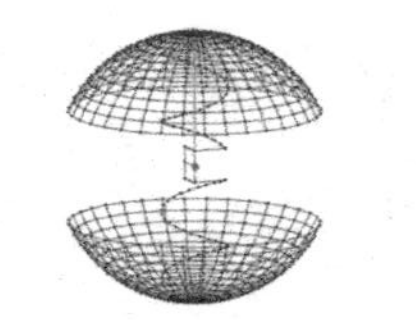 | Spherical-cap | Self-resonance, resonance with spherical top-loading and helical winding, $Q = 1.5Q_{chu}$ achievable | [91] 7.2.3.1.2 Figure 7.169 |
| | Crossbar loaded monopole | Small size (height $0.13\lambda_0$: $\lambda_0 = 3$ m), wide bandwidth (110~205 MHz and 420~445 MHz for −10 dB return loss), adjustable with the radius and length of the sleeve, almost omnidirectional pattern | [91] 7.2.3.1.2 Figure 7.176 |
| | Single slot line loaded with spiral | Miniaturized slot antenna, occupying area of $0.13\lambda_0 \times 0.15\lambda_0$ at 800 MHz band, narrow bandwidth (0.9%), gain 0.8 dB | [92] 7.2.3.1.3 Figure 7.179 |
| | Double slot line loaded with spiral | Nearly same occupying area as the single-element antenna ($0.165\lambda_0 \times 0.157\lambda_0$), improved bandwidth (2.4%) and gain ($1.5 \sim 1.7$ dB) at 800 MHz band | [92] 7.2.3.1.3 Figure 7.179 |
|  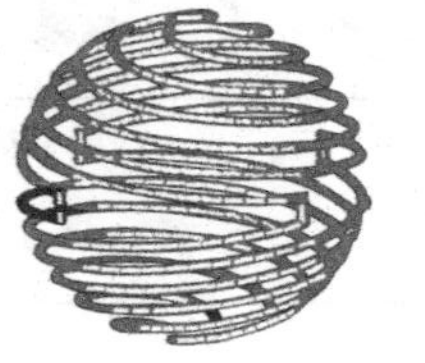 | Multi-arm spherical helix antenna | Electrically small size, four-folded arm, self-resonant with radius 4.2 cm, $ka = 0.266$ at 300 MHz, efficiency 97%, $Q = 84$ ($\approx 1.5Q_{chu}$) | [93] 7.2.2.2.3.1, 7.2.2.2.3.2 Figure 7.155 |

*(cont.)*

| Brief view | Antenna type | Main features | Applications | References and location in the text |
|---|---|---|---|---|
| | Electrically small spherical wire dipole antenna | Almost isotropic radiation pattern, diameter about 20 cm, resonance at 327 MHz, $Q$ about 12, close to the lower bound $Q_{chu}$, efficiency 93% | | [94] 7.2.2.2.3.3 Figure 7.152 |
| | Spherical split ring resonator (SSRR) antenna | Electrically small size (radius 21cm, resonance at 300 MHz), optimized with $Q \approx 3.5Q_{chu}$ | | [95] 7.2.2.2.3.4 Figure 7.160 |
| | Electrically small complementary pairs of thick monopole antennas | Electrically small size (height $\lambda/18$, (height/diameter) = 1), 3:1 VSWR bandwidth, gain 5.75 dB, minimum efficiency 25% | | [96] 7.2.4.2.1 Figure 7.194 |
| | Combined electric and magnetic type planar antenna | Small size (25×20 cm), unidirectional radiation over wide bandwidth, Combined electrical radiator (two dipole arms) with magnetic radiator (a loop cut on the ground plane) | | [97] 7.2.4.2.2 Figure 7.196 |

| | | | |
|---|---|---|---|
| | Dual-band and F-shaped antenna with inductive slot | Dual-band operation at 2.4 and 5.4 GHz, gain 1.5 dBi at 2.45 GHz and 4.9 dBi at 5.2 GHz, bandwidth 7.8% at 2.5 GHz and 11.4% at 5.37 GHz, a figure-eight pattern in the *E* plane and a nearly omnidirectional pattern in the *H* plane | [98]<br>7.2.4.3.1 |
| | **Miniaturized Composite Antenna** | | |
| | Modified, cavity-backed loop, sectionalized into six and looped with spiral and corrugated stub | Small dimensions ($<\lambda_0/10$), low-profile, similar radiation as a short monopole, constituted of sectionalized slot loop embedded on a metallic cavity, loaded with a spiral-like slot-line on the edge | [99]<br>7.2.4.3.2<br>Figure 7.204 |
| | **Metamaterial Integrated Antenna** | | |
| | Mu-negative material (MNG) loaded antenna | Miniaturized circular patch antenna designed with resonance mode depending on the MNG material parameters | [100]<br>7.2.5.1.1<br>[100] |
| 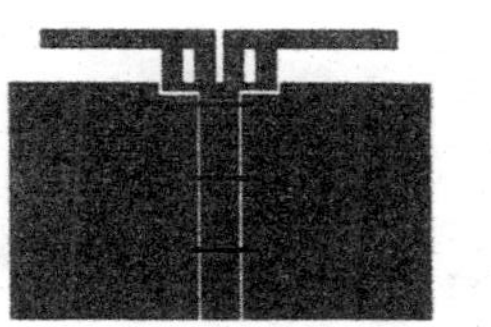 | • Circular patch antenna | | 7.2.5.1.1<br>[101]<br>7.2.5.1.1<br>Figure 7.210 |

(*cont.*)

| Brief view | Antenna type | Main features | Applications | References and location in the text |
|---|---|---|---|---|
| | • MNG material constituted of single ring array | | | Figure 7.212 |
| | • Elliptical patch antenna loaded with MNG material | | | Figure 7.214 |
| | Epsilon-negative material (ENG) loaded antenna Monopole loaded with a half-sphere ENG material | Miniaturized monopole antenna; example: ENG half-sphere radius of $\lambda/18$ and the monopole length $\lambda/50$ at $\lambda = 14.8$ cm, $Q = 42$ (about $1.5Q_{chu}$) | | [102] 7.2.5.1.2 Figure 7.218 |
| | Loop antenna loaded with an equivalent MNG structure placed on a finite ground plane | Miniaturized loop antenna with $ka = 0.11$, resonance at 300 MHz, narrow bandwidth | | [103] 7.2.5.1.1.3 Figure 7.217 |
| | Rectangular loop antenna loaded with an equivalent MNG structure with inter-digital capacitor | MNG structure is synthesized by either 3D extended capacitor-loaded loop (CLL) or 2D planar CLL with inter-digital capacitor or a lumped capacitor | | [103] 7.2.5.1.1 Figure 7.217 |

| | | | |
|---|---|---|---|
| 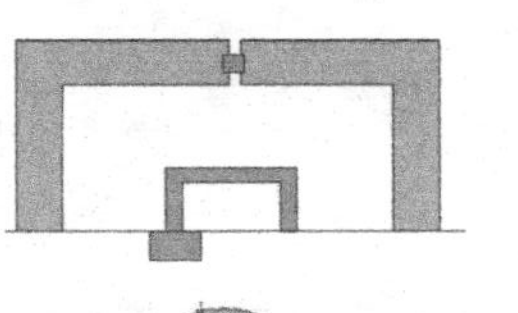 | Rectangular loop antenna loaded with equivalent MNG structure | Matching (resonance) in space for small antenna, using MNG structure implemented by capacitor-loaded loop (inter-digital capacitor can be used) | [103] 7.2.5.1.1 Figure 7.217 |
| | Monopole antenna loaded with equivalent ENG structure of a helical strip | Matching (resonance) in space for small antenna, using ENG structure implemented by cylindrical helix strip, greater bandwidth than a small monopole with the same length | [103] 7.2.5.1.2 Figure 7.226 |
| | **Metamaterial Integrated Antenna II** | | |
| 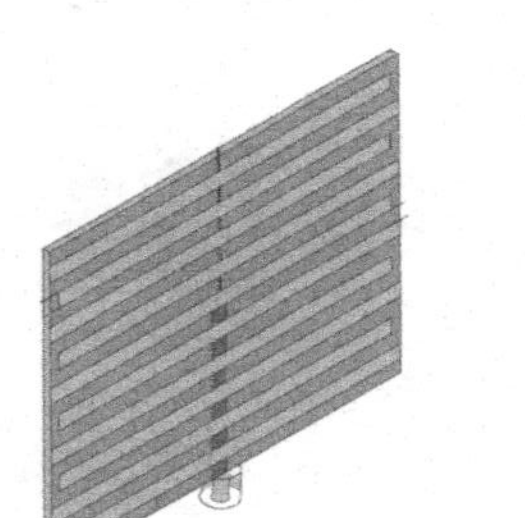 | Monopole antenna loaded with equivalent ENG structure incorporating meander line surface placed on a finite ground plane | Matching (resonance) in space for small antenna, using ENG structure implemented by meander line surface having large inductance | [103] 7.2.5.1.2 Figure 7.225 |

(*cont.*)

| Brief view | Antenna type | Main features | Applications | References and location in the text |
|---|---|---|---|---|
| | Backfire-to-endfire fan beam CRLH Leaky Wave antenna | Antenna constituted of a 1D CRLH metamaterial structure incorporating series-resonant tank with inter-digital capacitor and shunt anti-resonant tank with stub inductor | | [104] 7.2.5.2.1 Figure 7.227 |
| | Conical beam antenna using 2D CRLH LW antenna | Opening angle of conical beam varies with frequency, following the dispersion relation ($\beta/\omega$), leading to beam scanning | | [104] 7.2.5.2.1 Figure 7.227 |
| | 3D CRLH (rotated TL-matrix based) material | | | [104] 7.2.5.2.1 Figure 7.227 |
| | CRLH ZOR mode microstrip resonator antennas (three models for different frequencies) | With ZOR property, operating frequency not depending on the size, but only on the lumped LC values, leading to create ESA. Increasing size increases gain with fixed frequency, high efficiency and high directivity | | [104] 7.2.5.2.2 Figure 7.230 |
| | A CRLH loop resonant microstrip antenna | A rectilinear CRLH structure with compactly spaced stubs in a radial manner, shorted at the center of the structure, Monopole-like radiation with ZOR mode | | [104] Figure 7.231 |

| | | | |
|---|---|---|---|
| | **Metamaterial Integrated Antenna III** | | |
| 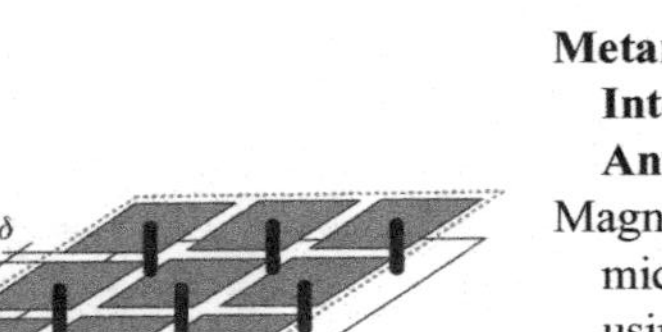 | Magnetic monopole microstrip antenna using CRLH ZOR mode in a mushroom structure | Monopole behavior due to magnetic current loop created around the mushroom patch | [104] Figure 7.232 |
| | Metamaterial applied antenna | Compact ($\lambda_0/11$) and low profile ($\lambda_0/28$), consisting of two 1D zero-degree phase shifting line metamaterials arranged in a ring structure for 1.7 GHz | [104] Figure 7.233 |
| | NRI TL metamaterial antenna composed with two unit cells | Four microstrip ZOR mode NRI TL unit cells with dimensions $\lambda_0/10 \times \lambda_0/10 \times \lambda_0/20$ over a $0.45\lambda_0$ ground plane, efficiency over 50%~70% at 3.1 GHz | [105] Figure 7.235 |
| | Dual monopole loaded with NRI TL metamaterial | Small size (22 mm×30 mm), broadband (−10 dB BW from 3.14 GHz to 7.2 GHz), printed monopole loaded with NRI TL unit cell, lowering the operating frequency and extending bandwidth | [106] Figure 7.236 |

*(cont.)*

| Brief view | Antenna type | Main features | Applications | References and location in the text |
|---|---|---|---|---|
| | Two-arm TL metamaterial antenna loaded with spiral inductors | Wideband (100 MHz for −10 dB BW), compact size ($\lambda_0/4 \times \lambda_0/7 \times \lambda_0/29$) at 3 GHz band, consisting of two TL metamaterial arms, each can operate at different frequency | | [107] Figure 7.223 |
| | Tri-band monopole antenna consisting of cascaded RH/LH TL metamaterial | Compact size (20×23.5×1.59 mm), multimode (dipole mode at 2.4 GHz, monopole mode at 5 GHz band and the 3rd mode formed by an L-shaped slot on the ground plane covering 3 GHz band), efficiency 67% at 2.45 GHz, 86% at 3.5 GHz and 85% at 5.5 GHz | Wireless systems (Wi-Fi, WiMAX, etc) | [108] Figure 7.239 |
| | Microstrip antenna partially filled with CRLH cells | Miniturized by using mushroom HIS structure of LH operation<br>Compact size ($\lambda_0/11$), simple structure and insensitive against ground plane size | | [109] [124] Figure 7.242 |

| | | | | |
|---|---|---|---|---|
| 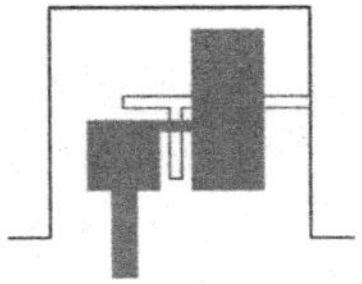 | Microstrip antenna loaded with CRLH TL metamaterial structure | Composed with cascaded CRLH TL structure in zero phase, bandwidth ($-10$ dB $S_{11}$) 4.5% at 2 GHz band | | [110]<br>[125]<br>Figure 7.243 |
| | **Planar patch antenna** | | | |
|  | E-shape patch (two open-end parallel slots on the patch) | Wideband operation, optimized by adjusting slot length, width, and position. About 30% bandwidth at 1.9–2.4 GHz. Simple structure, small size | Wireless communication systems | [111–113]<br>8.1.2.1.1<br>Figure 8.1 |
| 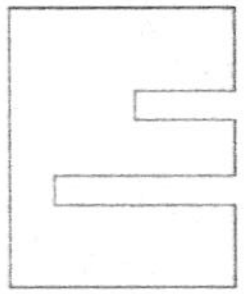 | E-shape patch with unequal arm slots | Modified from E-shape patch to generate circular polarization with relatively wideband axial ratio | GPS, RFID, WLAN, etc. | [114]<br>8.1.2.1.1.1.1<br>Figure 8.5 |

(*cont.*)

| Brief view | Antenna type | Main features | Applications | References and location in the text |
|---|---|---|---|---|
| 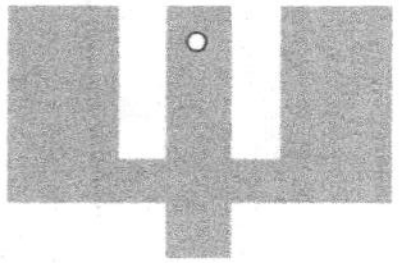 | $\psi$-shape patch | Modified from E-shape patch to form $\psi$-shape. Wider bandwidth than E-shape. Over 50% bandwidth | Wideband wireless communication systems | [115, 116] 8.1.2.1.1.1.2 Figure 8.9 |
| 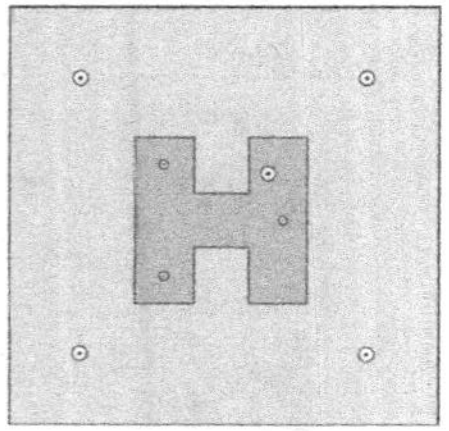 | H-shape patch | Circular polarization with wide bandwidth in both axial ratio and impedance | | [117–120] 8.1.2.1.1.1.3 Figure 8.11 |
| 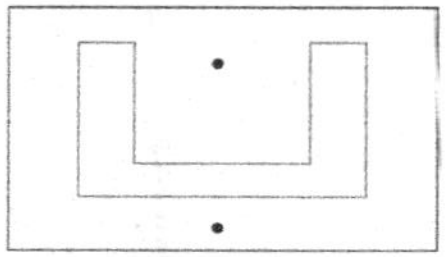 | U-slot patch | Wideband as well as multiband operation. Impedance bandwidth in excess of 30% | | [121–125] 8.1.2.1.1.1.3 Figure 8.29 |
| 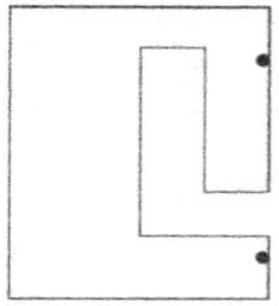 | Half-U-slot patch (L-shape open-end slot) on the patch | Reduced size U-slot patch, yet similar impedance bandwidth as the full-U-slot patch | | [126] 8.1.2.1.1.2.2 Figure 8.29 |

| | | | |
|---|---|---|---|
| 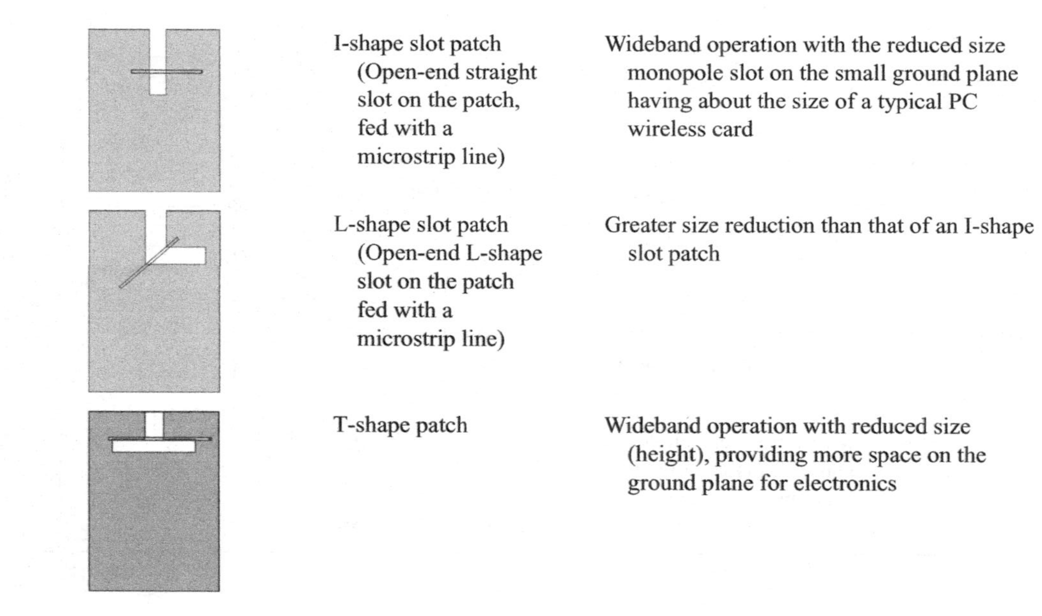 | I-shape slot patch (Open-end straight slot on the patch, fed with a microstrip line) | Wideband operation with the reduced size monopole slot on the small ground plane having about the size of a typical PC wireless card | [127] 8.1.2.1.1.2.1 Figure 8.16 |
| | L-shape slot patch (Open-end L-shape slot on the patch fed with a microstrip line) | Greater size reduction than that of an I-shape slot patch | [127] 8.1.2.1.1.2.1 Figure 8.17 |
| | T-shape patch | Wideband operation with reduced size (height), providing more space on the ground plane for electronics | [127] 8.1.2.1.1.2.1 Figure 8.18 |

(*cont.*)

| Brief view | Antenna type | Main features | Applications | References and location in the text |
|---|---|---|---|---|
| 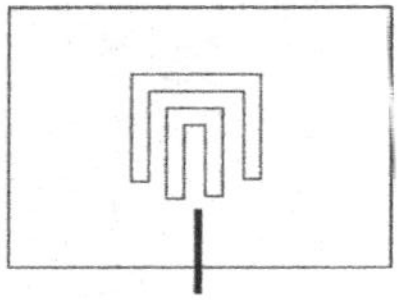 | Double-U-slot patch | Triple-band operation. Interaction of two slots excites center frequency; the shorter and longer slots excite higher and lower frequencies, respectively | Multiband systems | [128] 8.1.2.1.1.2.2 Figure 8.31 |
| 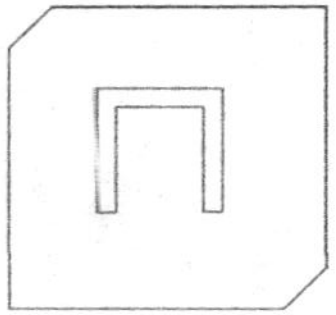 | U-slot patch with truncated corner | Wider bandwidths for both impedance and axial ratio than for patch without truncation | | [128] 8.1.2.1.1.2.2 Figure 8.31 |
| 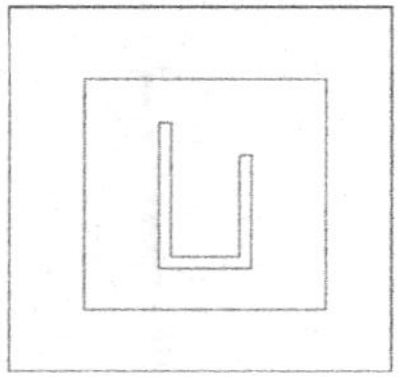 | Unequal-arm U-slot patch | Circular polarization with bandwidth performance similar to a U-slot patch with truncation | | [129] 8.1.2.1.1.2.2 Figure 8.31 |

| | | | | |
|---|---|---|---|---|
| | Square-slot patch (wide slot on the patch) | Wide bandwidth, 1:1.5 VSWR bandwidth of 1 GHz at 2 GHz, 0.5 GHz bandwidth for 2 dB gain | Wideband wireless systems | [130] 8.1.2.1.1.2.2 Figure 8.32 |
| | Rotated square-slot patch | Bandwidth enhanced compared with un-rotated square patch; 10 dB return loss bandwidth 2.2 GHz at 4.5 GHz | Wideband wireless systems | [131] 8.1.2.1.1.2.2 Figure 8.34 |
| | Octagonal wide slot with a square ring resonator | Ultra-wide-band operation with band-notched frequency, high, sharp band-rejection over 20 dB | UWB systems | [132] 8.1.2.1.2.3.4 |
| | $\lambda/8$ PIFA | Penta-band operation with two folded radiating slots generating two $\lambda/8$ modes to cover two lower bands, and two $\lambda/4$ modes to cover three higher bands | Mobile phone, GSM, UMTS, and WLAN systems | [133] 8.1.2.1.2.1.1 Figure 8.35 |

(*cont.*)

| Brief view | Antenna type | Main features | Applications | References and location in the text |
|---|---|---|---|---|
| | Bent-monopole antenna (BMA) | Penta-band operation with bent-monopole fed by coaxial cable with a thin ground line. Total monopole length determines the lower bands, and the ground-line length determines the higher bands | Mobile phones, CDMA, GSM, DCS, PCS and WCDMA | [133] 8.1.2.1.2.1.1 Figure 8.37 |
| | A printed loop with U-slot for wideband tuning | One-wavelength loop with a middle line provides four-frequency resonance. Bandwidth of two resonant modes increased by U-slot tuning element to cover two other frequency bands, resulting in hepta-band operation | Laptop computer, GSM/DCS/PCS/UMTS/WLAN | [134] 8.1.2.1.2.1.2 Figure 8.39 |
| | A printed strip monopole with closely coupled shorted strip | Eight-band operation covering three wide lower bands (698–960 MHz) and five higher bands (1719–2690 MHz) | LTE/GSM/UMTS Mobile phone | [135] 8.1.2.1.2.1.3 Figure 8.41 |
| | Basic planar square monopole antenna | Wide impedance bandwidth, with the square length about $0.21\lambda$ of the lower edge frequency | Wideband wireless systems and UWB systems | [136] 8.1.2.1.2.1.4 Figure 8.43 |

| | | | | |
|---|---|---|---|---|
| 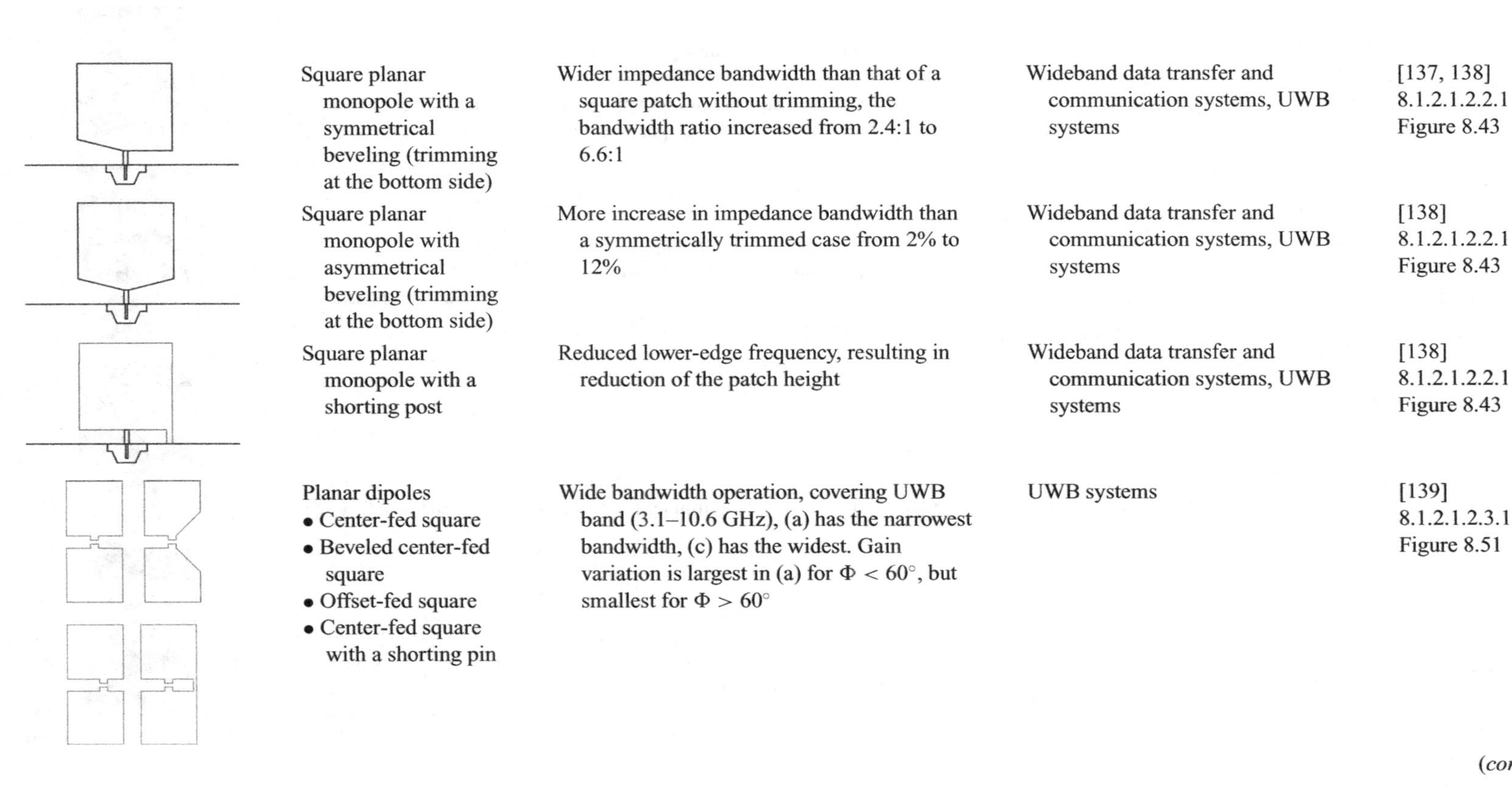 | Square planar monopole with a symmetrical beveling (trimming at the bottom side) | Wider impedance bandwidth than that of a square patch without trimming, the bandwidth ratio increased from 2.4:1 to 6.6:1 | Wideband data transfer and communication systems, UWB systems | [137, 138]<br>8.1.2.1.2.2.1<br>Figure 8.43 |
| | Square planar monopole with asymmetrical beveling (trimming at the bottom side) | More increase in impedance bandwidth than a symmetrically trimmed case from 2% to 12% | Wideband data transfer and communication systems, UWB systems | [138]<br>8.1.2.1.2.2.1<br>Figure 8.43 |
| | Square planar monopole with a shorting post | Reduced lower-edge frequency, resulting in reduction of the patch height | Wideband data transfer and communication systems, UWB systems | [138]<br>8.1.2.1.2.2.1<br>Figure 8.43 |
| | Planar dipoles<br>• Center-fed square<br>• Beveled center-fed square<br>• Offset-fed square<br>• Center-fed square with a shorting pin | Wide bandwidth operation, covering UWB band (3.1–10.6 GHz), (a) has the narrowest bandwidth, (c) has the widest. Gain variation is largest in (a) for $\Phi < 60^\circ$, but smallest for $\Phi > 60^\circ$ | UWB systems | [139]<br>8.1.2.1.2.3.1<br>Figure 8.51 |

*(cont.)*

| Brief view | Antenna type | Main features | Applications | References and location in the text |
|---|---|---|---|---|
| | Planar flare dipole with spline-defined outer curve | Wideband operation with optimized shape, delivering 5:1 (190–1000 MHz) bandwidth, gain greater than 0 dBi, small size (reduction from $\lambda/2$ dipole in excess of 30%) | UWB systems | [140, 141] 8.1.2.1.2.3.2.2 Figure 8.55 |
| | CPW-fed square monopole with symmetrical beveling | Wide bandwidth operation, covering UWB band | UWB systems | [142] 8.1.2.1.2.3 Figure 8.55 |
| | Planar binomial curved monopole | UWB operation. Impedance bandwidth depends on the order of the binomial function and the gap between the monopole and the ground plane | UWB systems | [143, 144] 8.1.2.1.2.3.2.1 Figure 8.55 |
| | Staircase-profile printed monopole | Impedance bandwidth and angle range for stable radiation can be designed by selecting number of stairs. Narrow angle range can be applied to specific UWB links | UWB systems | [145] 8.1.2.1.2.3 Figure 8.55 |

| | | | | |
|---|---|---|---|---|
|  | A printed beveled planar monopole (and a half of the monopole) | UWB operation with reduced size. With antenna structure half of the monopole (40% reduction in size), bandwidth is 8.25 GHz (2.85–11.1 GHz) and gain from −5.6 to 2.3 dBi | UWB systems | [146]<br>8.1.2.1.2.3 |
| 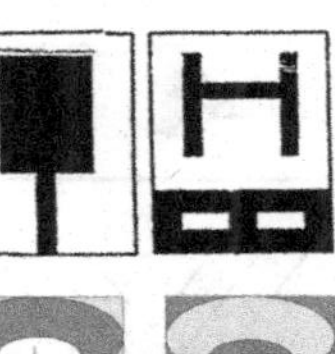 | Planar monopole with variable frequency band-notch function | UWB operation with variable band-stop frequency by varying parameters of H-shaped conductor back plane | UWB systems, high data-rate transfer and communication systems | [147]<br>8.1.2.1.2.3.3.2 |
| 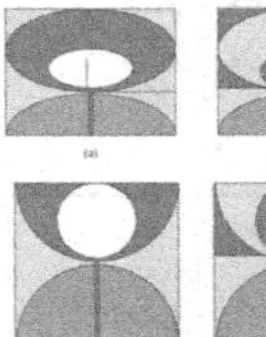 | Elliptical-shape planar monopole (complementary monopole) | UWB operation | The same as above | [148, 149]<br>8.1.2.1.2.3.3.2<br>Figure 8.67 |
| 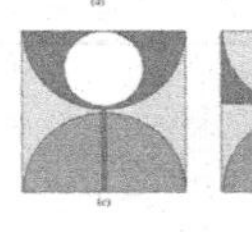 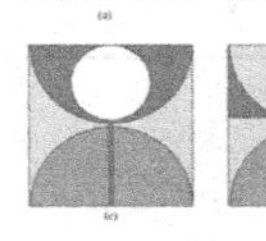 | Circular-shape planar monopole (complementary monopole) | UWB operation | | [148, 149]<br>8.1.2.1.2.3.3.2<br>Figure 8.67 |
| 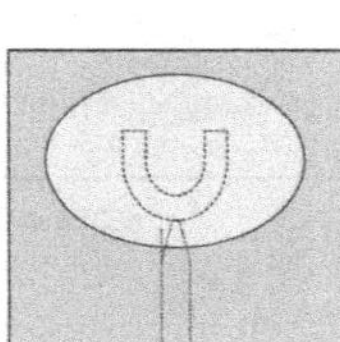 | Elliptical or Circular slot with U-shape tuning stub | UWB operation. Lower edge frequency depends on U-shape tuning stub and shape factor | The same as above | [150]<br>8.1.2.1.2.3.3.2<br>Figure 8.68 |

(*cont.*)

| Brief view | Antenna type | Main features | Applications | References and location in the text |
|---|---|---|---|---|
| | Beveled square planar monopole with a quasi-square slot ring on the planar surface | UWB operation with a band-notch at WLAN band | The same as above | [151] 8.1.2.1.2.3.3.1 Figure 8.63 |
| | V-shape planar patch with unequal arms fed through a triangle-PIFA coupled with V-shape patch | Wideband and multiband operation, determined by the feed position. Operating frequency depends on the length of the slot | UWB systems and multiband wireless systems | [152] 8.1.2.1.2.3.4 |
| | Sectorial loop composed of an arch and two sectors | Wideband operation with 8.5:1 frequency range for VSWR $\leq 2.2$. The size at the lowest frequency is $0.39\lambda_0$ | UWB systems | [153] 8.1.2.1.2.3.4.2 Figure 8.73 |
| | Printed rectangular monopole with a notch and a strip at the upper side corner | UWB operation with reduced ground plane by cutting a notch from the radiator and attaching a strip to the radiator; 3D omnidirectional radiation | UWB systems | [154] 8.1.2.1.2.3.3.3 Figure 8.70 |

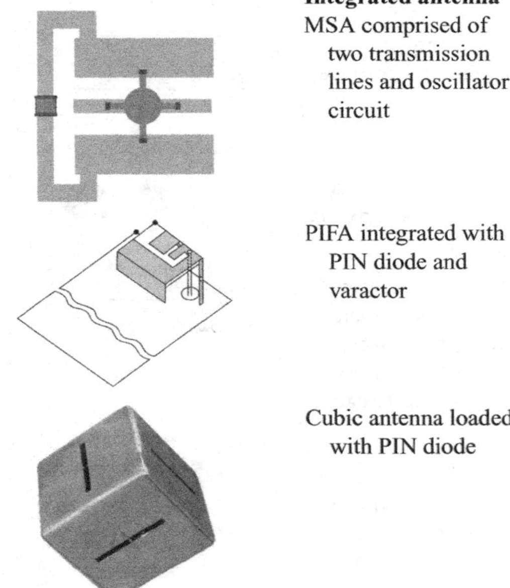

| | | | | |
|---|---|---|---|---|
| | **Integrated antenna** | | | |
| | MSA comprised of two transmission lines and oscillator circuit | MSA loaded with HF FET enhanced radiation power, gain 10 dBi, EIRP 11.2 dBm, at 8.5 GHz | Active integrated antenna | [132] 8.2.1.2 Figure 8.74 |
| | PIFA integrated with PIN diode and varactor | Frequency reconfigurable, multiband operation, four-band switching by PIN diode | Wireless communication and mobile phone systems (PCS, WCDMA, WiMAX, WLAN, etc.) | [133] 8.2.1.2 Figure 8.76 |
| | Cubic antenna loaded with PIN diode | Pattern reconfigurable, pattern is varied by on–off operation of slots on the cube surfaces by PIN diode switching | Pattern-control antenna diversity system | [134] 8.2.1.2 Figure 8.77 |
| | **Antenna on HIS (High Impedance Surface)** | | | |
| | Mushroom type surface | Periodically arrayed mushroom-like elements to compose HIS surface | Low-profile antenna, surface-wave suppression, reduction of backward radiation, mutual-coupling reduction | [135] 8.3.2 Figure 8.79 |

(*cont.*)

| Brief view | Antenna type | Main features | Applications | References and location in the text |
|---|---|---|---|---|
| | Corrugated surface | Periodically corrugated surface, wide impedance bandwidth | The same as above, low-profile UWB antenna | [136] 8.3.2 Figure 8.81 |
| | Combination of Jerusalem cross and fractal patch | HIS surface to allow low-profile antenna with enhanced gain | Low-profile antenna gain enhancement | [137] 8.3.2 Figure 8.82 |
| | **Antenna on Electromagnetic Bandgap (EBG)** | | | |
| | A dipole on EBG surface | Low-profile and gain increase | Low-profile antenna gain enhancement | [138] 8.3.3.2 Figure 8.84 |
| | Sleeve dipole on EBG surface | Low-profile and bandwidth increase | Low-profile antenna wide return-loss bandwidth | [139] 8.3.3.3 Figure 8.87 |
| | Two MSAs separated by EBG belt | Reduction of mutual coupling, allowing close spacing between two antennas | Low-profile antenna, mounting a few antennas in a small limited area, antenna array in a narrow space, antenna in MIMO system | [140] 8.3.3.4 Figure 8.88 |

| | | | | |
|---|---|---|---|---|
| | **Antenna on a Defected Ground Surface (DGS)** | Frequency band-stop performance and slow-wave property, use of single unit, limited number of units, or a periodic repetition of units available | Antenna, filters, power amplifier, oscillator, etc. | [141] 8.3.4 |
| | **DGS unit** | | | |
| | Ring | | | Figure 8.90 |
|  | Meander line | | | Figure 8.90 |
|  | Square | | | Figure 8.90 |
|  | Triangle | | | Figure 8.90 |
|  | H-shape | | | Figure 8.90 |
| | U-shape | | | Figure 8.90 |

(*cont.*)

| Brief view | Antenna type | Main features | Applications | References and location in the text |
|---|---|---|---|---|
| 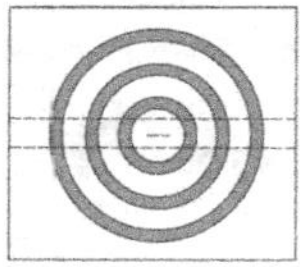 | DGS unit (circles) | Suppressing interference, phase noise, harmonics in active MSA | | [142] Figure 8.91 |
| 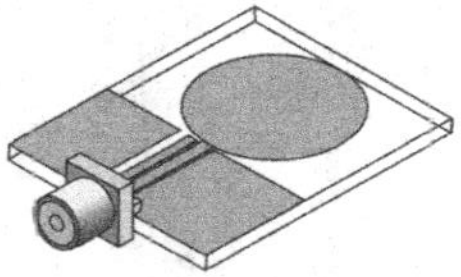 | Disk monopole patch with L-shape cut on the ground plane | Multiband operation due to DGS, creating a new current path and an additional resonance, small sizing | Card-size module antenna in MIMO systems, multiband antenna | [142] Figure 8.93 |
| | **Antenna with DBE (Degenerated Band Edge)** | | | |
| 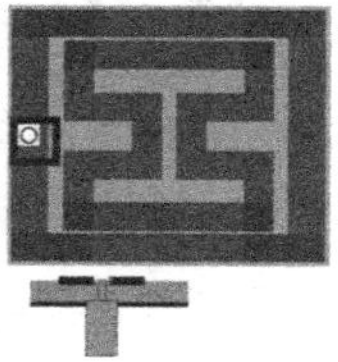 | Microstrip DBE antenna | Size reduction and yet wideband, sharp beam | Small antenna with wide bandwidth array antenna with high gain (narrow beam) | [181–186] Figure 8.98 |

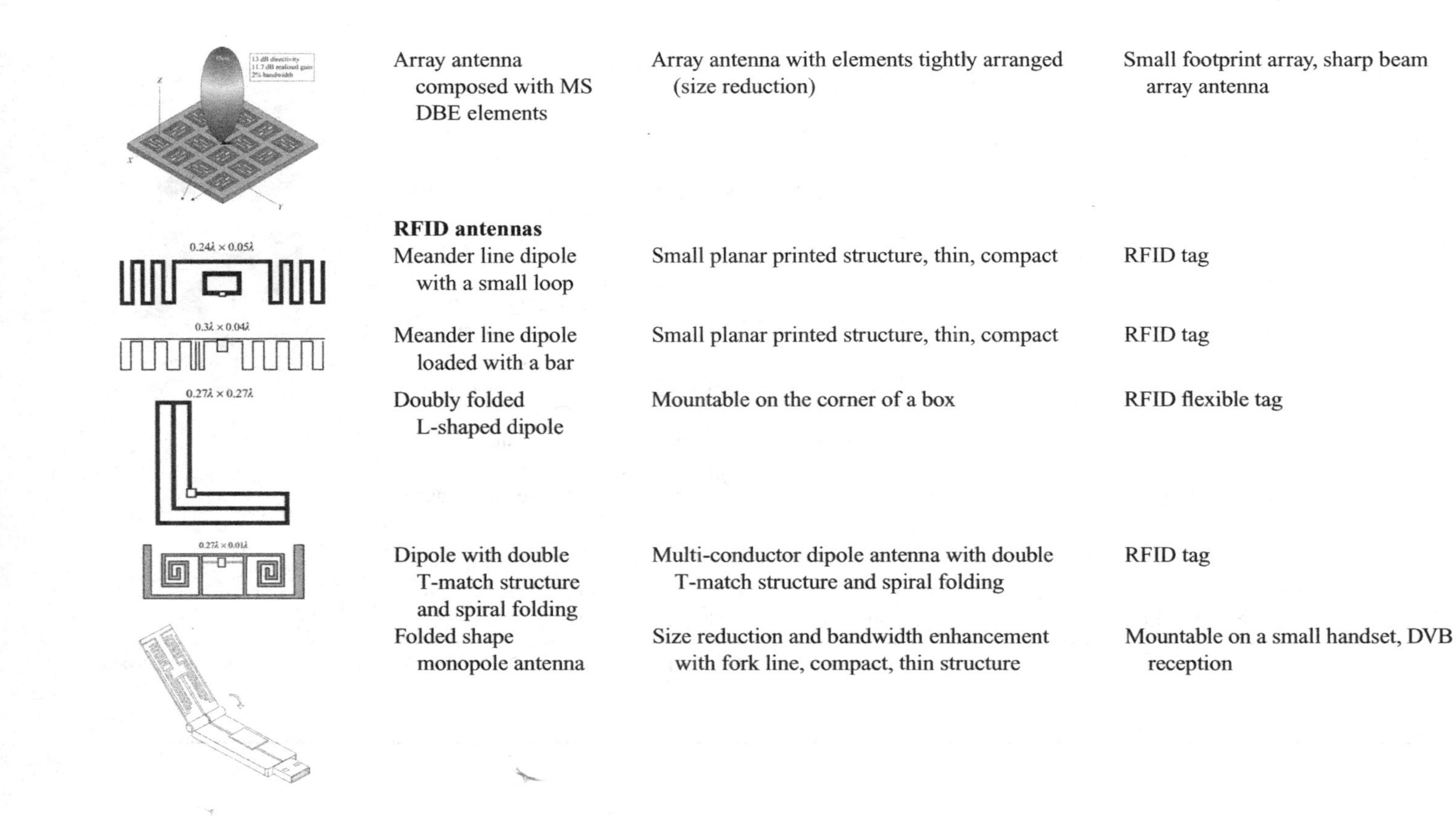

| | | | |
|---|---|---|---|
| Array antenna composed with MS DBE elements | Array antenna with elements tightly arranged (size reduction) | Small footprint array, sharp beam array antenna | [181–186] Figure 8.99 |
| **RFID antennas** | | | |
| Meander line dipole with a small loop | Small planar printed structure, thin, compact | RFID tag | [144] Figure 8.104 |
| Meander line dipole loaded with a bar | Small planar printed structure, thin, compact | RFID tag | [144, 145] Figure 8.104 |
| Doubly folded L-shaped dipole | Mountable on the corner of a box | RFID flexible tag | [144] [194] Figure 8.104 |
| Dipole with double T-match structure and spiral folding | Multi-conductor dipole antenna with double T-match structure and spiral folding | RFID tag | [145] [147] Figure 8.104 |
| Folded shape monopole antenna | Size reduction and bandwidth enhancement with fork line, compact, thin structure | Mountable on a small handset, DVB reception | [196] Figure 8.105 |

(*cont.*)

| Brief view | Antenna type | Main features | Applications | References and location in the text |
|---|---|---|---|---|
| 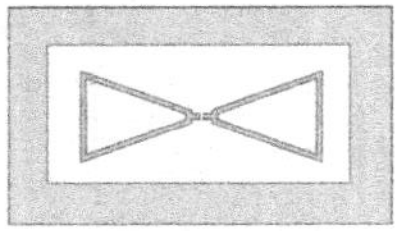  | Wired bow-tie dipole antenna recessed in a cavity | Mountable on a metal surface | RFID tag installation in a metal container, vehicle, aircraft, etc. | [197]<br>Figure 8.106 |
| 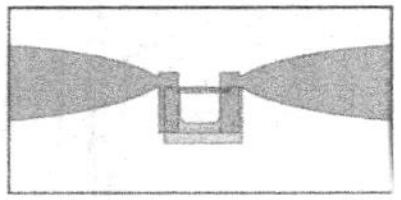  | Planar flared dipole loaded with an RFID sensor chip | Mountable directly on materials as paper block, plastic, wood, a bottle of a tap water, a glass bottle, etc. | RFID tag, sensor tag | [198]<br>Figure 8.107 |
|  | Conventional PIFA with a square conductor loaded with an RFID chip | Thin, planar, small size ($0.19\lambda \times 0.19\lambda$) | RFID tag | [144, 152]<br>8.4.2<br>Figure 8.108 |
| 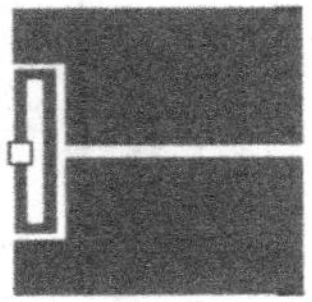  | Two-layer double PIFA fed by a small rectangular loop | Thin, planar, small size ($0.13\lambda \times 0.16\lambda$) | RFID tag | [144, 153]<br>8.4.2<br>Figure 8.108 |

| | | | | |
|---|---|---|---|---|
|  | Coplanar IFA with an additional horizontal stub | Thin, planar, small size ($0.09\lambda \times 0.18\lambda$) | RFID tag | [144, 154] 8.4.2 Figure 8.108 |
|  | Coplanar IFA with multiple folded elements | Thin, planar, small size ($0.11\lambda \times 0.08\lambda$) | RFID tag | [144, 155] 8.4.2 Figure 8.108 |
|  | IFA loaded with a Hilbert trace at the end | Thin, planar, small size, operating frequency and bandwidth adjustable by the length of the Hilbert trace | RFID tag | [201] 8.4.2 Figure 8.109 |
|  | Folded rectangular patch antenna loaded with H-shape slot | Thin, planar, small size, comparable to a credit card size | RFID tag, card type, with sensor chip medical applications | [202] 8.4.2 Figure 8.110 |
| | **Slot loaded patch antenna** | | | |
|  | Series connected OCSRR antenna (OCSRR: Open Complementary Split Ring Resonator) | Antenna composed of series OCSRR units, compact, small size, OCSRR comprised with a circular slot and a conductor loop inside the slot, creating higher input resistance and increase of bandwidth | RFID tag | [202] Figure 8.113 |

(*cont.*)

| Brief view | Antenna type | Main features | Applications | References and location in the text |
|---|---|---|---|---|
| | L-shaped slot dipole, loaded with a meander line at the end of the dipole | Thin, planar, small size | RFID tag, card type | [203]<br>8.4.2<br>Figure 8.114 |
| | Square patch antenna loaded with circular slots | Circular polarization | RFID tag | [204]<br>8.4.2<br>Figure 8.115 |
| | **Coil antenna** | | | |
| | Small coil antenna in a radio-controlled wristwatch | Very small coil antenna installed inside a wristwatch | Radio-controlled wristwatch | [203]<br>8.4.2<br>Figure 8.101 |

| | | | | |
|---|---|---|---|---|
| | Coil antenna (laminated amorphous metal core) | Very small-size coil antenna for radio-controlled wristwatch, to receive standard time signal at LF band (40, 60 kHz) | Radio-controlled wristwatch | [203]<br>8.4.2<br>Figure 8.103 |
| | **Loop antenna** | | | |
| | Circular loop antenna with two-line feeding stub | Resonance frequency and impedance adjustable by varying the feeding stub | RFID | [204]<br>8.4.2<br>Figure 8.116 |
| | Two-element defected loop (three cornered) | Directional performance by two elements, applicable to handheld device | Handheld RFID reader | [205, 206]<br>8.4.2<br>Figure 8.117 |

## References

[1] R. W. P. King, *Theory of Linear Antennas*, Harvard University Press, 1956, pp. 180–192.

[2] W. L. Stutzman and G. A. Thiele, *Antenna Theory and Design*, 2nd edn., John Wiley and Sons, 1998, 1.9 and 2.1.

[3] C. A. Balanis, *Antenna Theory, Analysis, and Design*, 2nd edn., John Wiley and Sons, 1997, 4.3.

[4] C. A. Balanis (ed.), *Modern Antenna Handbook*, John Wiley and Sons, 2008, 10.5.

[5] K. Fujimoto (ed.), *Mobile Antenna Systems Handbook*, 3rd edn., Artech House, 2008, 5.3.1.

[6] C. A. Balanis, *Antenna Theory, Analysis, and Design*, 2nd edn., John Wiley and Sons, 2008, 23.6.3.1 and 2.2.1.

[7] K. Fujimoto *et al.*, *Small Antennas*, Research Studies Press, UK, 1986, 2.4.

[8] R. W. King and C. H. Harrison, *Antennas and Waves, A Modern Approach*, MIT Press, 1969, 6.12.

[9] R. W. King and C. H. Harrison, *Antennas and Waves, A Modern Approach*, MIT Press, 1969, 6.14.

[10] C. A. Balanis, *Antenna Theory, Analysis, and Design*, 2nd edn., John Wiley and Sons, 1997, 5.4.

[11] R. W. King and C. H. Harrison, *Antennas and Waves, A Modern Approach*, MIT Press, 1969, 9.2.

[12] W. L. Stutzman and G. A. Thiele, *Antenna Theory and Design*, 2nd edn., John Wiley and Sons, 1998, 2.4.2.

[13] C. A. Balanis, *Antenna Theory, Analysis, and Design*, 2nd edn., John Wiley and Sons, 1997, 5.6.2.

[14] K. Fujimoto, *et al.*, *Small Antennas*, Research Studies Press, UK, 1986, 2.3.1.

[15] H. Morishita *et al.*, A Balance-Fed Loop Antenna System for Handset, *IEICE Transactions on Fundamentals of Electronics, Communications and Computer Sciences*, vol. E82-A, 1999, no. 7, pp. 1138–1143.

[16] S. Hayashida, Analysis and Design of Folded Loop Antennas, Doctoral Dissertation, National Defense Academy, Japan, 2006 (in Japanese).

[17] S. Hayashida, H. Morishita, and K. Fujimoto, A Wideband Folded Loop Antenna for Handsets, *IEICE*, vol. J86-B, 2003, no. 9, pp. 1799–1805 (in Japanese).

[18] Y. Kim *et al.*, A Folded Loop Antenna System for Handsets Developed and Based on the Advanced Design Concept, *IEICE Transactions on Communications*, vol. E84-B, 2001, no. 9, pp. 2468–2475.

[19] S. Hayashida, H. Morishita, and K. Fujimoto, Self-Balanced Wideband Folded Loop Antenna, *IEE Proceedings, Microwaves, Antennas & Propagation*, vol. 153, 2006, no. 1, pp. 7–12.

[20] S. Hayashida *et al.*, Characteristics of Built-in Folded Monopole Antenna for Handsets, *IEICE Transactions on Communications*, vol. E88-B, 2005, no. 6, pp. 2275–2283.

[21] H. Nakano, H. Taguchi, A. Yoshizawa, and J. Yamauchi, Shortening Ratios of Modified Dipole Antennas, *IEEE Transactions of Antennas and Propagation*, vol. 32, 1984, no. 4, pp. 385–386.

[22] P. E. Mayes, Balanced Backfire Zigzag Antennas, *1964 IEEE Int Conference Record*, pt. 1, pp. 158–165.

[23] S. H. Lee, Theory of Zigzag Antennas, Ph.D. Dissertation, Dept. of Electrical Engineering University of California, Berkeley, 1968, pp. 20, 31–33.

[24] S. H. Lee and K. K. Mei, Analysis of Zigzag Antennas, *IEEE Transactions on Antennas and Propagation*, vol. 18, 1970, no. 6, pp. 760–764.
[25] H. Nakano, H. Tagami, A. Yoshizawa, and J. Yamauchi, Shortening Ratios of Modified Dipole Antennas, *IEEE Transactions on Antennas and Propagation*, vol. 32, 1984, no. 4, pp. 385–386.
[26] K. Noguchi *et al.*, Increasing the Bandwidth of Meander Line Antennas Consisting of Two Strips, *Transactions of IEICE*, vol. JB2-B, 1999, no. 3, pp. 402–409.
[27] L. C. Godara (ed.), *Handbook of Antennas in Wireless Communications*, CRC Press, 2002, Chapter 12.
[28] S. R. Best and D. L. Hanna, A Performance Comparison of Fundamental Small-Antenna Designs, *IEEE Antennas and Propagation Magazine*, vol. 52, 2010, no. 1, pp. 47–70.
[29] C. A. Balanis (ed.), *Modern Antenna Handbook*, John Wiley and Sons, 2008, 10.6.1.
[30] H. Nakano, H. Tagami, A. Yoshizawa, and J. Yamauchi, Shortening Ratios of Modified Dipole Antennas, *IEEE Transactions on Antennas and Propagation*, vol. 32, 1984, no. 4, pp. 385–386.
[31] S. R. Best and D. L. Hanna, A Performance Comparison of Fundamental Small-Antenna Designs, *IEEE Antennas and Propagation Magazine*, vol. 52, 2010, no. 1, pp. 47–70.
[32] T. Endo, Y. Sunahara, and Y. Hoshihara, Resonance Frequency of Dielectric Loaded Normal Mode Helical Antenna, *IEICE Technical Report*, vol. 95, 1995, no. 535, pp. 1–6 and L. C. Godara (ed.), *Handbook of Antennas in Wireless Communications*, CRC Press, 2000, 12.2.2.2.
[33] K. Noguchi *et al.*, Impedance Characteristics of a Meander Line Antenna Mounted on a Conducting Plane, *IEICE National Convention*, B-1–106, 1999, p. 106.
[34] H. Choo and H. Ling, Design of Planar, Electrically Small Antennas with Inductively Coupled Feed Using a Genetic Algorithm, *IEEE APS International Symposium 2003*, 22.1.
[35] C. W. P. Huang *et al.*, FDTD Characterization on Meander Line Antennas for RF and Wireless Communications, *Progress in Electromagnetics Research PIER*, 24, 1991, pp. 185–199.
[36] K. Noguchi *et al.*, Impedance Characteristics of a Small Meander Line Antenna, *Transactions of IEICE*, vol. JB, BII, 1998, no. 2, pp. 183–184.
[37] K. Noguchi, *et al.*, Increasing the Bandwidth of Meander Line Antennas Consisting of Two Strips, *Transactions of IEICE*, vol. JB2-B, 1999, no. 3, pp. 402–409.
[38] C. A. Balanis (ed.), *Modern Antenna Handbook*, John Wiley and Sons, 2008, 9.3.2.
[39] W. L. Stutzman and G. A. Thiele, *Antenna Theory and Design*, 2nd edn., John Wiley and Sons, 1998, 6.2.1.
[40] J. D. Kraus and R. J. Marhefka, *Antennas*, 3rd edn., McGraw-Hill, 2002, pp. 8–22.
[41] N. Inagaki, K. Tamura, and K. Fujimoto, Theoretical Investigation on the Resonance Length of Normal Mode Helical Antennas, *Technical Report of Nagoya Institute of Technology*, vol. 23, 1971, pp. 335–341.
[42] K. Fujimoto *et al.*, *Small Antennas*, Research Studies Press, UK, 1987, pp. 59–75.
N. Inagaki, T. Marui, and K. Fujii, Newly Devised MoM Analysis and Design Data for NMHA, *Technical Report of IEICE*, AP2007–194 (2008–03), pp. 123–128.
[43] C. A. Balanis (ed.), *Modern Antenna Handbook*, John Wiley and Sons, 2008, 9.3.2.
[44] C. A. Balanis (ed.), *Modern Antenna Handbook*, John Wiley and Sons, 2008, 10.6.1.
[45] H. Choo and H. Ling, Design of Planar, Electrically Small Antennas with Inductively Coupled Feed Using a Genetic Algorithm, *IEEE APS International Symposium 2003*, 5.3.
[46] C. P. Baliada, J. Romeu, and A. Cardama, The Koch Monopole: A Small Fractal Antenna, *IEEE Transactions on Antennas and Propagation*, vol. 48, 2000, no. 11, pp. 1773–1781.

[47] Z. N. Chen (ed.), *Antennas for Portable Devices*, 7.3.2 John Wiley and Sons, 2007, pp. 248–258.
[48] D. Guha and Y. M. M. Antar, *Microstrip and Printed Antennas*, John Wiley and Sons, 2011.
[49] S. Honda *et al.*, On a Broadband Disk Monopole Antenna, *Technical Report of Television Society Japan*, ROFT 91–55 (1991–10), 1991.
[50] Z. N. Chen and M. Y. W. Chia, Impedance Characteristics of EMC Triangular Planar Monopoles, *Electronics Letters*, vol. 37, 2001, no. 21, pp. 1271–1272.
[51] Z. N. Chen, Impedance Characteristics of Planar Bow-Tie Like Monopole Antennas, *Electronics Letters*, vol. 36, 2000, no. 13, pp. 1100–1101.
[52] D. H. Werner and S. Ganguly, An Overview of Fractal Antenna Engineering Research, *IEEE Antennas and Propagation Magazine*, vol. 45, February 2003, no. 1, pp. 39–40.
[53] J. R-Mohassel, A. Mehdipour, and H. Aliakbarian, New Schemes of Size Reduction in Space Filling Resonant Dipole Antennas, *3rd European Conference on Antennas and Propagation*, vol. 23–27, 2009, pp. 2430–2432.
[54] J. L. Volakis, C-C Chen and K. Fujimoto, *Small Antennas*, McGraw-Hill, 2010, 3.2.4.
[55] N. Engheta and R. W. Ziolkowsky, *Metamaterials-Physics and Engineering Explorations*, John Wiley and Sons, 2006.
[56] J. Zhu, A. Hoorfar, and N. Engheta, Peano Antennas, *Antennas and Wireless Propagation Letters*, vol. 3, 2004, pp. 71–74.
[57] X. Chen, S. S-Naemi, and Y. Liu, A Down-Sized Hilbert Antenna for UHF Band, *IEEE International Symposium on Antennas and Propagation* 2003, pp. 581–584.
[58] M. Takiguchi and Y. Yamada, Radiation and Ohmic Resistances in Very Small Meander Line Antennas of Less than 0.1 Wavelength, *Transactions of IEICE*, vol. J87-B, 2004, no. 9, pp. 1336–1346.
[59] Y. Yamada and N. Michishita, Efficiency Improvement of a Miniaturized Meander Line Antenna by Loading a High $\varepsilon r$ Material, *IEEE iWAT*, 2005.
[60] F. Kuroki and H. Ohta, Miniaturized Cross Meander-Line Antenna Etched on Both Sides of Dielectric Substrate, *International Symposium on Antennas and Propagation (ISAP)* 2006, Singapore.
[61] J. L. Volakis, C-C Chen and K. Fujimoto, *Small Antennas*, McGraw-Hill, 2010, p. 142.
[62] N. Engheta and R. W. Ziolkowsky, *Metamaterials-Physics and Engineering Explorations*, John Wiley and Sons, 2006.
[63] J. Zhu, A. Hoorfar, and N. Engheta, Peano Antennas, *Antennas and Wireless Propagation Letters*, vol. 3, 2004, pp. 71–74.
[64] X. Chen, S. S-Naemi, and Y. Liu, A Down-Sized Hilbert Antenna for UHF Band, *IEEE International Symposium on Antennas and Propagation*, 2003, pp. 581–584.
[65] J. P. Gianvittorio and Y. Rahmat-Samii, Fractal Antenna: A Novel Antenna Miniaturization Technique and Applications, *IEEE Antennas and Propagation Magazine*, vol. 44, 2002, no. 1, pp. 20–36.
[66] S. R. Best, A Comparison of the Resonant Properties of Small Space-Filling Fractal Antennas, *IEEE Antennas and Wireless Propagation Letters*, vol. 2, 2003, pp. 197–200.
[67] C. P. Baliada, J. Romeu, and A. Cardama, The Koch Monopole: A Small Fractal Antenna, *IEEE Transactions on Antennas and Propagation*, vol. 48, 2000, no. 11, pp. 1773–1781.
[68] D. H. Werner and S. Ganguly, An Overview of Fractal Antenna Engineering Research, *IEEE Antennas and Propagation Magazine*, vol. 45, 2003, no. 1, pp. 38–57.

[69] J. P. Gianvittorio and Y. Rahmat-Samii, Fractal Antenna: A Novel Antenna Miniaturization Technique and Applications, *IEEE Antennas and Propagation Magazine*, vol. 44, 2002, no. 1, pp. 20–36.

[70] W. J. Krzysztofik, Modified Sierpinsky Fractal Monopole for ISM-Bands Handset Applications, *IEEE Transactions on Antennas and Propagation*, vol. 57, 2009, no. 3, pp. 606–615.

[71] H. Huang and A. Hoorfer, Miniaturization of Dual-Band Planar Inverted-F Antennas using Peano-Curve Elements, *International Symposium on Antennas and Propagation (ISAP)* 2006, a292 r206.

[72] W. L Stutzman and G. A. Thiele, *Antenna Theory and Design*, 2nd edn., John Wiley and Sons, pp. 252–258.

[73] M. McFadden and W. R. Scott, Analysis of the Equiangular Spiral Antenna on a Dielectric Substrate, *IEEE Transactions on Antennas and Propagation*, vol. 55, 2007, no.11, pp. 3163–3171.

[74] H. Nakano *et al.*, Equiangular Spiral Antenna Backed by a Shallow Cavity With Absorbing Strips, *IEEE Transactions on Antennas and Propagation*, vol. 56, 2008, no. 8, pp. 2742–2747.

[75] J. L. Volakis, C-C Chen and K. Fujimoto *Small Antennas*, McGraw-Hill, 2010, Chapter 5.

[76] J. L. Volakis, C-C Chen and K. Fujimoto, *Small Antennas*, McGraw-Hill, 2010, 5.4.

[77] J. L. Volakis, C-C Chen and K. Fujimoto, *Small Antennas*, McGraw-Hill, 2010, 5.6.2.

[78] J. L. Volakis, C-C Chen and K. Fujimoto, *Small Antennas*, McGraw-Hill, 2010, 6.4.

[79] J. L. Volakis, N. W. Nurnberger, and D. S. Filipovic, A Broadband Cavity-Backed Slot Spiral Antenna, *IEEE Antennas and Propagation Magazine*, vol. 43, 2001, no. 6, pp. 15–26.

[80] J. L. Volakis, C.-C. Chen and K. Fujimoto, *Small Antennas*, McGraw-Hill, 2010, 5.3 and J. L. Volakis, N. W. Numberger, and D. S. Filipouic, A Broadband Cavity-Backed Slot Spiral Antenna, *IEEE Antennas and Propagation Magazine*, vol. 43, 2001, no. 6, 5.3.

[81] H. Fujishima, Inverted-L Antenna with Self-Complementary Structure, *Technical Report of IEICE, A-P94-24*, 1994, pp. 23–28.

[82] P. Xu, K. Fujimoto, and L. Shiming, Performance of Quasi-self-complementary Antenna, *IEEE Antennas and Propagation Society International Symposium*, vol. 40, 2002, pp. 464–467.

[83] P. Xu and K. Fujimoto, L-shaped Self-complementary Antenna, *IEEE APS International Symposium*, vol. 3, 2003, pp. 95–98.

[84] R. Azadegan and K. Sarabandi, Bandwidth Enhancement of Miniaturized Slot Antennas Using Folded, Complementary, and Self-Complementary Realization, *IEEE Transactions on Antennas and Propagation*, vol. 55, 2007, no. 9, 2435–2444.

[85] Y. Mushiake, *Self-Complementary Antennas*, Springer Verlag, 1996.

[86] M. C. Buck and D. S. Filipovic, Two-Arm Sinuous Antennas, *IEEE Transactions on Antennas and Propagation*, vol. 56, 2008, no. 5, pp. 1229–1235.

[87] C. W. Harrison, Monopole with Inductive Loading, *IEEE Transactions on Antennas and Propagation*, AP-11, 1963, pp. 394–400.

[88] T. L. Simpson, The Disk Loaded Monopole Antenna, *IEEE Transactions on Antennas and Propagation* vol. 52, 2008, no. 2, pp. 542–550.

[89] S. R. Best and D. L. Hanna, A Performance Comparison of Fundamental Small-Antenna Designs, *IEEE Antennas and Propagation Magazine*, vol. 52, 2010, no. 1, pp. 47–70.

[90] C. W. Harrison, Monopole with Inductive Loading, *IEEE Transaction on Antennas and Propagation*, AP-11, 1963, pp. 394–400.

[91] L. J.-Ying and G. Y.-Beng, Characteristics of Broadband Top-Loaded Open-Sleeve Monopole, *IEEE APS International Symposium* 2006, 157.7, pp. 635–638.

[92] K. Surabandi and R. Azadegan, Design of an Efficient Miniaturized UHF Planar Antenna, *IEEE Transactions on Antennas and Propagation*, vol. 51, 2003, no. 6, pp. 1270–1276.

[93] S. R. Best, Low Q Electrically Small Linear and Elliptical Polarized Spherical Dipole Antennas, *IEEE Transactions on Antennas and Propagation*, vol. 53, 2003, no. 3, pp. 1047–1053.

[94] A. Mehdipour, H. Aliakbarian, and J. Rashed-Mohassel, A Novel Electrically Small Spherical Wire Antenna With Almost Isotropic Radiation Pattern, *IEEE Antennas and Wireless Propagation Letters*, vol. 7, 2009, pp. 396–399.

[95] O. S. Kim, Low-Q Electrically Small Spherical Magnetic Dipole Antennas, *IEEE Transactions on Antennas and Propagation*, vol. 58, 2010, no. 7, pp. 2210–2217.

[96] K. G. Schroeder and K. M. S. Hoo, Electrically Small Complementary Pair (ESCP) with Interelement Coupling, *IEEE Transactions on Antennas and Propagation*, AP-24, 1976, no. 4, pp. 411–418.

[97] D.-H. Kwon *et al.*, Small Printed Combined Electric-Magnetic Type Ultrawideband Antenna With Directive Radiation Characteristics, *IEEE Transactions on Antennas and Propagation*, vol. 56, 2008, no. 1, pp. 237–241.

[98] X.-C. Lin and C.-C. Yu, A Dual-band Slot-Monopole Hybrid Antenna, *IEEE Transactions on Antenna and Propagation*, vol. 56, 2008, no. 1, pp. 282–285.

[99] W. Hong and K. Sarabandi, Low Profile Miniaturized Planar Antenna With Omnidirectional Vertically Polarized Radiation, *IEEE Transactions on Antennas and Propagation*, vol. 24, 1976, no. 4, pp. 411–418.

[100] F. Bilotti, A. Alu, and L. Vegni, Design of Miniaturized Metamaterial Patch Antennas With μ–Negative Loading, *IEEE Transactions on Antennas and Propagation*, vol. 56, 2008, no. 6, pp. 1640–1647.

[101] P. Y. Chen and A. Alu, Sub-Wavelength Elliptical Patch Antenna Loaded With μ–Negative Metamaterials, *IEEE Transactions on Antennas and Propagation*, vol. 58, 2010, no. 9, pp. 2909–2919.

[102] H. R. Stuart and A. Pidwerbetsky, Electrically Small Antenna Elements Using Negative Permittivity Resonators, *IEEE Transactions on Antennas and Propagation*, vol. 54, 2006, no. 6, pp. 1644–1653.

[103] A. Erentok and R. W. Ziolkowsky, Metamaterial-Inspired Efficient Electrically Small Antennas, *IEEE Transactions on Antennas and Propagation*, vol. 56, 2008, no. 3, pp. 691–707.

[104] C. Caloz, T. Itoh, and A. Rennings, CRLH Metamaterial Leaky-Wave and Resonant Antennas, *IEEE Antennas and Propagation Magazine*, vol. 50, 2008, no. 5, pp. 26–39.

[105] M. A. Antoniades and G. V. Eleftheriades, A Folded-Monopole Model for Electrically Small NRI-TL Metamaterial Antennas, *IEEE Antennas and Wireless Propagation Letters*, vol. 7, 2008, pp. 425–428.

[106] M. Antoniades and G. V. Eleftheriades, A Broadband Dual-Mode Monopole Antenna Using NRI-TL Metamaterial Loading, *IEEE Antennas and Wireless Propagation Letters*, vol. 8, 2009, pp. 258–261.

[107] J. Zhu and G. V. Eleftheriades, A Compact Transmission-Line Metamaterial Antenna With Extended Bandwidth, *IEEE Antennas and Wireless Propagation Letters*, vol. 8, 2009, pp. 295–298.

[108] J. Zhu, M. A. Antoniades, and G. V. Eleftheriades, A Compact Tri-band Monopole Antenna With Single-Cell Metamaterial Loading, *IEEE Transactions on Antennas and Propagation*, vol. 58, 2010, no. 4, pp. 1031–1038.

[109] P. J. Herritz-Martinez *et al.*, Multifrequency and Dual-Mode Patch Antennas Filled With Left-Handed Structures, *IEEE Transactions on Antennas and Propagation*, vol. 56, 2008, no. 8, pp. 2527–2539.

[110] Y.-S. Wang, M.-Feng Hsu, and S.-J. Chung, A Compact Slot Antenna Utilizing a Right/Left-Handed Transmission Line Feed, *IEEE Transactions on Antennas and Propagation*, vol. 56, 2008, no. 3, pp. 675–682.

[111] F. Yang, X.-X. Zhang, and Y. Rahmat-Samii, Wide-Band E-Shaped Patch Antennas for Wireless Communications, *IEEE Transactions on Antennas and Propagation*, vol. 49, 2001, no. 7, pp. 1094–1100.

[112] Y. Chen, S. Yang, and Z. Nie, Bandwidth Enhancement Method for Low Profile E-Shaped Microstrip Patch Antennas, *IEEE Transactions on Antennas and Propagation*, vol. 58, 2010, no. 7, pp. 2442–2447.

[113] A. Khidre, K. F. Lee, F. Yang, and A. Elsherbeni, Wideband Circularly Polarized E-Shaped Patch Antenna for Wireless Applications, *IEEE Antennas and Propagation Magazine*, vol. 52, 2010, no. 5, pp. 219–229.

[114] A. Khidre, K. F. Lee, F. Yang, and A. Elsherbeni, Wideband Circularly Polarized E-Shaped Patch Antenna for Wireless Applications, *IEEE Antennas and Propagation Magazine*, vol. 52, 2010, no. 5, pp. 219–229.

[115] S. K. Sharma and L. Shafai, Performance of a Novel $\psi$-shape Microstrip Patch Antenna with Wide Bandwidth, *IEEE Antennas and Wireless Propagation Letters*, vol. 8, 2009, pp. 468–471.

[116] S. K. Sharma and L. Shafai, Investigation of a Novel $\psi$-shape Microstrip Patch Antenna With Wide Impedance Bandwidth, *IEEE APS International Symposium* June 2007, Digest vol. 45, pp. 881–884.

[117] K. L. Chung, A Wideband Circularly Polarized H-Shaped Patch Antenna, *IEEE Transactions on Antennas and Propagation*, vol. 58, 2010, pp. 3379–3383.

[118] K. F. Lee *et al.*, The Versatile U-Slot Antenna, *IEEE Antennas and Propagation Magazine*, vol. 52, 2010, no. 1, pp. 71–88.

[119] S. Weigand *et al.*, Analysis and Design of Broad-Band Single-Layer Rectangular U-Slot Microstrip Patch Antennas, *IEEE Transactions on Antennas and Propagation*, vol. 51, 2003, no. 3, pp. 457–468.

[120] K.-L. Wong, *Compact and Broadband Microstrip Antennas*, John Wiley and Sons, 2002, p. 239.

[121] A. K. Shackelford, K. F. Lee, and K. M. Luk, Design of Small-Size Wide-Bandwidth Microstrip-Patch Antennas, *IEEE Antennas and Propagation Magazine*, vol. 45, 2000, no. 1, pp. 75–83.

[122] C. L. Mak, K. M. Luk, K. F. Lee, and Y. L. Chow, Experimental Study of a Microstrip Patch Antenna with an L-shaped Probe, *IEEE Transactions on Antennas and Propagation*, vol. 48, 2000, no. 5, pp. 777–783.

[123] R. Chair *et al.*, Miniature Wide-Band Half U-Slot and Half E-Shaped Patch Antennas, *IEEE Transactions on Antennas and Propagation*, vol. 53, 2005, no. 8, pp. 2645–2652.

[124] C. L. Mak, R. Chair, K. F. Lee, K. M. Luk, and A. A. Kishk, Half-U-slot Patch Antenna with Shorting Wall, *International Symposium USNC/CNC/URSI National Radio Science Meeting*, vol. 2, 2003, pp. 876–879.

[125] T. Huynh and K. F. Lee, Single-Layer Single-Patch Wideband Microstrip Antenna, *Electronics Letters*, vol. 31, 1997, no. 16, pp. 1310–1312.

[126] A. K. Shackelford, K. F. Lee, K. M. Luk, and R. Chair, U-Slot Patch Antenna with Shorting Pin, *Electronics Letters*, vol. 37, 2001, no. 12, pp. 729–730.

[127] S. I. Latif, L. Shafai, and S. K. Sharma, Bandwidth Enhancement and Size Reduction of Microstrip Slot Antennas, *IEEE Transactions on Antennas and Propagation*, vol. 53, 2005, no. 3, pp. 994–1003.

[128] K. F. Lee, S. L. Steven, and A. Kishk, Dual- and Multiband U-Slot Patch Antennas, *IEEE Antennas and Wireless Propagation Letters*, vol. 2, 2008, p. 64.

[129] K. F. Tong and T. P. Wong, Circular Polarized U-Slot Antenna, *IEEE Transactions on Antennas and Propagation*, vol. 55, 2007, no. 8, pp. 2382–2385.

[130] J.-Y. Sze and K.-L. Wong, Bandwidth Enhancement of a Microstrip-Line-Fed Printed Wide-Slot Antenna, *IEEE Transactions on Antennas and Propagation*, vol. 49, 2001, no. 7, pp. 1020–1024.

[131] J.-Y. Jan and J.-W. Su, Bandwidth Enhancement of a Printed Wide-Slot Antenna With a Rotated Square, *IEEE Transactions on Antennas and Propagation*, vol. 53, 2005, no. 6, pp. 2111–2114.

[132] S. R. Best, A Comparison of the Resonant Properties of Small Space-Filling Fractal Antennas, *IEEE Antennas and Wireless Propagation Letters*, vol. 2, 2003, pp. 197–200.

[133] C.-H. Chang and K.-L. Wong, Printed λ/8-PIFA for Penta-Band WWAN Operation in the Mobile Phone, *IEEE Transactions on Antennas and Propagation*, vol. 57, 2009, no. 5, pp. 1373–1381.

[134] C.-M. Peng *et al.*, Bandwidth Enhancement of Internal Antenna by Using Reactive Loading for Penta-Band Mobile Handset Application, *IEEE Transactions on Antennas and Propagation*, vol. 59, 2011, no. 5, pp. 1728–1733.

[135] C.-W. Chiu and Y.-J. Chi, Printed Loop Antenna With a U-Shaped Tuning Element for Hepta-Band Laptop Applications, *IEEE Transactions on Antennas and Propagation*, vol. 58, 2010, no. 11, pp. 3464–3470.

[136] F.-H. Chu and K.-Lu Wong, Planar Printed Strip Monopole With a Closely-Coupled Parasitic Shorted Strip for Eight-Band LTE/GSM/UMTS Mobile Phone, *IEEE Transactions on Antennas and Propagation*, vol. 58, 2010, no. 10, pp. 3426–3431.

[137] G. Dubost and S. Zisler, *Antennas a Large Bande*, Masson, 1976, pp. 128–129.
S. Honda *et al.*, On a Broadband Disk Monopole Antenna, *Technical Report of Television Society Japan*, ROFT 91–55 (1991–10) 1991.

[138] M. J. Ammann and Z. N. Chen, Wideband Monopole Antennas for Multi-Band Wireless Systems, *IEEE Antennas and Propagation Magazine*, vol. 45, 2003, no. 2, pp. 146–150.

[139] X. H. Wu and Z. N. Chen, Comparison of Planar Dipoles in UWB Applications, *IEEE Transactions on Antennas and Propagation*, vol. 53, 2005, no. 6, pp. 1973–1983.

[140] S. Koulouridis and J. L. Volakis, A Novel Planar Conformal Antenna Designed With Splines, *IEEE Antennas and Wireless Propagation Letters*, vol. 8, 2009, pp. 34–36.

[141] S. Koulouridis and J. L. Volakis, Miniaturization of Flare Dipole via Shape Optimization and Matching Circuits, *IEEE APS International Symposium* 2007, pp. 4785–4788.

[142] W. Dullaert and H. Rogier, Novel Compact Model for the Radiation Pattern of UWB Antennas Using Vector Spherical and Slepian, *IEEE Transactions on Antennas and Propagation*, vol. 58, 2010, no. 2, pp. 287–298.

[143] D. Valderas *et al.*, UWB Staircase-Profile Printed Monopole Design, *IEEE Antennas and Wireless Propagation Letters*, vol. 7, 2008, pp. 255–259.

[144] X.-L. Liang, *et al.*, Printed Binomial-Curved Slot Antennas for Various Wideband Applications, *IEEE Transactions on Antennas and Propagation*, vol. 59, 2011, no. 4, pp. 1058–1065.

[145] C.-W. Ling *et al.*, Planar Binomial Curved Monopole Antennas for Ultra-wideband Communication, *IEEE Transactions on Antennas and Propagation*, vol. 55, 2007, no. 9, pp. 2622–2624.

[146] M. Sun, Y. P. Zhang, and Y. Lu, Miniaturization of Planar Monopole Antenna for Ultra-wideband Radios, *IEEE Transactions on Antennas and Propagation*, vol. 58, 2010, no. 7, pp. 2420–2425.

[147] R. Zaker and J. Nourinia, Novel Modified UWB Planar Monopole Antenna with Variable Frequency Band-Notch Function, *IEEE Antennas and Wireless Propagation Letters*, vol. 7, 2008, pp. 112–116.

[148] E. S. Angelopoulos *et al.*, Circular and Elliptical CPE-Fed Slot and Microstrip-Fed Antennas for Ultrawideband Applications, *IEEE Antennas and Wireless Propagation Letters*, vol. 5, 2006, pp. 294–297.

[149] A. M. Abbosh and E. Bialkowsky, Design of Ultrawideband Planar Monopole Antenna of Circular and Elliptical Shape, *IEEE Antennas and Wireless Propagation Letters*, vol. 7, 2009, pp. 17–23.

[150] P. Li and X. Chen, Study of Printed Elliptical/Circular Slot Antennas for Ultrawideband Applications, *IEEE Transactions on Antennas and Propagation*, vol. 4, 2006, no. 6. pp. 1670–1675.

[151] E. A-Daviu *et al.*, Modal Analysis and Design of Band-Notched UWB Planar Monopole Antennas, *IEEE Transactions on Antennas and Propagation*, vol. 58, 2010, no. 5, pp. 1457–1467.

[152] H. Elsadek and D. Nashaat, Multiband and UWB V-Shaped Antenna Configuration for Wireless Communications Applications, *IEEE Antennas and Wireless Propagation Letters*, vol. 7, 2008, pp. 89–91.

[153] N. Behdad and K. Sarabandi, A Compact Antenna for Ultrawide-Band Applications, *IEEE Transactions on Antennas and Propagation*, vol. 53, 2005, no. 7, pp. 2185–2191.

[154] Z. N. Chen, T. S. P. See, and X. Qing, Small Printed Ultrawideband Antenna with Reduced Ground Plane, *IEEE Transactions on Antennas and Propagation*, vol. 55, 2007, no. 2, pp. 383–388.

[155] K. Fujimoto, Integrated Antenna Systems, in K. Chang (ed.) *Encyclopedia of RF and Microwave Engineering*, vol. 3, John Wiley and Sons, 2005, pp. 2113–2147.
A. Mortazawi, T. Itoh, and J. Harvey, *Active Antennas and Quasi-Optical Arrays*, IEEE Press, 1999.

[156] A. Navarrow and J. K. Chang, *Integrated Active Antennas and Spatial Power Combining*, John Wiley and Sons, 1996.

[157] K. Chang, R. A. York, P. S. Hall, and T. Itoh, Active Integrated Antennas, *IEEE Transactions on Antennas and Propagation*, vol. 50, 2002, no. 3, pp. 937–943.

[158] C. H Mueller *et al.*, Small-Size X-Band Active Integrated Antenna with Feedback Loop, *IEEE Transactions on Antennas and Propagation*, vol. 56, 2008, no. 5, pp. 1236–1241.

[159] J.-H. Lim *et al.*, A Reconfigurable PIFA Using a Switchable PIN-Diode and a Fine-Tuning Varactor for USPCS/WCDMA/m-WiMAX/WLAN, *IEEE Transactions on Antennas and Propagation*, vol. 58, 2010, no. 7, pp. 2404–2411.

[160] G. Elli and S. Liw, Active Planar Inverted-F Antenna for Wireless Applications, *IEEE Transactions on Antennas and Propagation*, vol. 57, 2009, no. 10, pp. 2899–2906.

[161] L. M. Feldner *et al.*, Electrically Small Frequency-Agile PIFA-as-a Package for Portable Wireless Devices, *IEEE Transactions on Antennas and Propagation*, vol. 55, 2007, no. 11, pp. 3310–3319.

[162] S.-L. S. Yang and A. A. Kishk, Frequency Reconfigurable U-Slot Microstrip Patch Antenna, *IEEE Antennas and Wireless Propagation Letters*, vol. 7, 2008, pp. 127–129.

[163] J. Sarrazin, Y. Mahe, and S. Avrillon, Pattern Reconfigurable Cubic Antenna, *IEEE Transactions on Antennas and Propagation*, vol. 57, 2009, no. 2, pp. 310–317.

[164] D. Sievenpiper *et al.*, High-Impedance Electromagnetic Surfaces with a Forbidden Frequency Band, *IEEE Transactions on Microwave Theory and Techniques*, vol. 47, 1999, no. 11, pp. 2059–2074.

[165] S. Clavijo, R. E. Diaz, and W. E. Mckinzie, Design Methodology for Sievenpiper High-Impedance Surfaces: an Artificial Magnetic Conductor for Positive Gain Electrically Small Antennas, *IEEE Transactions on Antennas and Propagation*, vol. 51, 2003, no. 10, pp. 2678–2690.

[166] Q. Wu, J. Geng, and D. Su, On the Performance of Printed Dipole Antenna With Novel Composite Corrugated-Reflectors for Low-Profile Ultrawideband Applications, *IEEE Transactions on Antennas and Propagation*, vol. 58, 2010, no. 12, pp. 3829–3846.

[167] G. Gampala, R. Sammeta, and C. J. Reddy, A Thin Low-Profile Antenna Using a Novel High Impedance Ground Plane, *Microwave Journal*, July 2010, pp. 70–80.

[168] A. Suntives and R. Abhari, Design of a Compact Miniaturized Probe-Fed Patch Antennas Using Electro-magnetic Bandgap Structure, *IEEE APS International Symposium*, 2010.

[169] Z. Li and Y. Rahmat-Samii, PBC, PMC, and PEC Ground Plane: A Case Study for Dipole Antenna, *IEEE APS International Symposium 2000*, vol. 4, pp. 2258–2261.

[170] M. Z. Azad and M. Ali, Novel Wideband Directional Dipole Antenna on a Mushroom Like EBG Structure, *IEEE Transactions on Antennas and Propagation*, vol. 56, 2008, no. 5, pp. 1242–1250.

[171] F. Yang and Y. Rahmat Samii, Reflection Phase Characterizations of the EBG Ground Plane for Low Profile Wire Antenna Applications, *IEEE Transactions on Antennas and Propagation*, vol. 51, 2003, no. 10, pp. 2691–2703.

[172] S. R. Best and D. L. Hanna, Design of a Broadband Dipole in Proximity to an EBG Ground Plane, *IEEE Antennas and Propagation Magazine*, vol. 50, 2008, no. 6, pp. 52–64.

[173] L. Akhoondzadeh-Asl *et al.*, Wideband Dipoles on Electromagnetic Bandgap Ground Plane, *IEEE Transactions on Antennas and Propagation*, vol. 55, 2007, no. 9, pp. 2426–2434.

[174] F. Yang and Y. Rahmat-Samii, Microwave Antennas Integrated with Electromagnetic Band-Gap (EBG) Structure; a Low Mutual Coupling Design for Array Applications, *IEEE Transactions on Antennas and Propagation*, vol. 51, 2003, no. 10, pp. 2936–2946.

[175] D. Guha, S. Biswas, and Y. M. M. Antar, Defected Ground Structure for Microstrip Antennas, in D. Guha and Y. M. M. Antar (eds.) *Microstrip and Printed Antennas*, John Wiley and Sons, 2011, Chapter 12.

[176] D. Guha *et al.*, Concentric Ring-shaped Defected Ground Structures for Microstrip Applications, *IEEE Antennas and Wireless Propagation Letters*, vol. 5, 2006, pp. 402–405.

[177] M. A. Antoniades and G. V. Eleftheriades, A Compact Multiband Monopole Antenna With a Defected Ground Plane, *IEEE Antenna and Wireless Propagation Letters*, vol. 7, 2008, pp. 652–655.

[178] D. Guha, M. Biswas, and Y. M. M. Antar, Microstrip Patch Antenna with Defected Ground Structure for Cross Polarization Suppression, *IEEE Antennas and Wireless Propagation Letters*, vol. 4, 2005, pp. 455–458.

[179] Y. J. Sung, M. Kim, and Y. S. Kim, Harmonic Reduction with Defected Ground Structure for a Microstrip Patch Antenna, *IEEE Antennas and Wireless Propagation Letters*, vol. 2, 2003, pp. 111–113.

[180] H. B. El-Shaarawy *et al.*, Novel Reconfigurable Defected Ground Structure Resonator on Coplanar Waveguide, *IEEE Transactions on Antennas and Propagation*, vol. 58, 2010, no. 11, pp. 3622–3628.

[181] J. L. Volakis *et al.*, Antenna Miniaturization Using Magnetic Photonic and Degenerated Band-Edge Crystals, *IEEE Antennas and Propagation Magazine*, vol. 48, 2008, no. 5, pp. 12–27.

[182] J. L Volakis, C. Chen, and K. Fujimoto, *Small Antennas: Miniaturization Technique & Applications*, McGraw-Hill, 2010, Chapter 7.

[183] S. Yarga, K. Sertel, and J. L. Volakis, Degenerate Band Edge Crystals for Directive Antennas, *IEEE Transactions on Antennas and Propagation*, vol. 56, 2008, no. 1, pp. 119–126.

[184] S. Yarga, K. Sertel, and J. L. Volakis, A Directive Resonator Antenna Using Degenerate Band Edge Crystals, *IEEE Transactions on Antennas and Propagation*, vol. 57, 2009, no. 3, pp. 799–803.

[185] S. Yarga, K. Sertel, and J. L. Volakis, Multilayer Directive Resonator Antenna Operating at Degenerate Band Edge Modes, *IEEE Antennas and Wireless Propagation Letters*, vol. 8, 2009, pp. 287–290.

[186] G. Mumcu, K. Sertel, and J. L. Volakis, Miniature Antenna Using Printed Coupled Lines Emulating Degenerate Band Edge Crystals, *IEEE Transactions on Antennas and Propagation*, vol. 57, 2009, no. 6, pp. 1618–1623.

[187] W. Choi *et al.*, RFID Tag Antenna with a Meandered Dipole and Inductively Coupled Feed, *IEEE APS International Symposium 2006*, pp. 1347–1350.

[188] G. Marrocco, The Art of UHF RFID Antenna Design: Impedance-Matching and Size-Reduction Techniques, *IEEE Antennas and Propagation Magazine*, vol. 50, 2008, no. 1 pp. 66–79.

[189] A. Toccafondi and P. Braconi, Compact Load-Bars Meander Line Antenna for UHF RDID Transponder, *First European Conference on Antennas and Propagation*, Nice, France, 2006, p. 804.

[190] S. A. Delichatsios *et al.*, Albano Multidimensional UHD Passive RFID Tag Antenna Design, *International Journal of Radio Frequency Identification Technology and Applications*, 1.1 January 2006, pp. 24–40.

[191] H. Hirvonenen *et al.*, Planar Inverted-F Antenna for Radio Frequency Identification, *Electronics Letters*, vol. 40, 2004, no. 4, pp. 848–850.

[192] B. Yu *et al.*, Balanced RFID Tag Antenna Mountable on Metallic Plates, *IEEE APS International Symposium 2006*, pp. 3237–3240.

[193] C. H. Cheng and R. D. Murch, Asymmetric RFID Tag Antenna, *IEEE APS International Symposium 2006*, digest, pp. 1363–1366.

[194] S. A. Delichatsios *et al.*, Albano Multidimensional UHF Passive RFID Tag Antenna Design, *International Journal of Radio Frequency Identification Technology and Applications*, 1.1 January 2006, pp. 24–40.

[195] C. Cho, H. Cho, and I. Park, Design of Novel RFID Tag Antenna for Metallic Objects, *IEEE APS International Symposium 2006*, pp. 3245–3248.

[196] C.-K. Hsu and S.-J. Chung, A Wideband DVB Forked Shape Monopole Antenna with Coupling Effect for USB Dongle Application, *IEEE Transactions on Antennas and Propagation*, vol. 58, 2010, no. 9, pp. 2029–3036.

[197] D. Kim and J. Yeo, A Passive RFID Tag Antenna Installed in a Recessed Cavity in a Metallic Platform, *IEEE Transactions on Antennas and Propagation*, vol. 58, 2010, no. 12, pp. 3814–3820.

[198] T. Deleruyelle *et al.*, An RFID Tag Antenna Tolerant to Mounted Materials, *IEEE Antennas and Propagation Magazine*, vol. 52, 2010, no. 4, pp. 14–19.

[199] G. Marrocco, The Art of UHF RFID Antenna Design: Impedance-Matching and Size Reduction Techniques, *IEEE Antennas and Propagation Magazine*, vol. 50, 2008, no. 1, pp. 66–70.

[200] J.-T. Huang, J.-H. Shiao, and J.-M. Wu, A Miniaturized Hilbert Inverted-F Antenna for Wireless Sensor Network Applications, *IEEE Transactions on Antennas and Propagation*, vol. 58, 2010, no. 9, pp. 3100–3103.

[201] C. Occhuzzi, S. Cippitelli, and G. Marrocco, Modeling, Design and Experimentation of Wearable RFID Sensor Tag, *IEEE Transactions on Antennas and Propagation*, vol. 58, 2010, no. 8, pp. 2490–2498.

[202] J. Kim *et al.*, Design of a Meandered Slot Antenna for UHF RFID Applications, *IEEE APS International Symposium 2010*, 206.5.

[203] Z. N. C. Nasimuddin and X. Qing, Asymmetric-Circular Shaped Slotted Microstrip Antennas for Circular Polarization and RFID Applications, *IEEE Transactions on Antennas and Propagation*, vol. 58, 2010, no. 12, pp. 3821–3828.

[204] B. D. Braaten, A Novel Compact UHF RFID Tag Antenna Designed With Series connected Open Complementary Split Ring Resonator (OCSRR) Particles, *IEEE Transactions on Antennas and Propagation*, vol. 58, 2010, no. 11, pp. 3728–3733.

[205] K. Abe, Analysis and Design of Very Small Antennas for Receiving Long Wave Signals, Doctoral Dissertation, Tokyo Institute of Technology, 2007.

[206] K. Abe and J. Takada, Performance Evaluation of a Very Small Magnetic Core Loop Antenna for an LF Receiver, *Proceedings of Asia-Pacific Microwave Conference*, 2006, TH3C-4, pp. 935–938.

[207] H.-K. Ryu and J.-M. Woo, Small Circular Loop antenna for RFID Tag, *International Symposium on Antennas and Propagation*, (ISAP) 2006, a341, r315.

[208] J. J. Yu and S. Lim, A Miniaturized Circularly Polarized Antenna for an Active 433.92 MHz RFID Handheld Reader, *IEEE APS International Symposium 2010*, 206.1.

# Index

Printed in the United States
by Baker & Taylor Publisher Services

Printed in the United States
by Baker & Taylor Publisher Services